AF577049

Grundlehren der mathematischen Wissenschaften 260

A Series of Comprehensive Studies in Mathematics

Editors

S.S. Chern B. Eckmann P. de la Harpe
H. Hironaka F. Hirzebruch N. Hitchin
L. Hörmander M.-A. Knus A. Kupiainen
J. Lannes G. Lebeau M. Ratner D. Serre
Ya.G. Sinai N.J.A. Sloane J. Tits
M. Waldschmidt S. Watanabe

Managing Editors

M. Berger J. Coates S.R.S. Varadhan

Springer
New York
Berlin
Heidelberg
Barcelona
Budapest
Hong Kong
London
Milan
Paris
Singapore
Tokyo

Grundlehren der mathematischen Wissenschaften

A Series of Comprehensive Studies in Mathematics

A Selection

200. Doid: Lectures on Algebraic Topology
201. Beck: Continuous Flows in the Plane
202. Schmetterer: Introduction to Mathematical Statistics
203. Schoeneberg: Elliptic Modular Functions
204. Popov: Hyperstability of Control Systems
205. Nikol'skiĭ: Approximation of Functions of Several Variables and Imbedding Theorems
206. André: Homologie des Algèbres Commutatives
207. Donoghue: Monotone Matrix Functions and Analytic Continuation
208. Lacey: The Isometric Theory of Classical Banach Spaces
209. Ringel: Map Color Theorem
210. Gihman/Skorohod: The Theory of Stochastic Processes I
211. Comfort/Negrepontis: The Theory of Ultrafilters
212. Switzer: Algebraic Topology—Homotopy and Homology
213. Shafarevich: Basic Algebraic Geometry
214. van der Waerden: Group Theory and Quantum Mechanics
215. Schaefer: Banach Lattices and Positive Operators
216. Pólya/Szegö: Problems and Theorems in Analysis II
217. Stenström: Rings of Quotients
218. Gihman/Skorohod: The Theory of Stochastic Processes II
219. Duvant/Lions: Inequalities in Mechanics and Physics
220. Kirikov: Elements of the Theory of Representations
221. Mumford: Algebraic Geometry I: Complex Projective Varieties
222. Lang: Introduction to Modular Forms
223. Bergh/Löfström: Interpolation Spaces. An Introduction
224. Gilbarg/Trudinger: Elliptic Partial Differential Equations of Second Order
225. Schütte: Proof Theory
226. Karoubi: K-Theory. An Introduction
227. Grauert/Remmert: Theorie der Steinschen Räume
228. Segal/Kunze: Integrals and Operators
229. Hasse: Number Theory
230. Klingenberg: Lectures on Closed Geodesics
231. Lang: Elliptic Curves: Diophantine Analysis
232. Gihman/Skorohod: The Theory of Stochastic Processes III
233. Stroock/Varadhan: Multidimensional Diffusion Processes
234. Aigner: Combinatorial Theory
235. Dynkin/Yushkevich: Controlled Markov Processes
236. Grauert/Remmert: Theory of Stein Spaces
237. Köthe: Topological Vector Spaces II
238. Graham/McGehee: Essays in Commutative Harmonic Analysis
239. Elliott: Probabilistic Number Theory I
240. Elliott: Probabilistic Number Theory II
241. Rudin: Function Theory in the Unit Ball of C^R
242. Huppert/Blackburn: Finite Groups II
243. Huppert/Blackburn: Finite Groups III
244. Kubert/Lang: Modular Units
245. Cornfeld/Fomin/Sinai: Ergodic Theory
246. Naimark/Stern: Theory of Group Representations
247. Suzuki: Group Theory I
248. Suzuki: Group Theory II
249. Chung: Lectures from Markov Processes to Brownian Motion
250. Arnold: Geometrical Methods in the Theory of Ordinary Differential Equations
251. Chow/Hale: Methods of Bifurcation Theory
252. Aubin: Nonlinear Analysis on Manifolds. Monge-Ampère Equations
253. Dwork: Lectures on ρ-adic Differential Equations
254. Freitag: Siegelsche Modulfunktionen
255. Lang: Complex Multiplication
256. Hörmander: The Analysis of Linear Partial Differential Operators I
257. Hörmander: The Analysis of Linear Partial Differential Operators II

(continued after index)

M.I. Freidlin   A.D. Wentzell

Random Perturbations of Dynamical Systems

Second Edition

Translated by Joseph Szücs

With 33 Illustrations

Springer

M.I. Freidlin
Department of Mathematics
University of Maryland
College Park, MD 20742
USA

A.D. Wentzell
Department of Mathematics
Tulane University
New Orleans, LA 70118
USA

Joseph Szücs (*Translator*)
Texas A&M University at Galveston
P.O. Box 1675
Galveston, TX 77553
USA

Mathematics Subject Classification (1991): 58F30, 60F10, 60H25

Library of Congress Cataloging-in-Publication Data
Freĭdlin, M.I. (Mark Iosifovich)
Random perturbations of dynamical systems / Mark I. Freidlin, Alexander D. Wentzell ; [Joseph Szücs (translator)]. — 2nd ed.
p. cm. — (Grundlehren der mathematischen Wissenschaften ; 260)
"[Based on the] original Russian edition: Fluktuat͡sii v dinamicheskikh sistemakh pod deĭstviem malykh sluchaĭnykh vozmushcheniĭ, Nauka : Moscow, 1979"—T.p. verso.
Includes bibliographical references and index.
ISBN 0-387-98362-7 (hardcover)
1. Stochastic processes. 2. Perturbation (Mathematics)
I. Vent͡tsel', A.D. II. Vent͡tsel', A.D. Fluktuat͡sii v dinamicheskikh sistemakh pod deĭstviem malykh sluchaĭnykh vozmushcheniĭ. III. Title. IV. Series.
QA274.F73 1998
519.2—dc21 98-4686

Printed on acid-free paper.

Original Russian edition: *Fluktuatsii v Dinamicheskikh Sistemakh Pod Deistviem Malykh Sluchainykh Vozmushchenii*, Nauka: Moscow, 1979.

© 1998, 1984 Springer-Verlag New York, Inc.
All rights reserved. This work may not be translated or copied in whole or in part without the written permission of the publisher (Springer-Verlag New York, Inc., 175 Fifth Avenue, New York, NY 10010, USA), except for brief excerpts in connection with reviews or scholarly analysis. Use in connection with any form of information storage and retrieval, electronic adaptation, computer software, or by similar or dissimilar methodology now known or hereafter developed is forbidden.
The use of general descriptive names, trade names, trademarks, etc., in this publication, even if the former are not especially identified, is not to be taken as a sign that such names, as understood by the Trade Marks and Merchandise Marks Act, may accordingly be used freely be anyone.

Production managed by Anthony Guardiola; manufacturing supervised by Joe Quatela.
Typeset by Asco Trade Typesetting Ltd., Hong Kong.
Printed and bound by Edwards Brothers Inc., Ann Arbor, MI.
Printed in the United States of America.

9 8 7 6 5 4 3 2 1

ISBN 0-387-98362-7 Springer-Verlag New York Berlin Heidelberg SPIN 10647749

Preface to the Second Edition

The first edition of this book was published in 1979 in Russian. Most of the material presented was related to large-deviation theory for stochastic processes. This theory was developed more or less at the same time by different authors in different countries. This book was the first monograph in which large-deviation theory for stochastic processes was presented. Since then a number of books specially dedicated to large-deviation theory have been published, including S. R. S. Varadhan [4], A. D. Wentzell [9], J.-D. Deuschel and D. W. Stroock [1], A. Dembo and O. Zeitouni [1]. Just a few changes were made for this edition in the part where large deviations are treated. The most essential is the addition of two new sections in the last chapter. Large deviations for infinite-dimensional systems are briefly considered in one new section, and the applications of large-deviation theory to wave front propagation for reaction-diffusion equations are considered in another one.

Large-deviation theory is not the only class of limit theorems arising in the context of random perturbations of dynamical systems. We therefore included in the second edition a number of new results related to the averaging principle. Random perturbations of classical dynamical systems under certain conditions lead to diffusion processes on graphs. Such problems are considered in the new Chapter 8. Some new results concerning fast oscillating perturbations of dynamical systems with conservation laws are included in Chapter 7. A few small additions and corrections were made in the other chapters as well. We would like to thank Ruth Pfeiffer and Fred Torcaso for their help in the preparation of the second edition of this book.

College Park, Maryland — M.I. Freidlin
New Orleans, Louisiana — A.D. Wentzell

Preface

Asymptotical problems have always played an important role in probability theory. In classical probability theory dealing mainly with sequences of independent variables, theorems of the type of laws of large numbers, theorems of the type of the central limit theorem, and theorems on large deviations constitute a major part of all investigations. In recent years, when random processes have become the main subject of study, asymptotic investigations have continued to play a major role. We can say that in the theory of random processes such investigations play an even greater role than in classical probability theory, because it is apparently impossible to obtain simple exact formulas in problems connected with large classes of random processes.

Asymptotical investigations in the theory of random processes include results of the types of both the laws of large numbers and the central limit theorem and, in the past decade, theorems on large deviations. Of course, all these problems have acquired new aspects and new interpretations in the theory of random processes.

One of the important schemes leading to the study of various limit theorems for random processes is dynamical systems subject to the effect of random perturbations. Several theoretical and applied problems lead to this scheme. It is often natural to assume that, in one sense or another, the random perturbations are small compared to the deterministic constituents of the motion. The problem of studying small random perturbations of dynamical systems has been posed in the paper by Pontrjagin, Andronov, and Vitt [1]. The results obtained in this article relate to one-dimensional and partly two-dimensional dynamical systems and perturbations leading to diffusion processes. Other types of random perturbations may also be considered; in particular, those arising in connection with the averaging principle. Here the smallness of the effect of perturbations is ensured by the fact that they oscillate quickly.

The contents of the book consists of various asymptotic problems arising as the parameter characterizing the smallness of random perturbations converges to zero. Of course, the authors could not consider all conceivable schemes of small random perturbations of dynamical systems. In particular, the book does not consider at all dynamical systems generated by random vector fields. Much attention is given to the study of the effect of perturbations on large time intervals. On such intervals small perturbations essentially influence the behavior of the system in general. In order to take account of this influence, we have to be able to estimate the probabilities of rare events, i.e., we need theorems on the asymptotics of probabilities of large deviations for random processes. The book studies these asymptotics and their applications to problems of the behavior of a random process on large time intervals, such as the problem of the limit behavior of the invariant measure, the problem of exit of a random process from a domain, and the problem of stability under random perturbations. Some of these problems have been formulated for a long time and others are comparatively new.

The problems being studied can be considered as problems of the asymptotic study of integrals in a function space, and the fundamental method used can be considered as an infinite-dimensional generalization of the well-known method of Laplace. These constructions are linked to contemporary research in asymptotic methods. In the cases where, as a result of the effect of perturbations, diffusion processes are obtained, we arrive at problems closely connected with elliptic and parabolic differential equations with a small parameter. Our investigations imply some new results concerning such equations. We are interested in these connections and as a rule include the corresponding formulations in terms of differential equations.

We would like to note that this book is being written when the theory of large deviations for random processes is just being created. There have been a series of achievements but there is still much to be done. Therefore, the book treats some topics that have not yet taken their final form (part of the material is presented in a survey form). At the same time, some new research is not reflected at all in the book. The authors attempted to minimize the deficiencies connected with this.

The book is written for mathematicians but can also be used by specialists of adjacent fields. The fact is that although the proofs use quite intricate mathematical constructions, the results admit a simple formulation as a rule.

Contents

Preface to the Second Edition v

Preface vii

Introduction 1

CHAPTER 1
Random Perturbations 15
§1. Probabilities and Random Variables 15
§2. Random Processes. General Properties 17
§3. Wiener Process. Stochastic Integral 24
§4. Markov Processes and Semigroups 29
§5. Diffusion Processes and Differential Equations 34

CHAPTER 2
Small Random Perturbations on a Finite Time Interval 44
§1. Zeroth Approximation 44
§2. Expansion in Powers of a Small Parameter 51
§3. Elliptic and Parabolic Differential Equations with a Small Parameter at the Derivatives of Highest Order 59

CHAPTER 3
Action Functional 70
§1. Laplace's Method in a Function Space 70
§2. Exponential Estimates 74
§3. Action Functional. General Properties 79
§4. Action Functional for Gaussian Random Processes and Fields 92

CHAPTER 4
Gaussian Perturbations of Dynamical Systems. Neighborhood of an Equilibrium Point 103
§1. Action Functional 103
§2. The Problem of Exit from a Domain 108
§3. Properties of the Quasipotential. Examples 118
§4. Asymptotics of the Mean Exit Time and Invariant Measure for the Neighborhood of an Equilibrium Position 123
§5. Gaussian Perturbations of General Form 132

CHAPTER 5
Perturbations Leading to Markov Processes 136
§1. Legendre Transformation 136
§2. Locally Infinitely Divisible Processes 143
§3. Special Cases. Generalizations 153
§4. Consequences. Generalization of Results of Chapter 4 157

CHAPTER 6
Markov Perturbations on Large Time Intervals 161
§1. Auxiliary Results. Equivalence Relation 161
§2. Markov Chains Connected with the Process $(X_t^\varepsilon, \mathsf{P}_x^\varepsilon)$ 168
§3. Lemmas on Markov Chains 176
§4. The Problem of the Invariant Measure 185
§5. The Problem of Exit from a Domain 192
§6. Decomposition into Cycles. Sublimit Distributions 198
§7. Eigenvalue Problems 203

CHAPTER 7
The Averaging Principle. Fluctuations in Dynamical Systems with Averaging 212
§1. The Averaging Principle in the Theory of Ordinary Differential Equations 212
§2. The Averaging Principle when the Fast Motion is a Random Process 216
§3. Normal Deviations from an Averaged System 219
§4. Large Deviations from an Averaged System 233
§5. Large Deviations Continued 241
§6. The Behavior of the System on Large Time Intervals 249
§7. Not Very Large Deviations 253
§8. Examples 257
§9. The Averaging Principle for Stochastic Differential Equations 268

CHAPTER 8
Random Perturbations of Hamiltonian Systems 283
§1. Introduction 283
§2. Main Results 295
§3. Proof of Theorem 2.2 301
§4. Proofs of Lemmas 3.1 to 3.4 312
§5. Proof of Lemma 3.5 328
§6. Proof of Lemma 3.6 338
§7. Remarks and Generalizations 344

CHAPTER 9
Stability Under Random Perturbations 361
§1. Formulation of the Problem 361
§2. The Problem of Optimal Stabilization 367
§3. Examples 373

CHAPTER 10
Sharpenings and Generalizations 377
§1. Local Theorems and Sharp Asymptotics 377
§2. Large Deviations for Random Measures 385
§3. Processes with Small Diffusion with Reflection at the Boundary 392
§4. Wave Fronts in Semilinear PDEs and Large Deviations 397
§5. Random Perturbations of Infinite-Dimensional Systems 408

References 417

Index 429

Introduction

Let $b(x)$ be a continuous vector field in R^r. First we discuss nonrandom perturbations of a dynamical system

$$\dot{x}_t = b(x_t). \tag{1}$$

We may consider the perturbed system

$$\dot{X}_t = b(X_t, \psi_t), \tag{2}$$

where $b(x, y)$ is a function jointly continuous in its two arguments and turning into $b(x)$ for $y = 0$. We shall speak of *small* perturbations if the function ψ giving the perturbing effect is small in one sense or another.

We may speak of problems of the following kind: the convergence of the solution X_t of the perturbed system to the solution x_t of the unperturbed system as the effect of the perturbation decreases, approximate expressions of various accuracies for the deviations $X_t - x_t$ caused by the perturbations, and the same problems for various functionals of a solution (for example, the first exit time from a given domain D).

To solve the kind of problems related to a finite time interval we require less of the function $b(x, y)$ than in problems connected with an infinite interval (or a finite interval growing unboundedly as the perturbing effect decreases). The simplest result related to a finite interval is the following: if the solution of the system (1) with initial condition x_0 at $t = 0$ is unique, then the solution X_t of system (2) with initial condition X_0 converges to x_t uniformly in $t \in [0, T]$ as $X_0 \to x_0$ and $\|\psi\|_{0T} = \sup_{0 \le t \le T} |\psi_t| \to 0$. If the function $b(x, y)$ is differentiable with respect to the pair of its arguments, then we can linearize it near the point $x = x_t$, $y = 0$ and obtain a linear approximation δ_t of $X_t - x_t$ as the solution of the linear system

$$\dot{\delta}_t^i = \sum_j \frac{\partial b^i}{\partial x^j}(x_t, 0)\delta_t^j + \sum_k \frac{\partial b^i}{\partial y^k}(x_t, 0) \cdot \psi_t^k; \tag{3}$$

under sufficiently weak conditions, the norm $\sup_{0 \le t \le T} |X_t - x_t - \delta_t|$ of the remainder will be $o(|X_0 - x_0| + \|\psi\|_{0T})$. If $b(x, y)$ is still smoother, then we have the decomposition

$$X_t = x_t + \delta_t + \gamma_t + o(|X_0 - x_0|^2 + \|\psi\|_{0T}^2), \tag{4}$$

in which γ_t depends quadratically on perturbations of the initial conditions and the right side (the function γ_t can be determined from a system of linear differential equations with a quadratic function of ψ_t, δ_t on the right side), etc.

We may consider a scheme

$$\dot{X}_t^\varepsilon = b(X_t^\varepsilon, \varepsilon\psi_t) \tag{5}$$

depending on a small parameter ε, where ψ_t is a given function. In this case for the solution X_t^ε with initial condition $X_0^\varepsilon = x_0$ we can obtain a decomposition

$$x_t + \varepsilon Y_t^{(1)} + \varepsilon^2 Y_t^{(2)} + \cdots + \varepsilon^n Y_t^{(n)} \tag{6}$$

in powers of ε with the remainder infinitely small compared with ε^n, uniformly on any finite interval $[0, T]$.

Under more stringent restrictions on the function $b(x, y)$, results of this kind can be obtained for perturbations ψ_t which are not small in the norm of uniform convergence but rather, for example, in some $\mathbf{L}^p$-norm or another.

As far as results connected with an infinite time interval are concerned, stability properties of the unperturbed system (1) as $t \to \infty$ are essential.

Let x_* be an equilibrium position of system (1), i.e., let $b(x_*) = 0$. Let this equilibrium position be asymptotically stable, i.e., for any neighborhood $U \ni x_*$ let there exist a small neighborhood V of x_* such that for any $x_0 \in V$ the trajectory x_t starting at x_0 does not leave U for $t \geq 0$ and converges to x_* as $t \to \infty$. Denote by G_* the set of initial points x_0 from which there start solutions converging to x_* as $t \to \infty$. For any neighborhood U of x_* and any point $x_0 \in G_*$ there exist $\delta > 0$ and $T > 0$ such that for

$$|X_0 - x_0| < \delta, \sup\nolimits_{0 \leq t \leq \infty} |\psi_t| < \delta$$

the solution X_t of system (2) with initial condition x_0 does not go out of U for $t \geq T$. This holds uniformly in x_0 within any compact subset of G_* (i.e., δ and T can be chosen the same for all points x_0 of this compactum). This also implies the uniform convergence of X_t to x_t on the infinite interval $[0, \infty)$ provided that $X_0 \to x_0$, $\sup_{0 \leq t < \infty} |\psi_t| \to 0$.

On the other hand, if the equilibrium position x_* does not have the indicated stability properties, then by means of arbitrarily small perturbations, the solution X_t of the perturbed system can be "carried away" from x_* for sufficiently large t even if the initial point X_0 equals x_*. In particular, there are cases where the solution x_t of the unperturbed system cannot leave some domain D for $t \geq 0$, but the solution X_t of the system obtained from the initial one by an arbitrarily small perturbation leaves the domain in finite time.

Some of these results also hold for trajectories attracted not to a point x_* but rather a compact set of limit points, for example, for trajectories winding on a limit cycle.

There are situations where besides the fact that the perturbations are small, we have sufficient information on their statistical character. In this case it is appropriate to develop various mathematical models of small random perturbations.

The consideration of random perturbations extends the notion of perturbations considered in classical settings at least in two directions. Firstly, the requirements of smallness become less stringent: instead of absolute smallness for all t (or in integral norm) it may be assumed that the perturbations are small only in mean over the ensemble of all possible perturbations. Small random perturbations may assume large values but the probability of these large values is small. Secondly, the consideration of random processes as perturbations extends the notion of the stationarity character of perturbations. Instead of assuming that the perturbations themselves do not change with time, we may assume that the factors which form the statistical structure of the perturbations are constant, i.e., the perturbations are stationary as random processes.

Such an extension of the notion of a perturbation leads to effects not characteristic of small deterministic perturbations. Especially important new properties occur in considering a long lasting effect of small random perturbations.

We shall see what models of small random perturbations may be like and what problems are natural to consider concerning them. We begin with perturbations of the form

$$\dot{X}_t^\varepsilon = b(X_t^\varepsilon, \varepsilon\psi_t), \tag{7}$$

where ψ_t is a given random process, for example, a stationary Gaussian process with known correlation function. (Nonparametric problems connected with arbitrarily random processes which belong to certain classes and are small in some sense are by far more complicated.) For the sake of simplicity, let the initial point X_0 not depend on ε: $X_0^\varepsilon = x_0$. If the solution of system (7) is unique, then the random perturbation $\psi(t)$ leads to a random process X_t^ε.

The first problem which arises is the following: Will X_t^ε converge to the solution x_t of the unperturbed system as $\varepsilon \to 0$? We may consider various kinds of probabilistic convergence: convergence with probability 1, in probability, and in mean. If $\sup_{0 \leq t \leq T} |\psi_t| < \infty$ with probability 1, then, ignoring the fact that the realization of ψ_t is random, we may apply the results presented above to perturbations of the form $\varepsilon\psi_t$ and obtain, under various conditions on $b(x, y)$, that $X_t^\varepsilon \to x_t$ with probability 1, uniformly in $t \in [0, T]$ and that

$$X_t^\varepsilon = x_t + \varepsilon Y_t^{(1)} + o(\varepsilon) \tag{8}$$

or

$$X_t^\varepsilon = x_t + \varepsilon Y_t^{(1)} + \cdots + \varepsilon^n Y_t^{(n)} + o(\varepsilon^n) \tag{9}$$

($o(\varepsilon)$ and $o(\varepsilon^n)$ are understood as being satisfied with probability 1 uniformly in $t \in [0, T]$ as $\varepsilon \to 0$).

Nevertheless, it is not convergence with probability 1 which represents the main interest from the point of view of possible applications. In considering small random perturbations, perhaps we shall not have to do with X_t^ε for various ε simultaneously but only for one small ε. We shall be interested in questions such as: Can we guarantee with practical certainty that for a small ε the value of X_t^ε is close to x_t? What will the order of the deviation $X_t^\varepsilon - x_t$ be? What can be said about the distribution of the values of the random process X_t^ε and functionals thereof? etc. Fortunately, convergence with probability 1 implies convergence in probability, so that X_t^ε will converge to x_t in probability, uniformly in $t \in [0, T]$ as $\varepsilon \to 0$:

$$\mathsf{P}\left\{\sup_{0 \le t \le T} |X_t^\varepsilon - x_t| \ge \delta\right\} \to 0 \tag{10}$$

for any $\delta > 0$.

For convergence in mean we have to impose still further restrictions on $b(x, y)$ and ψ_t; we shall not discuss this.

From the sharper result (8) it follows that the random process

$$Y_t^\varepsilon = \frac{X_t^\varepsilon - x_t}{\varepsilon}$$

converges to a random process $Y_t^{(1)}$ in the sense of distributions as $\varepsilon \to 0$ (this latter process is connected with the random perturbing effect ψ_t through linear differential equations). In particular, this implies that if ψ_t is a Gaussian process, then in first approximation, the random process X_t^ε will be Gaussian with mean x_t and correlation function proportional to ε^2. This implies the following result: if f is a smooth scalar-valued function in R^r and $\operatorname{grad} f(x_{t_0}) \neq 0$, then

$$\mathsf{P}\left\{\frac{f(X_{t_0}^\varepsilon) - f(x_{t_0})}{\varepsilon} \le x\right\} = \Phi\left(\frac{x}{\sigma}\right) + o(1) \tag{11}$$

as $\varepsilon \to 0$, where $\Phi(y) = \int_{-\infty}^{y} (1/\sqrt{2\pi})e^{-z^2/2}dz$ is the Laplace function and σ is determined from grad $f(x_{t_0})$ and the value of the correlation function of $Y_t^{(1)}$ at the point (t_0, t_0). We may obtain sharper results from (9): an expansion of the remainder $o(1)$ in powers of ε. We may also obtain results relative to asymptotic distributions of functionals of Y_t^ε, $0 \le t \le T$, and sharpenings of them, connected with asymptotic expansions.

Hence for random perturbations of the form (7) we may pose and solve a series of problems characteristic of the limit theorems of probability theory. Results on the convergence in probability of a random solution of the perturbed system to a nonrandom function correspond to laws of large numbers for sums of independent random variables. We can speak of the limit distribution under a suitable normalization; this corresponds to results of the type of the central limit theorem. Also as in sharpenings of the central limit theorem, we may obtain asymptotic expansions in powers of the parameter.

In the limit theorems for sums of independent random variables there is still another direction: the study of probabilities of *large deviations* (after normalization) of a sum from the mean. Of course, all these probabilities converge to zero. Nevertheless we may study the problem of finding simple expressions equivalent to them or the problem of sharper (or rougher) asymptotics of them. The first general results concerning large deviations for sums of independent random variables have been obtained by Cramér [1]. These results have to do with asymptotics, up to equivalence, of probabilities of the form

$$\mathsf{P}\left\{\frac{\xi_1 + \cdots + \xi_n - nm}{\sigma\sqrt{n}} > x\right\} \tag{12}$$

as $n \to \infty$, $x \to \infty$ and also asymptotic expansions for such probabilities (under more stringent restrictions).

We may be interested in analogous problems for a family of random processes X_t^ε arising as a result of small random perturbations of a dynamical system. For example, let A be a set in a function space on the interval $[0, T]$, which does not contain the unperturbed trajectory x_t (and is at a positive distance from it). Then the probability

$$\mathsf{P}\{X^\varepsilon \in A\} \tag{13}$$

of the event that the perturbed trajectory X_t^ε belongs to A, of course, converges to 0 as $\varepsilon \to 0$, but what is the asymptotics of this infinitely small probability?

It may seem that such digging into extremely rare events contradicts the general spirit of probability theory, which ignores events of small probability. Nevertheless, it is exactly this determination of which almost unlikely events related to the random process X_t^ε on a finite interval are "more improbable" and which are "less improbable," that, in several cases, serves as a key to the question of what the behavior, with probability close to 1, of the process X_t^ε will be on an infinite time interval (or on an interval growing with decreasing ε).

Indeed, for the sake of definiteness, we consider the particular case of perturbations of the form (7):

$$\dot{X}_t^\varepsilon = b(X_t^\varepsilon) + \varepsilon\psi_t. \tag{14}$$

Furthermore, let ψ_t be a stationary Gaussian process. Assume that the trajectories of the unperturbed system (1), beginning at points of a bounded domain D, do not leave this domain for $t \geq 0$ and are attracted to a stable equilibrium position x_* as $t \to \infty$. Will the trajectories of the perturbed system (14) also have this property with probability near 1? The results above related to small nonrandom perturbations cannot help us answer this question, since the supremum of $|\psi_t|$ for $t \in [0, \infty)$ is infinite with probability 1 (if we do not consider the case of "very degenerate" processes ψ_t). We have to approach this question differently. We divide the time axis $[0, \infty)$ into a countable number of intervals of length T. On each of these intervals, for small ε, the most likely behavior of X_t^ε is such that the supremum of $|X_t^\varepsilon - x_t|$ over the interval is small. (For intervals with large indices, X_t^ε will be simply close to x_* with overwhelming probability.) All other ways of behavior, in particular, the exit of X_t^ε from D on a given time interval, will have small probabilities for small ε. Nonetheless, these probabilities are positive for any $\varepsilon > 0$. (Again, we exclude from our considerations the class of "very degenerate" random processes ψ_t.) For a given $\varepsilon > 0$ the probability

$$\mathsf{P}\{X_t^\varepsilon \notin D \text{ for some } t \in [kT, (k+1)T]\} \tag{15}$$

will be almost the same for all intervals with large indices. If the events involving the behavior of our random process on different time intervals were independent, we would obtain from this that sooner or later, with probability 1, the process X_t^ε leaves D and the first exit time τ^ε has an approximately exponential distribution with parameter $T^{-1}\mathsf{P}\{X_t^\varepsilon$ exits from D for some $t \in [kT, (k+1)T]\}$. The same will happen if these events are not exactly independent but the dependence between them decreases for distant intervals in a certain manner. This can be ensured by some weak dependence properties of the perturbing random process ψ_t.

Hence for problems connected with the exit of X_t^ε from a domain for small ε, it is essential to know the asymptotics of the probabilities of improbable events ("large deviations") involving the behavior of X_t^ε on finite time intervals. In the case of small Gaussian perturbations it turns out that these probabilities have asymptotics of the form $\exp\{-C\varepsilon^{-2}\}$ as $\varepsilon \to 0$ (rough asymptotics, i.e., not up to equivalence but logarithmic equivalence). It turns out that we can introduce a functional $S(\varphi)$ defined on smooth functions (which are smoother than the trajectories of X_t^ε), such that

$$\mathsf{P}\{\rho(X^\varepsilon, \varphi) < \delta\} \approx \exp\{-\varepsilon^{-2}S(\varphi)\} \tag{16}$$

for small positive δ and ε, where ρ is the distance in a function space (say, in the space of continuous functions on the interval from T_1 to T_2; for the precise meaning of formula (16), cf. Ch. 3). The value of the functional at a given function characterizes the difficulty of the passage of X_t^ε near the function. The probability of an unlikely event consists of the contributions $\exp\{-\varepsilon^{-2}S(\varphi)\}$ corresponding to neighborhoods of separate functions φ; as $\varepsilon \to 0$, only the summand with smallest $S(\varphi)$ becomes essential. Therefore, it is natural that the constant C providing the asymptotics is determined as the infimum of $S(\varphi)$ over the corresponding set of functions φ. Thus for the probability in formula (15) the infimum has to be taken over smooth functions φ_t leaving D for $t \in [kT, (k+1)T]$. (Exact formulations and the form of the functional $S(\varphi)$ may be found in §5, Ch. 4; there we discuss its application to finding the asymptotics of the exit time τ^ε as $\varepsilon \to 0$.)

Another problem related to the behavior of X_t^ε on an infinite time interval is the problem of the limit behavior of the stationary distribution μ^ε of X_t^ε as $\varepsilon \to 0$. This limit behavior is connected with the limit sets of the dynamical system (1). Indeed, the stationary distribution shows how much time the process spends in one set or another. It is plausible to expect that for small ε the process X_t^ε will spend an overwhelming amount of time near limit sets of the dynamical system and, most likely, near stable limit sets. If system (1) has only one stable limit set K, then the measure μ^ε converges weakly to a measure concentrated on K as $\varepsilon \to 0$ (we do not formulate our assertions in so precise a way that we take account of the possibility of the existence of distinct limits μ^{ε_i} for different sequences $\varepsilon_i \to 0$). However, if there are several stable sets, even if there are at least two, K_1 and K_2, then the situation becomes unclear; it depends on the exact form of small perturbations.

The problem of what happens to the stationary distribution of a random process arising as an effect of random perturbations of a dynamical system when these perturbations decrease has been posed in the paper of Pontrjagin, Andronov, and Vitt [1]. The approach applied in this article does not relate to perturbations of the form (14) but rather perturbations under whose influence there arise diffusion processes (given by formulas (19) and (20) below). This approach is based on solving the Fokker–Planck differential equation; in the one-dimensional case the problem of finding the asymptotics of the stationary distribution has been solved completely (cf. also Bernstein's article [1] which appeared in the same period). Some results involving the stationary distribution in the two-dimensional case have also been obtained.

Our approach is not based on equations for the probability density of the stationary distribution but rather the study of probabilities of improbable events. We outline the scheme of application of this approach to the problem of asymptotics of the stationary distribution.

The process X_t^ε spends most of the time in neighborhoods of the stable limit sets K_1 and K_2, it occasionally moves to a significant distance from K_1 or K_2 and returns to the same set, and it very seldom passes from K_1 to K_2 or conversely. If we establish that the probability of the passage of X_t^ε from K_1 to K_2 over a long time T (not depending on ε) converges to 0 with rate

$$\exp\{-V_{12}\varepsilon^{-2}\}$$

as $\varepsilon \to 0$, and the probability of passage from K_2 to K_1 has the order

$$\exp\{-V_{21}\varepsilon^{-2}\}$$

and $V_{12} < V_{21}$, then it becomes plausible that for small ε the process spends most of the time in the neighborhood of K_2. This is so since a successful "attempt" at passage from K_1 to K_2 will fall on a smaller number of time intervals $[kT, (k+1)T]$ spent by the process near K_1, than a successful attempt at passage from K_2 to K_1 with respect to the number of time intervals of length T spent near K_2. Then μ^ε will converge to a measure concentrated on K_2. The constants V_{12} and V_{21} can be determined as the infima of the functional $S(\varphi)$ over the smooth functions φ passing from K_1 to K_2 and conversely on an interval of length T (more precisely, they can be determined as the limits of these infima as $T \to \infty$).

The program of the study limit behavior which we have outlined here is carried out not for random perturbations of the form (14) but rather perturbations leading to Markov processes; the exact formulations and results are given in §4, Ch. 6.

As we have already noted, random perturbations of the form (14) do not represent the only scheme of random perturbations which we shall consider (and not even the scheme to which we shall pay the greatest attention). An immediate generalization of it may be considered, in which the random process ψ_t is replaced by a generalized random process, a "white noise," which can be defined as the derivative (in the sense of distributions) of the Wiener process w_t:

$$\dot{X}_t^\varepsilon = b(X_t^\varepsilon) + \varepsilon \dot{w}_t. \tag{17}$$

Upon integrating equation (17), it takes the following form which does not contain distributions:

$$X_t^\varepsilon = X_0 + \int_0^t b(X_s^\varepsilon)\,ds + \varepsilon(w_t - w_0). \tag{18}$$

For perturbations of this form we can solve a larger number of interesting problems than for perturbations of the form (14), since they lead to a Markov process X_t^ε.

A further generalization is perturbations which depend on the point of the space and are of the form

$$X_t^\varepsilon = b(X_t^\varepsilon) + \varepsilon\sigma(X_t^\varepsilon)\dot{w}_t. \tag{19}$$

where $\sigma(x)$ is a matrix-valued function. The precise meaning of equation (19) can be formulated in the language of stochastic integrals in the following way:

$$X_t^\varepsilon = X_0 + \int_0^t b(X_s^\varepsilon)\, ds + \varepsilon \int_0^t \sigma(X_s^\varepsilon)\, dw_s. \tag{20}$$

Every solution of equation (20) is also a Markov process (a diffusion process with drift vector $b(x)$ and diffusion matrix $\varepsilon^2\sigma(x)\sigma^*(x)$). For perturbations of the white noise type, given by formulas (19), (20), we can also obtain results on convergence to the trajectories of the unperturbed system, of the the type (10), and results on expansions of the type (9) in powers of ε, from which we can obtain results on asymptotic Gaussian character (for example, of the type (11)). Of course, since the white noise is a generalized process whose realizations are not bounded functions in any sense, these results cannot be obtained from the results concerning nonrandom perturbations mentioned at the beginning of the introduction; they have to be obtained independently (cf. §2, Ch. 2).

For perturbations of the white noise type we establish results concerning probabilities of large deviations of the trajectory X_t^ε from the trajectory x_t of the dynamical system (cf. §1, Ch. 4 and §3, Ch. 5). Moreover, because of the Markovian character of the processes, they become even simpler; in particular, the functional $S(\varphi)$ indicating the difficulty of passage of a trajectory near a function takes the following simple form:

$$S(\varphi) = \frac{1}{2}\int \sum_{i,j} a_{ij}(\varphi_t)(\dot\varphi_t^i - b^i(\varphi_t))(\dot\varphi_t^j - b^j(\varphi_t))\, dt,$$

where $(a_{ij}(x)) = (\sigma(x)\sigma^*(x))^{-1}$.

What other schemes of small random perturbations of dynamical systems shall we consider? What families of random processes will arise in our study? The generalizations may go in several directions and it is not clear which of these directions are preferred to others. Nevertheless, the problem may be posed in a different way: In what case may a given family of random processes be considered as a result of a random perturbation of the dynamical system (1)?

First, in the same way as we may consider the trajectory of a dynamical system, issued from any point, we have to be able to begin the random process at any point x of the space at any time t_0. Further the random process under consideration should depend on a parameter h characterizing the smallness of perturbations. For the sake of simplicity, we shall assume h is a positive numerical parameter converging to zero (in §3, Ch. 5 families depending on a two-dimensional parameter are considered). Hence for every real t_0, $x \in R^r$ and $h > 0$, $X_t^{t_0, x; h}$ is a random process with values in R^r, such that $X_{t_0}^{t_0, x; h} = x$. We shall say that $X_t^{t_0, x; h}$ is a result of small random perturbations of system (1) if $X_t^{t_0, x; h}$ converges in probability to the solution $x_t^{t_0, x}$ of the unperturbed system (1) with the initial condition $x_{t_0}^{t_0, x} = x$ as $h \downarrow 0$.

This scheme incorporates many families of random processes, arising in various problems naturally but not necessarily as a result of the "distortion" of some initial dynamical system.

EXAMPLE 1. Let $\{\xi_n\}$ be a sequence of independent identically distributed r-dimensional random vectors. For $t_0 \in R^1$, $x \in R^r$, $h > 0$ we put

$$X_t^{t_0, x; h} = x + h \sum_{k=[h^{-1}t_0]}^{[h^{-1}t]-1} \xi_k. \tag{21}$$

It is easy to see that $X^{t_0, x; h}$ converges in probability to $x^{t_0, x} = x + (t - t_0)m$, uniformly on every finite time interval as $h \downarrow 0$ (provided that the mathematical expectation $m = \mathsf{M}\xi_k$ exists), i.e., it converges to the trajectory of the dynamical system (1) with $b(x) \equiv m$.

EXAMPLE 2. For every $h > 0$ we construct a Markov process on the real line in the following way. Let two nonnegative continuous functions $l(x)$ and $r(x)$ on the real line be given. Our process, beginning at a point x, jumps to the point $x - h$ with probability $h^{-1}l(x)\,dt$ over time dt, to the point $x + h$ with probability $h^{-1}r(x)\,dt$, and it remains at x with the complementary probability. An approximate calculation of the mathematical expectation and variance of the increment of the process over a small time interval Δt shows that as $h \downarrow 0$, the random process converges to the deterministic, nonrandom process described by equation (1) with $b(x) = r(x) - l(x)$ (the exact results are in §2, Ch. 5).

Still another class of examples: ξ_t is a stationary random process and $X_t^h = X_t^{t_0, x; h}$ is the solution of the system

$$\dot{X}_t^h = b(X_t^h, \xi_{h^{-1}t}) \tag{22}$$

with initial condition x at time t_0. It can be proved under sufficiently weak assumptions that X_t^h converges to a solution of (1) with $b(x) = \mathsf{M}b(r, \xi_s)$ as $h \downarrow 0$ ($\mathsf{M}b(x, \xi_s)$ does not depend on s; the exact results may be found in §2, Ch. 7).

In the first example, the convergence in probability of $X_t^{t_0, x; h}$ as $h \downarrow 0$ is a law of large numbers for the sequence $\{\xi_n\}$. Therefore, in general we shall speak of results establishing the convergence in probability of random processes of a given family to the trajectories of a dynamical system as of results of the type of the law of large numbers. Similarly, results involving the convergence, in the sense of distributions, of a family of random processes $X_t^{t_0, x; h} - x_t^{t_0, x}$ after an appropriate normalization to a Gaussian process are results of the type of the central limit theorem. Results involving large deviations are results involving the asymptotics of probabilities of events that the realization of a random process falls in some sets of functions, not containing the trajectory $x_t^{t_0, x}$ of the unperturbed dynamical system. We say a few words on results of the last kind.

For the random step function (21) constructed from the independent random variables ξ_k, the results of the type of large deviations are connected, of course, with the asymptotics, as $n \to \infty$, of probabilities of the form

$$\mathsf{P}\left\{\frac{\xi_1 + \cdots + \xi_n}{n} > x\right\}. \tag{23}$$

The results concerning the asymptotics of probabilities (23) can be divided into two groups: for rapidly decreasing "tails" of the distribution of the terms ξ_i, the principal term of the probability is due to uniformly not too large summands and the asymptotics has the form $\exp\{-Cn\}$ (up to logarithmic equivalence); if, on the other hand, the "tails" of the ξ_i decrease slowly, then the principal part of probability (23) is due to one or more summands of order nx and the probability has the same order as $n\mathsf{P}\{\xi_i > nx\}$. The first general results concerning large deviations were obtained by Cramér under the assumption that the exponential moments $\mathsf{M}e^{z\xi_i}$ are finite, at least for all sufficiently small z; they belong to the first group of results. The results, considered in this book, on large deviations for families of random processes are also generalizations of results belonging to the first group. The assumptions under which they are obtained include analogues of the Cramér condition $\mathsf{M}e^{z\xi_i} < \infty$. Moreover, approximately half of the devices used in obtaining these results is a generalization of Cramér's method (cf. §§2 and 3, Ch. 3 and §§1 and 2, Ch. 5).

Furthermore, in this book we only consider rough results on large deviations, which hold up to logarithmic equivalence. In connection with this we introduce a notation for rough (logarithmic) equivalence:

$$A_h \asymp B_h \qquad (h \downarrow 0), \tag{24}$$

if $\ln A_h \sim \ln B_h$ as $h \downarrow 0$.

Cramér's results and a great many subsequent results are not rough but sharp (up to equivalence and even sharper). Nevertheless, we have to take into consideration that random processes are more complicated objects than sums of independent variables. One may try to obtain sharp results on the asymptotics of large deviations for families of random processes; some results have indeed been obtained in this direction. However, in this respect there is an essentially different direction of research: from theorems on large deviations one tries to obtain various other interesting results on the asymptotic behavior of families of random processes which are deterministic in the limit (which may be considered as a result of small random perturbations of a dynamical system). In the authors' opinion, one can deduce more interesting rough consequences from rough theorems on large deviations than sharp consequences from sharp theorems.

Hence we shall consider results of three kinds: results of the type of the law of large numbers, of the type of the central limit theorem, and rough results of the type of large deviations (and, of course, all sorts of consequences of these results). The results of the first type are the weakest; they follow from results of the second or third type. Sometimes we shall speak of them in the first place because it is easier to obtain them and because they are a sort of test of the correctness of a family of random processes to appear in general as a result of small perturbations of a dynamical system.

The results of the second and third types are independent of each other and neither is stronger than the other. Therefore, in some cases we do not consider results of the type of the central limit theorem but rather discuss large deviations immediately (and in the process of obtaining results in this area, we obtain results of the type of the law of large numbers automatically).

The random perturbations are said to be homogeneous in time if the distributions of the values of the arising random process at any finite number of moments of time does not change if we simultaneously shift these moments and the initial moment t_0 along the time axis. In this case all that can be said about perturbations can be formulated naturally in terms of the family $X_t^{x,h}$ of random processes beginning at the point x at time 0: $X_0^{x,h} = x$. Among the schemes of random perturbations we consider, only (21) is not homogeneous in time.

We discuss the content of the book briefly. First we note that we consider problems in probability theory in close connection with problems of the theory of partial differential equations. To the random processes arising as a result of small random perturbations there correspond problems connected with equations containing a small parameter. We study the random perturbations by direct probabilistic methods and then deduce consequences concerning the corresponding problems for partial differential equations. The problems involving the connection between the theory of Markov processes and that of partial differential equations are discussed in Chapter 1. There we recall the necessary information from the theory of random processes.

In Chapter 2 we consider mainly schemes of random perturbations of the form $\dot{X}_t^\varepsilon = b(X_t^\varepsilon, \varepsilon\xi_t)$ or $\dot{X}_t^\varepsilon = b(X_t^\varepsilon) + \varepsilon\sigma(X_t^\varepsilon)\dot{w}_t$, where $\dot{w}_t$ is a white noise process. We discuss results of the type of the law of large numbers in §1, we discuss sharper results, connected with asymptotic expansions in §2, and the application of these results to partial differential equations in §3.

In Chapter 3, for the first time in this book, we consider results involving large deviations for a very simple family of random processes, namely, for the Wiener process w_t multiplied by a small parameter ε. The rough asymptotics of probabilities of large deviations can be described by means of the action functional. The action functional appears in all subsequent chapters. The general questions involving the description of large deviations by means of such functionals constitute the content of §3 of this chapter. We calculate the action functional for families of Gaussian processes in §4.

Chapter 4 is devoted mainly to the study of perturbations of dynamical systems by a white noise process. We determine the action functional for the corresponding family of random processes. We study the problem of exit from a neighborhood of a stable equilibrium position of a dynamical system, due to random perturbations, and we determine the asymptotics of the average exit time of the neighborhood and the position at the first exit time. In the same chapter we study the asymptotics of the invariant measure for a dynamical system with one equilibrium position. The problems to be considered are closely connected with the behavior, as $\varepsilon \to 0$, of the solution of problems for elliptic equations with a small parameter at the derivatives of the highest order. The limit behavior of the solution of Dirichlet's problem for an elliptic equation of the second order with a small parameter at the derivatives of the highest order in the case where the characteristics of the corresponding degenerate equation go out to the boundary was studied by Levinson [1]. In Chapter 4 this limit behavior is studied in the case where the characteristics are attracted to a stable equilibrium position inside the domain. (The case of a more complicated behavior of the characteristics is considered in Chapter 6.) We consider Gaussian perturbations of the general form in the last section of Chapter 4.

In Chapter 5 we generalize results of Chapter 4 to a sufficiently large class of families of Markov processes (including processes with discontinuous trajectories). Here the connection with theorems on large deviations for sums of independent random variables becomes clearer; in particular, there appears the apparatus of Legendre transforms of convex functions, which is a natural tool in this area (a separate section is devoted to Legendre transforms).

In Chapter 6 the generalization goes in a different direction: from problems for systems with one equilibrium position to systems with a more complicated structure of equilibrium positions, limit sets, etc. Here an essential role is played by sets of points equivalent to each other in the sense of a certain equivalence relation connected with the system and the perturbations. In the case of a finite number of critical sets, the perturbed system can be approximated in some sense by a finite Markov chain with transition probabilities depending on the small parameter. For the description of the limit behavior of such chains a peculiar apparatus of discrete character, connected with graphs, is developed. A large portion of the results of this chapter admits a formulation in the language of differential equations.

In Chapter 7 we consider problems connected with the averaging principle. Principally, we consider random processes defined by equations of the form $\dot{X}_t^\varepsilon = b(X_t^\varepsilon, \xi_{t/\varepsilon})$, where ξ_t is a stationary process with sufficiently good mixing properties. For the family of random processes X_t^ε we establish theorems of the type of the law of large numbers, the central limit theorem, and finally, of large deviations. Special attention is paid to the last group of questions. In §6, Ch. 7 we study the behavior of X_t^ε on large time intervals. Here we also consider examples and the corresponding problems of the theory of partial differential equations. In Chapter 7 we also consider systems of differential equations in which the velocity of the fast motion depends on the "slow" variables.

White noise perturbations of Hamiltonian systems are considered in Chapter 8. Let $b(x)$ in (17) be a Hamiltonian vector field in $R^2 : b(x) = (\partial H(x)/\partial x^2, -\partial H(x)/\partial x^1)$, where $H(x)$, $x \in R^2$, is a smooth function. Let $\lim_{|x|\to\infty} H(x) = \infty$. Then all the points of the phase space R^2 are equivalent for the process X_t^ε defined by (17) from the large-deviation point of view. This means, roughly speaking, that for any $x \in R^2$ and any open set $\mathscr{E} \subset R^2$ one can find a nonrandom $t = t(\varepsilon)$ such that the probability that X_t^ε goes from x to $\mathscr{E}$ in the time $t(\varepsilon)$ is, at least, not exponentially small as $\varepsilon \downarrow 0$. Here the averaging principle rather than large-deviation theory allows us to calculate asymptotic behavior of many interesting characteristics of the process X_t^ε as $\varepsilon \downarrow 0$. One can single out a fast and a slow component of X_t^ε as $\varepsilon \downarrow 0$: The fast component is close to the deterministic motion with a fixed slow component. The slow component $H(X_{t/\varepsilon}^\varepsilon)$, at least locally, is close to the one-dimensional diffusion governed by an operator with the coefficients defined by averaging with respect to the fast component. But if the Hamiltonian $H(x)$ has more than one critical point, the slow component $H(X_{t/\varepsilon}^\varepsilon)$ does not converge in general to a Markov process as $\varepsilon \downarrow 0$. To have a Markov process the limit should consider a projection of X_t^ε on the graph homeomorphic to the set of all connected components of the level sets of $H(x)$, provided with the natural topology. We describe in Chapter 8 the diffusion processes on graphs and calculate the process which is the limit of the slow component of X_t^ε as $\varepsilon \downarrow 0$. As usual, such a result concerning the diffusion process implies a new result concerning the PDEs with a small diffusion coefficient and the Hamiltonian field as a drift. The limit of the solutions of a Dirichlet problem for such an equation is found as the solution of an appropriate boundary problem in a domain on the graph.

Chapter 9 contains the applications of the results obtained in the preceding chapters to the study of stability with respect to small random perturbations. We introduce a certain numerical characteristic of stability, which is connected with the action functional. A series of optimal stabilization problems is considered.

The last, tenth, chapter has the character of a survey. We discuss sharpenings of theorems on large deviations, large deviations for random measures, results concerning the action functional for diffusion processes with reflection at the boundary, random perturbations of infinite-dimensional systems, and applications of large-deviation theory to asymptotic problems for reaction-diffusion equations.

Since the first edition of this book was published, many papers have appeared that generalize and refine some of the results included there. Some further results and references can be found in Day [1], [2] (exit problem for processes with small diffusion). The averaging principle is studied in Kifer [4], [5], Liptser [1], [2], and Gulinskii and Veretennikov [1]. Problems concerning infinite-dimensional systems are considered in Da Prato and Zabczyk [1]. Further references can be found in these publications and in the monographs on large-deviation theory mentioned in the Foreword.

Chapter 1

Random Perturbations

§1. Probabilities and Random Variables

We shall assume known the basic facts of the Lebesgue integral and measure theory, as well as probability theory. The necessary information concerning these topics is contained, for example, in the corresponding chapters of the book by Kolmogorov and Fomin [1] and in the book by Gikhman and Skorokhod [1]. In this chapter we introduce notation and recall some information from the theory of stochastic processes in an appropriate form. We shall not provide proofs but rather references to the pertinent literature.

According to Kolmogorov's axiomatics, at the base of all probability theory is a triple $\{\Omega, \mathscr{F}, \mathsf{P}\}$ of objects called a probability field or probability space. Here Ω is a nonempty set, which is interpreted as the space of elementary events. The second object, $\mathscr{F}$, is a σ-algebra of subsets of Ω. Finally, P is a probability measure on the σ-algebra $\mathscr{F}$, i.e., a countably additive nonnegative set function normalized by the condition $\mathsf{P}(\Omega) = 1$. The elements of the σ-algebra $\mathscr{F}$ are called events.

The most important objects of probability theory are random variables, i.e., functions $\xi(\omega)$ defined on Ω with values on the real line R^1 such that $\{\omega: \xi(\omega) < x\} \in \mathscr{F}$ for every $x \in R^1$. In general, a random variable $\xi(\omega)$ with values in a measurable space $(X, \mathscr{B})$ is a measurable mapping of $(\Omega, \mathscr{F})$ into $(X, \mathscr{B})$[1].

If as $(X, \mathscr{B})$ we take the r-dimensional space R^r with the σ-algebra $\mathscr{B}^r$ of Borel sets, then the corresponding mapping $\xi(\omega)$ is called an r-dimensional random variable. The probability measure defined by the equality

$$\mu(D) = \mathsf{P}\{\xi(\omega) \in D\}, \qquad D \in \mathscr{B}$$

on the σ-algebra $\mathscr{B}$ is called the distribution of the random variable $\xi(\omega)$.

For random variables $\xi(\omega)$ with values in R^1 the mathematical expectation $\mathsf{M}\xi(\omega) = \int_\Omega \xi(\omega)\mathsf{P}(d\omega)$ is defined provided that this integral exists as a

[1] A measurable space is a set X together with a σ-algebra $\mathscr{B}$ of subsets of X. The measurability of a mapping means that the inverse image of every measurable set is measurable.

Lebesgue integral. In this book we shall use repeatedly Chebyshev's inequality

$$\mathsf{P}\{\xi(\omega) \geq a\} \leq \frac{\mathsf{M}f(\xi)}{f(a)}$$

for any nonnegative monotone increasing function $f(\cdot)$ on R^1 provided that $\mathsf{M}f(\xi) < \infty$.

For integrals on parts of Ω we shall sometimes use the notation

$$\int_A \xi(\omega)\mathsf{P}(d\omega) = \mathsf{M}(A;\xi).$$

If in the space $(X, \mathscr{B})$ a topology is given and the open sets are measurable, then we can speak of the convergence of random variables with values in $(X, \mathscr{B})$. Various kinds of convergence can be considered. A sequence of r-dimensional random variables $\xi_n(\omega)$ is said to converge in probability to an r-dimensional random variable $\xi(\omega)$ if $\lim_{n\to\infty} \mathsf{P}\{|\xi_n(\omega) - \xi(\omega)| > \delta\} = 0$ for any $\delta > 0$, where $|\xi_n(\omega) - \xi(\omega)|$ is the Euclidean length of the vector $\xi_n(\omega) - \xi(\omega)$. If $\lim_{n\to\infty} \mathsf{M}|\xi_n(\omega) - \xi(\omega)|^2 = 0$, then we say that ξ_n converges to ξ in mean square. Finally, a sequence ξ_n converges to ξ with probability 1 or almost surely if $\mathsf{P}\{\lim_{n\to\infty} \xi_n(\omega) = \xi(\omega)\} = 1$.

If $\xi_n \to \xi$ in mean square, then $\mathsf{M}\xi_n \to \mathsf{M}\xi$, as follows from the Cauchy–Bunyakovsky inequality. If $\xi_n \to \xi$ almost surely or in probability, then for the convergence of $\mathsf{M}\xi_n$ to $\mathsf{M}\xi$ we must impose additional assumptions. For example, it is sufficient to assume that the absolute values of variables ξ_n do not exceed a certain random variable $\eta(\omega)$ having finite mathematical expectation (Lebesgue's theorem). We shall use repeatedly this and other theorems on the passage to the limit under the mathematical expectation sign. All the needed information concerning this question can be found in the book by Kolmogorov and Fomin [1].

Let $\mathscr{G}$ be a σ-subalgebra of the σ-algebra $\mathscr{F}$ and suppose $\mathscr{G}$ is complete with respect to the measure P (this means that the σ-algebra $\mathscr{G}$ contains, together with every set A, all sets from $\mathscr{F}$ differing from A by a set of probability 0).

Let η be a one-dimensional random variable having finite mathematical expectation. The conditional mathematical expectation of η with respect to the σ-algebra $\mathscr{G}$, denoted by $\mathsf{M}(\eta|\mathscr{G})$, is defined as the function on Ω measurable with respect to the σ-algebra $\mathscr{G}$ for which

$$\int_\Lambda \mathsf{M}(\eta|\mathscr{G})\mathsf{P}(d\omega) = \int_\Lambda \eta(\omega)\mathsf{P}(d\omega).$$

for every $\Lambda \in \mathscr{G}$. The existence of the random variable $\mathsf{M}(\eta|\mathscr{G})$ follows from the Radon–Nikodym theorem. The same theorem implies that any two such random variables coincide everywhere except, maybe, on a set of measure 0. If the random variable $\chi_A(\omega)$ is equal to 1 on some set $A \in \mathscr{F}$ and 0 for $\omega \notin A$, then $\mathsf{M}(\chi_A \mid \mathscr{G})$ is called the conditional probability of the event A with respect to the σ-algebra $\mathscr{G}$ and is denoted by $\mathsf{P}(A \mid \mathscr{G})$. We list the basic properties of conditional mathematical expectations.

1. $\mathsf{M}(\eta \mid \mathscr{G}) \geq 0$ if $\eta \geq 0$.
2. $\mathsf{M}(\xi + \eta \mid \mathscr{G}) = \mathsf{M}(\xi \mid \mathscr{G}) + \mathsf{M}(\eta \mid \mathscr{G})$ if each term on the right exists.
3. $\mathsf{M}(\xi\eta \mid \mathscr{G}) = \xi \mathsf{M}(\eta \mid \mathscr{G})$ if $\mathsf{M}\xi\eta$ and $\mathsf{M}\eta$ are defined and ξ is measurable with respect to the σ-algebra $\mathscr{G}$.
4. Let $\mathscr{G}^1$ and $\mathscr{G}^2$ be two σ-algebras such that $\mathscr{G}^1 \subset \mathscr{G}^2 \subseteq \mathscr{F}$. We have $\mathsf{M}(\xi \mid \mathscr{G}^1) = \mathsf{M}(\mathsf{M}\xi \mid \mathscr{G}^2) \mid \mathscr{G}^1)$.
5. Assume that the random variable ξ does not depend on the σ-algebra $\mathscr{G}$, i.e., $\mathsf{P}(\{\xi \in D\} \cap A) = \mathsf{P}\{\xi \in D\} \cdot \mathsf{P}(A)$ for any Borel set D and any $A \in \mathscr{G}$. Then $\mathsf{M}(\xi \mid \mathscr{G}) = \mathsf{M}\xi$ provided that the latter mathematical expectation exists.

We note that the conditional mathematical expectation is defined up to values on a set of probability 0, and that all equalities between conditional mathematical expectations are satisfied everywhere with the possible exception of a subset of the space Ω having probability 0. In those cases which do not lead to misunderstanding, we shall not state this explicitly.

The proofs of Properties 1–5 and other properties of conditional mathematical expectations and conditional probabilities may be found in the book by Gikhman and Skorokhod [1].

§2. Random Processes. General Properties

Let us consider a probability space $\{\Omega, \mathscr{F}, \mathsf{P}\}$, a measurable space $(X, \mathscr{B})$, and a set T on the real line. A family of random variables $\xi_t(\omega)$, $t \in T$ with values in $(X, \mathscr{B})$ is called a random process. The parameter t is usually called time and the space X the phase space of the random process $\xi_t(\omega)$. Usually, we shall consider random processes whose phase space is either the Euclidean space R^r or a smooth manifold. For every fixed $\omega \in \Omega$ we obtain a function ξ_t, $t \in T$ with values in X, which is called a trajectory, realization, or sample function of the process $\xi_t(\omega)$.

The collection of distributions $\mu_{t_1, \dots, t_r}$ of the random variables $(\xi_{t_1}, \xi_{t_2}, \dots, \xi_{t_r})$ for all $r = 1, 2, 3, \dots$ and $t_1, \dots, t_r \in T$ is called the family of finite-dimensional distributions of ξ_t. If T is a countable set, then the finite-dimensional distributions determine the random process to the degree of uniqueness usual in probability theory. In the case where T is an interval of the real line, as is known, there is essential nonuniqueness. For example, one can have two processes with the same finite-dimensional distributions yet one has continuous trajectories for almost all ω and the other has discontinuous trajectories. To avoid this nonuniqueness, the requirement of separability of processes is introduced in the general theory. For all processes to be considered in this book there exist with probability one continuous variants or right continuous variants. Such a right continuous process is determined essentially uniquely by its finite-dimensional distributions. We shall always consider continuous or right continuous modifications, without stating this explicitly.

With every random process ξ_t, $t \in T$ there are associated the σ-algebras $\mathscr{F}_{\leq t} = \mathscr{F}^{\xi}_{\leq t} = \sigma\{\xi_s, s \leq t\}$ and $\mathscr{F}_{\geq t} = \mathscr{F}^{\xi}_{\geq t} = \sigma\{\xi_s, s \geq t\}$, which are the smallest σ-algebras with respect to which the random variables $\xi_s(\omega)$ are measurable for $s \leq t$ and $s \geq t$, respectively. It is clear that for $t_1 < t_2$ we have the inclusion $\mathscr{F}^{\xi}_{\leq t_1} \subseteq \mathscr{F}^{\xi}_{\leq t_2}$. In what follows we often consider the conditional mathematical expectations $\mathsf{M}(\eta | \mathscr{F}^{\xi}_{\leq t})$, which will sometimes be denoted by $\mathsf{M}(\eta | \xi_s, s \leq t)$. By $\mathsf{M}(\eta | \xi_t)$ we denote the conditional mathematical expectation with respect to the σ-algebra generated by the random variable ξ_t. Analogous notation will be used for conditional probabilities.

Assume given a nondecreasing family of σ-algebras $\mathscr{N}_t$: $\mathscr{N}_{t_1} \subseteq \mathscr{N}_{t_2}$ for $0 \leq t_1 \leq t_2; t_1, t_2 \in T$. We denote by $\mathscr{N}$ the smallest σ-algebra containing all σ-algebras $\mathscr{N}_t$ for $t \geq 0$. A random variable $\tau(\omega)$ assuming nonnegative values and the value $+\infty$ is called a Markov time (or a random variable not depending on the future) with respect to the family of σ-algebras $\mathscr{N}_t$ if $\{\omega: \tau(\omega) \leq t\} \in \mathscr{N}_t$ for every $t \in T$. An important example of Markov time is the first hitting time of a closed set in R^r by a process ξ_t continuous with probability 1. Here the role of the σ-algebras $\mathscr{N}_t$ is played by the nondecreasing family of σ-algebras $\mathscr{F}^{\xi}_{\leq t} = \sigma(\xi_s, s \leq t)$.

We denote by $\mathscr{N}_\tau$ the collection of the sets $A \subseteq \mathscr{N}$ for which

$$A \cap \{\tau \leq t\} \in \mathscr{N}_t$$

for all $t \in T$. It is easy to verify that $\mathscr{N}_\tau$ is a σ-algebra and τ is measurable with respect to this σ-algebra. If a process ξ_t is right continuous, and τ is a Markov time with respect to the σ-algebras $\mathscr{N}_t = \mathscr{F}^{\xi}_{\leq t}$, then the value ξ_τ is also measurable with respect to $\mathscr{N}_\tau = \mathscr{F}_{\leq \tau}$. The proofs of these assertions and a series of other properties of Markov times can be found in Wentzell's book [1].

The description of a random process by means of its finite-dimensional distributions is very cumbersome and is usually employed only in problems connected with the foundations of the theory. Interesting results can be obtained for special classes of random processes. We restrict ourselves to those basic classes of processes which occur in the book.

Gaussian Processes. We recall that an r-dimensional random variable $\xi = (\xi^1, \ldots, \xi^r)$ is said to be Gaussian if its characteristic function

$$f^{\xi}(z) = \mathsf{M} \exp\{i(z, \xi)\}, z \in R^r,$$

has the form $f^{\xi}(z) = \exp\{i(z, m) - (Rz, z)/2\}$, where $z = (z_1, \ldots, z_r) \in R^r$. $(z, \xi) = \sum_1^r z_k \xi^k, m = (m^1, \ldots, m^r)$ is the vector of mathematical expectations, i.e., $m^k = \mathsf{M}\xi^k$ and $R = (R^{ij})$ is the covariance matrix, i.e.,

$$R^{ij} = \mathsf{M}(\xi^i - m^i)(\xi^j - m^j).$$

A random process $\xi_t, t \in T$ is said to be Gaussian if all of its finite-dimensional distributions are Gaussian. Since a Gaussian distribution is determined by its mathematical expectation and covariance matrix, all finite-dimensional distributions of a Gaussian process ξ_t are completely determined by two functions: $m(t) = \mathsf{M}\xi_t$ and the correlation function

$$R(s, t) = \mathsf{M}(\xi_s - m(s))(\xi_t - m(t)).$$

We shall usually consider Gaussian processes for $t \in T = [0, T_0]$ and assume that the functions $m(t)$ and $R(s, t)$ are continuous for $s, t \in [0, T_0]$. Under these assumptions, the process ξ_t is continuous in mean square and it can be assumed that $\int_0^{T_0} |\xi_s|^2 \, ds < \infty$ for almost all ω. If in addition to the continuity of $m(t)$, we assume that the function $R(s, t)$ has some smoothness properties, for example, that it has a mixed second derivative for $s = t$, then there exists a continuous modification of ξ_t. This means that on the same probability space on which $\xi_t(\omega)$ is defined, there exists a random process $\xi_t^*(\omega)$ such that $\mathsf{P}\{\xi_t(\omega) = \xi_t^*(\omega)\} = 1$ for $t \in [0, T_0]$ and the functions $\xi_t^*(\omega)$ are continuous on the interval $[0, T_0]$ for almost all $\omega \in \Omega$ (cf. Gikhman and Skorokhod [1]).

With a process $\xi_t, t \in [0, T_0]$ there is associated the operator

$$A: (A\varphi)_t = \int_0^{T_0} R(s, t)\varphi_s \, ds,$$

called the correlation operator. If the function $R(s, t)$ is continuous, then A is a symmetric completely continuous operator in the Hilbert space $\mathbf{L}^2_{[0, T_0]}$ of functions defined on $[0, T_0]$ and with values in R^1. As usual, the norm in this space is given by the equality $\|\varphi\| = (\int_0^{T_0} |\varphi_s|^2 \, ds)^{1/2}$.

We shall also consider multidimensional Gaussian processes

$$\xi_t = (\xi_t^1, \ldots, \xi_t^r).$$

In this case $m(t) = \mathsf{M}\xi_t$ is a vector-valued function and $R(s, t) - (R^{ij}(s, t))$, where $R^{ij}(s, t) = \mathsf{M}(\xi_s^i - m^i(s))(\xi_j^i - m^j(t))$; $i, j = 1, 2, \ldots, r$. The correlation operator acts in the space of functions with values in R^r.

Markov processes are, roughly speaking, random processes whose behavior after a fixed time t under the condition that the behavior of the process is given until time t (inclusive) is the same as if the process began at the point X_t

at time t. This phrase can be turned into a precise definition in several ways. For this we need a certain technique connected with the circumstance that we have to stipulate the possibility of "emitting" the process from every point of the space in which it takes place.

Let $(\Omega, \mathscr{F})$ and $(X, \mathscr{B})$ be measurable spaces and in Ω let a nondecreasing system of σ-algebras $\mathscr{N}_t \subseteq \mathscr{F}$, $t \in T$ be chosen, where T is either the set $\{0, 1, 2, \ldots\}$ of nonnegative integers or the right half-line $[0, \infty)$.

A Markov process (more precisely, a homogeneous Markov process) with respect to the system of σ-algebras $\mathscr{N}_t$ is, by definition, a collection of the following objects:

(A) a random process $X_t(\omega)$, $t \in T$, $\omega \in \Omega$ with values in X;
(B) a collection of probability measures $\mathsf{P}_x(A)$ defined for $x \in X$ and $A \in \mathscr{F}$.

The following conditions are assumed to be satisfied:

(1) for every $t \in T$ the random variable $X_t(\omega)$ is measurable with respect to the σ-algebra $\mathscr{N}_t$;
(2) for any $t \in T$ and $\Gamma \in B$ the function $\mathsf{P}_x\{X_t \in \Gamma\} = P(t, x, \Gamma)$ is measurable in the variable x with respect to the σ-algebra $\mathscr{B}$;
(3) $P(0, x, X \setminus \{x\}) = 0$;
(4) if $t, u \in T$, $t \leq u$, $x \in X$, $\Gamma \in \mathscr{B}$, then the equality

$$\mathsf{P}_x\{X_u \in \Gamma | \mathscr{N}_t\} = P(u - t, X_t, \Gamma)$$

holds almost surely with respect to the measure P_x.

The function $P(t, x, \Gamma)$ is called the transition function of the Markov process.

If there is a topology in X and $\mathscr{B}$ is the σ-algebra of Borel sets in X, i.e., the σ-algebra generated by the open sets, then we can speak of various continuity properties of the Markov process. The weakest of these properties is stochastic continuity.

A Markov process is said to be stochastically continuous if its transition function has the following property: $\lim_{t \downarrow 0} P(t, x, X \setminus U) = 0$ for every $x \in X$ and every neighborhood U of x. We shall consider only stochastically continuous Markov processes.

A Markov process X_t is considered on not only one probability space but rather a whole family of probability spaces $\{\Omega, \mathscr{F}, \mathsf{P}_x\}$. The concept of a process that it can be emitted from every point x of the space can be expressed by making the trajectories of the process depend on x and by making the probabilities independent of x. Let $\{\Omega, \mathscr{F}, \mathsf{P}\}$ be a probability space and $\mathscr{N}_t$, $t \in T$ a nondecreasing system of σ-algebras.

A Markov family (with respect to the above system of σ-algebras) is, by definition, a collection of random processes $X_t^x(\omega)$, $t \in T$, $x \in X$ which satisfies the following conditions:

(1) for every $t \in T$ and $x \in X$ the random variable $X_t^x(\omega)$ is measurable with respect to $\mathcal{N}_t$;
(2) $\mathsf{P}\{X_t^x \in \Gamma\} = P(t, x, \Gamma)$ is a $\mathscr{B}$-measurable function of x;
(3) $\mathsf{P}(0, x, X \quad \{x\}) = 0$;
(4) if $t, u \in T$, $t \leq u$, x fi X and $\Gamma \in B$, then

$$\mathsf{P}\{X_u^x \in \Gamma | \mathcal{N}_t\} - P(u \quad t, X_t^x, \Gamma)$$

almost surely with respect to P.

The notions of a Markov process and a Markov family introduced above are special cases of the corresponding concepts in Dynkin's book [2]. It is easy to construct a Markov process corresponding to a given Markov family $X_t^x(\omega)$ on a probability space $\{\Omega, \mathscr{F}, \mathsf{P}\}$. For this it is sufficient to set

$$\Omega' = \Omega \times X, \qquad X_t(\omega') = X_t^x(\omega) \quad \text{for } \omega' = (\omega, x);$$

$$\mathsf{P}_x\{X_t \in \Gamma\} = \mathsf{P}\{X_t^x \in \Gamma\}.$$

It can be proved (cf. Dynkin [2]) that the random process $X_t(\omega')$ and the collection of the probability measures P_x, $x \in X$ on Ω' form a Markov process. The Markov process (X_t, P_x) thus constructed will be called the Markov process corresponding to the family X_t^x.

In what follows we shall use both Markov families and Markov processes. If the index x appears in the trajectories of X_t^x, then a Markov family is considered and if the index appears in the probability P_x, then the corresponding process is considered. Mathematical expectation with respect to P_x will be denoted by M_x.

Usually, we shall consider random processes defined by differential equations of the form $\dot{X}_t = b(X_t, \xi_t(\omega))$, where $\xi_t(\omega)$ is a random process. The solutions of this equation are defined for all possible initial conditions $X_0 = x \in X$. Let X_t^x be the solution issued from the point x: $X_0^x = x$. Under certain conditions, the processes $X_t^x(\omega)$ form a Markov family with respect to the nondecreasing system of σ-algebras $\mathscr{F}_{\leq t}^{\xi} = \sigma\{\xi_s, s \leq t\}$. We shall use the notation P_x and M_x for probabilities and mathematical expectations connected with the process X_t^x in the case of non-Markovian processes, as well The index x will indicate the initial condition under which the differential equation is being solved.

It follows from the definition of a Markov process that if the position of a process is known at time t, then the events determined by the process before and after time t are independent. If we fix the position of a process at a random time $\tau(\omega)$, then the events determined by the behavior of the Markov process before and after time $\tau(\omega)$ may turn out to be dependent even if $\tau(\omega)$ is a Markov time. Those Markov processes for which these events are independent for every Markov time $\tau(\omega)$ are called strong Markov processes. For a precise definition, cf. Dynkin [2]. A Markov process (X_t, P) with respect to a nondecreasing system of σ-algebras $\mathcal{N}_t$ is said to be strong Markov if for every Markov time τ with respect to the σ-algebras $\mathcal{N}_t$ and for all $t \geq 0$, $x \in X$, and $\Gamma \in \mathscr{B}$ the relation

$$\mathsf{P}_x\{X_{\tau+t} \in \Gamma \mid \mathcal{N}_\tau\} = P(t, X_\tau, \Gamma)$$

is satisfied for almost all points of the set $\Omega_\tau = \{\omega \in \Omega: \tau(\omega) < \infty\}$ with respect to the measure P_x.

Conditions ensuring that a given Markov process is a strong Markov process together with various properties of strong Markov processes, are discussed in detail in Dynkin's book [2]. We note that all Markov processes considered in the present book are strong Markov.

As a simple important example of a Markov process may serve a Markov chain with a finite number of states. This is a Markov process in which the parameter t assumes the values $0, 1, 2, \ldots$ and the phase space X consists of a finite number of points: $X = \{e_1, \ldots, e_n\}$. A homogeneous Markov chain (we shall only encounter such chains in this book) is given by the square matrix $P = (p_{ij})(i, j = 1, \ldots, n)$ of one-step transition probabilities: $\mathsf{P}_{e_i}\{X_1 = e_j\} = p_{ij}$. It follows from the definition of a Markov process that if the row vector $q(s) = (q_1(s), \ldots, q_n(s))$ describes the distribution of $X_s(\omega)$ (i.e.,

$$\mathsf{P}_x\{X_s = e_i\} = q_i(s)),$$

then $q(t) = q(s)P^{t-s}$ for $t > s$. A row vector $\bar{q} = (\bar{q}_1, \ldots, \bar{q}_n)$, $\bar{q}_i \geq 0, \sum_i^n \bar{q}_i = 1$, for which $\bar{q}P = \bar{q}$ is called an invariant distribution of the Markov process. Every chain with a finite number of states admits an invariant distribution. If all entries of P (or a power of it) are different from zero, then the invariant distribution $\bar{q}$ is unique and $\lim_{t\to\infty} \mathsf{P}_x\{X_t(\omega) = e_i\} = \bar{q}_i$ for all x and $e_i \in X$. This assertion, called the ergodic theorem for Markov chains, can be carried over to Markov processes of the general kind.

In the forthcoming sections we shall return to Markov processes. Now we recall some more classes of processes.

We say that a random process $\xi_t(\omega)$, $t \geq 0$ in the phase space $(R^r, \mathscr{B}^r)$ is a process with independent increments if the increments

$$\xi_{t_n} - \xi_{t_{n-1}}, \xi_{t_{n-1}} - \xi_{t_{n-2}}, \ldots, \xi_{t_2} - \xi_{t_1}$$

are independent random variables for any $t_n > t_{n-1} > \cdots > t_1 \geq 0$.

The Poisson process is an example of such a process. It is a random process ν_t, $t \geq 0$ assuming nonnegative integral values, having independent increments and right continuous trajectories with probability one for which

$$\mathsf{P}\{\nu_t - \nu_s = k\} = \frac{[(t-s)\lambda]^k}{k!} e^{-(t-s)\lambda}; \qquad 0 \leq s < t; \quad k = 0, 1, \ldots,$$

where λ is a positive parameter.

In the next section we shall consider another process with independent increments, the Wiener process, which plays an important role in the theory of random processes.

For a process with independent increments we may associate the Markov family $X_t^x = x + \xi_t - \xi_0$ and the Markov process which corresponds to this family. Another class of processes, the class of martingales, is also closely connected with processes with independent increments.

A random process ξ_t, $t \in T$ is called a martingale with respect to a nondecreasing family of σ-algebras $\mathscr{N}_t$ if the random variable ξ_t is measurable with respect to $\mathscr{N}_t$ for every $t \in T$ and $\mathsf{M}\xi_t < \infty$ and $\mathsf{M}(\xi_t | \mathscr{N}_s) = \xi_s$ for $s, t \in T$, $s < t$. If $\mathsf{M}(\xi_t | \mathscr{N}_s) \leq \xi_s$, then the process ξ_t is called a supermartingale.

A detailed exposition of the theory of martingales can be found in Doob's book [1].

A random process $\xi_t(\omega)$, $-\infty < t < \infty$ is said to be stationary (in the strict sense) if for a given natural number r and given $t_1, \ldots, t_r$ the distribution of the random variable $(\xi_{t_1+h}, \xi_{t_2+h}, \ldots, \xi_{t_r+h})$ is the same for all real numbers h. It is clear that if $\mathsf{M}|\xi_t|^2$ exists for a stationary process ξ_t, then $\mathsf{M}\xi_t = m =$ const and the correlation function $R(s, t) = \mathsf{M}(\xi_s - m)(\xi_t - m)$ depends only on $t - s$.

Finally, we recall the notion of weak convergence of measures corresponding to a family of random processes. Every random process $X_t(\omega)$ defines a mapping of the probability space $\{\Omega, \mathscr{F}, \mathsf{P}\}$ into the space of trajectories. This mapping induces a probability measure μ in the space of trajectories. In many problems in probability theory, random processes are considered up to the distributions which the processes induce in the space of trajectories. In connection with this, an important role is played by those types of convergence of random processes which mean convergence, in one sense or another, of the distributions in the space of trajectories. For the sake of definiteness, let X_t^ε, $t \in [0, T]$, $\varepsilon > 0$, be a family of random processes whose trajectories are, with probability one, continuous functions defined on $[0, T]$ with values in R^r. As usual, we denote by $\mathbf{C}_{0T}(R^r)$ the space of such functions with the topology of uniform convergence. Let μ^ε be the family of measures corresponding to the processes X_t^ε in $\mathbf{C}_{0T}(R^r)$. We say that the measure μ^ε on $\mathbf{C}_{0T}(R^r)$ converges weakly to a measure μ as $\varepsilon \to 0$ if

$$\lim_{\varepsilon \to 0} \int f(x)\mu^\varepsilon(dx) = \int f(x)\mu(dx).$$

for every continuous bounded functional $f(x)$ on $\mathbf{C}_{0T}(R^r)$. In Prokhorov [1] (cf. also Gikhman and Skorokhod [1]) conditions of compactness in the topology of weak convergence are studied for a family of measures corresponding to random processes. If the family of measures μ^ε is weakly compact and the finite-dimensional distributions corresponding to the random processes X_t^ε converge to the distributions of some process X_t, then the measures μ^ε converge weakly to the measure μ corresponding to the process X_t.

§3. Wiener Process. Stochastic Integral

A Wiener process is, by definition, a Gaussian process w_t, $t \in [0, \infty)$ with values in R^1 having the following properties:

(1) $\mathsf{M} w_t = 0$ for $t \geq 0$;
(2) $\mathsf{M} w_s w_t = \min(s, t)$;
(3) for almost all ω, the trajectories of $w_t(\omega)$ are continuous in $t \in [0, \infty)$.

It can be proved (cf., for example, Gikhman and Skorokhod [1]) that a process with these properties exists on an appropriate probability space $\{\Omega, \mathscr{F}, \mathsf{P}\}$. From the Gaussianness of w_t it follows that for arbitrary moments of time $t_n > t_{n-1} > \cdots > t_1 \geq 0$ the random variables

$$w_{t_n} - w_{t_{n-1}}, w_{t_{n-1}} - w_{t_{n-2}}, \ldots, w_{t_2} - w_{t_1}$$

have a joint Gaussian distribution and from property (2) it follows that they are uncorrelated: $\mathsf{M}(w_{t_{i+1}} - w_{t_i})(w_{t_{j+1}} - w_{t_j}) = 0$ for $i, j = 1, 2, \ldots, n; i \neq j$. We conclude from this that the increments of a Wiener process are independent. We note that the increment of a Wiener process from time s to t, $s < t$ has a Gaussian distribution and $\mathsf{M}(w_t - w_s) = 0$, $\mathsf{M}(w_t - w_s)^2 = t - s$. It can be calculated that $\mathsf{M}|w_t - w_s| = \sqrt{2\pi^{-1}(t - s)}$.

We recall some properties of a Wiener process. The upper limits (lim sup) of almost all trajectories of a Wiener process are $+\infty$ and the lower limits (lim inf) are $-\infty$. From this it follows in particular that the trajectories of a Wiener process pass through zero infinitely many times with probability 1 and the set $\{t: w_t(\omega) = 0\}$ is unbounded for almost all ω. The realizations of a Wiener process are continuous by definition. Nevertheless, with probability 1 they are nowhere differentiable and have infinite variation on every time interval. It can be proved that with probability one the trajectories of a Wiener process satisfy a Hölder condition with any exponent $\alpha < 1/2$ but do not satisfy it with exponents $\alpha \geq 1/2$. We also note the following useful identity:

$$\mathsf{P}\left\{\sup_{0 \leq s \leq T} w_s > a\right\} = 2\mathsf{P}\{w_T > a\}.$$

Every random process $\xi_t(\omega)$, $t \in T$ with values in a measurable space $(X, \mathscr{B})$ can be considered as a mapping of the space $(\Omega, \mathscr{F})$ into a space of functions defined on T, with values in X. In particular, a Wiener process $w_t(\omega)$, $t \in [0, T]$ determines a mapping of Ω into the space $\mathbf{C}^0_{0T}(R^1)$ of continuous functions on $[0, T]$ which are zero at $t = 0$. This mapping determines a probability measure μ_w in $\mathbf{C}^0_{0T}(R^1)$, which is called the Wiener measure. The support of the Wiener measure is the whole space $\mathbf{C}^0_{0T}(R^1)$. This means that an arbitrary small neighborhood (in the uniform topology) of every function $\varphi \in \mathbf{C}^0_{0T}(R^1)$ has positive Wiener measure.

A collection of r independent Wiener processes $w_t^1(\omega), w_t^2(\omega), \ldots, w_t^r(\omega)$ is called an r-dimensional Wiener process.

The important role of the Wiener process in the theory of random processes can be explained to a large degree by the fact that many classes of random processes with continuous trajectories admit a convenient representation in terms of a Wiener process. This representation is given by means of the stochastic integral. We recall the construction and properties of the stochastic integral.

Let us given a probability space $\{\Omega, \mathscr{F}, \mathsf{P}\}$, a nondecreasing family of σ-algebras $\mathscr{N}_t$, $t \geq 0$, $\mathscr{N}_t \subseteq \mathscr{F}$, and a Wiener process w_t on $\{\Omega, \mathscr{F}, \mathsf{P}\}$. We assume that the σ-algebras $\mathscr{N}_t$ are such that $\mathscr{F}^w_{\leq t} \subseteq \mathscr{N}_t$ for every $t \geq 0$ and

$$\mathsf{M}(w_t - w_s \mid \mathscr{N}_s) = 0; \qquad \mathsf{M}(|w_t - w_s|^2 \mid \mathscr{N}_s) = t - s$$

for every $0 \leq s \leq t$. This will be so, at any rate, if $\mathscr{N}_t = \mathscr{F}^w_{\leq t}$.

We say that a random process $f(t, \omega)$, $t \geq 0$ measurable in the pair (t, ω) does not depend on the future (with respect to the family of σ-algebras $\mathscr{N}_t$) if $f(t, \omega)$ is measurable with respect to $\mathscr{N}_t$ for every $t > 0$. We denote by $\mathbf{H}^2_{a,b}$, $0 \leq a < b < \infty$, the set of functions $f(t, \omega)$ not depending on the future and such that $\int_a^b \mathsf{M}|f(t, \omega)|^2\, dt < \infty$. For such functions we define Itô's stochastic integral $\int_a^b f(s, \omega)\, dw_s$. We note that since the trajectories of a Wiener process have infinite variation over any interval, this integral cannot be defined as a Stieltjes integral. Itô's integral is first defined for step functions belonging to $\mathbf{H}^2_{a,b}$. If $t_0 = a < t_1 < t_2 < \cdots < t_n = b$, $f(s, w) = f_i(w)$ for $s \in [t_i, t_{i+1})$, $i = 0, 1, \ldots, n - 1$ and $f(s, \omega) \in \mathbf{H}^2_{a,b}$, then we set

$$\int_a^b f(s, \omega)\, dw_s = \sum_{i=0}^{n-1} f_i(\omega)(w_{t_{i+1}} - w_{t_i}).$$

Consequently, we assign the variable $\eta_f(\omega) = \int_a^b f(s, \omega)\, dw_s$ to the step function $f(s, \omega) \in \mathbf{H}^b_{a,b}$. If we introduce the norm

$$\|f\|_{\mathbf{H}^2} = \left(\int_a^b \mathsf{M}|f(s, \omega)|^2\, ds \right)^{1/2}$$

in $\mathbf{H}^2_{a,b}$ and the norm $\|\eta_f\| = (\mathsf{M}\eta_f^2)^{1/2}$ in the space of random variables, then the mapping $f \to \eta_f$ is norm-preserving, as is easy to see. This mapping, defined first only for step functions, can be extended to the closure with preservation of the norm. It can be proved that the closure of the set of step functions in $\mathbf{H}^2_{a,b}$ coincides with $\mathbf{H}^2_{a,b}$. Therefore, to every element $f \in \mathbf{H}^2_{a,b}$ there corresponds a random variable η_f, which is called Itô's stochastic integral of the function $f(s, \omega)$ and is denoted by $\int_a^b f(s, \omega)\, dw_s$. We list the basic properties of the stochastic integral ($f(s, \omega), g(s, \omega) \in \mathbf{H}^2_{a,b}$):

1. $\int_a^b (\alpha f(s, \omega) + \beta g(s, \omega))\, dw_s = \alpha \int_a^b f(s, \omega)\, dw_s + \beta \int_a^b g(s, \omega)\, dw_s$;
2. $\mathsf{M}(\int_a^b f(s, \omega)\, dw_s | \mathcal{N}_a) = 0$;
3. $\mathsf{M}(\int_a^b f(s, \omega)\, dw_s \int_a^b g(s, \omega)\, dw_s | \mathcal{N}_a) = \mathsf{M}(\int_a^b f(s, \omega) g(s, \omega)\, ds | \mathcal{N}_a)$,

in particular,

$$\mathsf{M}\left(\left(\int_a^b f(s, \omega)\, dw_a\right)^2 \middle| \mathcal{N}_a\right) = \mathsf{M}\left(\int_a^b f^2(s, \omega)\, ds \middle| \mathcal{N}_a\right).$$

We note that the stochastic integral is defined up to an ω-set of measure zero and all equalities listed above are satisfied almost surely with respect to the measure P on Ω.

Now we consider the stochastic integral as a function of the upper limit of the integral. We denote by $\chi_t(s)$ the function equal to 1 for $s \le t$ and zero for $s > t$. If $f(s, \omega) \in \mathbf{H}^2_{a,b}$, then $\chi_t(s) f(s, \omega) \in \mathbf{H}^2_{a,b}$ for every t. We define $\int_a^t f(s, \omega)\, dw_s$ for $t \in [a, b]$ by means of the equality

$$\int_a^t f(s, \omega)\, dw_s = \int_a^b \chi_t(s) f(s, \omega)\, dw_s.$$

Since for every t, the integral $\int_a^b \chi_t(s) f(s, \omega)\, dw_s$ is determined up to events of probability zero, we have some arbitrariness in the definition of the left side. It can be proved that the right side can be defined for every t in such a way that the stochastic integral on the left side will be a continuous function of the upper limit for almost all ω. Whenever in this book, we consider stochastic integrals with a varying upper limit, we always have in mind the variant which is continuous with probability 1.

It follows from the above properties of the stochastic integral that the stochastic process $\xi_t = \int_a^t f(s, \omega)\, dw_s$, together with the nondecreasing family of σ-algebras $\mathcal{N}_t$, forms a martingale. This martingale has continuous trajectories with probability 1 and

$$\mathsf{M}(\xi_t^2 | \mathcal{N}_a) = \mathsf{M}\left(\int_a^t f^2(s, \omega)\, ds | \mathcal{N}_a\right) < \infty.$$

For these martingales we have the following generalized Kolmogorov inequality:

$$\mathsf{P}\left\{\max_{a \le t \le b}\left|\int_a^t f(s, \omega)\, dw_s\right| > c \,\Big|\, \mathcal{N}_a\right\} \le \frac{1}{c^2}\mathsf{M}\left(\int_a^b f^2(s, \omega)\, ds \,\Big|\, \mathcal{N}_a\right).$$

Sometimes we have to consider the stochastic integral with a random time as the upper limit. Let τ be a Markov time with respect to a nondecreasing system of σ algebras $\mathcal{N}_t$, $t \geqq 0$ and let $\chi_\tau(s)$ be equal to 1 for $s \geqq \tau$ and 0 for $s > \tau$. If $\chi_\tau(s) f(s, \omega) \in \mathbf{H}^2_{0,\infty}$, then $\int_0^\tau f(s, \omega)\, dw_s = \int_0^\infty \chi_\tau(s) f(s, \omega)\, dw_s$ and $\mathsf{M} \int_0^\tau f(s, \omega)\, dw_s = 0$.

In particular, $\chi_\tau(s) f(s, \omega) \in H^2_{0,\infty}$, if $|f(s, \omega)| < c < \infty$ for all $s > 0$ for almost all ω and if $\mathsf{M}\tau < \infty$.

Let $w_t = \{w_t^i\}$ now be an r-dimensional Wiener process and let $\mathcal{N}_t$ be the σ-algebra generated by the random variables w_s for $s \leqq t$. The stochastic integral

$$\int_a^b \Phi(s, \omega)\, dw_s$$

is defined in a natural manner for matrix-valued functions $\Phi(s, \omega)$ with entries belonging to $\mathbf{H}^2_{a,b}$. Namely, if w_t is understood as an r-dimensional column vector and the matrix $\Phi(s, \omega) = (\Phi_{ij}(s, \omega))$ has r columns and l row, then $\int_a^b \Phi(s, \omega)\, dw_s$ is the l-dimensional random variable whose ith component is equal to the sum

$$\sum_{j=1}^r \int_a^b \Phi_{ij}(s, w)\, dw_s^j.$$

We consider the l-dimensional random process

$$X_t = \int_a^t \Phi(s, \omega)\, dw_s + \int_a^t \Psi(s, \omega)\, ds, \qquad t \in [a, b].$$

Here the first term is the integral with respect to the r-dimensional Wiener process which we have just described and in the second term,

$$\Psi(s, \omega) = \{\Psi^i(s, \omega)\}$$

is an l-dimensional random process. The relation defining X_t is sometimes written in the form

$$dX_t = \Phi(t, \omega)\, dw_t + \Psi(t, \omega)\, dt,$$

and the expression $\Phi(t, \omega)\, dw_t + \Psi(t, \omega)\, dt$ is called the stochastic differential of X_t.

Let a function $u(t, x)$, $t \in [a, b]$, $x \in R^l$ have a continuous first derivative with respect to t and continuous second derivatives with respect to the space variables. In the theory of the stochastic integral an important role is played by Itô's formula giving an expression for the stochastic differential of the random process $\eta_t = u(t, X_t)$:

$$d\eta_t = \sum_{k=1}^{r} \sum_{i=1}^{l} \frac{\partial u}{\partial x^i}(t, X_t)\Phi_{ik}(t, \omega)dw_t^k$$
$$+ \left[\frac{\partial u}{\partial t}(t, X_t) + \sum_{i=1}^{l} \frac{\partial u}{\partial x^i}(t, X_t)\Psi^i(t, \omega)\right.$$
$$\left. + \frac{1}{2}\sum_{k=1}^{r} \sum_{i,j=1}^{i} \Phi_{ik}(t, \omega)\Phi_{jk}(t, \omega)\frac{\partial^2 u}{\partial x^i \partial x^i}(t, X_t)\right] dt.$$

A detailed exposition of the construction and proofs of all properties listed here of the stochastic integral together with additional properties can be found in the books by Gikhman and Skorokhod [1] and McKean [1].

As we have already noted, by means of the stochastic integral we can obtain representations of some classes of random processes in terms of the Wiener process. Let us discuss the representation of Gaussian processes in detail. We obtain a representation of the Gaussian process with mean zero and correlation function $R(s, t)$, which we assume to be continuous for $s, t \in [0, T]$. If considered in the space $\mathbf{L}^2_{0,T}$, the correlation operator A of such a process is completely continuous, nonnegative definite, symmetric and of finite trace. Let $e_1(t), \ldots, e_n(t), \ldots$ be its eigenfunctions and let $\lambda_1, \lambda_2, \ldots, \lambda_n, \ldots$ be the corresponding eigenvalues. It is known (Riesz and Sz.-Nagy [1]) that the kernel of such an operator can be expressed in the form

$$R(s, t) = \sum_k \lambda_k e_k(s)e_k(t).$$

We set

$$G(s, t) = \sum_k \sqrt{\lambda_k}\, e_k(s)e_k(t).$$

It follows from the finiteness of the trace of A that this series is always convergent in the space $\mathbf{L}^2_{[0,T]\times[0,T]}$ of square integrable functions on $[0, T] \times [0, T]$. It follows easily from the definition of $G(s, t)$ that

$$\int_0^T G(s, t_1)G(s, t_2)\, ds = R(t_1, t_2).$$

We consider the stochastic integral

$$X_t = \int_0^T G(s, t)\, dw_s.$$

This stochastic integral with respect to the Wiener process w_t exists, since the integrand does not depend on chance and $\int_0^T G^2(s, t)\,ds = R(t, t) < \infty$. It follows from previously mentioned properties of the stochastic integral that X_t is a Gaussian process and

$$\mathsf{M}X_t = 0; \qquad \mathsf{M}X_{t_1}X_{t_2} = \mathsf{M}\left(\int_0^T G(s, t_1)\,dw_s \cdot \int_0^T G(s, t_2)\,dw_s\right)$$
$$= \int_0^T G(s, t_1)G(s, t_2)\,ds = R(t_1, t_2).$$

Consequently, we have obtained a representation for the Gaussian process with vanishing mean and correlation function $R(s, t)$ as a stochastic integral of the nonrandom function $G(s, t)$.

Sometimes we consider a so-called white noise process $\dot{w}_t$, the derivative of a Wiener process w_t. As we have already mentioned, the derivative of a Wiener process does not exist in the ordinary sense. Nevertheless, the stochastic integral enables us to give a meaning to some expressions containing $\dot{w}_t$. Having defined Itô's integral for functions $f(s, \omega) \in \mathbf{H}_{0,T}^2$, we may set

$$\int_0^T f(s, \omega)\,dw_s = \int_0^T f(s, \omega)\dot{w}_s\,ds,$$

assuming that the left side defines the right side. In particular, the formula defining the process X_t can be written in the form

$$X_t = \int_0^T G(s, t)\dot{w}_s\,ds$$

and we may say that the Gaussian process X_t is the result of applying the integral operator with kernel $G(s, t)$ to the white noise process (this kernel is sometimes called the impulse response of the operator).

A large class of martingales continuous in the time variable admits a representation in the form of a stochastic integral with respect to the Wiener process. In §5 we construct diffusion processes starting from a Wiener process.

§4. Markov Processes and Semigroups

Let (X_t, P_x) be a Markov process on a phase space $(X, \mathscr{B})$ and let $P(t, x, \Gamma)$ be its transition function. We denote by $\mathbf{B}$ the Banach space of bounded $\mathscr{B}$-measurable functions on X with the norm $\|f\| = \sup_{x \in X} |f(x)|$. With the

Markov process (or with its transition function) we may associate the family of operators T_t, $t \geqq 0$ acting in **B** according to the formula

$$(T_t f)(x) = \mathsf{M}_x f(X_t) = \int_X f(y)P(t, x, dy).$$

Since $P(t, x, \Gamma)$ is a probability measure as a function of Γ, the operators T_t preserve nonnegativity and do not increase norm: if $f(x) \geqq 0$, then

$$T_t f(x) \geq 0. \; \|T_t f\| \leq \|f\|.$$

It follows from the Markov property that

$$T_{t+s} f(x) = \mathsf{M}_x f(X_{t+s}) = \mathsf{M}_x(\mathsf{M}_{X_t} f(X_s)) = T_t(T_s f)(x),$$

i.e., the operators T_t form a semigroup: $T_t T_s = T_{t+s}$. Consequently, with every Markov process (X_t, P_x) there is associated the contraction semigroup T_t acting in the space **B** of bounded measurable functions on the phase space.

The contraction semigroup is, of course, also associated with the Markov family X_t^x (defined on the probability space $(\Omega, \mathscr{F}, \mathsf{P})$ not depending on x):

$$(T_t f)(x) = \mathsf{M} f(X_t^x) = \int_X f(y)P(t, x, dy).$$

If the function $f(x)$ is the indicator $\chi_\Gamma(x)$ of a set $\Gamma \subseteq X$ (i.e., the function equal to 1 on Γ and 0 outside Γ), then we obtain $T_t \chi_\Gamma(x) = P(t, x, \Gamma)$ and the semigroup property $T_{t+s}\chi_\Gamma(x) = T_t(T_s \chi_\Gamma)$ can be written in the form

$$P(t + s, x, \Gamma) = \int_x P(t, x, dy)P(s, y, \Gamma),$$

which is called the Chapman–Kolmogorov equation.

With the transition function $P(t, x, \Gamma)$ we may associate another operator semigroup U_t, $t \geqq 0$ acting in the Banach space **V** of finite countability additive set functions on $(X, \mathscr{B})$ with the norm $\|\mu\|^*$ defined as the total variation of the set function μ:

$$(U_t \mu)(\Gamma) = \int_X P(t, x, \Gamma)\mu(dx); \qquad \mu \in V, \quad \Gamma \in \mathscr{B}.$$

As is easy to see, the operators T_t and U_t are conjugate to each other in the sense that

$$\int_X T_t f(x)\mu(dx) = \int_X f(x)(U_t \mu)(dx); \quad f \in \mathbf{B}, \quad \mu \in V.$$

The semigroup U_t describes the evolution of the one-dimensional distributions of the Markov process. Namely, if we consider the Markov process as beginning not at a given point $x \in X$ but rather a random point X_0 with distribution μ: $\mathsf{P}\{X_0 \in \Gamma\} = \mu(\Gamma)$, the distribution at time t will be exactly $U_t\mu$:

$$\mathsf{P}\{X_t \in \Gamma\} = \int_X \mathsf{P}\{X_0 \in dx\}P(t, x, \Gamma) = (U_t\mu)(\Gamma), \qquad \Gamma \in \mathscr{B}.$$

A measure μ on $(X, \mathscr{B})$ is called an invariant measure of the Markov process if $U_t\mu = \mu$ for all $t \geqq 0$. It is clear that the invariant measures form a cone in $\mathbf{V}$. If $\mu(X) = 1$, then the invariant measure μ is also called a stationary probability distribution.

The infinitesimal generator A of the semigroup T_t (it is also the infinitesimal generator of the Markov process (X_t, P_x) or the Markov family X_t^x with the given transition function) is defined by the equality

$$Af = \lim_{t \downarrow 0} \frac{T_t f - f}{t}.$$

Here convergence is understood as convergence in norm, i.e., the equality means that $\lim_{t\downarrow 0}\|t^{-1}(T_t f - f) - Af\| = 0$. The operator A is not defined for all elements of $\mathbf{B}$ in general. The domain of A is a vector subspace, which is denoted by D_A. It is everywhere dense in the space

$$\mathbf{B}_0 = \{f \in \mathbf{B}: \lim_{t\downarrow 0} \|T_t f - f\| = 0\}.$$

The infinitesimal generator determines the semigroup T_t uniquely on $\mathbf{B}_0$. If the transition function is stochastically continuous, then the semigroup T_t considered only on $\mathbf{B}_0$ (and consequently, the infinitesimal generator A, as well) determines uniquely the transition function and all finite-dimensional distributions of the Markov process (Markov family).

In the theory of semigroups it is proved that for every $f \in D_A$ the function $u_t(x) = T_t f(x)$ is a solution of the abstract Cauchy problem

$$\frac{\partial u_t(x)}{\partial t} = Au_t(x), \qquad \lim_{t\downarrow 0} u_t(x) = f(x).$$

The solution of this problem is always unique in the class of bounded functions

The infinitesimal generator A^* of the semigroup U_t can be defined analogously. For μ belonging to the domain of A^*, the function

$$V_t(\Gamma) = U_t\mu(\Gamma), \Gamma \in \mathscr{B},$$

is a solution of the corresponding Cauchy problem. In particular, it can be verified easily that every invariant measure belongs to the domain of A^* and $A^*\mu = 0$.

A detailed exposition of the semigroup theory of Markov processes can be found in Dynkin's book [2].

We consider examples of Markov processes and their infinitesimal generators.

First Example. Let X be a finite set and let $\mathscr{B}$ be the collection of its subsets. A Markov process with such a phase space is called a Markov process with a finite number of states. With every such process there is associated a system of functions $p_{ij}(t)$ $(i, j \in X,\ t \geqq 0)$ satisfying the following conditions:

(1) $p_{ij}(t) \geq 0$ for $i, j \in X$, $t \geq 0$;
(2) $\sum_{j \in X} p_{ij}(t) = 1$;
(3) $p_{ij}(0) = 0$ for $i \neq j$, $p_{ii}(0) = 1$ for $i \in X$;
(4) $p_{ij}(s + t) = \sum_{k \in X} p_{ik}(t) p_{kj}(s)$.

The transition function of the process can be expressed in terms of the functions $p_{ij}(t)$ in the following way:

$$P(t, x, \Gamma) = \sum_{y \in \Gamma} p_{xy}(t); \qquad x \in X, \quad \Gamma \in \mathscr{B}, \quad t \geq 0.$$

We shall only consider stochastically continuous processes with a finite number of states. For these processes the functions $p_{ij}(t)$ satisfy the additional condition

(5) $\lim_{t \downarrow 0} p_{ij}(t) = p_{ij}(0)$.

It can be proved under conditions (1)–(5) that the right derivatives at zero $q_{ij} = p'_{ij}(0)$ exist. We introduce the matrix $P(t) = (p_{ij}(t))$; and use the notation $Q = (q_{ij})$.

We calculate the infinitesimal generator A of our Markov process and the infinitesimal generator A^* of the adjoint operator semigroup U_t.

The space **B** of bounded measurable functions on X and the space **V** of countably additive set functions on $\mathscr{B}$ are finite-dimensional linear spaces with dimension equal to the number of elements of X. We identify the elements of **B** with column vectors and the elements of **V** with row vectors. The semigroup T_t acts on vectors belonging to **B** according to the formula

$$T_t f = P(t) f.$$

The infinitesimal generator A of the semigroup is defined on the whole of **B** and is given by the formula

$$Af = \lim_{t \downarrow 0} \frac{T_t f - f}{t} = Qf.$$

The adjoint semigroup U_t is given by multiplication by the matrix $P(t)$ on the right and its infinitesimal generator is given by multiplication by the matrix Q on the right. It can be proved easily that Q has at least one left eigenvector $m \in \mathbf{V}$ with eigenvalue zero ($mQ = 0$) and with nonnegative components whose sum is equal to one. Every such vector defines a stationary distribution of the process with a finite number of states. If all entries of Q are different from zero, then the stationary distribution is unique.

SECOND EXAMPLE. Let v_t be a Poisson process starting at zero (cf. §2). The collection of the processes $v_t^x = x + v_t$, $x \in R^1$, forms a Markov family with respect to the σ-algebras $\mathcal{N}_t = \mathcal{F}^v_{\leq t}$ (as phase space we take the real line R^1). The corresponding semigroup T_t acts according to the formula

$$T_t f(x) = \sum_{k=0}^{\infty} f(x + k)e^{-\lambda t} \frac{(\lambda t)^k}{k!}$$

in the space of bounded measurable functions and its infinitesimal generator A has the form

$$Af(x) = \lambda[f(x + 1) - f(x)].$$

Intuitively, the Poisson process can be described in the following way. If at some moment the trajectory is at the point x, then it spends an additional random time τ at x and then jumps 1 to the right, arriving at the point $x + 1$, it spends some time in this position and then jumps to $x + 2$ and so on. The random variable τ follows an exponential distribution:

$$\mathsf{P}\{\tau > t\} = \exp\{-\lambda t\}.$$

For a Poisson process, the number λ is the same for all states and also the length of a jump is fixed. We obtain a jump-like Markov process of the general form if we allow λ (describing the distribution of the time until exit from the state x) to depend on x and consider jumps whose lengths are random with a distribution depending on the initial state. A jump-like process (which can be considered not only on a line but also in r-space R^r) can be described by the infinitesimal generator

$$Af(x) = \lambda(x) \int [f(x + u) - f(x)]\mu_x(du).$$

Here integration is carried out over all space except for the point 0, $\lambda(x)$ characterizes the distribution of the time until exit from x, and the measure $\mu_x(du)$ gives the distribution of the jump length.

In the next section we shall consider a large class of Markov processes (Markov families) with continuous trajectories and the corresponding infinitesimal generators.

§5. Diffusion Processes and Differential Equations

Let w_t be an l-dimensional Wiener process and let $\mathcal{N}_t$ be the σ-algebra generated by the random variables w_s for $s \leqq t$. We consider the stochastic differential equation

$$\dot{X}_t = b(X_t) + \sigma(X_t)\dot{w}_t, \qquad X_0 = x.$$

in R^r. Here $b(x) = (b^1(x), \ldots, b^r(x))$ is a vector field in R^r and $\sigma(x) = (\sigma^i_j(x))$ is a matrix having l columns and r rows. By a solution of this equation we understand a random process $X_t = X_t(\omega)$ which satisfies the relation

$$X_t - x = \int_0^t b(X_s)ds + \int_0^t \sigma(X_s)dw_s.$$

with probability 1 for every $t \geqq 0$. We shall usually assume that the coefficients $b^i(x)$, $\sigma^i_j(x)$ satisfy the following conditions:

(1) $\sum_i |b^i(x) - b^i(y)| + \sum_{i,j} |\sigma^i_j(x) - \sigma^i_j(y)| \leq K|x - y|; \quad x, y \in R^r$,

(2) $\sum_i |b^i(x)| + \sum_{i,j} |\sigma^i_j(x)| \leq K(|x| + 1)$,

where $|x|$ is the Euclidean length of the vector $x \in R^r$ and K is a positive constant.

Under these conditions it can be proved that the above stochastic differential equation has a solution $X^x_t(\omega)$, $t \geq 0$ which is continuous with probability 1, the random variable $X^x_t(\omega)$ is measurable with respect to the σ-algebra $\mathcal{N}_t$ for every $t \geq 0$, and $\int_a^b \mathsf{M}|X^x_t|^2\, dt < \infty$ for any $b > a \geqq 0$. For every $t \geqq 0$, any two solutions of a stochastic differential equation having these properties coincide for almost all $\omega \in \Omega$.

Using the independence of the increments of a Wiener process and the uniqueness of the solution, it can be proved (Dynkin [2]) that the set of processes X^x_t for all possible initial points $x \in R^r$ forms a Markov family with respect to the system of the σ-algebras $\mathcal{N}_t$. It can be proved that the Markov process corresponding to this family is a strong Markov process.

Consequently, a stochastic differential equation determines a strong Markov process. This process is called a random diffusion process.

We shall study the infinitesimal generator of the diffusion process. Let the function $u(x)$, $x \in R^r$ have bounded continuous derivatives up to the second order. By Itô's formula we obtain

$$u(X^x_t) - u(x) = \int_0^t (\nabla u(X^x_s), \sigma(X^x_s)\, dw_s) + \int_0^t Lu(X^x_s)\, ds.$$

Here $\nabla u(x)$ is the gradient of $u(x)$, the quantity under the first integral sign on the right side is the Euclidean scalar product of the vectors $\nabla u(X^x_s)$ and $\sigma(X^x_s)\, dw_s$, and the differential operator L has the form

$$Lu(x) = \frac{1}{2} \sum_{i,j=1}^{r} a^{ij}(x) \frac{\partial^2 u}{\partial x^i \partial x^j}(x) + \sum_{i=1}^{r} b^i(x) \frac{\partial u}{\partial x^i}(x),$$

where $a(x) = (a^{ij}(x)) = \sigma(x)\sigma^*(x)$ is a square matrix of order n. It follows from the above expression for $u(X_t^x) - u(x)$ that $u(x) \in D_A$ and

$$Au(x) = \lim_{t \downarrow 0} \frac{\mathsf{M}_x u(X_t^x) - u(x)}{t} = \lim_{t \downarrow 0} \frac{1}{t} \int_0^t \mathsf{M} Lu(X_s^x)\, ds = Lu(x).$$

Here we have used the continuity of the function $Lu(x)$ and the fact that the mathematical expectation of the stochastic integral is equal to zero.

Hence the infinitesimal generator of the diffusion process is defined and coincides with L for smooth functions. The operator L is sometimes called the differential generator of the diffusion process, the functions $a^{ij}(x)$ are called diffusion coefficients and $b(x)$ is called the drift vector. It is easy to see that the matrix $(a^{ij}(x))$ of diffusion coefficients is nonnegative definite, i.e.,

$$\sum_{i,j=1}^{r} a^{ij}(x)\lambda_i\lambda_j \geq 0$$

for any real $\lambda_1, \ldots, \lambda_r$. Conversely, if a nonnegative definite matrix $(a^{ij}(x))$ and a vector $b(x)$ are given with sufficiently smooth entries, then we can construct a diffusion process with diffusion coefficients $a^{ij}(x)$ and drift $b(x)$. This can be done, for example, by means of a stochastic differential equation: if the matrix $\sigma(x)$ is such that $\sigma(x)\sigma^*(x) = (a^{ij}(x))$, then the solutions of the equation $\dot{X}_t = b(X_t) - \sigma(X_t)\dot{w}_t$ form a diffusion process with diffusion coefficients $a^{ij}(x)$ and drift $b(x)$. For the existence and uniqueness of solutions of the stochastic differential equation it is necessary that the coefficients $\sigma(x)$ and $b(x)$ satisfy certain regularity requirements. For example, as has already been indicated, it is sufficient that $\sigma(x)$ and $b(x)$ satisfy a Lipschitz condition. A representation of $(a^{ij}(x))$ in the form $(a^{ij}(x)) = \sigma(x)\sigma^*(x)$ with entries $\sigma_j^i(x)$ satisfying a Lipschitz condition is always possible whenever the functions $a^{ij}(x)$ are twice continuously differentiable (Freidlin [5]). If $\det(a^{ij}(x)) \neq 0$, then for such a representation it is sufficient that the functions $a^{ij}(x)$ satisfy a Lipschitz condition.

Consequently, every operator

$$L = \frac{1}{2} \sum_{i,j=1}^{r} a^{ij}(x) \frac{\partial^2}{\partial x^i\, \partial x^j} + \sum_{i=1}^{r} b^i(x) \frac{\partial}{\partial x^i}$$

with nonnegative definite matrix $(a^{ij}(x))$ and sufficiently smooth coefficients has a corresponding diffusion process. This diffusion process is determined essentially unique by its differential generator: any two processes with a common differential generator induce the same distribution in the space of trajectories. This is true in all cases where the coefficients $a^{ij}(x)$ and $b^i(x)$ satisfy some weak regularity conditions, which are always satisfied in our investigations.

In a majority of problems in probability theory we are interested in those properties of a random process which are determined by the corresponding distribution in the space of trajectories and do not depend on the concrete representation of the process. In connection with this we shall often say: "Let us consider the diffusion process corresponding to the differential operator L," without specifying how this process is actually given.

A diffusion process corresponding to the operator L can be constructed without appealing to stochastic differential equations. For example, if the diffusion matrix is nondegenerate, then a corresponding process can be constructed starting from the existence theorem for solutions of the parabolic equation $\partial u/\partial t = Lu(t, x)$. Relying on results of the theory of differential equations, we can establish a series of important properties of diffusion processes. For example, we can give conditions under which the transition function has a density. In many problems connected with degeneracies in one way or another, it seems to be more convenient to use stochastic differential equations.

We mention some particular cases. If $a^{ij}(x) = 0$ for all $i, j = 1, 2, \ldots, r$, then L turns into an operator of the first order:

$$L = \sum_{i=1}^{r} b^i(x) \frac{\partial}{\partial x^i}.$$

In this case, the stochastic differential equations turn into the following system of ordinary differential equations:

$$\dot{x}_t = b(x_t), \qquad x_0 = x.$$

Consequently, to any differential operator of the first order there corresponds a Markov process which represents a deterministic motion given by solutions of an ordinary differential equation. In the theory of differential equations, this ordinary differential equation is called the equation of characteristics and its solutions are the characteristics of the operator L.

Another particular case is when all drift coefficients $b^j(x) \equiv 0$ and the diffusion coefficients form a unit matrix: $a^{ij}(x) = \delta^{ij}$. Then, $L = \Delta/2$, where Δ is the Laplace operator. The corresponding Markov family has the form

$$w_t^x = x + w_t,$$

i.e., to the operator $\Delta/2$ there corresponds the family of processes which are obtained by translating the Wiener process by the vector $x \in R^r$. For the sake of brevity, the Markov process (w_t, P_x) connected with this family will also be called a Wiener process. The index in the probability or the mathematical expectation will indicate that the trajectory $w_t^x = x + w_t$ is considered. For example, $\mathsf{P}_x\{w_t \in \Gamma\} = \mathsf{P}\{x + w_t \in \Gamma\}$, $\mathsf{M}_x f(w_t) = \mathsf{M} f(x + w_t)$.

It is easy to see that if all coefficients of L are constant, then the corresponding Markov family consists of Gaussian processes of the form

$$X_t^x = x - \sigma w_t + bt.$$

The diffusion process will also be Gaussian if the diffusion coefficients are constant and the drift depends linearly on x.

Now let (X_t, P_x) be a diffusion process, A the infinitesimal generator of the process, and L the corresponding differential operator. Let us consider the Cauchy problem

$$\frac{\partial u(t, x)}{\partial t} = Lu(t, x); \qquad u(0, x) = f(x),$$

$$x \in R^r, \quad t > 0.$$

A generalized solution of this problem is, by definition, a solution of the following Cauchy problem:

$$\frac{\partial u_t}{\partial t} = Au_t, \qquad u_o = f.$$

The operator A is an extension of L, so that this definition is unambiguous. As has been noted in §3, the solution of the abstract Cauchy problem exists in every case where $f \in D_A$ and can be written in the form

$$u_t(x) = \mathsf{M}_x f(X_t) = T_t f(x).$$

If the classical solution $u(t, x)$ of the Cauchy problem exists, then, since $Au = Lu$ for smooth functions $u = u(t, x)$, the function $u(t, x)$ is also a solution of the abstract Cauchy problem and $u(t, x) = T_t f(x)$ by the uniqueness of the solution of the abstract problem. This representation can be extended to solutions of the Cauchy problem with an arbitrary bounded continuous initial function not necessarily belonging to D_A. This follows from the maximum principle for parabolic equations.

If the matrix $(a^{ij}(x))$ is nondegenerate and the coefficients of L are sufficiently regular (for example, they satisfy a Lipschitz condition), then the equation $\partial u/\partial t = Lu$ has the fundamental solution $p(t, x, y)$, i.e., the solution of the Cauchy problem with initial function $\delta(x - y)$. As can be seen easily, this fundamental solution is the denity of the transition function:

$$P(t, x, \Gamma) = \int_\Gamma p(t, x, y)\, dy.$$

The equation $\partial u/\partial t = Lu$ is called the backward Kolmogorov equation of the diffusion process (X_t, P_x).

Let $c(x)$ be a bounded uniformly continuous function on R^r. Consider the family of operators

$$\tilde{T}_t f(x) = \mathsf{M}_x f(X_t) \exp\left\{\int_0^t c(X_s)\, ds\right\}, \qquad t \geq 0,$$

in the space of bounded measurable functions on R^r. The operators $\tilde{T}_t$ form a semigroup (cf., for example, Dynkin [2]). Taking into account that

$$\exp\left\{\int_0^t c(X_s)\, ds\right\} = 1 + \int_0^t c(X_s)\, ds + o(t)$$

as $t \downarrow 0$, it is easy to prove that if $f \in D_A$ and the coefficients of the operator L are uniformly bounded on R^r, then f belongs to the domain of the infinitesimal generator $\tilde{A}$ of the semigroup $\tilde{T}_t$ and $\tilde{A}f(x) = Af(x) + c(x)f(x)$, where A is the infinitesimal generator of the semigroup $T_t f(x) = \mathsf{M}_x f(X_t)$. Using this observation, it can be proved that for a bounded continuous function $f(x)$ the solution of the Cauchy problem

$$\frac{\partial v(t, x)}{\partial t} = Lv(t, x) + c(x)v(t, x), \qquad x \in R^r, \quad t > 0,$$

$$v(0, x) = f(x)$$

can be written in the form

$$v(t, x) = \tilde{T}_t f(x) = \mathsf{M}_x f(X_t) \exp\left\{\int_0^t c(X_s)\, ds\right\}.$$

A representation in the form of the expectation of a functional of the trajectories of the corresponding process can also be given for the solution of a nonhomogeneous equation: if

$$\frac{\partial w}{\partial t} = Lw + c(x)w + g(x), \qquad w(0, x) = 0,$$

then

$$w(t, x) = \mathsf{M}_x \int_0^t g(X_s) \exp\left\{\int_0^s c(X_u)\, du\right\} ds.$$

A probabilistic representation of solutions of an equation with coefficients depending on t and x can be given in terms of the mean value of functionals of trajectories of the process determined by the nonhomogeneous stochastic differential equation

$$\dot{X}_t = b(t, X_t) + \sigma(t, X_t)\dot{w}_t, \qquad X_{t_0} = x.$$

The solutions of this equation exist for any $x \in R^r$, $t_0 \geqq 0$ if the coefficients are continuous in t and x and satisfy a Lipschitz condition in x with a constant independent of t (Gikhman and Skorokhod [1]). The set of the processes $X_t^{t_0, x}$ for all $t_0 \geqq 0$ and $x \in R^r$ forms a nonhomogeneous Markov family (cf. Dynkin [1]).

In the phase space R^r of a diffusion process (X_t, P_x) let a bounded domain D with a smooth boundary ∂D be given. Denote by τ the first exit time of the process from the domain D: $\tau = \tau(\omega) = \inf\{t: X_t \notin D\}$. In many problems we are interested in the mean of functionals depending on the behavior of the process from time 0 to time τ: for example, expressions of the form

$$\mathsf{M}_x \int_0^\tau f(X_s)\, ds,\ \mathsf{M}_x \exp\left\{\int_0^\tau f(X_s)\, ds\right\}, \text{ etc.}$$

These expressions as functions of the initial point x are solutions of boundary value problems for the differential generator L of the process (X_t, P_x). In the domain D let us consider Dirichlet's problem

$$Lu(x) + c(x)u(x) = f(x), \qquad x \in D;$$

$$u(x)|_{x \in \partial D} = \psi(x).$$

It is assumed that $c(x)$, $f(x)$, for $x \in R^r$, and $\psi(x)$, for $x \in \partial D$, are bounded continuous functions and $c(x) \leqq 0$. Concerning the operator L we assume that it is uniformly nondegenerate in $D \cup \partial D$, i.e., $\sum a^{ij}(x)\lambda_i\lambda_j \geqq k \sum \lambda_i^2$, $k > 0$, and all coefficients satisfy a Lipschitz condition. Under these conditions, the function

$$\tilde{u}(x) = -\mathsf{M}_x \int_0^\tau f(X_t) \exp\left\{\int_0^t c(X_s)\, ds\right\} dt + \mathsf{M}_x \psi(X_\tau) \exp\left\{\int_0^\tau c(X_s)\, ds\right\}$$

is the unique solution of the above Dirichlet problem.

In order to prove this, first we assume that the solution $u(x)$ of Dirichlet's problem can be extended with preservation of smoothness to the whole space R^r. We write $Y_t = \int_0^t c(X_s)\, ds$ and apply Itô's formula to the function $u(X_t)e^{Y_t}$:

$$\begin{aligned} u(X_t) \exp\left[\int_0^t c(X_s)\, ds\right] - u(x) &= \int_0^t (e^{Y_s}\nabla u(X_s), \sigma(X_s)\, dw_s) + \int_0^t e^{Y_s} f(X_s)\, ds \\ &\quad + \int_0^t e^{Y_s}[Lu(X_s) - f(X_s)]\, ds \\ &\quad + \int_0^t u(X_s)e^{Y_s}c(X_s)\, ds. \end{aligned}$$

This equality is satisfied for all $t \geqq 0$ with probability 1. We now replace t by the random variable τ. For $s < \tau$ the trajectory X_s does not leave D, and therefore, $Lu(X_s) = f(X_s) - c(X_s)u(X_s)$. Hence for $t = \tau$ the last two terms on the right side of the equality obtained by means of Itô's formula cancel each other. The random variable τ is a Markov time with respect to the σ-algebras $\mathcal{N}_t$. Under the assumptions made concerning the domain and the process, we have $\mathsf{M}_x\tau < K < \infty$. Therefore

$$\mathsf{M}_x \int_0^\tau (e^{Y_s}\nabla u(X_s), \sigma(X_s)\, dw_s) = 0.$$

utilizing these remarks, we obtain

$$\mathsf{M}_x u(X_\tau) \exp\left[\int_0^\tau c(X_s)\, ds\right] - u(x) = \mathsf{M}_x \int_0^\tau f(X_s) \exp\left[\int_0^\tau c(X_v)\, dv\right] ds.$$

Our assertion follows from this in the case where $u(x)$ can be extended smoothly to the whole space R^r. In order to obtain a proof in the general case, we have to approximate the domain D with an increasing sequence of domains $D_n \subset D$ with sufficiently smooth boundaries ∂D_n. As boundary functions we have to choose the values, on ∂D_n, of the solution $u(x)$ of Dirichlet's problem in D.

We mention some special cases. If $c(x) \equiv 0$, $\psi(x) \equiv 0$ and $f(x) \equiv -1$, then $\tilde{u}(x) = \mathsf{M}_x\tau$. The function $\tilde{u}(x)$ is the unique solution of the problem

$$L\tilde{u}(x) = -1 \quad \text{for } x \in D, \qquad \tilde{u}(x)|_{\partial D} = 0.$$

If $c(x) \equiv 0$, $f(x) \equiv 0$, then for $\tilde{u}(x) = \mathsf{M}_x\, \psi(X_\tau)$ we obtain the problem

$$L\tilde{u}(x) = 0, \qquad x \in D; \qquad \tilde{u}(x)|_{\partial D} = \psi(x).$$

If $c(x) > 0$, then, as is well known, Dirichlet's problem can "go out to the spectrum"; the solution of the equation $Lu + c(x)u = 0$ with vanishing boundary values may not be unique in this case. On the other hand, if $c(x) \leq c_0 < \infty$, and $\mathsf{M}_x e^{c_0\tau} < \infty$ for $x \in D$, then it can be proved that the solution of Dirichlet's problem is unique and the formulas giving a representation of the solution in the form of the expectation of a functional of the corresponding process remain valid (Khasminskii [2]). It can be proved that $\sup\{c: \mathsf{M}_x e^{c\tau} < \infty\} = \lambda_1$ is the smallest eigenvalue of the problem

$$-Lu = \lambda_1 u, \; u|_{\partial D} = 0.$$

We will need not only equations for expectations associated with diffusion processes, but also certain inequalities.

If $u(x)$ is a function that is smooth in D and continuous in its closure

$D \cup \partial D$, $Lu(x) \le c_0 < 0$ for $x \in D$, then

$$\mathsf{M}_x \tau \le \frac{u(x) - \min\{u(x) : x \in \partial D\}}{c_0}; \tag{5.1}$$

if $u(x)$ is positive in $D \cup \partial D$, and $Lu(x) - \lambda u(x) \le 0$, $x \in D$, then

$$\mathsf{M}_x e^{\lambda\tau} \le \frac{u(x)}{\min\{u(x) : x \in \partial D\}}. \tag{5.2}$$

A representation in the form of the expectation of a functional of trajectories of the corresponding process can be given for the solution of a mixed problem for a parabolic equation. For example, the solution $w(t, x)$ of the problem

$$\frac{\partial w}{\partial t} = Lw, \qquad t > 0, \quad x \in D;$$
$$w(0, x) = f(x), \quad x \in D; \qquad w(t, x)|_{t>0,\, x\in\partial D} = \psi(x)$$

can be represented in the form

$$w(t, x) = \mathsf{M}_x\{\tau > t; f(X_t)\} + \mathsf{M}_x\{\tau \le t; \psi(X_\tau)\},$$

under some regularity assumptions on the coefficients of the operator, the boundary of the domain D and the functions $f(x)$ and $\psi(x)$.

On the one hand, we can view the formulas mentioned in this section which connect the expectations of certain functionals of a diffusion process with solutions of the corresponding boundary value problems as a method of calculating these expectations by solving differential equations. On the other hand, we can study the properties of the functionals and their mathematical expectations by methods of probability theory in order to use this information for the study of solutions of boundary value problems. In our book the second point of view will be predominant.

A representation in the form of the mathematical expectation of a functional of trajectories of the corresponding process can also be given for several other boundary value problems, for example, for Neumann's problem, the third boundary value problem (Freidlin [3], Ikeda [1]).

Now we turn to the behavior of a diffusion process as $t \to \infty$. For the sake of simplicity, we shall assume that the diffusion matrix and the drift vector consist of bounded entries satisfying a Lipschitz condition and

$$\sum_{i,j=1}^{r} a^{ij}(x)\lambda_i\lambda_j \ge k \sum_{i=1}^{r} \lambda_i^2,\ k > 0,$$

for $x \in R^r$ and for all real $\lambda_1, \ldots, \lambda_r$. Such a nondegenerate diffusion process may have trajectories of two different types: either trajectories going out to infinity as $t \to \infty$ with probability $\mathsf{P}_x = 1$ for any $x \in R^r$ or trajectories which return to a given bounded region after an arbitrary large t with probability $\mathsf{P}_x = 1$, $x \in R^r$, although $\mathsf{P}_x\{\overline{\lim}_{t\to\infty} |X_t| = \infty\} = 1$.

The diffusion processes which have trajectories of the second type are said to be recurrent. The processes for which $\mathsf{P}_x\{\lim_{t\to\infty} |X_t| = \infty\} = 1$, are said to be transient. It is easy to prove that the trajectories of a recurrent process hit every open set of the phase space with probability $\mathsf{P}_x = 1$ for any $x \in R^r$. We denote by $\tau = \inf\{t: |X_t| < 1\}$ the first entrance time of the unit ball with center at the origin. For a recurrent process, $\mathsf{P}_x\{\tau < \infty\} = 1$. If $\mathsf{M}_x \tau < \infty$ for any $x \in R^r$, then the process (X_t, P_x) is said to be positively recurrent, otherwise it is said to be null recurrent. The Wiener process in R^1 or R^2 serves as an example of a null recurrent process. The Wiener process in R^r is transient for $r \geqq 3$. If, uniformly for all $x \in R^r$ lying outside some ball, the projection of the drift $b(x)$ onto the radius vector connecting the origin of coordinates with the point x is negative and bounded from below in its absolute value, then the process (X_t, P_x) is positively recurrent. It is possible to give stronger sufficient conditions for recurrence and positive recurrence in terms of so-called barriers—nonnegative functions $V(x)$, $x \in R^r$ for which $LV(x)$ has a definite sign and which behave in a certain way at infinity. The recurrence or transience of a diffusion process is closely connected with the formulation of boundary value problems for the operator L in unbounded domains. For example, the exterior Dirichlet problem for the Laplace operator in R^2, where the corresponding process is recurrent, has a unique solution in the class of bounded functions while in order to select the unique solution of the exterior Dirichlet problem for the operator Δ in R^3, it is necessary to prescribe the limit of the solution as $|x| \to \infty$.

It can be proved that if a diffusion process (X_t, P_x) is positively recurrent, then it has a unique stationary probability distribution $\mu(\Gamma)$, $\Gamma \in \mathscr{B}^r$, i.e., a probability measure for which $U_t \mu(\Gamma) = \int_{R^r} \mu(dx) P(t, x, \Gamma) = \mu(\Gamma)$. This measure has a density $m(x)$ which is the unique solution of the problem

$$L^* m(x) = 0 \quad \text{for } x \in R^r, \quad m(x) > 0, \qquad \int_{R^r} m(x)\, dx = 1.$$

Here L^* is the formal adjoint of L:

$$L^* m(x) = \frac{1}{2} \sum_{i,j=1}^{r} \frac{\partial^2}{\partial x^j \partial x^i} (a^{ij}(x) m(x)) - \sum_{i=1}^{r} \frac{\partial}{\partial x^i} (b^i(x) m(x)).$$

For positively recurrent diffusion processes the law of large numbers holds in the following form:

$$\mathsf{P}_x\left\{ \lim_{T\to\infty} \frac{1}{T} \int_0^T f(X_s)\, ds = \int_{R^r} f(x) m(x)\, dx \right\} = 1$$

for arbitrary $x \in R^r$ and any bounded measurable function $f(x)$ on R^r. The process X_t^x, $t \in [0, T]$ defined by the stochastic differential equation

$$\dot{X}_t^x = b(X_t^x) + \sigma(X_t^x)\dot{w}_t, \qquad X_0^x = x,$$

as every other random process induces a probability distribution in the space of trajectories. Since the trajectories of X_t^x are continuous with probability 1, this distribution is concentrated in the space $\mathbf{C}_{0T}^x$ of continuous functions assuming the value x at $t = 0$. We denote by μ_X the measure corresponding to X_t^x in $\mathbf{C}_{0T}^x$. Together with X_t^x, we consider the process Y_t^x satisfying the stochastic differential equation

$$\dot{Y}_t^x = b(Y_t^x) + \sigma(Y_t^x)\dot{w}_t + f(t, Y_t^x), \qquad Y_0^x = x.$$

The processes Y_t^x and X_t^x coincide for $t = 0$; they differ by the drift vector $f(t, Y_t)$. Let μ_Y be the measure corresponding to the process Y_t in $\mathbf{C}_{0T}^x$. We will be particularly interested in the question of when the measures μ_X and μ_Y are absolutely continuous with respect to each other and what the density of one measure with respect to the other looks like. Suppose there exists an r-dimensional vector $\varphi(t, x)$ with components bounded by an absolute constant and such that $\sigma(x)\varphi(t, x) = f(t, x)$. Then μ_X and μ_Y are absolutely continuous with respect to each other and the density $d\mu_Y/d\mu_X$ has the form

$$\frac{d\mu_Y}{d\mu^X}(X^x) = \exp\left\{\int_0^T (\varphi(t, X_t^x), dw_t) - \frac{1}{2}\int_0^T |\varphi(t, X_t^x)|^2\, dt\right\}$$

(Girsanov [1], Gikhman and Skorokhod [1]). In particular, if the diffusion matrix $a(x) = \sigma(x)\sigma^*(x)$ is uniformly nondegenerate for $x \in R^r$, then μ_x and μ_Y are absolutely continuous with respect to each other for any bounded measurable $f(t, x)$. If $X_t^x = x + w_t$ is the Wiener process and

$$Y_t^x = x + w_t + \int_0^t f(s)\, ds,$$

then

$$\frac{d\mu_Y}{d\mu_X} = \exp\left\{\int_0^T (f(s), dw_s) - \frac{1}{2}\int_0^T |f(s)|^2\, ds\right\}.$$

The last equality holds provided that $\int_0^T |f(s)|^2\, ds < \infty$.

Chapter 2

Small Random Perturbations on a Finite Time Interval

§1. Zeroth Order Approximation

In the space R^r we consider the following system of ordinary differential equations:

$$\dot{X}_t^\varepsilon = b(X_t^\varepsilon, \varepsilon\xi_t), \qquad X_0^\varepsilon = x. \tag{1.1}$$

Here $\xi_t(\omega)$, $t \geq 0$ is a random process on a probability space $\{\Omega, \mathscr{F}, \mathsf{P}\}$ with values in R^1 and ε is a small numerical parameter. We assume that the trajectories of $\xi_t(\omega)$ are right continuous, bounded and have at most a finite number of points of discontinuity on every interval $[0, T]$, $T < \infty$. At the points of discontinuity of ξ_t, where as a rule, equation (1.1) cannot be satisfied, we impose the requirement of continuity of X_t^ε. The vector field $b(x, y) = (b^1(x, y), \ldots, b^r(x, y))$, $x \in R^r$, $y \in R^1$ is assumed to be jointly continuous in its variables. Under these conditions the solution of problem (1.1) exists for almost all $\omega \in \Omega$ on a sufficiently small interval $[0, T]$, $T = T(\omega)$.

Let $b(x, 0) = b(x)$. We consider the random process X_t^ε as a result of small perturbations of the system

$$\dot{x}_t = b(x_t), \qquad x_0 = x. \tag{1.2}$$

Theorem 1.1. *Assume that the vector field* $b(x, y)$, $x \in R^r$, $y \in R^l$ *is continuous and that equation* (1.2) *has a unique solution on the interval* $[0, T]$. *Then, for sufficiently small* ε, *the solution of equation* (1.1) *is defined for* $t \in [0, T]$ *and*

$$\mathsf{P}\left\{\lim_{\varepsilon \to 0} \max_{0 \leq t \leq T} |X_t^\varepsilon - x_t| = 0\right\} = 1.$$

Strictly speaking, this result does not have a probabilistic character and belongs to the theory of ordinary differential equations. We are not going to give a detailed proof but only note that the existence of the solution X_t^ε on the whole interval $[0, T]$ follows from the proof of Peano's theorem on the existence of the solution of an ordinary differential equation (cf., for example,

Coddington and Levinson [1]) and the convergence follows from Arzelà's theorem on compactness of sets in $\mathbf{C}_{0T}$ if we take into account that the solution of equation (1.2) is unique.

Now in R^r we consider the stochastic differential equation with a small parameter

$$\dot{X}_t^\varepsilon = b(X_t^\varepsilon) + \varepsilon\sigma(X_t^\varepsilon)\dot{w}_t, \qquad X_0^\varepsilon = x. \tag{1.3}$$

This equation might be considered as a special case of equation (1.1) with $b(x, y) = b(x) + \sigma(x)y$. Nevertheless, here for y we have substituted a white noise process, whose trajectories are not only discontinuous functions but distributions in the general case. Therefore the convergence of the solution of equation (1.3) to the solution of equation (1.2), which is obtained for $\varepsilon = 0$, has to be considered separately.

Theorem 1.2. *Assume that the coefficients of equation* (1.3) *satisfy a Lipschitz condition and increase no faster than linearly*:

$$\sum_i [b^i(x) - b^i(y)]^2 + \sum_{i,j} [\sigma_j^i(x) - \sigma_j^i(y)]^2 \leq K^2|x - y|^2,$$
$$\sum_i [b^i(x)]^2 + \sum_{i,j} [\sigma_j^i(x)]^2 \leq K^2(1 + |x|^2).$$

Then for all $t > 0$ *and* $\delta > 0$ *we have*

$$\mathsf{M}|X_t^\varepsilon - x_t|^2 \leq \varepsilon^2 a(t), \qquad \lim_{\varepsilon \to 0} \mathsf{P}\left\{\max_{0 \leq s \leq t} |X_s^\varepsilon - x_s| > \delta\right\} = 0,$$

where $a(t)$ *is a monotone increasing function, which is expressed in terms of* $|x|$ *and* K.

For the proof, we need the following lemma, which we shall use several times in what follows.

Lemma 1.1. *Let* $m(t)$, $t \in [0, T]$, *be a nonnegative function satisfying the relation*

$$m(t) \leq C + \alpha \int_0^t m(s)\, ds, \qquad t \in [0, T], \tag{1.4}$$

with $C, \alpha > 0$. *Then*

$$m(t) \leq Ce^{\alpha t}$$

for $t \in [0, T]$.

Proof. From inequality (1.4) we obtain

$$m(t)\left(C + \alpha \int_0^t m(s)\, ds\right)^{-1} \leq 1.$$

Integrating both sides from 0 to t, we obtain

$$\ln\left(C + \alpha \int_0^t m(s)\, ds\right) - \ln C \leq \alpha t,$$

which implies that

$$C + \alpha \int_0^t m(s)\, ds \leq Ce^{\alpha t}.$$

The last inequality and (1.4) imply the assertion of the lemma. □

Now we begin the proof of the theorem. We prove that $\mathsf{M}|X_t^\varepsilon|^2$ is bounded uniformly in $\varepsilon \in [0, 1]$. For this, we apply Itô's formula (cf. §3, Ch. 1) to the function $1 + |X_t^\varepsilon|^2$. Taking into account that the mathematical expectation of the stochastic integral in this formula vanishes, we obtain

$$1 + \mathsf{M}|X_t^\varepsilon|^2 = 1 + |x|^2 + 2\int_0^t \mathsf{M}\,(X_s^\varepsilon, b(X_s^\varepsilon))\, ds + \varepsilon^2 \int_0^t \mathsf{M} \sum_{i,j} [\sigma_j^i(X_s^\varepsilon)]^2\, ds.$$

Since the coefficients of equation (1.3) increase no faster than linearly, the last relation implies the estimate

$$\begin{aligned} 1 + \mathsf{M}|X_t^\varepsilon|^2 &\leq 1 + |x|^2 \\ &\quad + 2\int_0^t \mathsf{M}\sqrt{|X_s^\varepsilon|^2 K^2(1 + |X_s^\varepsilon|^2)}\, ds \\ &\quad + \varepsilon^2 K^2 \int_0^t (1 + M|X_s^\varepsilon|^2)\, ds \leq 1 + |x|^2 \\ &\quad + (2K + \varepsilon^2 K^2)\int_0^t (1 + \mathsf{M}|X_s^\varepsilon|^2)\, ds. \end{aligned}$$

Using Lemma 1.1, we conclude that

$$1 + \mathsf{M}|X_t^\varepsilon|^2 \leq (1 + |x|^2)\exp[(2K + \varepsilon^2 K^2)t]. \tag{1.5}$$

Now we apply Itô's formula to the function $|X_t^\varepsilon - x_t|^2$ and take the the mathematical expectation on both sides of the equality:

$$\mathsf{M}|X_t^\varepsilon - x_t|^2 = 2\int_0^t \mathsf{M}(X_s^\varepsilon - x_s, b(X_s^\varepsilon) - b(x_s))\,ds + \varepsilon^2 \int_0^t \mathsf{M}\sum_{i,j}[\sigma_j^i(X_s^\varepsilon)]^2\,ds$$

It follows from this relation that

$$\mathsf{M}|X_t^\varepsilon - x_t|^2 \le 2K\int_0^t \mathsf{M}|X_s^\varepsilon - x_s|^2\,ds + \varepsilon^2K^2\int_0^t (1 + \mathsf{M}|X_s^\varepsilon|^2)\,ds,$$

and using Lemma 1.1, we obtain

$$\mathsf{M}|X_t^\varepsilon - x_t|^2 \le e^{2Kt}\cdot\varepsilon^2K^2\int_0^t (1 + \mathsf{M}|X_s^\varepsilon|^2)\,ds.$$

Combining the last inequality and (1.5), we obtain the first assertion of the theorem:

$$\mathsf{M}|X_t^\varepsilon - x_t|^2 \le \varepsilon^2K^2e^{2Kt}(1 + |x|^2)\int_0^t \exp[(2K + \varepsilon^2K^2)s]\,ds \le \varepsilon^2 a(t).$$

Now we prove the second assertion of Theorem 1.2. It follows from the definition of X_t^ε and x_t that

$$\max_{0\le s\le t}|X_s^\varepsilon - x_s| \le \int_0^t |b(X_s^\varepsilon) - b(x_s)|\,ds + \varepsilon\max_{0\le s\le t}\left|\int_0^s \sigma(X_v^\varepsilon)\,dw_v\right|. \tag{1.6}$$

From Chebyshev's inequality and the first assertion of the theorem we obtain an estimate of the first term on the right side of (1.6):

$$\begin{aligned}\mathsf{P}\left\{\int_0^t |b(X_s^\varepsilon) - b(x_s)|\,ds > \frac{\delta}{2}\right\} &\le 4\delta^{-2}\mathsf{M}\left[\int_0^t |b(X_s^\varepsilon) - b(x_s)|\,ds\right]^2 \\ &\le 4tK^2\delta^{-2}\int_0^t \mathsf{M}|X_s^\varepsilon - x_s|^2\,ds \qquad (1.7)\\ &\le 4tK^2\delta^{-2}\varepsilon^2\int_0^t a(s)\,ds = \varepsilon^2\delta^{-2}a_1(t).\end{aligned}$$

The estimation of the second term in (1.6) can be accomplished with use of the generalized Kolmogorov inequality for stochastic integrals:

$$\mathsf{P}\left\{\varepsilon \max_{0 \le s \le t} \left| \int_0^s \sigma(X_v^\varepsilon)\, dw_v \right| > \frac{\delta}{2}\right\} \le 4\delta^{-2}\varepsilon^2 \int_0^t \sum_{i,j} \mathsf{M}[\sigma_j^i(X_s^\varepsilon)]^2\, ds = \varepsilon^2\delta^{-2}a_2(t) \tag{1.8}$$

Estimates (1.6)–(1.8) imply the last assertion of the theorem. □

In some respect we make more stringent assumptions in Theorem 1.2 than in Theorem 1.1. We assumed that the coefficients satisfied a Lipschitz condition instead of continuity. However, we obtained a stronger result in that not only did we prove that X_t^ε converges to x_t, but we also obtain estimates of the rate of convergence. If we make even more stringent assumptions concerning the smoothness of the coefficients, then the difference $X_t^\varepsilon - x_t$ can be estimated more accurately. We shall return to this question in the next section. Now we will obtain a result on the zeroth approximation for a differential equation with a right side of a sufficiently general form.

We consider the differential equation

$$\dot{X}_t^\varepsilon = b(\varepsilon, t, X_t^\varepsilon, \omega), \qquad X_0^\varepsilon = x$$

in R^n. Here $b(\varepsilon, t, x, \omega) = (b^1(\varepsilon, t, x, \omega), \ldots, b^r(\varepsilon, t, x, \omega))$ is an r-dimensional vector defined for $x \in R^r$, $t \geqq 0$, $\varepsilon > 0$ and $\omega \in \Omega$.

We assume that the field $b(\varepsilon, t, x, \omega)$ is continuous in t and x for almost all ω for any $\varepsilon > 0$,

$$\sup_{t \ge \Delta, \varepsilon \in (0,1]} \mathsf{M}|b(\varepsilon, t, x, \omega)|^2 < \infty,$$

and for some $K > 0$ we have

$$|b(\varepsilon, t, x, \omega) - b(\varepsilon, t, y, \omega)| \le K|x - y|$$

almost surely for any $x, y \in R^r$, $t \ge 0$, $\varepsilon > 0$. We note that continuity in ε for fixed t, x, ω is not assumed.

Theorem 1.3. *We assume that there exists a continuous function $\bar{b}(t, x)$, $t > 0$. $x \in R^r$ such that for any $\delta > 0$, $T > 0$, $x \in R^r$ we have*

$$\lim_{\varepsilon \to 0} \mathsf{P}\left\{\left| \int_{t_0}^{t_0+T} b(\varepsilon, t, x, \omega)\, dt - \int_{t_0}^{t_0+T} \bar{b}(t, x)\, dt \right| > \delta\right\} = 0 \tag{1.9}$$

uniformly in $t_0 \ge 0$. Then the equation

$$\dot{\bar{x}}_t = \bar{b}(t, \bar{x}_t), \qquad \bar{x}_0 = x \tag{1.10}$$

has a unique solution and

$$\lim_{\varepsilon\to 0} \mathsf{P}\left\{\max_{0\le t\le T} |X_t^\varepsilon - \bar{x}_t| > \delta\right\} = 0$$

for every $T > 0$ *and* $\delta > 0$.

Proof. First we note that the function $\bar{b}(t, x)$ satisfies a Lipschitz condition in x with the same constant as the function $b(\varepsilon, t, x, \omega)$. Indeed, since the function $\bar{b}(t, x)$ is continuous, by the mean value theorem we have

$$\int_t^{t+\Delta} \bar{b}(s, x)\, ds = \bar{b}(t, x)\,\Delta + o(\Delta), \qquad \Delta \to 0.$$

Taking account of (1.9), we obtain that

$$\begin{aligned}|\bar{b}(t, x) - \bar{b}(t, y)| &= \frac{1}{\Delta}\left|\int_t^{t+\Delta} \bar{b}(s, x)\, ds - \int_t^{t+\Delta} \bar{b}(s, y)\, ds\right| + \frac{o(\Delta)}{\Delta} \\ &\le \frac{1}{\Delta}\left|\int_t^{t+\Delta} b(\varepsilon, s, x, \omega)\, ds - \int_t^{t+\Delta} b(\varepsilon, s, y, \omega)\, ds\right| \\ &\quad + \frac{o(\Delta)}{\Delta} + \delta_\varepsilon \le K|x - y| + \frac{o(\Delta)}{\Delta} + \delta_\varepsilon,\end{aligned}$$

where $\delta_\varepsilon = \delta_\varepsilon(t, \omega) \to 0$ in probability as $\varepsilon \to 0$.

Since this inequality holds for arbitrary small ε and Δ we have

$$|\bar{b}(t, x) - \bar{b}(t, y)| \le K|x < y|. \tag{1.11}$$

It follows from (1.11) that equation (1.10) has a unique solution.

By the definition of X_t^ε and $\bar{x}_t$ we have

$$\begin{aligned}X_t^\varepsilon - \bar{x}_t &= \int_0^t [b(\varepsilon, s, X_s^\varepsilon, \omega) - \bar{b}(s, \bar{x}_s)]\, ds \\ &= \int_0^t [b(\varepsilon, s, X_s^\varepsilon, \omega) - b(\varepsilon, s, \bar{x}_s, \omega)]\, ds \\ &\quad + \int_0^t [b(\varepsilon, s, \bar{x}_s, \omega) - \bar{b}(s, \bar{x}_s)]\, ds.\end{aligned}$$

Define $m(t) = m^\varepsilon(t) = \max_{0\le s\le t}|X_s^\varepsilon - \bar{x}_s|$. Using the preceding formula, we obtain the inequality

$$m(t) \le K \cdot \int_0^t m(s)\, ds + \max_{0\le t_1\le t}\left|\int_0^{t_1} [b(\varepsilon, s, \bar{x}_s, \omega) - \bar{b}(s, \bar{x}_s)]\, ds\right|.$$

Then we obtain by Lemma 1.1 that

$$m(T) \le e^{KT} \max_{0 \le t \le T} \left| \int_0^t [b(\varepsilon, s, \bar{x}_s, \omega) - \bar{b}(s, \bar{x}_s)]\, ds \right|, \tag{1.12}$$

where T is an arbitrary positive number.

We now show that the maximum on the right side of (1.12) converges to zero in probability as $\varepsilon \to 0$. Let n be a large integer, which we will choose later. Using the Lipschitz condition we have for $t \in [0, T]$,

$$\begin{aligned}
&\int_0^t [b(\varepsilon, s, \bar{x}_s, \omega) - \bar{b}(s, \bar{x}_s)]\, ds \\
&\quad = \sum_{k=0}^{n-1} \int_{kt/n}^{(k+1)t/n} [b(\varepsilon, s, \bar{x}_s, \omega) - \bar{b}(s, \bar{x}_s)]\, ds \\
&\quad = \sum_{k=0}^{n-1} \int_{kt/n}^{(k+1)t/n} [b(\varepsilon, s, \bar{x}_{kt/n}, \omega) - \bar{b}(s, \bar{x}_{kt/n})]\, ds \\
&\qquad + \sum_{k=0}^{n-1} \int_{kt/n}^{(k+1)t/n} [b(\varepsilon, s, \bar{x}_s, \omega) - b(\varepsilon, s, \bar{x}_{kt/n}, \omega)]\, ds \\
&\qquad + \sum_{k=0}^{n-1} \int_{kt/n}^{(k+1)t/n} [\bar{b}(s, \bar{x}_{kt/n}) - \bar{b}(s, \bar{x}_s)]\, ds \\
&\quad = \sum_{k=0}^{n-1} \int_{kt/n}^{(k+1)t/n} [b(\varepsilon, s, \bar{x}_{kt/n} \omega) - \bar{b}(s, x_{kt/n})]\, ds + p_{n,t}^{\varepsilon},
\end{aligned} \tag{1.13}$$

where $|\rho_{n,t}^{\varepsilon}| < C/n$ and C is a constant depending on the Lipschitz constant K and T.

By condition (1.9), the sum on the last side of the formula converges to zero in probability for given n. Consequently, (1.13) implies that

$$\lim_{\varepsilon \to 0} \mathsf{P}\left\{ \max_{0 \le k \le n} \left| \int_0^{kT/n} [b(\varepsilon, s, \bar{x}_s, \omega) - \bar{b}(s, \bar{x}_s)]\, ds \right| > \frac{\delta}{2} \right\} = 0 \tag{1.14}$$

for $n > 4C/\delta$.

Moreover, we note that

$$\begin{aligned}
&\mathsf{P}\left\{ \max_{0 \le t \le T} \left| \int_0^t [b(\varepsilon, s, \bar{x}_s, \omega) - \bar{b}(s, \bar{x}_s)]\, ds \right| > \delta \right\} \\
&\quad \le \mathsf{P}\left\{ \max_{0 \le k \le n} \left| \int_0^{kT/n} [b(\varepsilon, s, \bar{x}_s, \omega) - \bar{b}(s, \bar{x}_s)] \right| ds > \frac{\delta}{2} \right\} \\
&\qquad + \mathsf{P}\left\{ \max_{0 \le k \le n} \int_{kT/n}^{(k+1)T/n} |b(\varepsilon, s, \bar{x}_s, \omega) - \bar{b}(s, \bar{x}_s)|\, ds > \frac{\delta}{2} \right\}.
\end{aligned} \tag{1.15}$$

We estimate the last term by means of Chebyshev's inequality:

$$\begin{aligned}
&\mathsf{P}\left\{\max_k \int_{kT/n}^{(k+1)T/n} |b(\varepsilon, s, \bar{x}_s, \omega) - \bar{b}(s, \bar{x}_s)|\, ds > \frac{\delta}{2}\right\} \\
&\quad \leq n \cdot \max_k \mathsf{P}\left\{\int_{kT/n}^{(k+1)T/n} |b(\varepsilon, s, \bar{x}_s, \omega) - \bar{b}(s, \bar{x}_s)|\, ds > \frac{\delta}{2}\right\} \\
&\quad \leq n \frac{4}{\delta^2}\frac{T^2}{n^2} \sup_{\varepsilon \geq 0,\, \varepsilon \in (0,1]} \mathsf{M}|b(\varepsilon, s, \bar{x}_s, \omega) - \bar{b}(s, \bar{x}_s)|^2 \\
&\quad \leq \frac{4T^2}{n\delta^2} \sup_{s \geq 0,\, \varepsilon \in (0,1]} \mathsf{M}[|\bar{b}(s, \bar{x}_s)| + |b(\varepsilon, s, 0, \omega)| + K|\bar{x}_s|]^2 \leq \frac{C_1 T}{n\delta^2},
\end{aligned} \tag{1.16}$$

where C_1 is a constant. Here we have used the fact that

$$\sup_{s \geq 0, \varepsilon \in (0,1]} \mathsf{M}|b(\varepsilon, s, 0, \omega)|^2 < \infty.$$

It follows from (1.14)–(1.16) that the right side of (1.12) converges to zero in probability as $\varepsilon \to 0$. This completes the proof of Theorem 1.3. □

The random process X_t^ε considered in Theorem 1.3 can be viewed as a result of random perturbations of system (1.10). We shall return to the study of similar perturbations in Ch. 7.

§2. Expansion in Powers of a Small Parameter

We return to the study of equations (1.1) and (1.3). In this section we obtain an expansion of X_t^ε in powers of the small parameter ε provided that the functions $b(x, y)$ are sufficiently smooth.

We follow the usual approach of perturbation theory to obtain an expansion

$$X_t^\varepsilon = X_t^{(0)} + \varepsilon X_t^{(1)} + \cdots + \varepsilon^k X_t^{(k)} + \cdots \tag{2.1}$$

of X_t^ε in powers of ε. We substitute this expansion with unknown coefficients $X_t^{(0)}, \ldots, X_t^{(k)}, \ldots$ into equation (1.1) and expand the right sides in powers of ε. Equating the coefficients of the same powers on the left and right, we obtain differential equations for the successive calculation of the coefficients $X_t^{(0)}, X_t^{(1)}, \ldots$ in (2.1).

We discuss how the right side of (1.1) is expanded in powers of ε. Let $X(\varepsilon)$ be any power series with coefficients from R^r:

$$X(\varepsilon) = c_0 + c_1 \varepsilon + \cdots + c_k \varepsilon^k + \cdots.$$

We write

$$\Phi_k = \Phi_k(c_0, c_1, \ldots, c_k, y) = \frac{1}{k!}\frac{d^k b(X(\varepsilon), \varepsilon y)}{d\varepsilon^k}\bigg|_{\varepsilon=0}$$

It is easy to see that Φ_k depends linearly on c_k for $k \geq 1$ and Φ_k is a polynomial of degree k in the variable y. In particular,

$$\Phi_0 = b(c_0, 0),$$

$$\Phi_1 = B_1(c_0, 0)c_1 + B_2(c_0, 0)y,$$

where $B_1(x, y) = (\partial b^i(x, y)/\partial x^k)$ is a square matrix of order r and $B_2(x, y) = (\partial b^i(x, y)/\partial y^k)$ is a matrix having r rows and l columns. It is clear from the definition of Φ_k that the difference $\Phi_k - B_1(c_0, 0)c_k = \Psi_k(c_0, c_1, \ldots, c_{k-1}, y)$ is independent of c_k.

Carrying out the above program, we expand both sides of (1.1) in powers of ε:

$$\dot{X}_t^{(0)} + \varepsilon\dot{X}_t^{(t)} + \cdots + \varepsilon^k\dot{X}_t^{(k)} + \cdots = \Phi_0(X_t^{(0)}, \xi_t) + \varepsilon\Phi_1(X_t^{(0)}, X_t^{(1)}, \xi_t) + \cdots + \varepsilon^k\Phi_k(X_t^{(0)}, \ldots, X_t^{(k)}, \xi_t) + \cdots.$$

Hence we obtain the differential equations

$$\begin{aligned}
\dot{X}_t^{(0)} &= \Phi_0(X_t^{(0)}, \xi_t) = b(X_t^{(0)}, 0),\\
\dot{X}_t^{(1)} &= \Phi_1(X_t^{(0)}, X_t^{(1)}, \xi_t) = B_1(X_t^{(0)}, 0)X_t^{(0)} + B_2(X_t^{(0)}, 0)\xi_t,\\
&\cdots\cdots\cdots\cdots\cdots\cdots\\
\dot{X}_t^{(k)} &= \Phi_k(X_t^{(0)}, \ldots, X_t^{(k)}, \xi_t) = B_1(X_t^{(0)}, 0)X_t^{(k)} + \Psi_k(X_t^{(0)}, \ldots, X_t^{(k-1)}, \xi_t),\\
&\cdots\cdots\cdots\cdots\cdots\cdots
\end{aligned} \tag{2.2}$$

To these differential equations we add the initial conditions $X_0^{(0)} = x$, $X_0^{(1)} = 0, \ldots, X_0^{(k)} = 0, \ldots$. If $b(x, y)$ is sufficiently smooth, then equations (2.2), together with the initial conditions, determine the functions $X_t^{(0)}$, $X_t^{(1)}, \ldots, X_t^{(k)}, \ldots$ uniquely. The zeroth approximation is determined from the first equation of system (2.2), which coincides with (1.2). If $X_t^{(0)}$ is known, then the second equation in (2.2) is a linear equation in $X_t^{(1)}$. In general, if the functions $X_t^{(0)}, \ldots, X_t^{(k-1)}$ are known, then the equation for $X_t^{(k)}$ will be a nonhomogeneous linear equation with coefficients depending on time.

Theorem 2.1. *Suppose the trajectories of a process $\xi_t(\omega)$ are continuous with probability 1 and that the function $b(x, y)$, $x \in R^r$, $y \in R^l$, has $k + 1$ bounded continuous partial derivatives with respect to x and y. Then*

$$X_t^\varepsilon = X_t^{(0)} + \varepsilon X_t^{(1)} + \cdots + \varepsilon^k X_t^{(k)} + R_{k+1}^\varepsilon(t),$$

where the functions $X_t^{(i)}$, $i = 0, 1, \ldots, k$, are determined from system (2.2) *and*

$$\sup_{0 \le t \le T} |R_{k+1}^{\varepsilon}(t)| < C(\omega)\varepsilon^{k+1}, \qquad \mathsf{P}\{C(\omega) < \infty\} = 1.$$

Proof. From the definition of X_t^{ε}, $X_t^{(i)}$, $i = 0, 1, \ldots, k$, it follows that the function $R_{k+1}^{\varepsilon}(t) = X_t^{\varepsilon} - \sum_{i=0}^{k} \varepsilon^i X_t^{(i)}$ satisfies the relation

$$\begin{aligned}\dot{R}_{k+1}^{\varepsilon}(t) &= b(X_t^{\varepsilon}, \varepsilon\xi_t) - \sum_{i=0}^{k} \varepsilon^i \Phi_i(X_t^{(0)}, \ldots, X_t^{(i)}, \xi_t) \\ &= \left[b(X_t^{\varepsilon}, \varepsilon\xi_t) - b\left(\sum_{i=0}^{k} \varepsilon^i X_t^{(i)}, \varepsilon\xi_t\right)\right] \\ &\quad + \left[b\left(\sum_{i=0}^{k} \varepsilon^i X_t^{(i)}, \varepsilon\xi_t\right) - \sum_{i=0}^{k} \varepsilon^i \Phi_i(X_t^{(0)}, \ldots, X_t^{(i)}, \xi_t)\right].\end{aligned} \tag{2.3}$$

Since the first derivatives of $b(x, y)$ are bounded then the first term on the right side of (2.3) can be estimated in the following way:

$$\left|b(X_t^{\varepsilon}, \varepsilon\xi_t) - b\left(\sum_{i=0}^{k} \varepsilon^i X_t^{(i)}, \varepsilon\xi_t\right)\right| \le K_1 |R_{k+1}^{\varepsilon}(t)|, \tag{2.4}$$

where K_1 is a constant.

In the Taylor series of $b(\sum_{i=0}^{k} \varepsilon^i X_t^{(i)}, \varepsilon\xi_t)$ about $(X_t^{(0)}, 0)$, the coefficients of ε^i are equal to Φ_i up to $i = k$. It follows that

$$\begin{aligned}&\left|b\left(\sum_{i=0}^{k} \varepsilon^i X_t^{(i)}, \varepsilon\xi_t\right) - \sum_{i=0}^{k} \varepsilon^i \Phi_i(X_t^{(0)}, \ldots, X_t^{(i)}, \xi_t)\right| \\ &\quad \le \sum_{\substack{0 \le j \le k-1 \\ 1 \le i_1 \le k; \ldots; 1 \le i_j \le k}} K_{i_1, \ldots, i_j} \varepsilon^{i_1 + \cdots + i_j + k + 1 - j} |X_t^{(i_1)}| \cdots |X_t^{(i_j)}| |\xi_t|^{k+1-j}.\end{aligned} \tag{2.5}$$

Here $K_{i_1, \ldots; i_j}$ are constants depending on the maximum absolute value of the $(k + 1)$st derivatives of $b(x, y)$, on $i_1, \ldots, i_j$ and on the dimension.

The following lemma provides an estimate for $|X_t^{(i)}|$.

Lemma 2.1. *There exist constants $C_i < \infty$ such that*

$$|X_t^{(i)}| \le C_i \cdot \left(\max_{0 \le s \le t} |\xi_s|\right)^i$$

for all $t \le T$.

The lemma can be proved by induction (using equations (2.2), of course).

If we integrate (2.3) from 0 to t and take into account that $R_{k+1}^{\varepsilon}(0) = 0$ and inequalities (2.4), (2.5) and Lemma 2.1, as well, we obtain

$$|R_{k+1}^{\varepsilon}(t)| \leq K_1 \int_0^t |R_{k+1}^{\varepsilon}(s)|\, ds + K_2 t \sum_{i=k+1}^{k(k+1)} \left[\varepsilon \cdot \max_{0 \leq s \leq t} |\xi_s|\right]^i,$$

where K_2 is a constant. For $\varepsilon \leq (2 \max_{0 \leq s \leq T} |\xi_s|)^{-1}$ the sum on the right side does not exceed $2\varepsilon^{k+1} (\max_{0 \leq s \leq t} |\xi_s|)^{k+1}$. Using Lemma 1.1, we obtain

$$|R_{k+1}^{\varepsilon}(t)| \leq \varepsilon^{k+1} 2K_2 e^{K_1 t} \left(\max_{0 \leq s \leq t} |\xi_s| \right)^{k+1}.$$

This completes the proof of the theorem. □

Consequently, if $b(x, y)$ is sufficiently smooth, then X_t^{ε} can be calculated to any accuracy. For this we have to integrate the equations for $X_t^{(i)}$. All these equations are linear and have approximately the same structure. The zeroth approximation $X_t^{(0)}$ is a nonrandom function while all approximations beginning with the first one are random processes. We remark that $X_t^{(1)}$ is determined from the equation

$$\dot{X}_t^{(1)} = B_1(X_t^{(0)}, 0) X_t^{(1)} + B_2(X_t^{(0)}, 0)\xi_t; \qquad X_0^{(1)} = 0.$$

It is clear from this that $X_t^{(1)}$ can be obtained from ξ_t by means of a linear (nonrandom) transformation. In particular, if ξ_t is a Gaussian process, then $X_t^{(1)}$ is also Gaussian, and consequently, the approximation $X_t^{(0)} + \varepsilon X_t^{(1)}$ of X_t^{ε} to within values of order ε^2 is a Gaussian process.

We discuss the one-dimensional case in more detail: X_t^{ε} is a process in R^1 and ξ_t is a one-dimensional process. Then the equation for $X_t^{(1)}$ can be solved by quadratures:

$$X_t^{(1)} = \int_0^t b_2'(X_s^{(0)}, 0)\xi_s \exp\left\{ \int_s^t b_1'(X_u^{(0)}, 0)\, du \right\} ds.$$

We write out the equation for $X_t^{(2)}$:

$$\begin{aligned} \dot{X}_t^{(2)} = b_1'(X_t^{(0)}, 0) X_t^{(2)} &+ \tfrac{1}{2}[b_{11}''(X_t^{(0)}, 0)(X_t^{(1)})^2 \\ &+ 2b_{12}''(X_t^{(0)}, 0) X_t^{(1)} \xi_t + b_{22}''(X_t^{(0)}, 0)\xi_t^2], \qquad X_0^{(2)} = 0. \end{aligned}$$

Here b_1' and b_2' are the derivatives of $b(x, y)$ with respect to x and y and the b_{ij}'' are the second derivatives of the same function. The equation for $X_t^{(2)}$ can also be solved by quadratures. $X_t^{(2)}$ is a quadratic functional of the process ξ_t. The equations for $X_t^{(i)}$ look similar for $i = 3, 4, \ldots$ and can be integrated successively by quadratures.

These equations become especially simple if $X_0 = x$ is an equilibrium position of the unperturbed system. In this case the functions $X_t^{(i)}$ can be found as solutions of nonhomogeneous linear equations with constant coefficients.

Theorem 2.1 can be used to calculate the expansions, in powers of a small parameter, of smooth functions of X_t^ε and their mathematical expectations. For example, if the first and second derivatives of a function $f(x)$ are bounded, then

$$\begin{aligned} \mathsf{M}f(X_t^\varepsilon) &= \mathsf{M}[f(X_t^{(0)}) + (\nabla f(X_t^{(0)}), X_t^{(1)})\varepsilon + O(\varepsilon^2)] \\ &= f(X_t^{(0)}) + \varepsilon(\nabla f(X_t^{(0)}), \mathsf{M}X_t^{(1)}) + O(\varepsilon^2), \end{aligned}$$

where the function $m(t) = \mathsf{M}X_t^{(1)}$ is a solution of the differential equation

$$\dot{m}(t) = B_1(X_t^{(0)}, 0)m(t) + B_2(X_t^{(0)}, 0)\mathsf{M}\xi_t.$$

We now consider equation (1.3). Formally we can consider equation (1.3) as a special case of (1.1) with $b(x, y) = b(x) + \sigma(x)y$. Assuming $b(x)$ and $\sigma(x)$ are sufficiently smooth, we can write down equations (2.2) for this case:

$$\begin{aligned} &\dot{X}_t^{(0)} = b(X_t^{(0)}), \qquad X_0^{(0)} = X_0^\varepsilon = x, \\ &\dot{X}_t^{(1)} = B(X_t^{(0)})X_t^{(1)} + \sigma(X_t^{(0)})\dot{w}_t, \qquad X_0^{(1)} = 0, \\ &\cdots\cdots\cdots\cdots\cdots\cdots\cdots\cdots\cdots \\ &\dot{X}_t^{(k)} = \Phi_k(X_t^{(0)}, \ldots, X_t^{(k)}, \dot{w}_t), \qquad X_0^{(k)} = 0. \\ &\cdots\cdots\cdots\cdots\cdots\cdots\cdots\cdots\cdots \end{aligned} \tag{2.6}$$

Here $B(x) = (\partial b^i(x)/\partial x^j)$. These equations are all linear in $\dot{w}_t$. The set of the first k equations of system (2.6) may be considered as a stochastic differential equation for the process $X_t^{k+1} = (X_t^{(0)}, X_t^{(1)}, \ldots, X_t^{(k)})$. If $b(x)$ and $\sigma(x)$ have bounded derivatives up to order $k + 1$, then this stochastic differential equation has a unique solution and determines a $(k + 1)$-dimensional Markov process. As in the case of equation (1.1), the zeroth approximation is a deterministic motion along the trajectories of the unperturbed dynamical system (1.2). The process $X_t^{(1)}$ can be determined from a stochastic differential equation whose drift vector depends linearly on $X_t^{(1)}$ and the diffusion coefficients depend only on t. It can be verified easily that $X_t^{(1)}$ is a Gaussian process. (It follows, for example, from the fact that the solution of a stochastic differential equation can be constructed by the method of successive approximations). Therefore, the solution of equation (1.3) to within values of order ε^2 is a Gaussian Markov process $X_t^{(0)} + \varepsilon X_t^{(1)}$ which is nonhomogeneous in time.

As an example, we consider the one-dimensional stochastic differential equation

$$\dot{X}_t^\varepsilon = b(X_t^\varepsilon) + \varepsilon\dot{w}_t, \qquad X_0^\varepsilon = x.$$

The zeroth approximation $X_t^{(0)}$ is the solution of the equation $\dot{X}_t^{(0)} = b(X_t^{(0)})$, $X_0^{(0)} = x$. For $X_t^{(1)}$ we obtain the equation

$$\dot{X}_t^{(1)} = b'(X_t^{(0)})X_t^{(1)} + \dot{w}_t.$$

If we consider $X_t^{(0)}$ known, then the solution of this equation can be written in the form

$$X_t^{(1)} = \int_0^t \exp\left\{\int_s^t b'(X_u^{(0)})\,du\right\} dw_s.$$

Therefore

$$X_t^\varepsilon = X_t^{(0)} + \varepsilon \int_0^t \exp\left\{\int_s^t b'(X_u^0)\,du\right\} dw_s + o(\varepsilon).$$

We formulate the following theorem concerning the expansion of the solution of the stochastic differential equation in powers of a small parameter ε.

Theorem 2.2. *Suppose the coefficients $b^i(x)$ and $\sigma_j^i(x)$ have bounded partial derivatives up to order $k + 1$ inclusive.*

Then for the solution X_t^ε of equation (1.3) *we have the expansion*

$$X_t^\varepsilon = X_t^{(0)} + \varepsilon X_t^{(1)} + \cdots + \varepsilon^k X_t^{(k)} + R_{k+1}^\varepsilon(t), \tag{2.7}$$

where $X_t^{(0)}, X_t^{(1)}, \ldots, X_t^{(k)}$ are determined from equations (2.6). *The random process $X_t^{k+1} = (X_t^{(0)}, \ldots, X_t^{(k)})$ is determined by the first $k + 1$ equations of system* (2.6). *The process X_t^ε is approximated to within values of order ε^2 by the Gaussian process $X_t^{(0)} + \varepsilon X_t^{(1)}$. The remainder in* (2.7) *satisfies the inequality*

$$\sup_{0 \le t \le T} (\mathsf{M}|R_{k+1}^\varepsilon(t)|^2)^{1/2} \le C\varepsilon^{k+1}, \qquad C < \infty.$$

The proof of this theorem differs from that of Theorem 2.1 in some technical details only. However, these technical details require tedious calculations connected with the proof of the differentiability of a solution of a stochastic differential equation with respect to a parameter. These questions are outside the scope of our main theme, and therefore, we do not give the proof of Theorem 2.5 here. The proof can be found in Blagoveshchenskii and Freidlin [1] and Blagoveshchenskii [1].

From the expansion

$$X_t^\varepsilon = X_t^{(0)} + \varepsilon X_t^{(1)} + \cdots + \varepsilon^k X_t^{(k)} + o(\varepsilon^k) \tag{2.8}$$

of the realizations of X_t^ε it is easy to obtain expansions in powers of ε for smooth functionals of realizations. Let the functional F be Fréchet differentiable at the point $X^{(0)}$. The derivative of F at this point is a linear functional $F'(X^{(0)}; h)$. In this case we have

$$F(X^\varepsilon) = F(X^{(0)}) + \varepsilon F'(X^{(0)}; X^{(1)}) + o(\varepsilon) \tag{2.9}$$

as $\varepsilon \to 0$, where $o(\varepsilon)$ is understood in the same way as in (2.8) (uniformly in $t \subset [0, T]$ for almost all ω or in the sense of convergence in probability uniformly in t).

In the case where $X_t^{(1)}$ is a Gaussian random process, from the expansion (2.9) we obtain that the value of the functional $F(X^\varepsilon)$ is asymptotically normal with standard deviation proportional to ε^2. The coefficient of proportionality in this asymptotic standard deviation can be expressed in terms of the derivative $F'(X^{(0)}; h)$ and the correlation function of $X_t^{(1)}$ (the asymptotic mean is equal to $F(X^{(0)})$ provided that $\mathsf{M}X_t^{(1)} = 0$).

Now consider the case where F is twice differentiable, the second derivative being a bilinear functional $F''(X^{(0)}; h_1, h_2)$. We obtain the following expansion to within values of order $o(\varepsilon^2)$ as $\varepsilon \to 0$:

$$\begin{aligned} F(X^\varepsilon) = {} & F(X^{(0)}) + \varepsilon F'(X^{(0)}; X^{(1)}) \\ & + \varepsilon^2[\tfrac{1}{2}F''(X^{(0)}; X^{(1)}, X^{(1)}) + F'(X^{(0)}; X^{(2)})] + o(\varepsilon^2); \end{aligned} \tag{2.10}$$

and so on.

For example, if the functional F has the form

$$F(\varphi) = \int_0^T g(\varphi_t)\, dt, \tag{2.11}$$

then formulas (2.9) and (2.10) take the forms (for the sake of simple notation, we consider the one-dimensional case):

$$\int_0^T g(X_t^\varepsilon)\, dt = \int_0^T g(X_t^{(0)})\, dt + \varepsilon \int_0^T g'(X_t^{(0)})X_t^{(1)}\, dt + o(\varepsilon); \tag{2.12}$$

$$\begin{aligned} \int_0^T g(X_t^\varepsilon)\, dt = {} & \int_0^T g(X_t^{(0)})\, dt + \varepsilon \int_0^T g'(X_t^{(0)})X_t^{(1)}\, dt \\ & + \varepsilon^2\left[\frac{1}{2}\int_0^T g''(X_t^{(0)})(X_t^{(1)})^2\, dt + \int_0^T g'(X_t^{(0)})X_t^{(2)}\, dt\right] + o(\varepsilon^2). \end{aligned} \tag{2.13}$$

In problems connected with random processes, we often have to consider functionals defined in terms of the first exit time of a domain D. In the case of a domain with a smooth boundary, the functional $\tau(\varphi) = \min\{t : \varphi_t \notin D\}$ will not be Fréchet differentiable or even continuous at all points φ of the space of continuous functions. Nevertheless, it will be differentiable at all points φ for which φ_t has a derivative for $t = \tau(\varphi)$ whose direction is not tangent to the boundary. We shall not prove this in the language of derivatives of functionals but rather formulate it directly in the language of expansions in powers of ε.

Theorem 2.3. *Suppose* (2.8) *holds with* $k = 1$. *Let* t_0 *be the first time of exit of* $X_t^{(0)}$ *from a domain D and let* τ^ε *be the first time of exit of* X_t^ε *from D. Let the boundary* ∂D *of D be once differentiable at the point* $X_{t_0}^{(0)}$ *and let n be the exterior normal at this point. Suppose that* $(\dot{X}_{t_0}^{(0)}, n) > 0$. *Then we have*

$$\tau^\varepsilon = t_0 - \varepsilon \frac{(X_{t_0}^{(1)}, n)}{(\dot{X}_{t_0}^{(0)}, n)} + o(\varepsilon), \tag{2.14}$$

$$X_{\tau^\varepsilon}^\varepsilon = X_{t_0}^{(0)} + \varepsilon \left[X_{t_0}^{(1)} - \dot{X}_{t_0}^{(0)} \frac{(X_{t_0}^{(1)}, n)}{(\dot{X}_{t_0}^{(0)}, n)} \right] + o(\varepsilon) \tag{2.15}$$

as $\varepsilon \to 0$ (*here* $o(\varepsilon)$ *is understood in the sense of convergence with probability* 1 *or convergence in probability depending on how* $o(\varepsilon)$ *is interpreted in the expansion* (2.8)).

Proof. We use the expansion (2.8) on the interval $[0, T]$, where $T > t_0$. First we obtain that $\tau^\varepsilon \to t_0$ as $\varepsilon \to 0$. From this we obtain

$$\begin{aligned} X_{\tau^\varepsilon}^\varepsilon &= X_{\tau^\varepsilon}^{(0)} + \varepsilon X_{\tau^\varepsilon}^{(1)} + o(\varepsilon) \\ &= X_{t_0}^{(0)} + (\tau^\varepsilon - t_0)\dot{X}_{t_0}^{(0)} + o(\tau^\varepsilon - t_0) + \varepsilon X_{t_0}^{(1)} + o(\varepsilon). \end{aligned} \tag{2.16}$$

Taking the scalar product of (2.16) and n, we obtain

$$(X_{\tau^\varepsilon}^\varepsilon - X_{t_0}^{(0)}, n) = (\tau^\varepsilon - t_0)(\dot{X}_{t_0}^{(0)}, n) + o(\tau^\varepsilon - t_0) + \varepsilon(X_{t_0}^{(1)}, n) + o(\varepsilon) \tag{2.17}$$

On the other hand, because of the smoothness of ∂D at the point $X_{t_0}^{(0)}$, the scalar product on the left side of (2.17) will be infinitesimal compared to $X_{\tau^\varepsilon}^\varepsilon - X_{t_0}^{(0)}$. It follows from this and from (2.16) that

$$(X_{\tau^\varepsilon}^\varepsilon - X_{t_0}^{(0)}, n) = o(\tau^\varepsilon - t_0) + o(\varepsilon). \tag{2.18}$$

From (2.17) and (2.19) we obtain the expansion (2.14) for τ^ε. Substituting the expansion in (2.16) again, we obtain (2.15). □

The coefficient of ε in the expansion (2.15) can be obtained by projecting $X_{t_0}^{(1)}$ parallel to $\dot{X}_{t_0}^{(0)}$ onto the tangent hyperplane at $X_{t_0}^{(0)}$.

If the expansion (2.8) holds with $k = 2$, the function $X_t^{(0)}$ is twice differentiable and the random function $X_t^{(1)}$ is once differentiable, then we can obtain an expansion of τ^ε and $X_{\tau^\varepsilon}^\varepsilon$ to within $o(\varepsilon^2)$ (although the corresponding functional is twice differentiable only on some subspace). On the other hand, if $X_t^{(1)}$ is not differentiable (this happens in the case of diffusion processes with small diffusion, considered in Theorem 2.2) then we do not obtain an expansion for τ^ε to within $o(\varepsilon^2)$. We explain why this is so.

The fact is that in the proof of Theorem 2.3 we did not use the circumstance that τ^ε is exactly the first time of reaching the boundary but only that it is a time when X_t^ε is on the boundary, converging to t_0. If we consider a process X_t^ε of a simple form: $X_t^\varepsilon = x_0 + t + \varepsilon w_t$, then the first time τ^ε of reaching a point $x_1 > x_0$ and the last time σ^ε of being at x_1 differ by a quantity of order ε^2. Indeed, by virtue of the strong Markov property with respect to the Markov time τ^ε, we obtain that the distribution of $\sigma^\varepsilon - \tau^\varepsilon$ is the same as that of the random variable $\zeta^\varepsilon = \max\{t : t + \varepsilon w_t = 0\}$. Then, we use the fact that $\varepsilon^{-2}(t\varepsilon^2 + \varepsilon w_{t\varepsilon^2}) = t + \varepsilon^{-1} w_{t\varepsilon^2} = t + \tilde{w}_t$, where $\tilde{w}_t$ is again a Wiener process issued from zero and $\zeta^\varepsilon = \varepsilon^2 \tilde{\zeta}$, where $\tilde{\zeta} = \max\{t : t + \tilde{w}_t = 0\}$.

§3. Elliptic and Parabolic Differential Equations with a Small Parameter at the Derivatives of Highest Order

In the theory of differential equations of elliptic or parabolic type, much attention is devoted to the study of the behavior, as $\varepsilon \to 0$, of solutions of boundary value problems for equations of the form $L^\varepsilon u^\varepsilon + c(x)u^\varepsilon = f(x)$ or $\partial v^\varepsilon / \partial t = L^\varepsilon v^\varepsilon + c(x)v^\varepsilon + g(x)$, where L^ε is an elliptic differential operator with a small parameter at the derivatives of highest order:

$$L^\varepsilon = \frac{\varepsilon^2}{2} \sum_{i,j=1}^{r} a^{ij}(x) \frac{\partial^2}{\partial x^i \partial x^j} + \sum_{i=1}^{r} b^i(x) \frac{\partial}{\partial x^i}.$$

As was said in Chapter 1, with every such operator L^ε (whose coefficients are assumed to be sufficiently regular) there is associated a diffusion process $X_t^{\varepsilon,x}$. This diffusion process can be given by means of the stochastic equation

$$\dot{X}_t^{\varepsilon,x} = b(X_t^{\varepsilon,x}) + \varepsilon\sigma(X_t^{\varepsilon,x})\dot{w}_t, \qquad X_0^{\varepsilon,x} = x, \tag{3.1}$$

where $\sigma(x)\sigma^*(x) = (a^{ij}(x)), b(x) = (b^1(x), \ldots, b^r(x))$. For this process we shall sometimes use the notation $X_t^{\varepsilon,x}$, sometimes $X_t^\varepsilon(x)$ (in the framework of the notion of a Markov family), sometimes X_t^ε and in which case we shall write the index x in the probability and consider the Markov process $(X_t^\varepsilon, \mathsf{P}_x)$.

In the preceding two sections of this chapter we obtained several results concerning the behavior of solutions $X_t^{\varepsilon,x}(\omega)$ of equations (3.1) as $\varepsilon \to 0$. Since the solutions of the boundary value problems for L^ε can be written as mean values of some functionals of the trajectories of the family $(X_t^{\varepsilon,x}, \mathsf{P})$ results concerning the behavior of solutions of boundary value problems as $\varepsilon \to 0$ can be obtained from the behavior of $X_t^{\varepsilon,x}(\omega)$ as $\varepsilon \to 0$. The present section is devoted to these questions.

We consider the Cauchy problem

$$\frac{\partial v^\varepsilon(t,x)}{\partial t} = L^\varepsilon v^\varepsilon(t,x) + c(x)v^\varepsilon(t,x) + g(x); \qquad t>0, \quad x \in R^r,$$

$$v^\varepsilon(0,x) = f(x) \tag{3.2}$$

for $\varepsilon > 0$ and together with it the problem for the first-order operator which is obtained for $\varepsilon = 0$:

$$\frac{\partial v^0(t,x)}{\partial t} = L^0 v^0 + c(x)v^0 + g(x); \qquad t>0, \quad x\in R^r, \quad v^0(0,x) = f(x). \tag{3.3}$$

We assume that the following conditions are satisfied.

(1) the functions $c(x)$ are uniformly continuous and bounded for $x \in R^r$;
(2) the coefficients of L^1 satisfy a Lipschitz condition;
(3) $k^{-2}\sum \lambda_i^2 \le \sum_{i,j=1}^r a^{ij}(x)\lambda_i\lambda_j \le k^2 \sum \lambda_i^2$ for any real $\lambda_1, \lambda_2, \ldots, \lambda_r$ and $x \in R^r$, where k^2 is a positive constant.

Under these conditions, the solutions of problems (3.2) and (3.3) exists and are unique.

All results of this paragraph remain valid in the case where the form $\sum a^{ij}(x)\lambda_i\lambda_j$ is only nonnegative definite. However, in the case of degeneracies the formulation of boundary value problems has to be adjusted and the notion of a generalized solution has to be introduced. We shall make the adjustments necessary in the case of degeneracies after an analysis of the non-degenerate case.

Theorem 3.1. *If conditions* (1)–(3) *are satisfied, then the limit* $\lim_{\varepsilon\to 0} v^\varepsilon(t,x) = v^0(t,x)$ *exists for every bounded continuous initial function* $f(x)$, $x \in R^r$. *The function* $v^0(t,x)$ *is a solution of problem* (3.3).

For the proof we note first of all that if condition (3) is satisfied, then there exists a matrix $\sigma(x)$ with entries satisfying a Lipschitz condition for which $\sigma(x)\sigma^*(x) = (a^{ij}(x))$ (cf. §5, Ch. 1).

The solution of equation (3.2) can be represented in the following way:

$$v^\varepsilon(t,x) = \mathsf{M} f(X_t^{\varepsilon,x}) \exp\left[\int_0^t c(X_s^{\varepsilon,x})\,ds\right] + \mathsf{M}\int_0^t g(X_s^{\varepsilon,x}) \exp\left[\int_0^s c(X_u^{\varepsilon,x})\,du\right] ds, \tag{3.4}$$

where $X_t^{\varepsilon,x}$ is the Markov family constructed by means of equation (3.1). It follows from Theorem 1.2 that the processes $X_s^{\varepsilon,x}(\omega)$ converge to $X_s^{0,x}$ (the solution of equation (1.2) with initial condition $X_0^{0,x} = x$) in probability uniformly on the interval $[0, t]$ as $\varepsilon \to 0$. Taking into account that there is a bounded continuous functional of $X_s^{\varepsilon,x}(\omega)$ under the sign of mathematical expectation in (3.4), by the Lebesgue dominated convergence theorem we conclude that

$$\lim_{\varepsilon \downarrow 0} v^\varepsilon(t,x) = f(X_t^{0,x}) \exp\left[\int_0^t c(X_s^{0,x})\,ds\right] + \int_0^t g(X_s^{0,x}) \exp\left[\int_0^s c(X_u^{0,x})\,du\right] ds.$$

An easy substitution shows that the function on the right side of the equality is a solution of problem (3.3). Theorem 3.1 is proved. □

If we assume that the coefficients of L^ε have bounded derivatives up to order $k + 1$ inclusive, then the matrix $\sigma(x)$ can be chosen so that its entries also have $k + 1$ bounded derivatives. In this case, by virtue of Theorem 2.2 we can write down an expansion for $X_t^{\varepsilon,x}$ in powers of ε up to order k. If the functions $f(x)$, $c(x)$, and $g(x)$ have $k + 1$ bounded derivatives, then, as follows from (2.7), we have an expansion in powers of ε up to order k with remainder of order ε^{k+1}.

Hence, for example, if $g(x) \equiv c(x) \equiv 0$ and $r = 1$, then the solution of problem (3.2) can be written in the form

$$\begin{aligned} v^\varepsilon(t, x) &= \mathsf{M}_x f(X_t^\varepsilon) \\ &= \mathsf{M}_x f(X_t^{(0)} + \varepsilon X_t^{(1)} + \cdots + \varepsilon^k X_t^{(k)} + R_{k+1}^\varepsilon(t)) \\ &= \sum_{i=0}^{k} \varepsilon^i \mathsf{M}_x G_i + O(\varepsilon^{k+1}), \end{aligned} \tag{3.5}$$

where $X_t^{(0)}, X_t^{(1)}, \ldots, X_t^{(k)}$ are the coefficients mentioned in Theorem 2.2 of the expansion of X_t in powers of the small parameter;

$$G_i = G_i(X_t^{(0)}, \ldots, X_t^{(i)}) = \frac{1}{i!} \frac{d^i}{d\varepsilon^i} f(X_t^{(0)} + \varepsilon X_t^{(1)} + \cdots + \varepsilon^k X_t^{(k)}) \big|_{\varepsilon = 0}.$$

We can derive from formula (3.5) and the equations defining the processes $X_t^{(i)}$ that the coefficients of the odd powers of ε vanish. The coefficients of ε^{2m} are the solutions of some first-order partial differential equations; they can, of course, be found by solving systems of ordinary differential equations.

We illustrate the method of finding the coefficients of the expansion of $v^\varepsilon(t, x)$ in the simplest case, i.e., for dimension 1 and up to terms of order ε^2. For the coefficients of the expansion in powers of ε of the solution of the stochastic differential equation

$$\dot{X}_t^\varepsilon = b(X_t^\varepsilon) + \varepsilon\sigma(X_t^\varepsilon)\dot{w}_t, \qquad X_0^\varepsilon = x, \tag{3.6}$$

we write out the first three equations in (2.6):

$$\dot{X}_t^{(0)} = b(X_t^{(0)}), \qquad X_0^{(0)} = x; \tag{3.7}$$

$$\dot{X}_t^{(1)} = b'(X_t^{(0)})X_t^{(1)} + \sigma(X_t^{(0)})\dot{w}_t, \qquad X_0^{(1)} = 0, \tag{3.8}$$

$$\dot{X}_t^{(2)} = b'(X_t^{(0)})X_t^{(2)} + \tfrac{1}{2}b''(X_t^{(0)})(X_t^{(1)})^2 + \sigma'(X_t^{(0)})X_t^{(1)}\dot{w}_t, \qquad X_0^{(2)} = 0. \tag{3.9}$$

The function $X_t^{(0)}$ is nonrandom and another notation for it is $x_t(x)$.

If f is a twice continuously differentiable function, then we have the expansion

$$\begin{aligned} f(X_t^\varepsilon) = f(X_t^{(0)}) &+ \varepsilon f'(X_t^{(0)})X_t^{(1)} \\ &+ \varepsilon^2[f'(X_t^{(0)})X_t^{(2)} + \tfrac{1}{2}f''(X_t^{(0)})(X_t^{(1)})^2] + o(\varepsilon^2). \end{aligned} \tag{3.10}$$

We take mathematical expectation on both sides:

$$\begin{aligned} v^\varepsilon(t, x) = \mathsf{M}_x f(x_t^\varepsilon) = f(x_t(x)) &+ \varepsilon f'(x_t(x))\mathsf{M}_x X_t^{(1)} \\ &+ \varepsilon^2[f'(x_t(x))\,\mathsf{M}_x X_t^{(2)} + \tfrac{1}{2}f''(x_t(x))\,\mathsf{M}_x(X_t^{(1)})^2] + o(\varepsilon^2). \end{aligned} \tag{3.11}$$

Since the process $X_t^{(1)}$ is Gaussian with zero mean, the coefficient of ε vanishes. To obtain $\mathsf{M}_x(X_t^{(1)})^2$, we apply formula (3.8) and Itô's formula:

$$\frac{d}{dt}(X_t^{(1)})^2 = 2b'(X_t^{(0)})(X_t^{(1)})^2 + 2\sigma(X_t^{(0)})b'(X_t^{(0)})\dot{w}_t + \sigma(X_t^{(0)})^2. \tag{3.12}$$

Taking mathematical expectation on both sides, we obtain the following nonhomogeneous linear differential equation for $\mathsf{M}_x(X_t^{(1)})^2$ with initial condition $\mathsf{M}_x(X_0^{(1)})^2 = 0$:

$$\frac{d}{dt}\mathsf{M}(X_t^{(1)})^2 = 2b'(x_t(x))\mathsf{M}_x(X_t^{(1)})^2 + a(x_t(x)). \tag{3.13}$$

Solving this equation, we also find the solution of the equation for $\mathsf{M}_x X_t^{(2)}$, which can be obtained by taking the mathematical expectation of (3.9):

$$\frac{d}{dt}\mathsf{M}_x X_t^{(2)} = b'(x_t(x))\mathsf{M}_x X_t^{(2)} + \tfrac{1}{2}b''(x_t(x))\mathsf{M}_x(X_t^{(1)})^2, \qquad \mathsf{M}_x X_0^{(2)} = 0. \tag{3.14}$$

Hence for the determination of the coefficients of the expansion of $v^\varepsilon(t, x)$ in powers of ε to within order 2, it is sufficient to solve the nonlinear equation $\dot{x}_t(x) = b(x_t(x))$, and the two linear equations (3.13) and (3.14).

The same result can be obtained in a simpler way by using standard methods of the theory of differential equations. Nevertheless, methods of probability theory can also be applied to less standard asymptotic problems. For example, suppose the function f is not smooth at a point $y = x_t(x)$ but has a power-like "corner": $f(z) = f(y) + C|z - y|^\alpha + o(|z - y|^\alpha)$ as $z \to y$, $0 < \alpha \leq 1$. We use the expansion

$$X_s^\varepsilon = x_s(x) + \varepsilon X_s^{(1)} + o(\varepsilon)$$

of X_s^ε. For $s = t$ we obtain

$$\begin{aligned} X_t^\varepsilon &= y + \varepsilon X_t^{(1)} + o(\varepsilon), \\ f(X_t^\varepsilon) &= f(y) + \varepsilon^\alpha C |X_t^{(1)}|^\alpha + o(\varepsilon^\alpha), \\ v_\varepsilon(t, x) &= \mathsf{M}_x f(X_t^\varepsilon) = f(y) + \varepsilon^\alpha C \mathsf{M}_x |X_t^{(1)}|^\alpha + o(\varepsilon^\alpha). \end{aligned}$$

We obtain the mathematical expectation above by using the fact that $X_t^{(1)}$ is Gaussian; it is equal to

$$\begin{aligned} &\int_{-\infty}^{\infty} |u|^\alpha \frac{1}{\sqrt{2\pi \mathsf{M}_x (X_t^{(1)})^2}} \exp[-u^2/2\mathsf{M}_x (X_t^{(1)})^2]\, du \\ &\qquad = \frac{(2\mathsf{M}_x (X_t^{(1)})^2)^{\alpha/2}}{\sqrt{\pi}} \cdot \Gamma\left(\frac{\alpha}{2} + \frac{1}{2}\right). \end{aligned}$$

If f vanishes in the neighborhood of $X_t^{(0)}$, the position of the unperturbed dynamical system (1.2) at time t, then all terms of (3.5) vanish. It turns out that in this case $v^\varepsilon(t, x)$ is logarithmically equivalent to $\exp\{-C\varepsilon^{-2}\}$, where C is a constant. We return to this case in the following chapter.

Now we consider Dirichlet's problem for the elliptic equation with a small parameter

$$\begin{aligned} \frac{\varepsilon^2}{2} \sum_{i,j}^{r} a^{ij}(x) \frac{\partial^2 u^\varepsilon}{\partial x^i \partial x^j} + \sum_{i=1}^{r} b^i(x) \frac{\partial u^\varepsilon}{\partial x^i} + c(x) u^\varepsilon(x) &= L^\varepsilon u^\varepsilon + c(x) u^\varepsilon \\ &= g(x), \qquad u^\varepsilon(x)|_{\partial D} = \psi(x). \end{aligned} \tag{3.15}$$

in a bounded domain $D \subset R^r$ with boundary ∂D.

We assume that the coefficients satisfy conditions (1)–(3) and $c(x) \leq 0$. For the sake of simplicity, the boundary ∂D of D is assumed to be smooth and the function $\psi(x)$, $x \in \partial D$ continuous. Under these conditions, there exists a unique solution of problem (3.15) for every $\varepsilon \neq 0$. This solution can be written in the form (cf. §5, Ch. 1)

$$u^\varepsilon(x) = \mathsf{M}_x\left[\psi(X^\varepsilon_{\tau^\varepsilon}) \exp\left[\int_0^{\tau^\varepsilon} c(X^\varepsilon_s)\, ds\right] - \int_0^{\tau^\varepsilon} g(x^\varepsilon_s) \exp\left[\int_0^s c(X^\varepsilon_v)\, dv\right] ds\right], \tag{3.16}$$

where (X_t, P_x) is the Markov process defined by equation (3.1) and $\tau^\varepsilon = \min\{t\colon X^\varepsilon_t \notin D\}$. In case we use the notation $X^\varepsilon_t(x)$, we also write $\tau^\varepsilon(x)$.

We shall say that a trajectory $x_t(x)$, $x \in D$ of system (1.2) leaves D in a *regular manner* if $T(x) = \min\{t\colon x_t(x) \notin D\} < \infty$ and $x_{T(x)+\delta}(x) \notin D \cup \partial D$ for sufficiently small $\delta > 0$.

Theorem 3.2. *Suppose conditions* (1)–(3) *are satisfied and the domain D is bounded and has a smooth boundary. If $c(x) < 0$ for all $x \in D \cup \partial D$ and for a given x, the trajectory $x_t(x)$, $t \geq 0$ does not leave D, then* $\lim_{\varepsilon\to 0} u^\varepsilon(x) = u^0(x)$ *exists and*

$$u^0(x) = -\int_0^\infty g(x_s(x)) \exp\left[\int_0^s c(x_v(x))\, dv\right] ds.$$

If $c(x) \leq 0$ for all $x \in D \cup \partial D$ and for a given x, the trajectory $x_t(x)$ leaves D in a regular manner, then

$$\lim_{\varepsilon\to 0} u^\varepsilon(x) = u^0(x) = \psi(x_{T(x)}(x)) \exp\left[\int_0^{T(x)} c(x_s(x))\, ds\right] - \int_0^{T(x)} g(x_s(x)) \exp\left[\int_0^s c(x_v(x))\, dv\right] ds.$$

Proof. First let $T(x) = +\infty$. For every $T < \infty$, the distance of the trajectory segment $x_s(x)$, $0 \leq s \leq T$ from ∂D is positive. We denote this distance by δ_T. For every $\alpha > 0$ and a sufficiently small $\varepsilon_0 > 0$ we have

$$\mathsf{M}\left\{\max_{0\leq s\leq T} |X^\varepsilon_s(x) - x_s(x)| > \frac{\delta_T}{2}\right\} < \alpha \tag{3.17}$$

for $\varepsilon < \varepsilon_0$. This follows from the second assertion of Theorem 1.2. From the definition of δ_T and (3.17) it follows that

$$\mathsf{P}\{\tau^\varepsilon(x) < T\} < \alpha. \tag{3.18}$$

We write

$$c_0 = \min_{x\in D\cup\partial D} |c(x)|, \qquad \psi_0 = \max_{x\in\partial D} |\psi(x)|, \qquad g_0 = \max_{x\in D\cup\partial D} |g(x)|.$$

On the basis of (3.18) we arrive at the following estimation:

$$\left| u^\varepsilon(x) + \int_0^\infty g(x_s(x)) \exp\left[\int_0^s c(x_v(x))\, dv\right] ds \right|$$

$$\leq \psi_0 e^{-c_0 T} + \int_T^\infty g_0 e^{-c_0 s}\, ds + \alpha(\psi_0 + g_0 c_0^{-1})$$

$$+ \mathsf{M} \int_0^T \left| g(X_s^\varepsilon(x)) \exp\left[\int_0^s c(X_v^\varepsilon(x))\, dv\right]\right.$$

$$\left. - g(x_s(x)) \exp\left[\int_0^s c(x_v(x))\, dv\right] \right| ds.$$

Since α and $e^{-c_0 T}$ can be chosen arbitrarily small for ε sufficently small and $\sup_{0 \leq s \leq T} |X_s^\varepsilon(x) - x_s(x)| \to 0$ in probability as $\varepsilon \to 0$, the first assertion of the theorem follows from the last inequality.

Now let $x_t(x)$ leave D in a regular manner (Figure 1). We have $\tau^\varepsilon(x) \to T(x)$ in probability as $\varepsilon \to 0$. Indeed, for every sufficiently small $\delta > 0$ we have

$$x_{T(x)-\delta}(x) \in D, \qquad x_{T(x)+\delta}(x) \notin D \cup \partial D.$$

Let δ_1 be the distance of the trajectory segment $x_s(x)$, $s \in [0, T(x) - \delta]$ from ∂D, let δ_2 be the distance of $x_{T(x)+\delta}(x)$ from ∂D, and let $\bar{\delta} = \min(\delta_1, \delta_2)$. By Theorem 1.2 we have

$$\lim_{\varepsilon \to 0} \mathsf{P}\left\{ \sup_{0 \leq s \leq T(x)+\delta} |X_s^\varepsilon(x) - x_s(x)| > \bar{\delta} \right\} = 0.$$

This implies that $\tau^\varepsilon(x) \in [T(x) - \delta, T(x) + \delta]$ with probability converging to 1 as $\varepsilon \to 0$. This means that $\tau^\varepsilon(\varepsilon) \to T(x)$ in probability. Using this circumstance and Theorem 1.2, the last assertion of the theorem follows from (3.16).

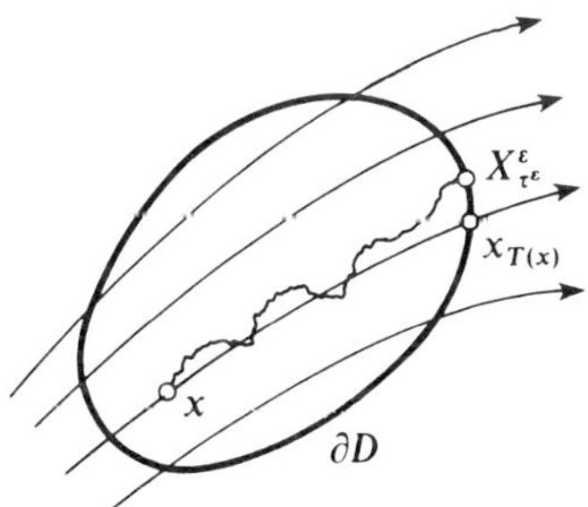

Figure 1.

The passage to the limit under the sign of mathematical expectation is legitimate by virtue of the uniform boundedness of the expression under the sign of mathematical expectation. □

Now let $c(x)$ be of an arbitrary sign. We only assume that it is continuous. In this case problem (3.15) may go out to the spectrum in general: its solution may not exist for every right side and may not be unique. As was discussed in §5, Ch. 1, in order that this does not occur it is sufficient that $c(x) \le c_0$ for $x \in D$ and $\mathsf{M}_x e^{c_0 \tau} < \infty$.

Lemma 3.1. *Suppose that for every $x \in D$, the trajectory $x_t(x)$ leaves D in a regular manner and $T(x) \le T_0 < \infty$ for $x \in D$. For some $\delta > 0$, let*

$$\max_{T(x) \le t \le T(x)+\delta} \rho(x_t(x), D \cup \partial D) \ge c > 0$$

for all $x \in D$. Then for any λ there exist $A(\lambda)$ and $\varepsilon(\lambda) > 0$ such that

$$\sup_{x \in D} \mathsf{M}_x e^{\lambda \tau^\varepsilon} \le A(\lambda) < \infty$$

for $\varepsilon \le \varepsilon(\lambda)$.

Proof. As follows from the analysis carried out in the proof of Theorem 3.2, if $x_t(x)$ leaves D in a regular manner, then $\tau^\varepsilon(x) \to T(x)$ in probability as $\varepsilon \to 0$. The conditions $T(x) \le T_0$ and $\max \rho(x_t(x), D \cup \partial D) \ge c$ imply that for every $\delta > 0$ there exists $\varepsilon_0 > 0$ such that for $\varepsilon < \varepsilon_0$ we have

$$\mathsf{P}\{|\tau^\varepsilon(x) - T(x)| > \delta\} < \delta$$

for all $x \in D$. This implies that

$$\sup_{x \in D} \mathsf{P}_x\{\tau^\varepsilon > 2T_0\} < \delta. \tag{3.19}$$

Moreover, using (3.19) and the Markov property of $(X_t^\varepsilon, \mathsf{P}_x)$, we obtain

$$\sup_{x \in D} \mathsf{P}_x\{\tau^\varepsilon > n \cdot 2T_0\} = \sup_{x \in D} \mathsf{M}_x\{\tau^\varepsilon > (n-1) \cdot 2T_0;\ \mathsf{P}_{X^\varepsilon_{(n-1)2T_0}}\{\tau^\varepsilon > 2T_0\}\}$$

$$\le \delta \cdot \sup_{x \in D} \mathsf{P}_x\{\tau^\varepsilon > (n-1)2T_0\}.$$

It follows from this inequality that

$$\mathsf{P}_x\{\tau^\varepsilon > n \cdot 2T_0\} < \delta^n$$

for every integer n and $x \in D$.

Since δ can be chosen arbitrarily small for ε sufficiently small, from the last inequality we obtain the assertion of the lemma:

$$\begin{aligned} \mathsf{M}_x e^{\lambda\tau^\varepsilon} &\le \sum_{n=0}^{\infty} e^{\lambda\cdot 2T_0(n+1)} \mathsf{P}_x\{\tau^\varepsilon > 2T_0 n\} \\ &\le e^{\lambda\cdot 2T_0}\cdot \sum_{n=0}^{\infty} (e^{\lambda\cdot 2T_0}\delta)^n = A(\lambda) < \infty. \end{aligned} \qquad \square$$

Corollary. *For every $k > 0$ there exists a constant $B = B(k)$ such that $t^k \le Be^t$. By Lemma 3.1 this implies that*

$$\mathsf{M}_x(\tau^\varepsilon)^k \le B(k) M_x e^{\tau^\varepsilon} \le B(k)A(1) = \tilde{A} < \infty.$$

Theorem 3.3. *Suppose that conditions (1)–(3) are satisfied, the domain D is bounded and has a smooth boundary ∂D and the function $\psi(x)$ is continuous on ∂D. Suppose furthermore that for all $x \in D$, the trajectories $x_t(x)$ leave D in a regular manner,*

$$\sup\nolimits_{x\in D} T(x) \le T_0 < \infty,$$

and

$$\max_{T(x)\le t\le T(x)+\delta} \rho(x_t(x).\ D \cup \partial D) \ge c > 0.$$

Then for every continuous function $c(x)$, $x \in D \cup \partial D$, the problem (3.15) has a unique solution for sufficiently small ε and

$$\begin{aligned} \lim_{\varepsilon\to 0} u^\varepsilon(x) = u^0(x) &= \psi(x_{T(x)}(x)) \exp\left[\int_0^{T(x)} c(x_s(x))\, ds\right] \\ &\quad - \int_0^{T(x)} g(x_s(x)) \exp\left[\int_0^s c(x_v(x))\, dv\right] ds. \end{aligned}$$

Proof. As was indicated in §1.5, the existence of a unique solution of problem (3.15) and the validity of (3.16) are guaranteed if

$$\mathsf{M}_x \exp\left\{\tau^\varepsilon \cdot \max_{x\subset D\cup\partial D} c(x)\right\} \le A < \infty.$$

Therefore, the first assertion of the theorem follows from Lemma 3.1.

The second assertion follows from (3.16) if we note that for any $t > 0$ and $x \in D$ we have

$$\tau^\varepsilon(x) \to T(x), \qquad \sup_{0\le s\le t} |X_s^\varepsilon(x) - x_s(x)| \to 0$$

in probability as $\varepsilon \to 0$ and note that the mathematical expectation of the square of the random variable in (3.16) under the sign of mathematical expectation is bounded uniformly in $\varepsilon < \varepsilon_0$ provided that ε_0 is sufficiently small. □

Remark 1. If $T(x) < \infty$ but the trajectory $x_t(x)$ does not leave D in a regular manner, then, as follows from simple examples, the limit function may have discontinuities on this trajectory.

Remark 2. It is easy to verify that the limit function $u^0(x)$ in Theorem 3.3 satisfies the following first-order equation obtained for $\varepsilon = 0$:

$$L^0u^0(x) + c(x)u^0(x) = \sum_{i=1}^{r} b^i(x)\frac{\partial u^0}{\partial x^i} + c(x)u^0(x) = g(x).$$

The function $u^0(x)$ is chosen from the solutions of this equation by the condition that it coincides with $\psi(x)$ at those points of the boundary of D through which the trajectories $x_t(x)$ leave D.

Remark 3. Now let us allow the matrix $(a^{ij}(x))$ to have degeneracies. In this case problem (3.15) must be modified. First, we cannot prescribe boundary conditions at all points of the boundary. This is easily seen from the example of first-order equations; boundary conditions will not be assumed at some points of the boundary. Second, a classical solution may not exist even in the case of infinitely differentiable coefficients, and it is necessary to introduce the notion of a generalized solution. Third and finally, a generalized solution may not be unique without additional assumptions. To construct a theory of such equations with a nonnegative characteristic form, we can use methods of probability theory. The first results in this area were actually obtained in this way (Freidlin [1], [4], [6]). Some of these results were subsequently obtained by traditional methods of the theory of differential equations. If the entries of $(a^{ij}(x))$ have bounded second derivatives, then there exists a factorization $(a^{ij}(x)) = \sigma(x)\sigma^*(x)$, where the entries of $\sigma(x)$ satisfy a Lipschitz condition. In this case the process $(X^\varepsilon, \mathsf{P}_x)$ corresponding to the operator L^ε is constructed by means of equations (3.1). In Freidlin's publications [1] and [4], this process is used to make precise the formulation of boundary value problems for L^ε, to introduce the notion of a generalized solution, to prove existence and uniqueness theorems, and to study the smoothness of a generalized solution.

In particular, if the functions $a^{ij}(x)$ have bounded second derivatives and satisfy the hypotheses of Theorem 3.2 or Theorem 3.3, respectively (with the exception of nondegeneracy), then for every sufficiently small ε, the generalized solution exists, is unique, and satisfies equality (3.16). In this case the assertion of Theorem 3.2 (Theorem 3.3) also holds if by $u^\varepsilon(x)$ we understand the generalized solution.

After similar adjustments, Theorem 3.1 also remains valid.

Theorems 3.2 and 3.3 used results concerning the limit behavior of X_t^ε which are of the type of law of large numbers. From finer results (expansions in powers of ε) we can obtain finer consequences concerning the asymptotics of the solution of Dirichlet's problem. Concerning the expansion of the solution in powers of a small parameter (in the case of smooth boundary conditions), the best results are not obtained by methods of pure probability theory but rather by purely analytical or combined (cf. Holland [1]) methods. We consider an example with nonsmooth boundary conditions.

Let the characteristic $x_t(x)$, $t \geq 0$ issued from an interior point x of a domain D with a smooth boundary leave the domain, intersecting its boundary for the value t_0 of the parameter; at the point $y = x_{t_0}(x)$ the vector $b(y)$ is directed strictly outside the domain. Let u^ε be a solution of the Dirichlet problem $L^\varepsilon u^\varepsilon = 0$, $u^\varepsilon \to 1$ as we approach some subdomain Γ_1 of the boundary and $u^\varepsilon \to 0$ as we approach the interior points of $\partial D \setminus \Gamma_1$ (and u^ε is assumed to be bounded everywhere). Suppose that the surface area of the boundary of Γ_1 is equal to zero. Then the solution $u^\varepsilon(x)$ is unique and can be represented in the form

$$u^\varepsilon(x) = \mathsf{M}_x \chi_{\Gamma_1}(X_{\tau^\varepsilon}^\varepsilon).$$

If y is an interior point of Γ_1 or $\partial D \setminus \Gamma_1$, then the value of u^ε at the point x converges to 1 or 0, respectively, as $\varepsilon \to 0$ (results concerning the rate of convergence must rely on results of the type of large deviations; cf. Ch. 6, Theorems 2.1 and 2.2). On the other hand, if y belongs to the boundary of the domain Γ_1, then the expansion (2.15) reduces the problem of asymptotics of $u^\varepsilon(x)$ to the problem of asymptotics of the probability that the Gaussian random vector $X_{t_0}^{(0)} - \dot{X}_{t_0}^{(0)}[(X_{t_0}^{(1)}, n)/(\dot{X}_{t_0}^{(0)}, n)]$ hits the ε^{-1} times magnified projection of Γ_1 onto the tangent plane (tangent line in the two-dimensional case). In particular, in the two-dimensional case if Γ_1 is a segment of an arc with y as one endpoint, then $\lim_{\varepsilon \to 0} u^\varepsilon(x) = \frac{1}{2}$. The same is true in the higher dimensional case provided that the boundary of Γ_1 is smooth at y. If this boundary has a "corner" at y, then the problem reduces to the problem of the probability that a normal random vector with mean zero falls into an angle (solid angle, cone) with vertex at zero. Using an affine transformation, one can calculate the angle (solid angle).

Chapter 3

Action Functional

§1. Laplace's Method in a Function Space

We consider a random process $X_t^\varepsilon = X_t^\varepsilon(x)$ in the space R^r defined by the stochastic differential equation

$$\dot{X}_t^\varepsilon = b(X_t^\varepsilon) + \varepsilon \dot{w}_t, \qquad X_0^\varepsilon = x. \tag{1.1}$$

Here, as usual, w_t is a Wiener process in R^r and the field $b(x)$ is assumed to be sufficiently smooth. As is shown in §1, Ch. 2, as $\varepsilon \to 0$, the trajectories of X_t^ε converge in probability to the solution of the unperturbed equation

$$\dot{x}_t = b(x_t), \qquad x_0 = x. \tag{1.2}$$

uniformly on every finite time interval. In the special case which we are considering it is easy to give an estimate of the probability

$$\mathsf{P}\left\{\sup_{0 \le t \le T} |X_t^\varepsilon(x) - x_t(x)| > \delta\right\}$$

which is sharper than that given in Ch. 2. Indeed, it follows from equations (1.1) and (1.2) that

$$X_t^\varepsilon(x) - x_t(x) = \int_0^t [b(X_s^\varepsilon(x)) - b(x_s(x))]\, ds + \varepsilon w_t. \tag{1.3}$$

Assuming that $b(x)$ satisfies a Lipschitz condition with constant K, we obtain from (1.3) that

$$\sup_{0 \le t \le T} |X_t^\varepsilon - x_t| \le \varepsilon e^{KT} \sup_{0 \le t \le T} |w_t|. \tag{1.4}$$

This implies that the probability of the deviation of $X_t^\varepsilon(x)$ from the trajectory of the dynamical system decreases exponentially with decreasing ε:

$$\begin{aligned}\mathsf{P}\left\{\sup_{0\le t\le T}|X_t^\varepsilon(x)-x_t(x)|>\delta\right\} &\le \mathsf{P}\left\{\sup_{0\le t\le T}|w_t|>\frac{\delta}{\varepsilon}e^{-KT}\right\}\\ &\le 2\mathsf{P}\left\{|w_T|>\frac{\delta}{2\varepsilon}e^{-KT}\right\}\\ &= O\left(\left(\frac{\delta^2}{\varepsilon^2 T}e^{-2KT}\right)^{(r/2)-1}\exp\left(-\frac{\delta^2}{8\varepsilon^2 T}e^{-2KT}\right)\right)\\ &= O(e^{-C\varepsilon^{-2}}),\end{aligned}$$

where C is a positive constant (depending on δ, K, and T).

This estimation means that if a subset A of the space of continuous functions on the interval from 0 to T contains a function $x_t(x)$ together with its δ-neighborhood in this space, then the main contribution to the probability $\mathsf{P}\{X^\varepsilon(x)\in A\}$ is given by this δ-neighborhood; the probability of the remaining part of A is exponentially small.

In many problems we are interested in probabilities $\mathsf{P}\{X^\varepsilon(x)\in A\}$ for sets A not containing the function $x_t(x)$ together with its neighborhood. Such problems arise, for example, in connection with the study of stability under random perturbations, when we are mainly interested in the probability of exit from a neighborhood of a stable equilibrium position or of a stable limit cycle in a given time or we are interested in the mean exit time from such a neighborhood. As we shall see, similar problems arise in the study of the limit behavior of an invariant measure of a diffusion process X_t^ε as $\varepsilon\to 0$, in connection with the study of elliptic differential equations with a small parameter at the derivatives of highest order and in other problems.

If the function $x_t(x)$ together with some neighborhood of it is not contained in A, then $\mathsf{P}\{X^\varepsilon(x)\in A\}\to 0$ as $\varepsilon\to 0$. It turns out that in this case, under certain assumptions on A, there exists a function $\varphi\in A$ such that the principal part of the probability measure of A is concentrated near φ; more precisely, for any neighborhood $U(\varphi)$ of φ we have

$$\mathsf{P}\{X^\varepsilon(x)\in A\setminus U(\varphi)\} = o(\mathsf{P}\{X^\varepsilon(x)\in U(\varphi)\})$$

as $\varepsilon\to 0$.

A similar situation arises in applying Laplace's method to calculate the asymptotics as $\varepsilon\to 0$ of integrals of the form $\int_a^b e^{-\varepsilon^{-1}f(x)}g(x)\,dx$. If x_0 is the only minimum point of the continuous function $f(x)$ on the interval $[a, b]$ and the function $g(x)$ is continuous and positive, then the major contribution to this integral is given by the neighborhood of x_0. Indeed, let U_1 be a neighborhood of x_0. Since x_0 is the only minimum point of f in [a, b], we have

$\min_{x \in [a,b] \setminus U_1} f(x) > f(x_0) + \gamma$, where γ is a positive number. Using this estimate, we obtain

$$\int_{[a,b]\setminus U_1} g(x) \exp\{-\varepsilon^{-1} f(x)\}\, dx < (b-a) \max_{x \in [a,b]} g(x) \exp\{-\varepsilon^{-1}(f(x_0) + \gamma)\}. \tag{1.5}$$

For the integral over the neighborhood of x_0, we obtain the following lower estimate:

$$\int_{U_1} g(x) \exp\{-\varepsilon^{-1} f(x)\}\, dx > \int_{xo+\delta}^{xo+\delta} g(x) \exp\{-\varepsilon^{-1}(f(x_0) + \gamma/2)\}\, dx$$
$$> 2\delta \min_{x \in [a,b]} g(x) \exp\{-\varepsilon^{-1}(f(x_0) + \gamma/2)\}, \tag{1.6}$$

where δ is chosen from the conditions: $\max_{|x - x_0| \le \delta} f(x) < f(x_0) + \gamma/2$, $\{x : |x - x_0| < \delta\} \subset U_1$. It follows from estimates (1.5) and (1.6) that the integral $\int_a^b g(x) \exp\{-\varepsilon^{-1} f(x)\}\, dx$ is logarithmically equivalent to $\exp\{-\varepsilon^{-1} f(x_0)\}$ as $\varepsilon \to 0$, i.e.,

$$\lim_{\varepsilon \to 0} \varepsilon \ln \int_a^b g(x) \exp\{-\varepsilon^{-1} f(x)\}\, dx = -f(x_0).$$

Using the Taylor expansion of f around x_0, we can obtain more accurate asymptotic formulas for the integral $\int_a^b g(x) \exp\{-\varepsilon^{-1} f(x)\}\, dx$.

The situation is analogous in the calculation of probabilities of various events connected with the process X_t^ε. It turns out that we can introduce a functional $S(\varphi)$ of a function on the interval $[0, T]$ such that for sufficiently small ε and δ we have

$$\mathsf{P}\{\|X^\varepsilon - \varphi\| < \delta\} \approx \exp\{-\varepsilon^{-2} S(\varphi)\}.$$

(The precise meaning of this formula will be explained in the following sections.)

If the minimum of $S(\varphi)$ on the set A is attained at $\tilde{\varphi}$, then in analogy with Laplace's method we may expect that for small ε, the basic contribution to $P\{X^\varepsilon \in A\}$ is given by a neighborhood of $\tilde{\varphi}$. In order to prove this, we need to obtain a lower estimate of $\mathsf{P}\{\|X^\varepsilon - \varphi\| < \delta\}$ of the type of estimate (1.6) and an upper estimate of the type of estimate (1.5) for the probability of the remaining portion of A. We carry out this program in the present chapter.

The idea of applying similar constructions in asymptotic problems in a function space goes back to R. Feynman's work in quantum mechanics (cf. Feynman and Hibbs [1]). If there exists a classical mechanical system for which the action on the trajectory φ_t is $S(\varphi)$, then, as is known, the motion of this system takes place along extremals of the functional $S(\varphi)$. The same functional can be used for a quantum mechanical description of a system. In a quantum mechanical motion various trajectories are possible and to every trajectory φ_t there is assigned the weight $C \exp\{(i/h)S(\varphi)\}$, called the probability amplitude. (To give this a precise meaning is a difficult problem.) Here h is the Planck constant and C is a normalizing factor. The probability amplitude of a set of trajectories can be calculated by summing (integrating) the contributions of the trajectories which constitute the set. The square of the absolute value of the probability amplitude corresponding to a set of trajectories is interpreted as the probability of the corresponding quantum mechanical motion. Such an approach is convenient in problems connected with the quasiclassical approximation in quantum mechanics, i.e., where the various characteristics of motion are approximated by their principal terms as $h \to 0$ and by successive corrections. Hence in Feynman's description of a quantum mechanical system, the correspondence principle, asserting that every quantum mechanical motion turns into a classical one as $h \to 0$, follows immediately from the circumstance that in summing probability amplitudes, for small h the major contribution to the sum is given by the trajectories φ which are extremals of the action functional, i.e., the classical trajectories. The contribution of the other trajectories will be significantly smaller because of the oscillation of the probability amplitude. This reasoning is an infinite-dimensional analogue of the principle of stationary phase, according to which the major contribution to the integral

$$\int g(x) \exp\{ih^{-1}S(x)\}\, dx$$

is given by the stationary points of the function $S(x)$.

The functional $S(\varphi)$ which we have introduced for the study of the behavior as $\varepsilon \to 0$ of probabilities of events connected with the process X_t^ε plays a role analogous to that of the action functional in Feynman's description of a quantum mechanical system; only our arguments are an infinite-dimensional analogue of Laplace's method rather than the method of stationary phase (and therefore, they are simpler).

In analogy with quantum mechanical problems, we shall call $\varepsilon^{-2}S(\varphi)$ the action functional for the corresponding family of random processes. Of course, this does not mean that we give this functional a mechanical interpretation; we only have in mind the analogy with the role played by the action in Feynman's approach to quantum mechanics.

In the next section we introduce the action functional and obtain the necessary estimates for the process $X_t^\varepsilon = \varepsilon w_t$, i.e., in the case where the vector field $b(x)$ is identically zero. The form of the action functional and the corresponding estimates for a process X_t with an arbitrary field $b(x)$ and also for some other processes will be established in §4 of this chapter and in §1 of the next chapter.

§2. Exponential Estimates

We denote by $\mathbf{C}_{T_1 T_2} = \mathbf{C}_{T_1 T_2}(R^r)$ the set of continuous functions on the interval $[T_1, T_2]$ with values in R^r. In this space we shall consider the metric $\rho_{T_1 T_2}(\varphi, \psi) = \sup_{T_1 \le t \le T_2} |\varphi_t - \psi_t|$. For absolutely continuous functions φ_t we define the functional

$$S(\varphi) = S_{T_1 T_2}(\varphi) = \frac{1}{2} \int_{T_1}^{T_2} |\dot{\varphi}_t|^2 \, ds;$$

if a function $\varphi \in \mathbf{C}_{T_1 T_2}$ is not absolutely continuous on $[T_1, T_2]$ or the integral is divergent, then we set $S(\varphi) = +\infty$.

Let w_t be a Wiener process in R^r, $w_0 = 0$.

The *action functional* for the family of random processes $X_t^\varepsilon = \varepsilon w_t$ is, by definition, the functional $I_{T_1 T_2}^\varepsilon(\varphi) = \varepsilon^{-2} S_{T_1 T_2}(\varphi)$; the functional $S_{T_1 T_2}(\varphi)$ will be called the normalized action functional for the family εw_t.

Theorem 2.1. *For any δ, γ, and K there exists an $\varepsilon_0 > 0$ such that*

$$\mathsf{P}\{\rho_{0T}(X^\varepsilon, \varphi) < \delta\} \ge \exp\{-\varepsilon^{-2}[S_{0T}(\varphi) + \gamma]\}$$

for $\varepsilon < \varepsilon_0$, where $T > 0$ and $\varphi \in \mathbf{C}_{0T}$ are such that $\varphi_0 = 0$ and $T + S_{0T}(\varphi) \le K$.

This theorem gives a lower estimate of the probability of "passing through the δ-tube about φ." In calculating $\mathsf{P}\{X^\varepsilon \in A\}$ for a set $A \subset \mathbf{C}_{0T}$, this theorem enables us to give a lower estimate of the neighborhood of an extremal $\tilde{\varphi} \in A$. In order to be able to apply Laplace's method, we also have to give an upper estimate of the probability that a trajectory of X^ε moves far from the "most probable" function $\tilde{\varphi}$. The necessary estimate is contained in the theorem below.

Theorem 2.2. *Let s be a positive number. Write*

$$\Phi(s) = \{\varphi \in \mathbf{C}_{0T}, \varphi_0 = 0, S_{0T}(\varphi) \le s\}.$$

For any $\delta > 0$, $\gamma > 0$, and $s_0 > 0$ there exists an $\varepsilon_0 > 0$ such that for $0 < \varepsilon \le \varepsilon_0$ and $s < s_0$ we have

$$\mathsf{P}\{\rho_{0T}(X^\varepsilon, \Phi(s)) \ge \delta\} \le \exp\{-\varepsilon^{-2}(s - \gamma)\}.$$

Proof of Theorem 2.1. If $S_{0T}(\varphi) \le K < \infty$, then φ is absolutely continuous and $\int_0^T |\dot{\varphi}_s|^2 \, ds < \infty$. We consider the random process $Y_t^\varepsilon = X_t^\varepsilon - \varphi_t$ obtained from $X_t^\varepsilon = \varepsilon w_t$ by the shift by φ. A shift by a function having a square integrable derivative induces an absolute continuous change of measures in $\mathbf{C}_{0T}$. If $\mu_{\varepsilon w}$ is the measure in $\mathbf{C}_{0T}$ corresponding to the process $X_t^\varepsilon = \varepsilon w_t$ and μ_{Y^ε} is the measure corresponding to Y_t^ε, then the density of the second measure with respect to the first one has the form

$$\frac{d\mu_{Y^\varepsilon}}{d\mu_{\varepsilon w}}(\varepsilon w) = \exp\left\{-\varepsilon^{-1}\int_0^T (\dot\varphi_s, dw_s) - \frac{\varepsilon^{-2}}{2}\int_0^T |\dot\varphi|^2\, ds\right\}.$$

Using this expression, we obtain

$$\begin{aligned}\mathsf{P}\{\rho_{0T}(X^\varepsilon, \varphi) < \delta\} \\ &= \mathsf{P}\{\rho_{0T}(Y^\varepsilon, 0) < \delta\} = \int_{\{\rho_{0T}(\varepsilon w, 0) < \delta\}} \frac{d\mu_{Y^\varepsilon}}{d\mu_{\varepsilon w}}(\varepsilon w)\mathsf{P}(dw) \\ &= \exp\left\{-\frac{\varepsilon^{-2}}{2}\int_0^T |\dot\varphi_s|^2\, ds\right\}\int_{\{\rho_{0T}(\varepsilon w, 0) < \delta\}} \exp\left\{-\varepsilon^{-1}\int_0^T (\dot\varphi_s, dw_s)\right\}\mathsf{P}(d\omega).\end{aligned} \tag{2.1}$$

It is easy to see that the probability of the set on which we integrate converges to 1 uniformly in $T \le K$ as $\varepsilon \to 0$. This follows, for example, from Kolmogorov's inequality. In particular, we can choose a positive number ε_1 such that

$$\mathsf{P}\{\rho_{0T}(\varepsilon w, 0) < \delta\} \ge \tfrac{3}{4}. \tag{2.2}$$

for $\varepsilon < \varepsilon_1$ and $T \le K$. Moreover, by Chebyshev's inequality we have

$$\begin{aligned}\mathsf{P}\left\{-\varepsilon^{-1}\int_0^T (\dot\varphi_s, dw_s) \le -2\sqrt{2}\varepsilon^{-1}\sqrt{S_{0T}(\dot\varphi)}\right\} \\ &\le \mathsf{P}\left\{\left|\varepsilon^{-1}\int_0^T (\dot\varphi_s, dw_s)\right| \ge 2\sqrt{2}\varepsilon^{-1}\sqrt{S_{0T}(\varphi)}\right\} \\ &\le \frac{\varepsilon^{-2}\mathsf{M}(\int_0^T (\dot\varphi_s, dw_s))^2}{8\varepsilon^{-2}S_{0T}(\varphi)} = 1/4,\end{aligned}$$

i.e.,

$$\mathsf{P}\left\{\exp\left\{-\varepsilon^{-1}\int_0^T (\dot\varphi_s, dw_s)\right\} \ge \exp\{-2\sqrt{2}\varepsilon^{-1}\sqrt{S_{0T}(\varphi)}\}\right\} \ge 3/4. \tag{2.3}$$

From estimates (2.2) and (2.3) we conclude that

$$\int_{\{\rho_{0T}(\varepsilon w, 0) < \delta\}} \exp\left\{-\varepsilon^{-1}\int_0^T (\dot\varphi_s, dw_s)\right\}\mathsf{P}(dw) > \tfrac{1}{2}\exp\{-2\sqrt{2}\varepsilon^{-1}\sqrt{S_{0T}(\varphi)}\},$$

and therefore,

$$\mathsf{P}\{\rho_{0T}(X^\varepsilon, \varphi) < \delta\} > \tfrac{1}{2}\exp\{-\varepsilon^{-2}S_{0T}(\varphi) - 2\sqrt{2}\varepsilon^{-1}\sqrt{S_{0T}(\varphi)}\}.$$

This implies the assertion of Theorem 2.1. □

Proof of Theorem 2.2. We have to estimate the probability that the trajectories of our process move far from the set of small values of the functional $S_{0T}(\varphi)$. On the trajectories of $X_t^\varepsilon = \varepsilon w_t$ themselves, the action functional is equal to $+\infty$ and we approximate the functions of X_t^ε by smoother functions. We denote by $l_t^\varepsilon, 0 \le t \le T$ the random polygon with vertices at the points $(0, 0), (\Delta, X_\Delta^\varepsilon), (2\Delta, X_{2\Delta}^\varepsilon), \ldots, (T, X_T^\varepsilon)$. We shall choose Δ later and now we only say that T/Δ is an integer. The event $\{\rho_{0T}(X^\varepsilon, \Phi(s)) \ge \delta\}$ may occur in two ways: either with $\rho_{0T}(X^\varepsilon, l^\varepsilon) < \delta$ or with $\rho_{0T}(X^\varepsilon, l^\varepsilon) \ge \delta$. In the first case we certainly have $l^\varepsilon \notin \Phi(s)$, i.e., $S_{0T}(l^\varepsilon) > s$. From this we obtain

$$\mathsf{P}\{\rho_{0T}(X^\varepsilon, \Phi(s)) \ge \delta\} \le \mathsf{P}\{S_{0T}(l^\varepsilon) > s\} + \mathsf{P}\{\rho_{0T}(X^\varepsilon, l^\varepsilon) \ge \delta\}. \quad (2.4)$$

To estimate the first probability, we transform $S_{0T}(l^\varepsilon)$:

$$S_{0T}(l^\varepsilon) = \frac{1}{2}\int_0^T |\dot{l}_t^\varepsilon|^2\,dt = \frac{\varepsilon^2}{2}\sum_{k=1}^{T/\Delta} \frac{|w_{k\Delta} - w_{(k-1)\Delta}|^2}{\Delta}.$$

As a consequence of self-similarity and independence of the increments of a Wiener process, the sum $\sum_{k=1}^{T/\Delta} \Delta^{-1}|w_{k\Delta} - w_{(k-1)\Delta}|^2$ is distributed in the same way as $\sum_{i=1}^{rT/\Delta} \xi_i^2$, where the ξ_i are independent random variables having normal distribution with parameters (0, 1):

$$\mathsf{P}\{S_{0T}(l^\varepsilon) > s\} = \mathsf{P}\left\{\sum_{i=1}^{rT/\Delta} \xi_i^2 > 2\varepsilon^{-2}s\right\}.$$

We estimate the right side by means of Chebyshev's exponential inequality. Since

$$\mathsf{M}\exp\left\{\frac{1-\alpha}{2}\xi_i^2\right\} = C_\alpha < \infty$$

for every $\alpha > 0$, we have

$$\mathsf{P}\{S_{0T}(l^\varepsilon) > s\} = \mathsf{P}\left\{\sum_{i=1}^{rT/\Delta} \xi_i^2 > 2\varepsilon^{-2}s\right\} \le \frac{\mathsf{M}\exp\left\{\dfrac{1-\alpha}{2}\displaystyle\sum_{i=1}^{rT/\Delta}\xi_i^2\right\}}{\exp\{\varepsilon^{-2}s(1-\alpha)\}}$$
$$= C_\alpha^{rT/\Delta}\exp\{-\varepsilon^{-2}s(1-\alpha)\}.$$

This implies that there exists an $\varepsilon_0 > 0$ such that

$$\mathsf{P}\{S_{0T}(l^\varepsilon) > s\} \le \tfrac{1}{2}\exp\{-\varepsilon^{-2}(s-\gamma)\} \tag{2.5}$$

for $\varepsilon < \varepsilon_0$ and $s \le s_0$.

Now we estimate the second term in inequality (2.4):

$$\begin{aligned}\mathsf{P}\left\{\rho_{0T}(X^\varepsilon, l^\varepsilon) \ge \delta\right\} &\le \sum_{k=1}^{T/\Delta} \mathsf{P}\left\{\max_{(k-1)\Delta \le t \le k\Delta} |X_t^\varepsilon - l_t^\varepsilon| \ge \delta\right\} \\ &= \frac{T}{\Delta}\mathsf{P}\left\{\max_{0\le t\le\Delta} |X_t^\varepsilon - l_t^\varepsilon| \ge \delta\right\} \\ &\le \frac{T}{\Delta}\mathsf{P}\left\{\max_{0\le t\le\Delta} |\varepsilon w_t| \ge \frac{\delta}{2}\right\}.\end{aligned} \tag{2.6}$$

Here we have used the facts that for distinct k's, the random variables $\max_{(k-1)\Delta \le t \le k\Delta} |X_t^\varepsilon - l_t^\varepsilon|$ are identically distributed and

$$\begin{aligned}\mathsf{P}\left\{\max_{0\le t\le\Delta} |X_t^\varepsilon - l_t^\varepsilon| \ge \delta\right\} &= \mathsf{P}\left\{\max_{0\le t\le\Delta} \left|\varepsilon w_t - \varepsilon\frac{t}{\Delta} w_\Delta\right| \ge \delta\right\} \\ &\le \mathsf{P}\left\{\max_{0\le t\le\Delta} |\varepsilon w_t| \ge \frac{\delta}{2}\right\}.\end{aligned}$$

Continuing estimation (2.6) and taking account of the inequality

$$\mathsf{P}\{w_\Delta^i > z\} \le \frac{\sqrt{\Delta}}{z\sqrt{2\pi}}\exp(-z^2/2\Delta)$$

true for a normal random variable w_Δ^i with mean 0 and variance Δ, we have

$$\begin{aligned}\mathsf{P}\{\rho_{0T}(X^\varepsilon, l^\varepsilon) \ge \delta\} &\le \frac{T}{\Lambda}\mathsf{P}\left\{\max_{0\le t\le\Delta} |\varepsilon w_t| > \frac{\delta}{2}\right\} \\ &\le \frac{4rT}{\Delta}\mathsf{P}\{w_\Delta^i > \delta/2r\varepsilon\} \le \frac{4rT}{\Delta}\cdot\frac{2r\varepsilon}{\delta\sqrt{2\pi}}\sqrt{\Delta}\exp\left(-\frac{\delta^2}{8r^2\Delta\varepsilon^2}\right).\end{aligned} \tag{2.7}$$

Now it is sufficient to take $\Delta < \delta^2/4r^2 s_0$, and the right side of (2.7) will be smaller than $\frac{1}{2}\exp\{-e^{-2}(s-\gamma)\}$ for ε sufficiently small and $s \le s_0$. This and inequalities (2.4) and (2.5) imply the assertion of the theorem. □

We establish some properties of the functional $S_{0T}(\varphi)$.

Lemma 2.1.

(a) *The functional $S_{0T}(\varphi)$ is lower semicontinuous in the sense of uniform convergence, i.e., if a sequence $\varphi^{(n)}$ converges to φ in $\mathbf{C}_{0T}$, then $S_{0T}(\varphi) \le \underline{\lim}_{n\to\infty} S_{0T}(\varphi^{(n)})$.*

(b) *The set of functions* φ_t, $0 \le t \le T$ *such that* φ_0 *belongs to some compact subset of* R^r *and* $S_{0T}(\varphi) \le s_0 < \infty$ *is compact.*

Proof. (a) It is sufficient to consider the case where the finite limit $\lim_{n\to\infty} S_{0T}(\varphi^{(n)})$ exists. We use the following fact (cf. Riesz and Sz.-Nagy [1], p. 86): a function φ_t is absolutely continuous and its derivative is square integrable if and only if

$$\sup_{0 \le t_0 < t_1 < \cdots < t_N \le T} \sum_{i=1}^{N} \frac{|\dot{\varphi}_{t_i} - \varphi_{t_{i-1}}|^2}{t_i - t_{i-1}} \tag{2.8}$$

is finite and in this case the supremum is equal to $\int_0^T |\dot{\varphi}_t|^2 \, dt$.

Expression (2.8) is equal to

$$\begin{aligned} &\sup_{0 \le t_0 < t_1 < \cdots < t_N \le T} \lim_{n\to\infty} \sum_{i=1}^{N} \frac{|\varphi_{t_i}^{(n)} - q_{t_{i-1}}^{(n)}|^2}{t_i - t_{i-1}} \\ &\le \varliminf_{n\to\infty} \sup_{0 \le t_0 < t_1 < \cdots < t_N \le T} \sum_{i=1}^{N} \frac{|\varphi_{t_i}^{(n)} - \varphi_{t_{i-1}}^{(n)}|^2}{t_i - t_{i-1}} \\ &= \lim_{n\to\infty} \int_0^T |\dot{\varphi}_s^{(n)}|^2 \, ds = 2 \lim_{n\to\infty} S(\varphi^{(n)}) < \infty. \end{aligned}$$

This implies that φ is absolutely continuous and $S(\varphi) \le \lim_{n\to\infty} S(\varphi^{(n)})$.

(b) From the estimation $\int_0^T|\dot{\varphi}_s|^2 \, ds = 2S_{0T}(\varphi) \le 2s_0$ we obtain that

$$|\varphi_t| = \left|\varphi_0 + \int_0^t \dot{\varphi}_s \, ds\right| \le |\varphi_0| + \sqrt{T \int_0^T |\dot{\varphi}_s|^2 \, ds} \le |\varphi_0| + \sqrt{2Ts_0}.$$

Consequently, all functions of our set are uniformly bounded. This estimation implies the equicontinuity of the functions φ:

$$\begin{aligned} |\varphi_{t+h} - \varphi_t| \le \int_t^{t+h} |\dot{\varphi}_s| \, ds &\le \sqrt{h \int_t^{t+h} |\dot{\varphi}_s|^2 \, ds} \\ &\le \sqrt{2S(\varphi)}\sqrt{h} \le \sqrt{2s_0}\sqrt{h}. \end{aligned}$$

The compactness follows from Arzelà's theorem. □

Corollary. *On every nonempty closed set in* $\mathbf{C}_{0T}$ *for which the initial values of* φ_0 *are contained in some compact set, the functional* $S_{0T}(\varphi)$ *attains its smallest value and values close to the smallest value are assumed by the functional only near functions at which the minimum is attained.*

Theorem 2.3. *Let a function* $\varphi \in \mathbf{C}_{0T}$ *be such that* $\varphi_0 = 0$, $S_{0T}(\varphi) < \infty$. *We have*

$$\lim_{\delta \downarrow 0} \overline{\lim_{\varepsilon \downarrow 0}} \varepsilon^2 \ln \mathsf{P}\{\rho_{0T}(X^\varepsilon, \varphi) < \delta\} = \lim_{\delta \downarrow 0} \underline{\lim_{\varepsilon \downarrow 0}} \varepsilon^2 \ln \mathsf{P}\{\rho_{0T}(X^\varepsilon, \varphi) < \delta\}$$
$$= -S_{0T}(\varphi).$$

For the proof of this theorem, it is sufficient to establish that for every $\gamma > 0$ and every sufficiently small $\delta_0 > 0$ there exists an $\varepsilon_0 > 0$ such that

$$\begin{aligned} \exp\{-\varepsilon^{-2}(S_{0T}(\varphi) - \gamma)\} &\geq \mathsf{P}\{\rho_{0T}(X^\varepsilon, \varphi) < \delta_0\} \\ &\geq \exp\{-\varepsilon^{-2}(S_{0T}(\varphi) + \gamma)\} \end{aligned} \tag{2.9}$$

for $0 < \varepsilon < \varepsilon_0$. The right-hand inequality constitutes the assertion in Theorem 2.1. We prove the left-hand inequality. We choose $\delta_0 > 0$ so small that

$$\inf_{\psi:\, \rho_{0T}(\varphi, \psi) \leq \delta_0} S_{0T}(\psi) > S_{0T}(\varphi) - \gamma/4.$$

This can be done relying on Lemma 2.1. By the corollary to the same lemma, we have

$$\delta_1 = \rho_{0T}(\{\psi: \rho_{0T}(\varphi, \psi) \leq \delta_0\}, \Phi(S_{0T}(\varphi) - \gamma/2)) > 0.$$

We apply Theorem 2.2:

$$\begin{aligned} \mathsf{P}\{\rho_{0T}(X^\varepsilon, \varphi) < \delta_0\} &\leq \mathsf{P}\{\rho_{0T}(X^\varepsilon, \Phi(S_{0T}(\varphi) - \gamma/2)) > \delta_1\} \\ &\leq \exp\{-\varepsilon^{-2}(S_{0T}(\varphi) - \gamma)\} \end{aligned}$$

whenever ε is sufficiently small. By the same token inequalities (2.9), and with them Theorem 2.3, are proved. □

The proofs of Theorems 2.1 and 2.2 reproduce in the simplest case of the process $X_t^\varepsilon = \varepsilon w_t$ the construction used in the articles [1] and [4] by Wentzell and Freidlin. In this special case, analogous results were obtained in Schilder's article [1].

§3. Action Functional. General Properties

The content of §2 may be divided into two parts: one which is concerned with the Wiener process and the other which relates in general to ways of describing the rough (to within logarithmic equivalence) asymptotics of families of measures in metric spaces. We consider these questions separately.

Let X be a metric space with metric ρ. On the σ-algebra of its Borel subsets let μ^h be a family of probability measures depending on a parameter $h > 0$. We shall be interested in the asymptotics of μ^h as $h \downarrow 0$ (the changes which have to be made if we consider convergence to another limit or to ∞ or if we consider a parameter not on the real line but in a more general set, are obvious).

Let $\lambda(h)$ be a positive real-valued function going to $+\infty$ as $h \downarrow 0$ and let $S(x)$ be a function on X assuming values in $[0, \infty]$. We shall say that $\lambda(h)S(x)$ is an *action function* for μ^h as $h \downarrow 0$ if the following assertions hold:

(0) the set $\Phi(s) = \{x: S(x) \leq s\}$ is compact for every $s \geq 0$;

(I) for any $\delta > 0$, any $\gamma > 0$ and any $x \in X$ there exists an $h_0 > 0$ such that

$$\mu^h\{y: \rho(x, y) < \delta\} \geq \exp\{-\lambda(h)[S(x) + \gamma]\} \tag{3.1}$$

for all $h \leq h_0$;

(II) for any $\delta > 0$, any $\gamma > 0$ and any $s > 0$ there exists an $h_0 > 0$ such that

$$\mu^h\{y: \rho(y, \Phi(s)) \geq \delta\} \leq \exp\{-\lambda(h)(s - \gamma)\} \tag{3.2}$$

for $h \leq h_0$.

If ξ^h is a family of random elements of X defined on the probability spaces $\{\Omega^h, \mathscr{F}^h, \mathsf{P}^h\}$, then the action function for the family of the distributions μ^h, $\mu^h(A) = \mathsf{P}^h\{\xi^h \in A\}$ is called the action function for the family ξ^h. In this case formulas (3.1) and (3.2) take the form

$$\mathsf{P}^h\{\rho(\xi^h, x) < \delta\} \leq \exp\{-\lambda(h)[S(x) + \gamma]\}, \tag{3.1'}$$

$$\mathsf{P}^h\{\rho(\xi^h, \Phi(s)) \geq \delta\} \leq \exp\{-\lambda(h)(s - \gamma)\}. \tag{3.2'}$$

Separately, the functions $S(x)$ and $\lambda(h)$ will be called the *normalized action function* and *normalizing coefficient*. It is clear that the decomposition of an action function into two factors $\lambda(h)$ and $S(x)$ is not unique; moreover, $\lambda(h)$ can always be replaced by a function $\lambda_1(h) \sim \lambda(h)$. Nevertheless, we shall prove below that for a given normalizing coefficient, the normalized action function is uniquely defined.

EXAMPLE 3.1. $X = R^1$, μ^h is the Poisson distribution with parameter h for every $h > 0$, and we are interested in the behavior of μ^h as $h \downarrow 0$. Here $\lambda(h) = -\ln h$; $S(x) = x$ for every nonnegative integer x and $S(x) = +\infty$ for the remaining x.

If X is a function space, we shall use the term *action functional.* Hence for the family of random processes εw_t, where w_t is a Wiener process, $t \in [0, T]$ and $w_0 = 0$, a normalized action functional as $\varepsilon \to 0$ is $S(\varphi) = \frac{1}{2}\int_0^T |\dot{\varphi}_t|\, dt$ for absolutely continuous φ_t, $0 \le t \le T$, $\varphi_0 = 0$, and $S(\varphi) = +\infty$ for all other φ and the normalizing coefficient is equal to ε^{-2} (as the space X we take the space of continuous functions on the interval $[0, T]$ with the metric corresponding to uniform convergence).

We note that condition (0) implies that $S(x)$ attains its minimum on every nonempty closed set. It is sufficient to consider only the case of a closed $A \subseteq X$ with $S_A = \inf\{S(x): x \in A\} < \infty$. We choose a sequence of points $x_n \in A$ such that $s_n = S(x_n) \downarrow s_A$. The nested compact sets $\Phi(s_n) \cap A$ are nonempty (since $\Phi(s_n) \cap A \ni x_n$), and therefore, their intersection is nonempty and contains a point x_A, $S(x_A) = s_A$.

It would be desirable to obtain immediately a large number of examples of families of random processes for which we could determine an action functional. The following result (Freidlin [7]) helps us with that.

Theorem 3.1. *Let $\lambda(h)S^{\mu}(x)$ be the action function for a family of measures μ^h on a space X (with metric ρ_X) as $h \downarrow 0$. Let φ be a continuous mapping of X into a space Y with metric ρ_Y and let a measure ν^h on Y be given by the formula $\nu^h(A) = \mu^h(\varphi^{-1}(A))$. The asymptotics of the family of measures ν^h as $h \downarrow 0$ is given by the action function $\lambda(h)S^{\nu}(y)$, where $S^{\nu}(y) = \min\{S^{\mu}(x): x \in \varphi^{-1}(y)\}$ (the minimum over the empty set is set to be equal to $+\infty$).*

Proof. We introduce the following notation:

$$\Phi^{\mu}(s) = \{x: S^{\mu}(x) \le s\}, \Phi^{\nu}(s) = \{y: S^{\nu}(y) \le s\}.$$

It is easy to see that $\Phi^{\nu}(s) = \varphi(\Phi^{\mu}(s))$, from which we obtain easily that S^{ν} satisfies condition (0).

We prove condition (I). We fix an arbitrary $y \in Y$ and a neighborhood of it. If $S^{\nu}(y) = \infty$, then there is nothing to prove. If $S^{\nu}(y) < \infty$, then there exists an x such that $\varphi(x) = y$, $S^{\nu}(y) = S^{\mu}(x)$. We choose a neighborhood of x whose image is contained in the selected neighborhood of y and thus obtain the condition to be proved.

Now we pass to condition (II). The pre-image of the set $\{y: \rho_Y(y, \Phi^{\nu}(s)) \ge \delta\}$ under φ is closed and does not intersect the compact set $\Phi^{\mu}(s)$. Therefore, we can choose a positive δ' such that the δ'-neighborhood of $\Phi^{\mu}(s)$ does not intersect $\varphi^{-1}\{y: \rho_Y(y, \Phi^{\nu}(s)) \ge \delta\}$. From inequality (3.2) with ρ_X, Φ^{μ} and δ' in place of δ, we obtain a similar inequality for ν^h, ρ_Y, Φ^{ν} and δ. □

Using this theorem in the special case where X and Y are the same space with distinct metrics, we obtain that if $\lambda(h)S(x)$ is an action function for a family of measures μ^h as $h \downarrow 0$ in a metric ρ_1 and another metric ρ_2 is such that $\rho_2(x, y) \to 0$ if $\rho_1(x, y) \to 0$, then $\lambda(h)S(x)$ is an action function in the metric ρ_2, as well. Of course, this simple assertion can be obtained directly. From this we obtain, in particular, that $\varepsilon^{-2}S_{0T}(\varphi) = (1/2\varepsilon^2)\int_0^T |\dot{\varphi}_t|^2\, dt$ remains an action functional for the family of the processes εw_t, $0 \le t \le T$ as $\varepsilon \downarrow 0$ if we consider the metric of the Hilbert space $\mathbf{L}^2_{0T}$.

The following examples are more interesting.

EXAMPLE 3.2. Let $G(s, t)$ be a k times continuously differentiable function on the square $[0, T] \times [0, T]$, $k \ge 1$. In the space $\mathbf{C}_{0T}$ we consider the operator $\tilde{G}$ defined by the formula

$$\tilde{G}\varphi_t = \int_0^T G(s, t)\, d\varphi_s.$$

Here the integral is understood in the sense of Stieltjes and by the assumed smoothness of G, integration by parts is allowed:

$$\tilde{G}\varphi_t = G(T, t)\varphi_T - G(0, t)\varphi_0 - \int_0^T \frac{\partial G(s, t)}{\partial s}\varphi_s\, ds.$$

This equality shows that $\tilde{G}$ is a continuous mapping of $\mathbf{C}_{0T}$ into the space $\mathbf{C}^{(k-1)}_{0T}$ of functions having $k - 1$ continuous derivatives with the metric

$$\rho^{k-1}(\varphi, \psi) = \max_{\substack{0 \le i \le k-1 \\ 0 \le t \le T}} \left| \frac{d^i(\varphi_t - \psi_t)}{dt^i} \right|.$$

We calculate the action functional for the family of the random processes

$$X_t^\varepsilon = \varepsilon \tilde{G} w_t = \varepsilon \int_0^T G(s, t)\, dw_s$$

in $\mathbf{C}^{(k-1)}_{0T}$ as $\varepsilon \to 0$. By Theorem 3.1, the normalizing coefficient remains the same and the normalized action functional is given by the equality

$$\begin{aligned} S^X(\varphi) = S^X_{0T}(\varphi) &= \min\{S^w_{0T}(\psi) : \tilde{G}\psi = \varphi\} \\ &= \min\left\{\frac{1}{2}\int_0^T |\dot{\psi}_s|^2\, ds : \tilde{G}\psi = \varphi\right\}; \end{aligned}$$

if there are no ψ for which $\tilde{G}\psi = \varphi$, then $S^X(\varphi) = +\infty$.

We introduce an auxiliary operator G in $\mathbf{L}^2_{0T}$, given by the formula

$$Gf(t) = \int_0^T G(s, t) f(s)\, ds,$$

and express S_{0T}^X in terms of the inverse of G. This will not be a one-to-one operator in general, since G vanishes on some subspace $\mathbf{L}_0 \subseteq \mathbf{L}_{0T}^2$, which may be nontrivial. We make the inverse operator one-to-one artificially, by setting $G^{-1}\varphi = \psi$, where ψ is the unique function in $\mathbf{L}_{0T}^2$ orthogonal to $\mathbf{L}_0$ and such that $G\psi = \varphi$. The operator G^{-1} is defined on the range of G.

If $S^X(\varphi) < \infty$, then there exists a function $\psi \in \mathbf{C}_{0T}$ such that $\tilde{G}\psi = \varphi$ and $S_{0T}^w(\psi) < \infty$. Then ψ is absolutely continuous and $\tilde{G}\psi = G\dot{\psi}$. Therefore,

$$S_{0T}^X(\varphi) = \min\left\{\frac{1}{2}\int_0^T |\dot{\psi}_s|^2\, ds \colon \tilde{G}\psi = \varphi\right\} = \min\left\{\frac{1}{2}\|f\|^2 \colon Gf = \varphi\right\},$$

where $\|f\|$ is the norm of f in $\mathbf{L}_{0T}^2$. Any element f for which $Gf = \varphi$ can be represented in the form $f = G^{-1}\varphi + f'$, where $f' \in \mathbf{L}_0$. Taking into account that $G^{-1}\varphi$ is orthogonal to $\mathbf{L}_0$, we obtain $\|f\|^2 = \|G^{-1}\varphi\|^2 + \|f'\|^2 \geq \|G^{-1}\varphi\|^2$. This means that $S_{0T}^X(\varphi) = \frac{1}{2}\|G^{-1}\varphi\|^2$ for φ in the range of G. For the remaining $\varphi \in \mathbf{C}_{0T}^{(k-1)}$ the functional assumes the value $+\infty$.

Example 3.3. We consider a random process X_t^ε on the interval $[0, T]$, satisfying the linear differential equation

$$P\left(\frac{d}{dt}\right)X_t^\varepsilon \equiv \sum_{k=0}^n a_k \frac{d^k X_t^\varepsilon}{dt^k} = \varepsilon \dot{w}_t,$$

where $\dot{w}_t$ is a one-dimensional white noise process. In order to choose a unique solution, we have to prescribe n boundary conditions; for the sake of simplicity, we assume them to be homogeneous linear and nonrandom. We denote by $G(s, t)$ the Green's function of the boundary value problem connected with the operator $P(d/dt)$ and our boundary conditions (cf. Coddington and Levinson [1]). The process X_t^ε can be represented in the form $X_t^\varepsilon = \varepsilon \int_0^T G(s, t)\, dw_s$, i.e., $X^\varepsilon = \tilde{G}(\varepsilon w)$. In this case the corresponding operator has the single-valued inverse $G^{-1} = P(d/dt)$ with domain consisting of the functions satisfying the boundary conditions. From this we conclude that the action functional for the family of the processes X_t^ε as $\varepsilon \to 0$ is $\varepsilon^{-2}S_{0T}^X(\varphi)$, where

$$S_{0T}^X(\varphi) = \frac{1}{2}\int_0^T \left|P\left(\frac{d}{dt}\right)\varphi\right|^2 dt,$$

and if φ does not satisfy the boundary conditions or the derivative $d^{n-1}\varphi_t/dt^{n-1}$ is not absolutely continuous, then $S_{0T}^X(\varphi) = +\infty$.

The action functional has an analogous form in the case of a family of multidimensional random processes which are the solution of a system of linear differential equations on the right side of which there is a white noise multiplied by a small parameter.

In the next chapter (§1), by means of the method based on Theorem 3.1, we establish that for a family of diffusion processes arising as a result of a perturbation of a dynamical system $\dot{x} = b(x_t)$ by adding to the right side a white noise multiplied by a small parameter, the action functional is equal to $(1/2\varepsilon^2) \int_0^T |\dot{\varphi}_t - b(\varphi_t)|^2 \, dt$.

We consider the conditions (0), (I), and (II) in more detail.

In the case of a complete space X, condition (0) can be split into two: lower semicontinuity of $S(x)$ on X (which is equivalent to closedness of $\Phi(s)$ for every s) and relative compactness of $\Phi(s)$. Such a splitting is convenient in the verification of the condition (cf. Lemma 2.1). The condition of semicontinuity of S is not stringent: it is easy to prove that if the functions $\lambda(h)$ and $S(x)$ satisfy conditions (I) and (II), then so do $\lambda(h)$ and the lower semicontinuous function $\underline{S}(x) = S(x) \wedge \underline{\lim}_{y \to x} S(y)$. (The method of redefining the normalized action functional by semicontinuity is used in a somewhat more refined form in the proof of Theorem 2.1 of Ch. 5.)

Theorem 3.2. *Condition* (I), *together with the condition of relative compactness of* $\Phi(s)$, *is equivalent to the condition*:

($\mathrm{I_{eq}}$) *for any* $\delta > 0$, $\gamma > 0$ *and* $s_0 > 0$ *there exists* $h_0 > 0$ *such that inequality* (3.1) *is satisfied for all* $h \le h_0$ *and all* $x \in \Phi(s_0)$.

Condition (II) *implies the following*:

($\mathrm{II_{eq}}$) *for any* $\delta > 0$, $\gamma > 0$ *and* $s_0 > 0$ *there exists an* $h_0 > 0$ *such that inequality* (3.2) *is satisfied for all* $h \le h_0$ *and* $s \le s_0$.

Conditions (0) *and* (II) *imply the following*:

($\mathrm{II_+}$) *for any* $\delta > 0$ *and* $s \ge 0$ *there exist* $\gamma > 0$ *and* $h_0 > 0$ *such that*

$$\mu^h\{y: \rho(y, \Phi(s)) \ge \delta\} \le \exp\{-\lambda(h)(s + \gamma)\} \tag{3.3}$$

for all $h \le h_0$.

We only prove the last assertion. The values of S on the closed set $A = \{y: \rho(y, \Phi(s)) \ge \delta\}$ are greater than s. Therefore, the infimum of $S(y)$ on this set is greater than s. We select a positive γ so that

$$\inf\{S(y): y \in A\} > s + 2\gamma;$$

then $A \cap \Phi(s + 2\gamma) = \varnothing$. We select a positive δ' not exceeding $\rho(A, \Phi(s + 2\gamma))$ (this distance is positive by virtue of the compactness of the second set) and use inequality (3.2) for δ' in place of δ and $s + 2\gamma$ in place of s. □

Conditions (I) and (II) were introduced in describing the rough asymptotics of probabilities of large deviations in the papers [1], [4] of Wentzell and Freidlin. There are other methods of description; however, under condition (0) they are equivalent to the one given here.

In Varadhan's paper [1] the following conditions occur instead of conditions (I) and (II):

(I′) for any open $A \subseteq X$ we have

$$\varliminf_{h \downarrow 0} \lambda(h)^{-1} \ln \mu^h(A) \geq -\inf\{S(x): x \in A\}; \tag{3.4}$$

(II′) for any closed $A \subseteq X$ we have

$$\varlimsup_{h \downarrow 0} \lambda(h)^{-1} \ln \mu^h(A) \leq -\inf\{S(x): x \in A\}. \tag{3.5}$$

Theorem 3.3. *Conditions* (I) *and* (I′) *are equivalent. Condition* (II′) *implies* (II) *and conditions* (0) *and* (II) *imply* (II′)

Consequently, (I) $\Leftrightarrow$ (I′) *and* (II) $\Leftrightarrow$ (II′) *under condition* (0).

Proof. The implications (I′) $\Rightarrow$ (I), (I) $\Rightarrow$ (I′) and (II′) $\Rightarrow$ (II) can be proved very simply. We prove, for example, the last one. The set $A = \{y: \rho(y, \Phi(s)) \geq \delta\}$ is closed and $S(y) > s$ in it. Therefore, $\inf\{S(y): y \in A\} \geq s$. From (II′) we obtain: $\varlimsup_{h \downarrow 0} \lambda(h)^{-1} \ln \mu^h(A) \leq -s$, which means that for every $\gamma > 0$ and h sufficiently small we have: $\lambda(h)^{-1} \ln \mu^h(A) \leq -s + \gamma$, i.e., (3.2) is satisfied.

Now let (0) and (II) be satisfied. We prove (II′). Choose an arbitrary $\gamma > 0$ and put $s = \inf\{S(y): y \in A\} - \gamma$. The closed set A does not intersect the compact set $\Phi(s)$. Therefore, $\delta = \rho(A, \Phi(s)) > 0$. We use inequality (3.2) and obtain that

$$\begin{aligned} \mu^h(A) &\leq \mu^h\{y: \rho(y, \Phi(s)) \geq \delta\} \\ &\leq \exp\{-\lambda(h)(s - \gamma)\} = \exp\{-\lambda(h)(\inf\{S(y): y \in A\} - 2\gamma)\}. \end{aligned}$$

for sufficiently small h. Taking logarithms, dividing by the normalizing coefficient $\lambda(h)$ and passing to the limit, we obtain that $\varlimsup_{h \downarrow 0} \lambda(h)^{-1} \ln \mu^h(A) \leq -\inf\{S(y): y \in A\} + 2\gamma$, which implies (3.5), since $\gamma > 0$ is arbitrary. □

In Borovkov's paper [1] the rough asymptotics of probabilities of large deviations is characterized by one condition instead of the two conditions (I) and (II) or (I′) and (II′).

We shall say that a set $A \subseteq X$ is *regular* (with respect to the function S) if the infimum of S on the closure of A coincides with the infimum of S on the set of interior points of A:

$$\inf\{S(x): x \in [A]\} = \inf\{S(x): x \in (A)\}.$$

We introduce the following condition:

(I $\frac{1}{2}$) for any regular Borel set $A \subseteq X$,

$$\lim_{h \downarrow 0} \lambda(h)^{-1} \ln \mu^h(A) = -\inf\{S(x): x \in A\}. \tag{3.6}$$

Theorem 3.4. *Conditions* (0), (I) *and* (II) *imply* (I $\frac{1}{2}$). *Moreover, if A is a regular set and* $\min\{S(x): x \in [A]\}$ *is attained at a unique point x_0, then*

$$\lim_{h \downarrow 0} \frac{\mu^h(A \cap \{x: \rho(x_0, x) < \delta\})}{\mu^h(A)} = 1 \tag{3.7}$$

for every $\delta > 0$.

Conversely, conditions (0) *and* (I $\frac{1}{2}$) *imply* (I) *and* (II).

We note that in terms of random elements of X, (3.7) can be rewritten in the form

$$\lim_{h \downarrow 0} \mathsf{P}^h\{\rho(x_0, \xi^h) < \delta \,|\, \xi^h \in A\} = 1. \tag{3.7$'$}$$

Proof. We use the equivalences (I) $\Leftrightarrow$ (I$'$) and (II) $\Leftrightarrow$ (II$'$) already established (under condition (0)). That (I$'$) and (II$'$) imply (I $\frac{1}{2}$) is obvious. Moreover, if A is a non-Borel regular set, then (3.6) is satisfied if $\mu^h(A)$ is replaced by the corresponding inner and outer measures.

To obtain relation (3.7), we observe that

$$\min\{S(x): x \in [A], \rho(x, x_0) \geq \delta\} > S(x_0).$$

We obtain from (3.5) that

$$\begin{aligned} &\overline{\lim_{h \downarrow 0}}\, \lambda(h)^{-1} \ln \mu^h\{x \in A: \rho(x_0, x) \geq \delta\} \\ &\qquad \leq \overline{\lim_{h \downarrow 0}}\, \lambda(h)^{-1} \ln \mu^h\{x \in [A]: \rho(x_0, x) \geq \delta\} < -S(x_0). \end{aligned}$$

This means that $\mu^h\{x \in A: \rho(x_0, x) \geq \delta\}$ converges to zero faster than $\mu^h(A) \asymp \exp\{-\lambda(h)S(x_0)\}$.

We show that (I $\frac{1}{2}$) implies (I$'$) and (II$'$). For any positive δ and any set $A \subset X$ we denote by $A_{+\delta}$ the δ-neighborhood of A and $A_{-\delta}$ the set of points which lie at a distance greater than δ from the complement of A. We set $s(\pm\delta) = \inf\{S(x): x \in A_{\pm\delta}\}$ (infimum over the empty set is assumed to be equal to $+\infty$). The function s is defined for all real values of the argument; at zero we define it to be $\inf\{S(x): x \in A\}$. It is easy to see that it is a non-increasing function which is continuous except at possibly a countable number of points.

If A is an open set, then s is left continuous at zero. Therefore, for an arbitrarily small $\gamma > 0$ there exists a $\delta > 0$ such that $s(-\delta) < s(0) - \gamma$, and s is continuous at $-\delta$. The latter ensures the applicability of (3.6) to $A_{-\delta}$. We obtain

$$\varliminf_{h \downarrow 0} \lambda(h)^{-1} \ln \mu^h(A) \geq \lim_{h \downarrow 0} \lambda(h)^{-1} \ln \mu^h(A_{-\delta}) > -s(0) - \gamma.$$

Since γ is arbitrary, this implies (3.4).

In the case of a closed set A we use condition (0) to establish the right continuity of s at zero and then we repeat the same reasoning with $A_{+\delta}$ replacing $A_{-\delta}$. □

EXAMPLE 3.4. Let A be the exterior of a ball in $\mathbf{L}^2_{0T}$: $A = \{\varphi \in \mathbf{L}^2_{0T}: \|\varphi\| > c\}$. This set is regular with respect to the functional S^X_{0T} considered in Example 3.3. To verify this, we multiply all elements of A by a number q smaller than but close to one. The open set $q \cdot A = q \cdot (A)$ absorbs $[A]$ and the infimum of the functional as well as all of its values change insignificantly (they are multiplied by q^2).

Since A is regular, we have

$$\lim_{\varepsilon \to 0} \varepsilon^2 \ln \mathsf{P}\{\|X^\varepsilon\| > c\} = \lim_{\varepsilon \to 0} \varepsilon^2 \ln \mathsf{P}\{\|X^\varepsilon\| \geq c\} = -\min\left\{\frac{1}{2}\left\|P\left(\frac{d}{dt}\right)\varphi\right\|^2\right\}, \tag{3.8}$$

where the minimum is taken over all functions φ satisfying the boundary conditions and equal to c in norm.

We consider the special case where the operator $P(d/dt)$ with the boundary conditions is self-adjoint in $\mathbf{L}^2_{0T}$. Then it has a complete orthonormal system of eigenfunctions $e_k(t)$, $k = 1, 2, \ldots$, corresponding to the eigenvalues λ_k, $k = 1, 2, \ldots$ (cf., for example, Coddington and Levinson [1]). If a function φ in $\mathbf{L}^2_{0T}$ is representable in the form $\sum c_k e_k$, then $P(d/dt)\varphi = \sum_{k=1}^{\infty} c_k \lambda_k e_k$ and $\|P(d/dt)\varphi\|^2 = \sum c_k^2 \lambda_k^2$. This implies that the minimum in (3.8) is equal to $c^2/2$ multiplied by λ_1^2, the square of the eigenvalue with the smallest absolute value. Consequently,

$$\lim_{\varepsilon \to 0} \varepsilon^2 \ln \mathsf{P}\{\|X^\varepsilon\| > c\} = -c^2\lambda_1^2/2. \tag{3.9}$$

The infimum of $S(\varphi)$ on the sphere of radius c in $\mathbf{L}^2_{0T}$ is attained on the eigenfunctions, multiplied by c, corresponding to the eigenvalue λ_1. If this eigenvalue is simple (and $-\lambda_1$ is not an eigenvalue), then there are only two such functions: ce_1 and $-ce_1$. Then for any $\delta > 0$ we have

$$\lim_{\varepsilon \downarrow 0} \mathsf{P}\{\|X^\varepsilon - ce_1\| < \delta \text{ or } \|X^\varepsilon + ce_1\| < \delta \,|\, \|X^\varepsilon\| > c\} = 1.$$

The same is true for the conditional probability under the condition $\|X^\varepsilon\| \geq c$.

A concrete example: $P(d/dt)\varphi = d^2\varphi/dt^2$, boundary conditions: $\varphi_0 = \varphi_T = 0$. This operator is self-adjoint. The equation $\varphi'' = \lambda\varphi, \varphi_0 = \varphi_T = 0$ for the eigenfunctions has the solution: $\lambda = \lambda_k = -k^2\pi^2/T^2$; $e_k(t) = (\sqrt{2}/T)\sin(k\pi t/T)$. The eigenvalues are simple. We have

$$\lim_{\varepsilon\to 0} \varepsilon^2 \ln \mathsf{P}\{\|X^\varepsilon\| > c\} = -\frac{c^2\pi^4}{2T^4}$$

for every $\delta > 0$ the conditional probability that X^ε is in the δ-neighborhood of one of the functions $\pm c(\sqrt{2}/T)\sin(\pi t/T)$, under the condition that $\|X^\varepsilon\|$ is greater (not smaller) than c, converges to 1 as $\varepsilon \to 0$.

If our differential operator with certain boundary conditions is not self-adjoint, then λ_1^2 in (3.9) must be replaced with the smallest eigenvalue of the product of the operator (with the boundary conditions) with its adjoint.

Example 3.5. Let $b(x)$ be a continuous function from R^r into R^r. On the space $\mathbf{C}_{0T}$ of continuous functions on the interval $[0, T]$ with values in R^r, consider the functional $S_{0T}(\varphi)$ equal to $\frac{1}{2}\int_0^T |\dot\varphi_s - b(\varphi_s)|^2\, ds$, if φ is absolutely continuous and $\varphi_0 = x_0$ to $+\infty$ on the rest of $\mathbf{C}_{0T}$. Let an open set $D \ni x_0$, $D \neq R^r$ be such that there exist interior points of the complement of D arbitrarily close to every point of the boundary ∂D of D (i.e., $\partial D = \partial[D]$). Let us denote by A_D the open set of continuous functions φ_t, $0 \leq t \leq T$ such that $\varphi_t \in D$ for all $t \in [0, T]$. We prove that $\bar{A}_D = \mathbf{C}_{0T} \setminus A_D$ is a regular set with respect to S_{0T}.

Let the minimum of S_{0T} on the closed set $\bar{A}_D$ be attained at the function φ_t, $0 \leq t \leq T$, $\varphi_0 = x_0$ (Figure 2). This function certainly reaches the boundary at some point $t_0 \neq 0$: $\varphi_{t_0} \in \partial D$. The minimum is finite, because there exist arbitrarily smooth functions issued from x_0 and leaving D in the time interval $[0, T]$; this implies that the function φ_t is absolutely continuous.

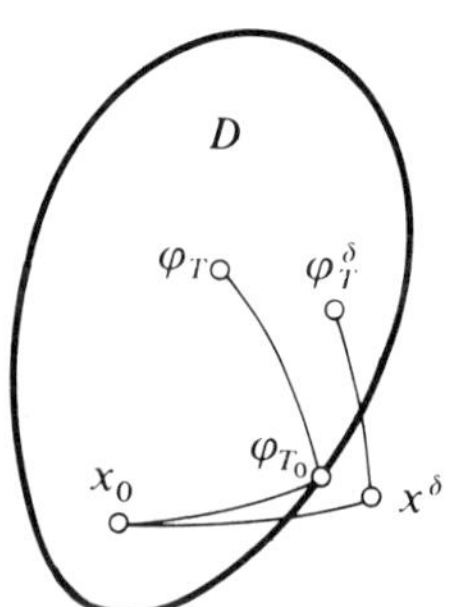

Figure 2.

For any $\delta > 0$ there exists an interior point x^δ of $R^r \setminus D$ in the δ-neighborhood of the point φ_{t_0}. We put

$$\varphi_t^\delta = \varphi_t + \frac{t}{t_0}(x^\delta - \varphi_{t_0}), \qquad 0 \le t \le T;$$

this function belongs to the interior of $\overline{A}_D$. We prove that $S_{0T}(\varphi^\delta) \to S_{0T}(\varphi)$ as $\delta \downarrow 0$. This implies the regularity of A_D. We have

$$\begin{aligned} S_{0T}(\varphi^\delta) - S_{0T}(\varphi) &= \frac{1}{2}\int_0^T [|\dot\varphi_t^\delta - b(\varphi_t^\delta)|^2 - |\dot\varphi_t - b(\varphi_t)|^2]\, dt \\ &= \int_0^T (\dot\varphi_t^\delta - b(\varphi_t^\delta) - \dot\varphi_t + b(\varphi_t), \dot\varphi_t - b(\varphi_t))\, dt \\ &\quad + \frac{1}{2}\int_0^T |\dot\varphi_t^\delta - b(\varphi_t^\delta) - \dot\varphi_t + b(\varphi_t)|^2\, dt. \end{aligned}$$

On the other hand, $\dot\varphi_t^\delta - b(\varphi_t^\delta) - \dot\varphi_t + b(\varphi_t) = t_0^{-1}(x^\delta - \varphi_{t_0}) + b(\varphi_t) - b(\varphi_t^\delta) \to 0$ uniformly in $t \in [0, T]$ as $\delta \downarrow 0$. Consequently, the scalar product of this function with $\dot\varphi_t - b(\varphi_t)$ and its scalar square in $\mathbf{L}_{0T}^2$ also converge to zero.

In Ch. 4 we prove that $\varepsilon^2 S_{0T}(\varphi)$ is the action functional for the family of diffusion processes X_t^ε described by the equation $\dot X_t^\varepsilon = b(\dot X_t^\varepsilon) + \varepsilon \dot w_t$, $X_0^\varepsilon = x_0$ (provided that b satisfies a Lipschitz condition). Then as $\varepsilon \to 0$ we have: $\mathsf{P}\{X_t^\varepsilon \text{ exits from } D \text{ for some value } t \in [0, T]\} \asymp \exp\{-\varepsilon^{-2} \min_{\varphi \in \overline{A}_D} S_{0T}(\varphi)\}$. If this minimum is attained at a unique function, then the trajectories of X_t^ε, going out of D lie near this function with an overwhelming probability for small ε.

We note that if D does not satisfy the condition $\partial D = \partial[D]$, then the corresponding set $\overline{A}_D$ may not be regular.

If the boundary ∂D is smooth, then we can prove that A_D is regular with respect to the same functional.

The last remark is concerned with the notion of regularity: if the action function is continuous (which was not the case in several examples related to function spaces), then a sufficient condition of regularity of the set A is the coincidence of $\partial(A)$ with $\partial[A]$.

Here is another form of the description of rough asymptotics (the integral description):

(III) If $F(x)$ is a bounded continuous function on X, then

$$\lim_{h \downarrow 0} \lambda(h)^{-1} \ln \int_X \exp\{\lambda(h)F(x)\}\mu^h(dx) = \max_X \{F(x) - S(x)\}. \tag{3.10}$$

This condition (under condition (0)) is also equivalent to conditions (I′) and (II′) (or (I) and (II)). A deduction of (III) (and even more complicated integral conditions) from (I′) and (II′) is contained in Varadhan's article [1].

We mention still another general assertion.

Theorem 3.5. *The value of the normalized action function at an element $x \in X$ can be expressed by either of the following two limits:*

$$S(x) = -\lim_{\delta \downarrow 0} \overline{\lim_{h \downarrow 0}}\, \lambda(h)^{-1} \ln \mu^h\{y: \rho(x, y) < \delta\}$$
$$= -\lim_{\delta \downarrow 0} \underline{\lim_{h \downarrow 0}}\, \lambda(h)^{-1} \ln \mu^h\{y: \rho(x, y) < \delta\}. \tag{3.11}$$

Proof. From condition (I) we deduce that $\underline{\lim}_{\delta \downarrow 0} \underline{\lim}_{h \downarrow 0} \geq -S(x)$, and from conditions (II) and (0) that $\overline{\lim}_{\delta \downarrow 0} \overline{\lim}_{h \downarrow 0} \leq -S(x)$. We prove the second one. For any $\gamma > 0$, the compact set $\Phi(S(x) - \gamma)$ does not contain x. We choose a positive δ smaller than one half of the distance of x from this compact set. We have

$$\mu^h\{y: \rho(x, y) < \delta\} \leq \mu^h\{y: \rho(y, \Phi(S(x) - \gamma)) \geq \delta\}.$$

By (3.2), this does not exceed $\exp\{-\lambda(h)(S(x) - 2\gamma)\}$ for h sufficiently small. This gives us the necessary assertion. □

We note that we cannot replace conditions (I), (II) (under condition (0)) by relations (3.11). Relations (3.11) do not imply (II), as the following example shows: $X = R^1$, μ^h is a mixture, with equal weights, of the normal distribution with mean 0 and variance h and the normal distribution with mean 0 and variance $\exp\{e^{1/h^2}\}$, and $\lambda(h) = h^{-1}$. Here $\lim_{\delta \downarrow 0} \lim_{h \downarrow 0} \lambda(h)^{-1} \ln \mu^h\{y: \rho(x, y) < \delta\} = -x^2/2$ for all x; the function $x^2/2$ satisfies (0) but (II) is not satisfied.

We formulate, in a general language, the methods used for the verification of conditions (I) and (II) in §2 (they will be used in §4 and §§1 and 2 of Ch. 5 as well). To deduce (I), we select a function $g_x(y)$ on X so that the family of measures $\tilde{\mu}^h(dy) = \exp\{\lambda(h) g_x(y)\}\mu^h(dy)$ converge to a measure concentrated at x; then we use the fact that

$$\mu^h\{y: \rho(x, y) < \delta\} = \int_{\{y: \rho(x, y) < \delta\}} \exp\{-\lambda(h) g_x(y)\}\tilde{\mu}^h(dy).$$

On a part of the domain of integration with a sufficiently large $\tilde{\mu}^h$-measure we estimate $g_x(y)$ from above: $g_x(y) \leq g_x(x) + \gamma$ (we use Chebyshev's inequality) and we take $g_x(x)$ as $S(x)$. This method was used by Cramér [1] to obtain precise rather than rough results in the study of distributions of sums of independent random variables.

We have obtained condition (II) in the following way: we chose a function $\tilde{x}(x)$ such that its values belong to $\{y: S(y) < \infty\}$ and a set A such that $\rho(\tilde{x}(x), x) \leq \delta$ for $x \in A$. Further we used the inequality

$$\mu^h\{x: \rho(x, \Phi(s)) \geq \delta\} \leq \mu^h(X \setminus A) + \mu^h\{x \in A: \tilde{x}(x) \notin \Phi(s)\}.$$

We estimated the first term by means of Kolmogorov's exponential inequality and the second term also by means of Chebyshev's exponential inequality:

$$\begin{aligned}\mu^h\{x \in A: \tilde{x}(x) \notin \Phi(s)\} &= \mu^h\{x \in A: S(\tilde{x}(x)) > s\} \\ &\leq \int_A \exp\{(1 - \varkappa)\lambda(h)S(\tilde{x}(x))\}\mu^h(dx) \\ &\quad \times \exp\{-(1 - \varkappa)\lambda(h)s\}.\end{aligned}$$

Of course, the problem of estimating the integral on A is solved separately in each case.

Finally, we discuss the peculiarities arising in considering families of measures depending, besides the main parameter h, on another parameter x assuming values in a space X (X will have the meaning of the point from which the trajectory of the perturbed dynamical system is issued; we restrict ourselves to the case of perturbations which are homogeneous in time). First, we shall consider not just one space of functions and one metric; instead for any $T > 0$, we shall consider a space of functions defined on the interval $[0, T]$ and a metric ρ_{0T}. Correspondingly, the normalized action functional will depend on the interval: $S = S_{0T}$. (In the case of perturbations considered in Chs. 4, 5, and 7, this functional turns out to have the form $S_{0T}(\varphi) = \int_0^T L(\varphi_t, \dot{\varphi}_t)\,dt$, and it is convenient to define it analogously for all intervals $[T_1, T_2]$, $-\infty \leq T_1 < T_2 \leq +\infty$ on the real line.)

Moreover, to every point $x \in X$ and every value of the parameter h there will correspond its own measure in the space of functions on the interval $[0, T]$. We can follow two routes: we either consider a whole family of functionals depending on the parameter x (the functional corresponding to x is assumed to be equal to $+\infty$ for all functions $\dot{\varphi}_t$ for which $\varphi_0 \neq x$) or introduce a new definition of the action functional suitable for the situation. We follow the second route.

Let $\{\Omega^h, \mathscr{F}^h, \mathsf{P}_x^h\}$ be a family of probability spaces depending on the parameters $h > 0$ and x running through the values of a metric space X and let X_t^h, $t \geq 0$ be a random process on this space with values in X. Let ρ_{0T} be a metric in the space of functions on $[0, T]$ with values in X and let $S_{0T}(\varphi)$ be a functional on this space. We say that $\lambda(h)S_{0T}(\varphi)$ is the action functional for the family of the random processes (X_t^h, P_x^h) uniformly in a class

$\mathscr{A}$ of subsets of X if we have:

(0_c) the functional S_{0T} is lower semicontinuous and the sum of the $\Phi_x(s)$ for $x \in K$ is compact for any compact set $K \subseteq X$, where $\Phi_x(s)$ is the set of functions on $[0, T]$ such that $\varphi_0 = x$ and $S_{0T}(\varphi) \leq s$;

(I_u) for any $\delta > 0$, any $\gamma > 0$, any $s_0 > 0$ and any $A \in \mathscr{A}$ there exists an $h_0 > 0$ such that

$$\mathsf{P}_x^h\{\rho_{0T}(X^h, \varphi) < \delta\} \geq \exp\{-\lambda(h)[S_{0T}(\varphi) + \gamma]\} \tag{3.12}$$

for all $h < h_0$, all $x \in A$ and all $\varphi \in \Phi_x(s_0)$;

(II_u) for any $\delta > 0$, $\gamma > 0$, $s_0 > 0$ and any $A \in \mathscr{A}$ there exists an $h_0 > 0$ such that

$$\mathsf{P}_x^h\{\rho_{0T}(X^h. \Phi_x(s)) \geq \delta\} \leq \exp\{-\lambda(h)(s - \gamma)\}. \tag{3.13}$$

for all $h \leq h_0$, $s \leq s_0$ and $x \in A$.

For example, in Ch. 4 we prove that $(1/2\varepsilon^2) \int_0^T |\dot{\varphi}_s - b(\varphi_s)|^2\, ds$ is an action functional for the family of processes given by the stochastic equation $\dot{X}_t^\varepsilon = b(X_t^\varepsilon) + \varepsilon \dot{w}_t$, uniformly on the whole space as $\varepsilon \to 0$.

§4. Action Functional for Gaussian Random Processes and Fields

Let X_t be a Gaussian random process defined for $0 \leq t \leq T$, with values in R^r and having mean zero and correlation matrix $a(s, t) = (a^{ij}(s, t))$, $a^{ij}(s, t) = \mathsf{M}X_s^i X_t^j$. The functions $a^{ij}(s, t)$ will be assumed to be square integrable on $[0, T] \times [0, T]$.

We denote by A the correlation operator of X_t acting in the Hilbert space $\mathbf{L}_{0T}^2$ of functions on $[0, T]$ with values in R^r: $Af_t = \int_0^T a(s, t) f_s\, ds$. The scalar product in this space is $(f, g) = \int_0^T \sum_{i=1}^r f^i(s) g^i(s)\, ds$, and the norm is $\|f\| = (f, f)^{1/2}$. We retain the notation (f, g) for the integral $\int_0^T (f_s, g_s)\, ds$ in the case where the functions do not belong to $\mathbf{L}_{0T}^2$ but the integral is defined in some sense. For example, if $\dot{w}_t$ is an r-dimensional white noise process, then $(f, \dot{w}) = \int_0^T f(s)\, dw_s$.

We assume that A has finite trace. As is known (Gikhman and Skorokhod [2], Ch. V, §5), this implies that the trajectories of X_t belong to $\mathbf{L}_{0T}^2$. We denote by $A^{1/2}$ the nonnegative symmetric square root of A. The operator $A^{1/2}$ as well as A are integral operators with a square integrable kernel. In order to reconstruct this kernel, we consider the eigenfunctions $e_1(t), \ldots, e_n(t), \ldots$ of A and the corresponding eigenvalues $\lambda_1, \ldots, \lambda_n, \ldots$. Since $a(s, t)$ is a correlation function, A is nonnegative definite, i.e., $\lambda_k \geq 0$. We put $G(s, t) = \sum \lambda_k^{1/2} e_k(s) e_k(t)$. The operator A has finite trace $\sum \lambda_k < \infty$. Therefore, the series $\sum \lambda_k^{1/2} e_k(s) e_k(t)$ is convergent in $\mathbf{L}_{[0,T]\times[0,T]}^2$. The function $G(s, t)$ is precisely the kernel of the integral operator $A^{1/2}$. In order to see this, it is sufficient to note that the eigenfunctions of the operator with kernel $G(s, t)$ coincide with the eigenfunctions of A and its eigenvalues are square roots of the corresponding eigenvalues of A.

The random process X_t admits an integral representation in terms of the kernel of $A^{1/2}$:

$$X_t = \int_0^T G(s, t)\, dw_s,$$

where w_s is an r-dimensional Wiener process. We shall write this equality in the form $X_t = A^{1/2}\dot{w}_t$. In order to prove it, it is sufficient to note that $\tilde{X} = A^{1/2}\dot{w}$ is a Gaussian process with $\mathsf{M}\tilde{X}_t = 0$ and

$$\mathsf{M}\tilde{X}_s\tilde{X}_t = \int_0^T G(s, u)G(t, u)\, du = a(s, t).$$

The operators A and $A^{1/2}$ nullify some subspace $\mathbf{L}_0 \subseteq \mathbf{L}^2_{0T}$. This subspace is nontrivial in general, and therefore, for a definition of the inverses A^{-1} and $A^{-1/2}$, additional agreements are needed. We define the inverses by setting $A^{-1}\varphi = \psi_1$, $A^{-1/2}\varphi = \psi_2$ if $A\psi_1 = \varphi$, $A^{1/2}\psi_2 = \varphi$ and ψ_1, ψ_2 are orthogonal to $\mathbf{L}_0$. Consequently, A^{-1} and $A^{-1/2}$ are defined uniquely on the ranges of A and $A^{1/2}$, respectively.

On $\mathbf{L}^2_{0T}$ we consider the functional $S(\varphi)$ equal to $\frac{1}{2}\|A^{-1/2}\varphi\|^2$; if $A^{-1/2}\varphi$ is not defined, we set $S(\varphi) = +\infty$. For functions φ for which A^{-1} is defined, we have $S(\varphi) = \frac{1}{2}(A^{-1}\varphi, \varphi)$, i.e., $S(\varphi)$ is an extension of the functional $\frac{1}{2}(A^{-1}\varphi, \varphi)$.

Theorem 4.1. *The functional $S(\varphi)$ is a normalized action functional for the Gaussian process $X_t^\varepsilon = \varepsilon X_t$ in $\mathbf{L}^2_{0T}$ as $\varepsilon \to 0$. The normalizing function is $\lambda(\varepsilon) = \varepsilon^{-2}$.*

Proof. Let $G_N(s, t)$ be a symmetric positive definite continuously differentiable kernel on the square $[0, T] \times [0, T]$ such that

$$\int_0^T \int_0^T [G(s, t) - G_N(s, t)]^2\, ds\, dt < \frac{1}{N},$$

where $G(s, t)$ is the kernel of $A^{1/2}$. We denote by G_N the operator with kernel $G_N(s, t)$. First we verify the inequality

$$\mathsf{P}\{\|X^\varepsilon - \varphi\| < \delta\} \geq \exp\{-\varepsilon^{-2}(S(\varphi) + \gamma)\} \tag{4.1}$$

for any $\delta, \gamma > 0$ for sufficiently small positive ε. If $S(\varphi) = +\infty$, then (4.1) is obvious. Let $S(\varphi) < \infty$. There exists a $\psi \in \mathbf{L}^2_{0T}$ orthogonal to $\mathbf{L}_0$ and such that $A^{1/2}\psi = \varphi$. We put $\varphi_N = G_N\psi$ and $X_N = G_N\dot{w}$. Choose N_0 so large that $\|\varphi_N - \varphi\| \le \|\psi\|(\int_0^T\int_0^T |G(s, t) - G_N(s, t)]^2\, ds\, dt)^{1/2} < \delta/3$ for $N \ge N_0$. For such N we have

$$\{\|\varepsilon X - \varphi\| < \delta\} \supseteq \{\|\varepsilon X - \varepsilon X_N\| < \delta/3, \|\varepsilon X_N - \varphi_N\| < \delta/3\}.$$

From this we obtain

$$\mathsf{P}\{\|\varepsilon X - \varphi\| < \delta\} \ge \mathsf{P}\left\{\|\varepsilon X_N - \varphi_N\| < \frac{\delta}{3}\right\} - \mathsf{P}\{\|\varepsilon X - \varepsilon X_N\| \ge \delta/3\}. \tag{4.2}$$

Since $G_N(s, t)$ is a continuously differentiable kernel, in estimating the first term of the right side, we can use the result of Example 3.2:

$$\mathsf{P}\{\|\varepsilon X_N - \varphi_N\| < \delta/3\} \ge \exp\{-\varepsilon^{-2}(S(\varphi) + \gamma)\} \tag{4.3}$$

for ε smaller than some ε_1. Here we have used the fact that

$$G_N^{-1}\varphi_N = \psi = A^{-1/2}\varphi, \qquad S(\varphi) = \tfrac{1}{2}\|\psi\|^2 = \tfrac{1}{2}\|G_N^{-1}\varphi_N\|^2.$$

Moreover, using Chebyshev's inequality, we obtain

$$\begin{aligned} \mathsf{P}\{\|\varepsilon X - \varepsilon X_N\| \ge \delta/3\} &= \mathsf{P}\left\{\int_0^T \left[\int_0^T (G(s, t) - G_N(s, t))\, dw_s\right]^2 dt \ge \frac{\delta^2}{9\varepsilon^2}\right\} \\ &\le e^{-a\varepsilon^{-2}}\mathsf{M}\exp\left\{\frac{9a}{\delta^2}\int_0^T \left[\int_0^T (G(s, t) - G_N(s, t))\, dw_s\right]^2 dt\right\} \end{aligned} \tag{4.4}$$

for an arbitrary $a > 0$. For sufficiently large N, the mathematical expectation on the right side is finite. To see this, we introduce the eigenfunctions φ_k and corresponding eigenvalues μ_k, $k = 1, 2, \ldots$ of the symmetric square integrable kernel $\Gamma_N(s, t) = G(s, t) - G_N(s, t)$. It is easy to verify the following equalities:

$$\begin{aligned} \Gamma_N(s, t) &= \sum \mu_k \varphi_k(s)\varphi_k(t), \\ \sum \mu_k^2 &= \int_0^T\int_0^T \Gamma_N(s, t)^2\, ds\, dt, \\ \int_0^T \Gamma_N(s, t)\, dw_s &= \sum_k \mu_k\varphi_k(t)\int_0^T \varphi_k(s)\, dw_s, \\ \int_0^T\left[\int_0^T \Gamma_N(s, t)\, dw_s\right]^2 dt &= \sum_k \mu_k^2\left[\int_0^T \varphi_k(s)\, dw_s\right]^2 \end{aligned}$$

The random variables $\xi_k = \int_0^T \varphi_k(s)\, dw_s$ have normal distribution with mean zero and variance one and they are independent for distinct k. Taking account of these properties, we obtain

$$\mathsf{M} \exp\left\{\frac{9a}{\delta^2} \int_0^T \left[\int_0^T \Gamma_N(s, t)\, dw_s\right]^2 dt\right\} = \mathsf{M} \exp\left\{\frac{9a}{\delta^2} \sum \mu_k^2 \xi_k^2\right\}$$

$$= \prod_{k=1}^{\infty} \mathsf{M} \exp\left\{\frac{9a}{\delta^2} \mu_k^2 \xi_k^2\right\} = \prod_{k=1}^{\infty} \left(1 - \frac{18a}{\delta^2} \mu_k^2\right)^{-1/2} \tag{4.5}$$

The last equality holds if $(18a/\delta^2)\mu_k^2 < 1$ for all k. Since

$$\sum \mu_k^2 = \int_0^T \int_0^T \Gamma_N(s, t)^2\, ds\, dt \to 0$$

as $N \to \infty$, relation (4.5) is satisfied for all N larger than some $N_1 = N_1(a, \delta)$. The convergence of the series $\sum \mu_k^2$ implies the convergence of the infinite product in (4.5). Consequently, if $N > N_1$, then the mathematical expectation on the right side of (4.4) is finite and for $a = S(\varphi) + \gamma$ and for ε sufficiently small we obtain

$$\mathsf{P}\{\|\varepsilon X - \varepsilon X_N\| \geq \delta/3\} \leq \text{const} \cdot \exp\{-\varepsilon^{-2}(S(\varphi) + \gamma)\}. \tag{4.6}$$

Combining estimates (4.2), (4.3), and (4.6), we obtain (4.1).

Now we prove that for any $s > 0$, $\gamma > 0$, $\delta > 0$ there exists an ε_0 such that

$$\mathsf{P}\{\rho(\varepsilon X, \Phi(s)) \geq \delta\} \leq \exp\{-\varepsilon^{-2}(s - \gamma)\} \tag{4.7}$$

for $\varepsilon \leq \varepsilon_0$, where $\Phi(s) = \{\varphi \in \mathbf{L}_{0T}^2 : S(\varphi) \leq s\}$.

Along with the image $\Phi(s)$ of the ball of radius $\sqrt{2s}$ in $\mathbf{L}_{0T}^2$ under the mapping $A^{1/2}$, we consider the image $\Phi_N(s)$ of the same ball under G_N. Let $N > 6s/\delta$. Taking account of the definition of $G_N(s, t)$, we obtain

$$\begin{aligned}\sup_{\varphi: \|\varphi\| \leq \sqrt{2s}} \|G\varphi - G_N \varphi\|^2 &= \sup_{\varphi: \|\varphi\| \leq \sqrt{2s}} \int_0^T \left[\int_0^T (G(s, t) - G_N(s, t))\varphi_s\, ds\right]^2 dt \\ &\leq \sup_{\varphi: \|\varphi\| \leq \sqrt{2s}} \int_0^T \int_0^T (G(s, t) - G_N(s, t))^2\, ds\, dt \|\varphi\|^2 \\ &\leq \delta/3.\end{aligned}$$

From this we obtain

$$\begin{aligned}\mathsf{P}\{\rho(\varepsilon X, \Phi(s)) \geq \delta\} \leq\ & \mathsf{P}\{\|\varepsilon X - \varepsilon X_N\| \geq \delta/3\} \\ & + \mathsf{P}\{\rho(\varepsilon X_N, \Phi_N(s)) \geq \delta/3\}\end{aligned} \tag{4.8}$$

for $N > 6s/\delta$. By virtue of (4.4), the first term on the right side can be made smaller than $\exp\{-\varepsilon^{-2}(s - \gamma/2)\}$ for N sufficiently large. The estimate

$$\mathsf{P}\{\rho(\varepsilon X_N, \Phi_N(s)) \geq \delta/3\} \leq \exp\{-\varepsilon^{-2}(s - \gamma/2)\}$$

of the second term follows from Example 3.2. Relying on these estimates, we can derive (4.7) from (4.8). Theorem 4.1 is proved. □

Remark. As was established in the preceding section, estimates (4.1) and (4.7) are satisfied for sufficiently small ε uniformly in all functions φ with $S(\varphi) \leq$ const and for all $s \leq s_0 < \infty$, respectively.

In the above proof, taken from Freidlin [1], the main role was played by the representation $X_t = A^{1/2}\dot{w}_t$ of the Gaussian process X_t. This representation enabled us to reduce the calculation of the action functional for the family of processes εX_t in $\mathbf{L}^2_{0T}$ to the estimates obtained in §2 of the corresponding probabilities for the Wiener process (for the uniform metric). We are going to formulate and prove a theorem very close to Theorem 4.1; nevertheless, we shall not rely on the estimates for the Wiener process in the uniform metric but rather reproduce proofs from §3.2 in a Hilbert space setting (cf. Wentzell [4]). This enables us to write out an expression of the action functional for Gaussian random processes and fields in various Hilbert norms. Of course, a Gaussian random field X_z can also be represented in the form $\int_0^T G(s, z)\, dw_s$ and we can use the arguments in the proof of Theorem 4.1. Nevertheless, this representation is not as natural for a random process.

Let $\mathbf{H}$ be a real Hilbert space. We preserve the notation (,) and $\| \;\|$ for the scalar product and norm of $\mathbf{H}$. In $\mathbf{H}$ we consider a Gaussian random element X with mean 0 and correlation functional $B(f, g) = \mathsf{M}(X, f)(X, g)$. This bilinear functional can be represented in the form $B(f, g) = (Af, g)$ where A is a self-adjoint linear operator which turns out to be automatically nonnegative definite and completely continuous with a finite trace (Gikhman and Skorokhod [2], Ch. V, §5). As earlier, we write $S(\varphi) = \frac{1}{2}\|A^{-1/2}\varphi\|^2$. If $A^{-1/2}\varphi$ is not defined, we set $S(\varphi) = +\infty$. In order to make $A^{-1/2}$ single-valued, as $A^{-1/2}\varphi$ we again choose that element ψ which is orthogonal to the null space of A and for which $A^{1/2}\psi = \varphi$.

Theorem 4.2. *Let s, δ and γ be arbitrary positive numbers. We have*

$$\mathsf{P}\{\|\varepsilon X - \varphi\| < \delta\} \geq \exp\{-\varepsilon^{-2}(S(\varphi) + \gamma)\} \tag{4.9}$$

for $\varepsilon > 0$ sufficiently small. Inequality (4.9) *is satisfied uniformly for all φ with $S(\varphi) \leq s < \infty$. If $\Phi(s) = \{\varphi \in H: S(\varphi) \leq s\}$*, then

$$\mathsf{P}\{\rho(\varepsilon X, \Phi(s)) \geq \delta\} \leq \exp\{-\varepsilon^{-2}(s - \gamma)\} \tag{4.10}$$

for $\varepsilon > 0$ *sufficiently small. Inequality* (4.10) *is satisfied uniformly for all* $s \le s_0 < \infty$.

Proof. Let e_i, $i = 1, 2, \ldots$ be orthonormal eigenfunctions of A and let λ_i be the corresponding eigenvalues. We denote by X_i and φ_i the coordinates of X and φ in the basis $e_1, e_2, \ldots$. Here $X_i = (X, e_i)$, $i = 1, 2, \ldots$, are independent Gaussian random variables with mean zero and variance $\mathsf{M}X_i^2 = \mathsf{M}(X, e_i)^2 = (Ae_i, e_i) = \lambda_i$. The functional $S(\varphi)$ can be represented in the following way: $S(\varphi) = \frac{1}{2}\|A^{-1/2}\varphi\|^2 = \frac{1}{2}\sum(\varphi_i^2/\lambda_i)$. We assume that $S(\varphi) < \infty$. Then the joint distribution of the Gaussian random variables $X_j - \varepsilon^{-1}\varphi_j$, $i = 1, 2, \ldots$, has a density p with respect to the distribution of the variables X_i, $i = 1, 2, \ldots$:

$$p(x_1, \ldots, x_n, \ldots) = \prod_{i=1}^{\infty} \exp\left\{-\lambda_i^{-1}\varepsilon^{-1}\varphi_i x_i - \frac{\lambda_i^{-1}}{2}\varepsilon^{-2}\varphi_i^2\right\}.$$

Therefore,

$$\begin{aligned}\mathsf{P}\{\|\varepsilon X - \varphi\| < \delta\} &= \mathsf{P}\left\{\sum_{i=1}^{\infty}(X_i - \varepsilon^{-1}\varphi_i)^2 < (\delta/\varepsilon)^2\right\} \\ &= \mathsf{M}\{\textstyle\sum X_i^2 < (\delta/\varepsilon)^2; p(X_1, X_2, \ldots)\} \\ &= \mathsf{M}\left\{\|X\|^2 < (\delta/\varepsilon)^2;\right. \\ &\qquad \left.\times \exp\left\{-\varepsilon^{-1}\sum_{i=1}^{\infty}\lambda_i^{-1}\varphi_i X_i - \varepsilon^2 S(\varphi)\right\}\right\}.\end{aligned} \tag{4.11}$$

Using Chebyshev's inequality, we find that

$$\mathsf{P}\{\|X\|^2 < (\delta/\varepsilon)^2\} \ge 1 - \varepsilon^2\delta^{-2}\mathsf{M}\|X\|^2 = 1 - \varepsilon^2\delta^{-2}\sum_{i=1}^{\infty}\lambda_i \ge 3/4$$

for $\varepsilon \le 2\delta(\sum_{i=1}^{\infty}\lambda_i)^{-1/2}$ and

$$\begin{aligned}\mathsf{P}\{|\textstyle\sum \lambda_i^{-1}\varphi_i X_i| < K\} &\ge 1 - K^{-2}\mathsf{M}\left(\displaystyle\sum_{i=1}^{\infty}\lambda_i^{-1}\varphi_i X_i\right)^2 \\ &= 1 - K^{-2}\sum_{i=1}^{\infty}\lambda_i^{-1}\varphi_i^2 = 1 - 2K^{-2}S(\varphi) \ge 3/4\end{aligned}$$

for $K \ge 2\sqrt{2}\sqrt{S(\varphi)}$. It follows from these inequalities that the random variable under the sign of mathematical expectation in (4.11) is greater than $\exp\{-\varepsilon^{-2}S(\varphi) - \varepsilon^{-1}K\}$ with probability not smaller than 3/4. This implies inequality (4.9).

Now we prove the second assertion of the theorem. We denote by $\tilde{X}$ the random vector with coordinates $(X_1, X_2, \ldots, X_{i_0}, 0, 0, \ldots)$. The choice of the index i_0 will be specified later. It is easy to verify that

$$\mathsf{P}\{\rho(\varepsilon X, \Phi(s)) \geq \delta\} \leq \mathsf{P}\{\varepsilon\tilde{X} \notin \Phi(s)\} + \mathsf{P}\{\rho(X, \tilde{X}) \geq \delta/\varepsilon\}. \quad (4.12)$$

The first probability is equal to

$$\mathsf{P}\{S(\varepsilon\tilde{X}) > s\} = \mathsf{P}\{S(\tilde{X}) > s\varepsilon^{-2}\} = \mathsf{P}\left\{\sum_{i=1}^{i_0} \lambda_i^{-1} X_i^2 > 2s\varepsilon^{-2}\right\}. \quad (4.13)$$

The random variable $\sum_{i=1}^{i_0} \lambda_i^{-1} X_i^2$ is the sum of squares of i_0 independent normal random variables with parameters (0, 1), and consequently, has a χ^2-distribution with i_0 degrees of freedom. Using the expression for the density of a χ^2-distribution, we obtain

$$\begin{aligned}\mathsf{P}\{\varepsilon\tilde{X} \notin \Phi(s)\} &= \mathsf{P}\left\{\sum_{i=1}^{i_0} \lambda_i^{-1} X_i^2 > 2s\varepsilon^{-2}\right\} \\ &= \int_{2s\varepsilon^{-2}}^{\infty} \frac{1}{\Gamma(i_0/2)2^{i_0/2}} x^{i_0/2-1} e^{-x/2}\, dx \\ &\leq \text{const} \cdot \varepsilon^{-i_0} \exp\{-\varepsilon^{-2} s\}\end{aligned} \quad (4.14)$$

for $\varepsilon > 0$ sufficiently small. The second probability in (4.12) can be estimated by means of Chebyshev's exponential inequality:

$$\mathsf{P}\{\rho(X, \tilde{X}) > \delta/\varepsilon\} \leq \exp\left\{-\frac{c}{2}(\delta/\varepsilon)^2\right\} \mathsf{M} \exp\left\{\frac{c}{2}\rho(X, \tilde{X})^2\right\}. \quad (4.15)$$

The mathematical expectation on the right side is finite for sufficiently large i_0. This can be proved in the same way as the finiteness of the mathematical expectation in (4.4); we have to take account of the convergence of the series $\sum_{i=1}^{\infty} \lambda_i$. Substituting $c = 2s\delta^{-2}$ in (4.15), we have

$$\mathsf{P}\{\rho(X, \tilde{X}) > \delta/\varepsilon\} \leq \text{const} \cdot \exp\{-s\varepsilon^{-2}\} \quad (4.16)$$

for sufficiently large i_0. Combining formulas (4.13), (4.14) and (4.16), we obtain the last assertion of the theorem. □

This theorem enables us to calculate the action functional for Gaussian random processes and fields in arbitrary Hilbert norms. It is only required that the realizations of the process belong to the corresponding Hilbert space. In many examples, for example, in problems concerning the crossing of a level by a random process or field, it is desirable to have estimations in the uniform norm. We can use imbedding theorems to obtain such estimations.

Let D be a bounded domain in R^r with smooth boundary. Let us denote by $\mathbf{W}_2^l$ the function space on D obtained from the space of infinitely differentiable functions in D by completing it in the norm

$$\|u\|_{W_2^l} = \left(\sum_{|q| \leq l} \int_D (u^{(q)}(x))^2 \, dx \right)^{1/2},$$

where $q = (q_1, \ldots, q_r)$, $|q| = \sum q_i$ and $u^{(q)} = \partial^{|q|} u / \partial x_1^{q_1} \cdots \partial x_r^{q_r}$. The space $\mathbf{W}_2^l$ with this norm is a separable Hilbert space (Sobolev [1]). Roughly speaking, $\mathbf{W}_2^l$ consists of functions having square integrable derivatives of order l. In order that the realizations of a Gaussian random field X_z, $z \in D \subset R^r$ with mean zero and correlation function $a(u, v) = \mathsf{M} X_u X_v$ belong to $\mathbf{W}_2^l(D)$, it is sufficient (cf., for example, Gikhman and Skorokhod [1]) that the correlation function have continuous derivatives up to order $2l$ inclusive. Let m be a multiindex $(m_1, \ldots, m_r)$, $m_t \geq 0$, and let $|m| = m_1 + m_2 + \cdots + m_r \leq l - r/2$. For all $x \in D$ we have the estimation

$$\left| \frac{\partial^{|m|} u(x)}{\partial x_1^{m_1} \cdots \partial x_r^{m_r}} \right| \leq \text{const} \cdot \|u\|_{W_2^l}.$$

This inequality comprises the content of an imbedding theorem (cf. Sobolev [1] or Ladyzhenskaya and Ural'tseva [1], Theorem 2.1). It follows easily from this that if the correlation function of a random field has continuous derivatives of order $2l$ in $D \cup \partial D$, then estimates (4.9) and (4.10) are satisfied in the metric of $\mathbf{C}_D^{(m)}$ for $m < l < r/2$. Moreover, the functional $S(\varphi)$ is defined by the equality $S(\varphi) = \frac{1}{2}\|A^{-1/2}\varphi\|^2$, where A is the correlation operator. As above, $A^{-1/2}$ is made single-valued by requiring the orthogonality of $A^{-1/2}\varphi$ to the null space of A. If $A^{-1/2}\varphi$ is not defined, then $S(\varphi) = +\infty$. In particular, for $2l > r$, estimates (4.9) and (4.10) hold in the norm of $\mathbf{C}_D$.

If we apply sharper imbedding theorems to obtain estimates in $\mathbf{C}_D$, we can reduce the requirements on the smoothness of the correlation function.

Lemma 4.1.

(a) *The set* $\Phi(s) = \{\varphi \in \mathbf{H}: S(\varphi) \leq s\}$, $s < \infty$ *is compact in* $\mathbf{H}$;
(b) *the functional* $S(\varphi)$ *is lower semicontinuous, i.e., if* $\|\varphi_n - \varphi\| \to 0$ *as* $n \to \infty$, *then* $S(\varphi) \leq \underline{\lim}_{n \to \infty} S(\varphi_n)$.

Proof. First we prove (b). We consider the Hilbert space $\mathbf{H}_1 \subset \mathbf{H}$ obtained by completing the domain $D_{A^{-1/2}}$ in the norm $\|f\|_1 = \|A^{-1/2} f\|$. This expression indeed defines a norm, since $A^{-1/2}$ is linear and does not vanish on nonzero elements. It is sufficient to prove our assertion for a sequence

φ_n such that $\lim_{n\to\infty} S(\varphi_n)$ exists and is finite. For such a sequence, $\|\varphi_n\|_1 = [2S(\varphi_n)]^{1/2} \le \text{const} < \infty$ for all n. Since the set $\{\varphi_n\}$ is bounded, it is weakly compact, i.e., there exists an element $\tilde{\varphi} \in \mathbf{H}_1$ such that some subsequence φ_{n_i} converges weakly to $\tilde{\varphi}$ in $\mathbf{H}_1$: $(\varphi_{n_i}, f)_{H_1} = (A^{-1/2}\varphi_{n_i}, A^{-1/2}f) \to (\tilde{\varphi}, f)_{H_1} = (A^{-1/2}\tilde{\varphi}, A^{-1/2}f)$ as $i \to \infty$ for any $f \in \mathbf{H}_1$. This implies that φ_{n_i} converges weakly to $\tilde{\varphi}$ in $\mathbf{H}$. Indeed, let $g \in \mathbf{H}$. We have $Ag \in \mathbf{H}_1$ and

$$
\begin{aligned}
(\varphi_{n_i}, g) &= (A^{1/2}A^{-1/2}\varphi_{ni}, g) = (A^{-1/2}\varphi_{ni}, A^{1/2}g) \\
&= (A^{-1/2}\varphi_{n_i}, A^{-1/2}Ag) = (\varphi_{n_i}, Ag)_{H_1} \to 0
\end{aligned}
$$

as $i \to \infty$. By virtue of the uniqueness of weak limits in $\mathbf{H}$ we obtain that $\tilde{\varphi} = \varphi$. Now we obtain assertion (b) of the lemma from the lower semicontinuity of the norm in the weak topology.

Assertion (a) of the lemma follows from (b) and from the fact that A, and thus $A^{1/2}$, are completely continuous. □

Remark. It follows from the proof that $S(\varphi)$ is lower semicontinuous in the topology of weak convergence in $\mathbf{H}$.

We indicate some more properties of $S(\varphi)$, clarifying its probabilistic meaning.

Theorem 4.3. *Let $\varphi \in D_{A^{-1/2}}$. We have*

$$
\lim_{\delta\downarrow 0} \lim_{\varepsilon\downarrow 0} \varepsilon^2 \ln \mathsf{P}\{\|\varepsilon X - \varphi\| < \delta\} = -S(\varphi); \tag{4.17}
$$

$$
\lim_{\delta\downarrow 0} \frac{\mathsf{P}\{\|\varepsilon X - \varphi\| < \delta\}}{\mathsf{P}\{\|\varepsilon X\| < \delta\}} = \exp\{-\varepsilon^{-2}S(\varphi)\}. \tag{4.18}
$$

Proof. The first assertion was proved in a general setting §3. We prove the second assertion only for $\varphi \in D_{A^{-1}}$. (Concerning a proof for $\varphi \in D_{A^{-1/2}}$, cf. Sytaya [1]; there is a sharpening of the first assertion of the theorem there.) Using the notation of Theorem 4.2, from equality (4.11) we obtain

$$
\begin{aligned}
\mathsf{P}\{\|\varepsilon X - \varphi\| < \delta\} &= \mathsf{M}\left\{\|\varepsilon X\| < \delta; \exp\left\{\varepsilon^{-1}\sum_{i=1}^{\infty}\lambda_i^{-1}\varphi_i x_i - \varepsilon^{-2}S(\varphi)\right\}\right\} \\
&= \exp\{-\varepsilon^{-2}S(\varphi)\}\mathsf{M}\left\{\|\varepsilon X\| < \delta; \exp\left\{\varepsilon^{-1}\sum_{i=1}^{\infty}\lambda_i^{-1}\varphi_i X_i\right\}\right\}.
\end{aligned}
\tag{4.19}
$$

If $\varphi \in D_{A^{-1}}$, then $|\sum \lambda_i^{-1}\varphi_i X_i| = |(A^{-1}\varphi, X)| \le \|A^{-1}\varphi\| \, \|X\|$. From this we conclude that

$$\lim_{\delta \downarrow 0} \frac{\mathsf{M}\{\|\varepsilon X\| < \delta; \exp\{\varepsilon^{-1}\sum \lambda_i^{-1}\varphi_i X_i\}\}}{\mathsf{P}\{\|\varepsilon X\| < \delta\}} = 1.$$

The last equality and (4.19) imply the second assertion of the theorem. □

Remark. Relation (4.18) is also interesting in problems not containing a small parameter. It can be interpreted as an assertion stating that for a Gaussian random field (process) X, the functional $\exp\{-S(\varphi)\}$ plays the role of a nonnormalized probability density with respect to "uniform distribution in the Hilbert space **H**."

EXAMPLE 4.1. Let D be a bounded domain in R^r with smooth boundary ∂D. Let X_z be a Gaussian random field defined for $z \in D \cup \partial D$ and having mean zero and correlation function $a(z_1, z_2)$. We assume that $a(z_1, z_2)$ has continuous derivatives through order $r + 2$. Then the realizations of X_z have square integrable partial derivatives up to order $[r/2] + 1$ on D with probability 1. In other words, these realizations belong to $\mathbf{W}_2^{[r/2]+1}(D)$, and consequently, by the above imbedding theorem, $X \in \mathbf{C}_{D \cup \partial D}$ and

$$\|X\|_c \le \text{const} \cdot \|X\|_{\mathbf{W}_2^{[r/2]+1}}.$$

This implies that the functional $S(\varphi)$ defined as $\frac{1}{2}\|A^{-1/2}\varphi\|^2$ if $A^{-1/2}\varphi$ is defined and as $+\infty$ for the remaining φ, is a normalized action functional for the random field εX_z as $\varepsilon \to 0$ not only in $\mathbf{W}_2^{[r/2]+1}$ but also in $\mathbf{C}_{D \cup \partial D}$.

We put $G = \{\varphi \in \mathbf{C}_{D \cup \partial D}: \max_{z \in D \cup \partial D} |\varphi(z)| \ge 1\}$. We study the behavior of $\mathsf{P}\{\varepsilon X \in G\}$ as $\varepsilon \to 0$. Every element of the boundary of the closed set G can be moved into the interior of G by multiplying it by a number arbitrarily close to 1. Since $S(\alpha\varphi) = \alpha^2 S(\varphi)$, this implies the regularity of G with respect to $S(\varphi)$. Therefore, by Theorem 3.4 we have

$$\lim_{\varepsilon \to 0} \varepsilon^2 \ln \mathsf{P}\{\varepsilon X \in G\} = -\inf_{\varphi \in G} S(\varphi).$$

We calculate the infimum.

We denote by $e_k(z)$ and λ_k the eigenfunctions and the corresponding eigenvalues of the symmetric kernel $a(z_1, z_2)$. It is known that $a(z_1, z_2) = \sum \lambda_k e_k(z_1)e_k(z_2)$. We calculate the value of S at the function $\varphi_{z_0}(z) = a(z_0, z) = \sum \lambda_k e_k(z_0)e_k(z)$. We have

$$S(\varphi_{z_0}) = \tfrac{1}{2}\|A^{-1/2}\varphi_{z_0}\|^2 = \tfrac{1}{2}\sum \lambda_k^{-1}(\varphi_{z_0}, e_k)$$
$$= \tfrac{1}{2}\sum \lambda_k e_k^2(z_0) = \tfrac{1}{2}a(z_0, z_0).$$

Let the function $a(z, z)$ attain its maximum on $D \cup \partial D$ at the point $\hat{z}$. Put

$$\hat{\varphi}(z) = a(\hat{z}, \hat{z})^{-1} \varphi_{\hat{z}}(z).$$

At $\hat{z}$ the function $\hat{\varphi}$ assumes the value 1; $S(\hat{\varphi}) = \frac{1}{2} a(\hat{z}, \hat{z})^{-1}$. We show that $S(\hat{\varphi})$ is the infimum of $S(\varphi)$ on G. Indeed, let $\varphi(z) = \sum c_k e_k(z)$ be any function assuming a value not smaller than 1 at some point $\bar{z}$. Using the Cauchy inequality, we obtain

$$\begin{aligned} S(\varphi) \cdot a(\bar{z}, \bar{z}) &= \tfrac{1}{2} \sum \lambda_k^{-1} c_k^2 \sum \lambda_k e_k^2(\bar{z}) \\ &\geq \tfrac{1}{2} \sum c_k e_k(\bar{z}) = \tfrac{1}{2} \varphi(\bar{z}) \geq \tfrac{1}{2}. \end{aligned}$$

This implies

$$S(\varphi) \geq \tfrac{1}{2} a(\bar{z}, \bar{z})^{-1} \geq \tfrac{1}{2} a(\hat{z}, \hat{z})^{-1}. \tag{4.20}$$

Hence $\inf_{\varphi \in G} S(\varphi) = \frac{1}{2} a(\hat{z}, \hat{z})^{-1}$.

If we assume that $\hat{z}$ is the only absolute maximum point of the function $a(z, z)$ on $D \cup \partial D$, then the equality in (4.20) can be attained only for $c_k = \lambda_k e_k(\hat{z}) a(\hat{z}, \hat{z})^{-1}$, $k = 1, 2, \ldots$, i.e., the infimum of $S(\varphi)$ is attained only for $\varphi(z) = \hat{\varphi}(z)$. Theorem 3.4 yields

$$\lim_{\varepsilon \to 0} \varepsilon^2 \ln \mathsf{P}\{\varepsilon X \in G\} = -\tfrac{1}{2} a(\hat{z}, \hat{z})^{-1} = -\frac{1}{2} \left(\max_{z \in D \cup \partial D} a(z, z) \right)^{-1}$$

If $a(\hat{z}, \hat{z}) > a(z, z)$ for all remaining points z of $D \cup \partial D$, then

$$\lim_{\varepsilon \to 0} \mathsf{P}\left\{ \sup_z |\varepsilon X_z - \hat{\varphi}(z)| < \delta \,|\, \varepsilon X \in G \right\} = 1$$

for every $\delta > 0$.

In the case of a homogeneous random field X_z there exists no function φ along which the realizations of εX_z reach the level 1 with overwhelming probability as $\varepsilon \to 0$ under the condition that they reach it at all. Nevertheless, in this case, the functions $a(z_0, z)$ for various z_0 can be obtained from each other by shifts and it can be proved that

$$\lim_{\varepsilon \to 0} \mathsf{P}\left\{ \max_{z \in D \cup \partial D} \left| \varepsilon X_z - \frac{a(z - z_0)}{a(0)} \right| < \delta \,|\, \varepsilon X_{z_0} \geq 1 \right\} = 1,$$

where $\delta > 0$, $a(z) = a(z_0, z_0 + z)$.

Chapter 4

Gaussian Perturbations of Dynamical Systems. Neighborhood of an Equilibrium Point

§1. Action Functional

In this chapter we shall consider perturbations of a dynamical system

$$\dot{x}_t = b(x_t), \qquad x_0 = x, \qquad b(x) = (b^1(x), \ldots, b^r(x)) \tag{1.1}$$

by a white noise process or by a Gaussian process in general. Unless otherwise stated, we shall assume that the functions b^i are bounded and satisfy a Lipschitz condition: $|b(x) - b(y)| \le K|x - y|$, $|b(x)| \le K < \infty$. Here we pay particular attention to the case where the perturbed process has the form

$$\dot{X}_t^\varepsilon = b(X_t^\varepsilon) + \varepsilon \dot{w}_t, \qquad X_0^\varepsilon = x, \tag{1.2}$$

where w_t is an r-dimensional Wiener process. In Ch. 2 we mentioned that as $\varepsilon \to 0$, the processes X_t^ε converge in probability to the trajectories of the dynamical system (1.1), uniformly on every finite interval $[0, T]$. This result can be viewed as a version of the law of large numbers. In Ch. 2 there is also an assertion, concerning the processes X_t^ε, of the type of the central limit theorem: the normalized difference $(X_t^\varepsilon - x_t)$ converges to a Gaussian process. This result characterizes deviations of order ε from the limiting dynamical system. In this chapter we study the asymptotics of probabilities of large (or order 1) deviations for the family X_t^ε of processes and consider a series of problems relating to the behavior of the perturbed process on large time intervals. In the last section we consider large deviations for the processes X_t^ε, which are defined by the equations

$$\dot{X}_t^\varepsilon = b(X_t^\varepsilon, \varepsilon\zeta_t), \qquad X_0^\varepsilon = x, \tag{1.3}$$

where ζ_t is a Gaussian process in R^l and (b, x, y), $x \in R^r$, $y \in R^l$ is a continuous function for which $b(x, 0) = b(x)$.

Let ψ_t be a continuous function on $[0, T]$ with values in R^r. In $\mathbf{C}_{0T}(R^r)$ we consider the operator $B_x\colon \psi \to v$, where $v = v_t$ is the solution of the equation

$$v_t = x + \int_0^t b(v_s)\, ds + \psi_t, \qquad t \in [0, T]. \tag{1.4}$$

It is easy to prove that under the assumptions made concerning $b(x)$, the solution of equation (1.4) exists and is unique for any continuous function ψ and any $x \in R^r$. The operator B_x has the inverse

$$(B_x^{-1}v)_t = \psi_t = v_t - x - \int_0^t b(v_s)\, ds.$$

Lemma 1.1. *Suppose the function $b(x)$ satisfies the Lipschitz condition*

$$|b(x) - b(y)| \leq K|x - y|.$$

Then the operator B_x in $\mathbf{C}_{0T}(R^r)$ satisfies the Lipschitz condition

$$\|B_x\varphi - B_x\psi\| \leq e^{KT}\|\varphi - \psi\|; \qquad \varphi, \psi \in C_{0T}(R^r).$$

Proof. By the definition of B_x, for $u = B_x\varphi$, $v = B_x\psi$ we have

$$|u_t - v_t| = \left|\int_0^t (b(u_s) - b(v_s))\, ds + (\varphi_t - \psi_t)\right|$$

$$\leq K \int_0^t |u_s - v_s|\, ds + \|\varphi - \psi\|, \qquad t \in [0, T].$$

Relying on Lemma 1.1 of Ch. 2, this implies the assertion of Lemma 1.1. □

On $\mathbf{C}_{0T}(R^r)$ we consider the functional $S(\varphi) = S_{0T}(\varphi)$ which is defined by the equality

$$S_{0T}(\varphi) = \frac{1}{2}\int_0^T |\dot{\varphi}_s - b(\varphi_s)|^2\, ds$$

for absolutely continuous functions and we set $S_{0T}(\varphi) = +\infty$ for the remaining $\varphi \in \mathbf{C}_{0T}(R^r)$.

Theorem 1.1. *The functional $\varepsilon^{-2}S(\varphi)$ is the action functional for the family X_t^ε of processes defined by equation* (1.2) *in $\mathbf{C}_{0T}(R^r)$ as $\varepsilon \to 0$ uniformly with respect to the initial point $x \in R^r$.*

Proof. For every fixed x, the assertion follows from Theorem 3.1 of Ch. 3 and Lemma 1.1 if we take account of the form of the actional functional for the family of processes εw_t.

Indeed, since B_x is continuous in $\mathbf{C}_{0T}(R^r)$ and has an inverse, the action functional for the family of the processes $X^\varepsilon = B_x(\varepsilon w)$ has the form $\varepsilon^{-2}S(\varphi)$, where

$$S(\varphi) = \frac{1}{2}\int_0^T \left|\frac{d}{dt}(B_x^{-1}\varphi)_t\right|^2 dt = \frac{1}{2}\int_0^T |\dot{\varphi}_t - b(\varphi_t)|^2 dt,$$

if the function $(B_x^{-1}\varphi)_t = \varphi_t - x - \int_0^t b(\varphi_s)\,ds$ is absolutely continuous. It is clear that this function is absolutely continuous if and only if φ_t is so.

Now we have to prove uniformity in x. The fulfillment of formulas (3.11), and (3.12) of Ch. 3 for small values of the parameter for all $x \in R^r$ follows immediately from the uniform continuity of B_x in x (a Lipschitz condition with a constant independent of x). It remains to prove (0_c) of §3 of Ch. 3: the lower semicontinuity of S_{0T} and the compactness of $\bigcup_{x\in K}\mathbf{\Phi}_x(s)$ for any compact set K. This can be deduced from the fact that $(x, \psi) \to B_x\psi$ is a homeomorphism if we consider only ψ with $\psi = 0$. □

This implies in particular that the infimum of $S_{0T}(\varphi)$ on any bounded closed subset of $\mathbf{C}_{0T}(R^r)$ is attained and S_{0T} assumes values close to the smallest value only near functions at which the minimum is attained.

We note that if $S_{0T}(\varphi) = 0$, then the function φ defined on $[0,T]$ is a trajectory of the dynamical system (1.1), since in this case, φ is absolutely continuous and satisfies the condition $\dot{\varphi}_t = b(\varphi_t)$ almost everywhere on $[0, T]$.

We consider some simple applications of Theorem 1.1. Let D be a domain in R^r, let ∂D be its boundary, let $c(x)$ be a bounded continuous function on R^r, and let $g(x)$ be a bounded continuous function defined on ∂D.

We write $\tau^\varepsilon = \min\{t: X_t^\varepsilon \notin D\}$, where X_t^ε is a solution of equation (1.6) and

$$H_D(t, x) = \{\varphi \in \mathbf{C}_{0T}(R^r): \varphi_0 = x, \varphi_t \in D \cup \partial D\},$$

$$\bar{H}_D(t, x) = \{\varphi \in \mathbf{C}_{0T}(R^r): \varphi_0 = x, x_s \notin D \quad \text{for some } s \in [0, t]\}.$$

Theorem 1.2. *Suppose that the boundary of D coincides with the boundary of its closure. We have*

$$\lim_{\varepsilon\to 0} \varepsilon^2 \ln \mathsf{P}_x\{X_t^\varepsilon \in D\} = -\min_{\varphi\in H_D(t,x)} S_{0T}(\varphi), \tag{1.5}$$

$$\lim_{\varepsilon\to 0} \varepsilon^2 \ln \mathsf{P}_x\{\tau^\varepsilon \le t\} = -\min_{\varphi\in \bar{H}_D(t,x)} S_{0T}(\varphi). \tag{1.6}$$

If the extremal $\hat{\varphi}_s$ *providing the minimum of* $S_{0t}(\varphi)$ *on* $\bar{H}_D(t, x)$ *is unique and it assumes a value in* ∂D *at only one value* $\hat{s} \in [0, t]$, *then*

$$\lim_{\varepsilon \to 0} \frac{\mathsf{M}_x\{\tau^\varepsilon \le t; g(X^\varepsilon_{\tau^\varepsilon}) \exp\{\int_0^{\tau^\varepsilon} c(X^\varepsilon_s)\, ds\}\}}{\mathsf{P}_x\{\tau^\varepsilon \le t\}} = g(\hat{\varphi}_{\hat{s}}) \exp\left\{\int_0^{\hat{s}} c(\hat{\varphi}_s)\, ds\right\}. \quad (1.7)$$

Proof. Since $\varepsilon^{-2} S_{0T}(\varphi)$ is the action functional for the family of processes X_t^ε, relations (1.5) and (1.6) follow from Theorem 3.4 of Ch. 3. In seeing this, we have to use the regularity of $H_D(t, x)$ and $\bar{H}_D(t, x)$, which was proved in Example 3.5 of Ch. 3.

To prove (1.7), we consider the events $A_1^\delta = \{\tau^\varepsilon \le t, \rho_{0t}(X^\varepsilon, \hat{\varphi}) < \delta\}$, $A_2^\delta = \{\tau^\varepsilon \le t, \rho_{0t}(X^\varepsilon, \hat{\varphi}) \ge \delta\}$. On the set A_1^δ we have

$$\left| g(\hat{\varphi}_{\hat{s}}) \exp\left\{\int_0^{\hat{s}} c(\hat{\varphi}_s)\, ds\right\} - g(X^\varepsilon_{\tau^\varepsilon}) \exp\left\{\int_0^{\tau^\varepsilon} c(X^\varepsilon_s)\, ds\right\} \right| < \lambda_\delta, \quad (1.8)$$

where $\lambda_\delta \to 0$ as $\delta \downarrow 0$. This estimate follows from the circumstance that the time spent by the curve $\hat{\varphi}_s$ in the δ-neighborhood of the point $\hat{\varphi}_{\hat{s}} \in \partial D$ converges to zero as $\delta \downarrow 0$ and after the time $\hat{s}$, the extremal $\hat{\varphi}_s$ does not hit ∂D. The relations $A_1^\delta \cup A_r^\delta = \{\tau^\varepsilon \le t\}$ and (1.8) imply the inequality

$$\begin{aligned} &\left(g(\hat{\varphi}_{\hat{s}}) \exp\left\{\int_0^{\hat{s}} c(\hat{\varphi}_s)\, ds\right\} - \lambda_\delta\right)[\mathsf{P}_x\{\tau^\varepsilon \le t\} - \mathsf{P}_x(A_2^\delta)] \\ &\qquad - \|g\| e^{\|c\| t} \mathsf{P}_x(A_2^\delta) \le \mathsf{M}_x\left\{\tau^\varepsilon \le t; g(X^\varepsilon_{\tau^\varepsilon}) \exp\left\{\int_0^{\tau^\varepsilon} c(X^\varepsilon_s)\, ds\right\}\right\} \qquad (1.9) \\ &\qquad \le \left(g(\hat{\varphi}_{\hat{s}}) \exp\left\{\int_0^{\hat{s}} c(\hat{\varphi}_s)\, ds\right\} + \lambda_\delta\right) \mathsf{P}_x\{\tau^\varepsilon \le t\} + \|g\| e^{\|c\| t} \mathsf{P}_x(A_2^\delta), \end{aligned}$$

where $\|c\| = \sup|c(x)|$, $\|g\| = \sup|g(x)|$. Since $\hat{\varphi}$ is the only extremal of the action functional on $\bar{H}_D(t, x)$, the value of S_{0t} at functions reaching ∂D and being at a distance not smaller than δ from $\hat{\varphi}$ is greater than $S_{0t}(\hat{\varphi}) + \gamma$, where γ is some positive number. Using Theorem 1.1 (the upper estimate with $\gamma/2$ instead of γ), we obtain that

$$\mathsf{P}_x(A_2^\delta) \le \exp\{-\varepsilon^{-2}(S_{0t}(\hat{\varphi}) + \gamma/2)\}$$

for ε sufficiently small.

Taking account of relation (1.6), from this we conclude that

$$\lim_{\varepsilon \to 0} \mathsf{P}_x(A_2^\delta)/\mathsf{P}_x\{\tau^\varepsilon \le t\} = 0. \quad (1.10)$$

If we now divide inequality (1.9) by $\mathsf{P}_x\{\tau^\varepsilon \le t\}$ and take account of (1.10) and the fact that $\lim_{\delta \downarrow 0} \lambda_\delta = 0$, we obtain (1.7). □

Thus, the calculation of the principal term of the logarithmic asymptotics of probabilities of events concerning the process X_t^ε has been reduced to the solution of some variational problems. These problems are of a standard character. For the extremals we have the ordinary Euler equation and the minimum itself can be found conveniently by means of the Hamilton–Jacobi equation (cf., for example, Gel'fand and Fomin [1]). If we write

$$V(t, x, y) = \min_{\varphi_0 = x,\, \varphi_t = y} S_{0t}(\varphi),$$

then

$$\min_{\varphi \in H_D(t, x)} S_{0t}(\varphi) = \min_{y \in D \cup \partial D} V(t, x, y), \qquad \min_{\varphi \in \bar{H}_D(t, x)} S_{0t}(\varphi) = \min_{\substack{0 \le s \le t \\ y \notin D}} V(s, x, y).$$

The Hamilton–Jacobi equation for $V(t, x, y)$ has the form

$$\frac{\partial V(t,x,y)}{\partial t} + \tfrac{1}{2}|\nabla_y V(t,x,y)|^2 + (b(y), \nabla_y V(t,x,y)) = 0, \tag{1.11}$$

where ∇_y is the gradient operator in the variable y. We have to add the conditions $V(0, x, x) = 0$, $V(t, x, y) \ge 0$ to equation (1.11).

Concluding this section, we note that the functions

$$u^\varepsilon(t, x) = \mathsf{P}_x\{X_t^\varepsilon \in D\}, \qquad v^\varepsilon(t, x) = \mathsf{P}_x\{\tau^\varepsilon \le t\},$$

$$w^\varepsilon(t, x) = \mathsf{M}_x\left\{\tau^\varepsilon \le t;\, g(X_{\tau^\varepsilon}^\varepsilon) \exp\left\{\int_0^{\tau^\varepsilon} c(X_s^\varepsilon)\, ds\right\}\right\}$$

are the solutions of the following problems:

$$\frac{\partial u^\varepsilon}{\partial t} = \frac{\varepsilon^2}{2}\Delta u^\varepsilon + (b(x), \nabla_x u^\varepsilon), \qquad x \in R^r, \quad t > 0;$$

$$u^\varepsilon(0, x) = 1 \quad \text{for } x \in D,$$

$$u^\varepsilon(0, x) = 0 \quad \text{for } x \notin D;$$

$$\frac{\partial v^\varepsilon}{\partial t} = \frac{\varepsilon^2}{2}\Delta v^\varepsilon + (b(x), \nabla_x v^\varepsilon), \qquad x \in D, \quad t > 0;$$

$$v^\varepsilon(0, x) = 0, \qquad v^\varepsilon(t, x)|_{x \in \partial D} = 1;$$

$$\frac{\partial w^\varepsilon}{\partial t} = \frac{\varepsilon^2}{2}\Delta w^\varepsilon + (b(x), \nabla_x w^\varepsilon) + c(x) w^\varepsilon, \qquad x \in D, t > 0;$$

$$w^\varepsilon(0, x) = 0,$$

$$w^\varepsilon(t, x)|_{x \in \partial D} = g(x);$$

so that Theorem 1.2 can be viewed as assertions, concerning the behavior as the parameter converges to zero, of solutions of differential equations with a small parameter at the derivatives of the highest order.

§2. The Problem of Exit from a Domain

Let D be a bounded domain in R^r and let ∂D be its boundary, which we assume to be smooth for the sake of simplicity. If the trajectory $x_t(x)$ of the system (1.1) issued from the point $x \in D$ leaves $D \cup \partial D$ within finite time, then for small ε, the trajectories of the process X_t^ε issued from x leave D within the same time with probability close to one and the first exit from D takes place near the exit point of $x_t(x)$ from D with overwhelming probability (cf. Ch. 2).

In this paragraph we shall assume that $(b(x), n(x)) < 0$ for $x \in \partial D$, where $n(x)$ is the exterior normal to the boundary of D, so that the curves $x_t(x)$ cannot leave D for $x \in D$. The trajectories of X_t^ε issued from a point $x \in D$ leave D with probability 1 (and in this case for every $\varepsilon \neq 0$). Nevertheless, the point of exit and the time necessary for reaching the boundary are not any more determined by the trajectory $x_t(x)$ of the dynamical system for small ε, but rather depend on the form of the field $b(x)$ in the whole domain D in general. Here we study the problem of exit from D for the simplest structure of $b(x)$ in D which is compatible with the condition $(b(x), n(x)) < 0$. A more general case will be considered in Ch. 6.

Let $O \in R^r$ be an asymptotically stable equilibrium position of system (1.1), i.e., for every neighborhood $\mathscr{E}_1$ of O let there exist a smaller neighborhood $\mathscr{E}_2$ such that the trajectories of system (1.1) starting in $\mathscr{E}_2$ converge to zero without leaving $\mathscr{E}_1$ as $t \to \infty$.

We say that D is attracted to O if the trajectories $x_t(x)$, $x \in D$ converge to the equilibrium position O without leaving D as $t \to \infty$.

The quasipotential of the dynamical system (1.1) with respect to the point O, is, by definition, the function $V(O, x)$ defined by the equality

$$V(O, x) = \inf\{S_{T_1T_2}(\varphi): \varphi \in \mathbf{C}_{T_1T_2}(R^r), \varphi_{T_1} = O, \varphi_{T_2} = x\}.$$

We note that the endpoints of the interval $[T_1, T_2]$ are not fixed. The meaning of the term "quasipotential" will be clarified in the next section. It is easy to verify that $V(O, x) \geq 0$, $V(O, O) = 0$ and the function $V(O, x)$ is continuous.

Theorem 2.1. *Let O be a stable equilibrium position of system* (1.1) *and suppose that the domain D is attracted to O and $(b(x), n(x)) < 0$ for $x \in \partial D$. Suppose furthermore that there exists a unique point $y_0 \in \partial D$ for which $V(O, y_0) = \min_{y \in \partial D} V(O, y)$. Then*

$$\lim_{\varepsilon \to 0} \mathsf{P}_x\{\rho(X_{\tau^\varepsilon}^\varepsilon, y_0) < \delta\} = 1,$$

for every $\delta > 0$ and $x \in D$, where $\tau^\varepsilon = \inf\{t: X_t^\varepsilon \in \partial D\}$.

The proof of this theorem will be done according to the following plan. First we show (by means of Lemma 2.1) that with probability converging to 1 as $\varepsilon \to 0$, the Markov trajectories X_t^ε issued from any point $x \in D$ hit a small neighborhood of the equilibrium position before they go out to ∂D (Fig. 3). Since X_t^ε is a strong Markov process, this implies that it is sufficient to study how the trajectories starting in a small neighborhood of O go out of the domain.

Let Γ and γ be two small spheres of radii μ and $\mu/2$ with their center at the equilibrium position O (Fig. 4). We introduce an increasing sequence of

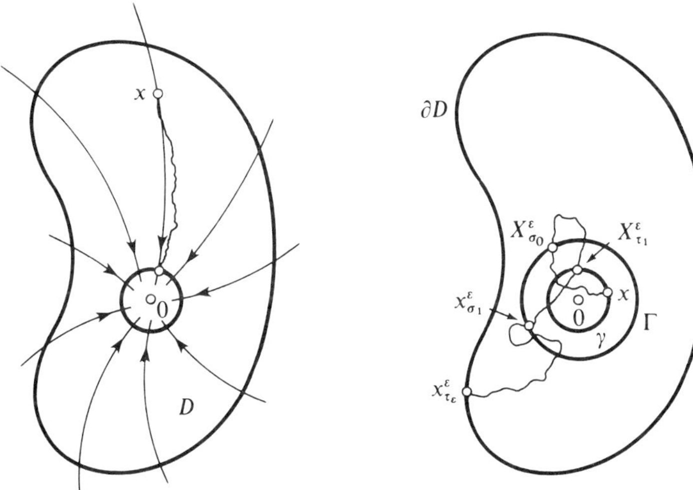

Figure 3.

Figure 4.

Markov times $\tau_0, \sigma_0, \tau_1, \sigma_1, \tau_2, \ldots$ in the following way: $\tau_0 = 0$ and $\sigma_n = \inf\{t > \tau_n : X_t^\varepsilon \in \Gamma\}$, $\tau_n = \inf\{t > \sigma_{n-1} : X_t^\varepsilon \in \gamma \cup \partial D\}$ (if at a certain step, the process X_t^ε does not reach the set Γ any more, we set the corresponding Markov time and all subsequent ones equal to $+\infty$. We note that infinite τ_n's or σ_n's can be avoided if we change the field $b(x)$ outside D in an appropriate way.) The sequence $Z_n = X_{\tau_n}^\varepsilon$ forms a Markov chain on the set $\gamma \cup \partial D$ (generally speaking, a nonconservative chain: Z_n is not defined if $\tau_n = \infty$ but this can happen only after exit to ∂D). For small ε, this Markov chain passes from any point $x \in \gamma \cup \partial D$ to the set γ in one step with overwhelming probability. On the other hand, it turns out that if the chain indeed passes from $x \in \gamma$ onto ∂D, then with probability converging to 1 as $\varepsilon \to 0$, this passage takes place into a point lying in the neighborhood of the minimum point y_0 of the quasipotential: for every $\delta > 0$ we have

$$\lim_{\varepsilon \to 0} \mathsf{P}_z\{|Z_1 - y_0| < \delta \,|\, Z_1 \in \partial D\} = 1 \tag{2.1}$$

uniformly in $z \in \gamma$. The assertion of the theorem can be deduced easily from this.

The proof of (2.1) will of course, use the action functional and its properties.

We formulate and prove the auxiliary assertions needed for the proof.

Lemma 2.1. *Let F be a compact set in R^r and let T and δ be positive numbers. There exist positive numbers ε_0 and β such that*

$$\mathsf{P}_x\{\rho_{0T}(X^\varepsilon, x(x)) \geq \delta\} \leq \exp\{-\varepsilon^{-2}\beta\},$$

for any $x \in F$ and $\varepsilon < \varepsilon_0$, where $x_t(x)$ is the trajectory of the dynamical system issued from x.

Proof. Put

$$G(x) = \{\varphi \in \mathbf{C}_{0T}: \varphi_0 = x, \rho_{0T}(\varphi, x(x)) \geq \delta\}.$$

By the corollary to Theorem 1.1, the infimum d of $S_{0T}(\varphi)$ on the closed set $\bigcup_{x\in F} G(x)$ is attained at some element of this set. The functional S_{0T} vanishes only on trajectories of the dynamical system. Therefore, $d > 0$.

For any $d' < d$, the sets $\bigcup_{x\in F} G(x)$ and $\bigcup_{x\in F} \Phi_x(d')$ are disjoint. Let us denote the distance between them by δ'. (δ' is positive because the first set is closed and the second is compact.) We use Theorem 1.1: for any $\gamma > 0$ we have

$$\begin{aligned}\mathsf{P}\{\rho_{0T}(X^\varepsilon, x(x)) \geq \delta\} &= \mathsf{P}_x\{X^\varepsilon \in G(x)\} \\ &\leq \mathsf{P}_x\{\rho_{0T}(X_0^\varepsilon, \Phi_x(d') \geq \delta'\} \leq \exp\{-\varepsilon^{-2}(d' - \gamma)\}\end{aligned}$$

for sufficiently small ε and for all $x \in F$. Hence the assertion of the lemma holds for $\beta = d' - \gamma$ (as β we can therefore choose any number smaller than d). □

In what follows, we denote by $\mathscr{E}_\delta(a)$ the δ-neighborhood of a point $a \in R^r$.

We shall need the following lemma in §4, as well.

Lemma 2.2. *Suppose that the point O is a stable equilibrium position of system (1.1), the domain D is attracted to O and $(b(x), n(x)) < 0$ for $x \in \partial D$. Then for any $\alpha > 0$ we have:*

(a) *there exist positive constants a and T_0 such that for any function φ_t assuming its values in the set $D \cup \partial D \setminus \mathscr{E}_\alpha(O)$ for $t \in [0, T]$, we have the inequality $S_{0T}(\varphi) > a(T - T_0)$;*
(b) *there exist positive constants c and T_0 such that for all sufficiently small $\varepsilon > 0$ and any $x \in D \cup \partial D \setminus \mathscr{E}_\alpha(O)$ we have the inequality*

$$\mathsf{P}_x\{\zeta_\alpha > T\} \leq \exp\{-\varepsilon^{-2}c(T - T_0)\},$$

where $\zeta_\alpha = \inf\{t: X_t^\varepsilon \notin D \setminus \mathscr{E}_\alpha(O)\}$.

Proof. (a) Let $\mathscr{E}_{\alpha'}(O)$ be a neighborhood of O such that the trajectories $x_t(x)$ of the dynamical system issued from $\mathscr{E}_{\alpha'}(O)$ never leave $\mathscr{E}_\alpha(O)$. We denote by $T(\alpha, x)$ the time spent by $x_t(x)$ until reaching $\mathscr{E}_{\alpha'}(O)$. Since D is attracted to O, we have $T(\alpha, x) < \infty$ for $x \in D \cup \partial D$. The function $T(\alpha, x)$ is upper semicontinuous in x (because $x_t(x)$ depends continuously on x). Consequently, it attains its largest value $T_0 = \max_{x \in D \cup \partial D} T(\alpha, x) < \infty$.

The set of functions from $\mathbf{C}_{0T_0}$, assuming their values in $D \cup \partial D \setminus \mathscr{E}_\alpha(O)$, is closed in $\mathbf{C}_{0T_0}$. By the corollary to Theorem 1.1, the functional S_{0T_0} attains its infimum on this set. This infimum is different from zero, since otherwise some trajectory of the dynamical system would belong to this set.

Hence for all such functions, $S_{0T_0}(\varphi) \geq A > 0$. By the additivity of S, for functions φ spending time T longer than T_0 in $D \cup \partial D \setminus \mathscr{E}_\alpha(O)$, we have $S_{0T}(\varphi) \geq A$; for functions spending time $T \geq 2T_0$ in $D \cup \partial D \setminus \mathscr{E}_\alpha(O)$, we have $S_{0T}(\varphi) \geq 2A$, etc. In general, we have

$$S_{0T}(\varphi) \geq A[T/T_0] > A(T/T_0 - 1) = a(T - T_0).$$

(b) From the circumstances that D is attracted to O and that $(b(x), n(x)) < 0$ on the boundary of D, it follows that the same properties will be enjoyed by the δ-neighborhood of D for sufficiently small $\delta > 0$. We shall assume that δ is smaller than $\alpha/2$. By assertion (a), there exist constants T_0 and A such that $S_{0T_0}(\varphi) > A$ for functions which do not leave the closed δ-neighborhood of D and do not get into $\mathscr{E}_{\alpha/2}(O)$. For $x \in D$, the functions in the set $\Phi_x(A) = \{\varphi: \varphi_0 = x, S_{0T_0}(\varphi) \leq A\}$ reach $\mathscr{E}_{\alpha/2}(O)$ or leave the δ-neighborhood of D during the time from 0 to T_0; the trajectories of X_t^ε for which $\zeta_\alpha > T_0$ are at a distance not smaller than δ from this set. By Theorem 1.1, this implies that for small ε and all $x \in D$ we have

$$\mathsf{P}_x\{\zeta_\alpha > T_0\} \leq \exp\{-\varepsilon^{-2}(A - \gamma)\}.$$

Then we use the Markov property:

$$\begin{aligned} \mathsf{P}_x\{\zeta_\alpha > (n + 1)T_0\} &= \mathsf{M}_x[\zeta_\alpha > nT_0; \mathsf{P}_{X_{nT_0}^\varepsilon}\{\zeta_\alpha > T_0\}] \\ &\leq \mathsf{P}_x\{\zeta_\alpha > nT_0\} \cdot \sup_{y \in D} \mathsf{P}_y\{\zeta_\alpha > T_0\}; \end{aligned}$$

and we obtain by induction that

$$\begin{aligned} \mathsf{P}_x\{\zeta_\alpha > T\} &\leq \mathsf{P}_x\left\{\zeta_\alpha > \left[\frac{T}{T_0}\right]T_0\right\} \leq \left[\sup_{y \in D} \mathsf{P}_y\{\zeta_\alpha > T_0\}\right]^{[T/T_0]} \\ &\leq \exp\left\{-\varepsilon^{-2}\left(\frac{T}{T_0} - 1\right)(A - \gamma)\right\}. \end{aligned}$$

Hence as c, we may take $(A - \gamma)T_0$, where γ is an arbitrarily small number. □

We formulate another simple lemma.

Lemma 2.3. *There exists a positive constant L such that for any x and* $y \in R^r$ *there exists a smooth function* φ_t, $\varphi_0 = x$, $\varphi_T = y$, $T = |x - y|$ *for which* $S_{0T}(\varphi) \le L \cdot |x - y|$.

Indeed, we may put $\varphi_t = x + [(y - x)/(|y - x|)]$.

We now pass to the proof of the theorem. Let $\delta > 0$. We write

$$d = \min\{V(O, y): y \in \partial D, |y - y_0| \ge \delta\} - V(O, y_0).$$

Since y_0 is the only minimum place of V, we have $d > 0$.

We choose a positive number $\mu < d/5L$ such that the sphere Γ of radius μ and center O is inside D (L is the constant from Lemma 2.3).

Lemma 2.4. *For sufficiently small* ε *we have*

$$\mathsf{P}_x\{Z_1 \in \partial D\} \ge \exp\{-\varepsilon^{-2}(V(O, y_0) + 0.45d)\}$$

for all $x \in \gamma$. *(We recall that* γ *is the sphere of radius* $\mu/2$ *and center O.)*

Proof. We choose a point y_1 outside $D \cup \partial D$ at a distance not greater than $\mu/2$ from y_0. There exists $T > 0$ such that for any point $x \in \gamma$ there exists a function φ_t^x, $0 \le t \le T$, $\varphi_0^x = x$, $\varphi_T^x = y_1$, $S_{0T}(\varphi^x) \le V(O, y_0) + 0.4d$.

Indeed, first of all we choose a function $\dot{\varphi}_t^{(1)}$, $0 \le t \le T_1$, $\varphi_0^{(1)} = O$, $\varphi_{T_1}^{(1)} = y_0$ such that $S_{0T_1}(\varphi^{(1)}) \le V(O, y_0) + 0.1d$. We cut off its first portion up to the point $x_1 = \varphi_{t_1}^{(1)}$ of the last intersection of $\varphi_t^{(1)}$ with Γ, i.e., we introduce the new function $\varphi_t^{(2)} = \varphi_{t_1+t}^{(1)}$, $0 \le t \le T_2 = T_1 - t_1$. We have $\varphi_0^{(2)} = x_1$, $\varphi_T^{(2)} = y_0$, $S_{0T_2}(\varphi^{(2)}) = S_{t_1T_1}(\varphi^{(1)}) \le V(O, y_0) + 0.1d$. Moreover, by Lemma 2.3, we choose functions $\varphi_t^{(3)}$, $0 \le t \le T_3 = \mu$, $\varphi_0^{(3)} = O$, $\varphi_{T_3}^{(3)} = x_1$, $S_{0T_3}(\varphi^{(3)}) \le 0.2d$; $\varphi_t^{(4)}$, $0 \le t \le T_4$, $\varphi_0^{(4)} = y_0$, $\varphi_{T_4}^{(4)} = y_1$, $S_{0T_4}(\varphi^{(4)}) \le 0.1d$. Finally, by the same lemma, for any $x \in \gamma$ we choose a function $\varphi_t^{(5)}$, $0 \le t \le T_5 = \mu/2$, $\varphi_0^{(5)} = x$, $\varphi_{T_5}^{(5)} = O$, $S_{0T_5}(\varphi^{(5)}) \le 0.1d$ depending on x. We construct the function φ_t^x out of pieces $\varphi^{(5)}$, $\varphi^{(3)}$, $\varphi^{(2)}$ and $\varphi^{(4)}$: $\varphi_t^x = \varphi_t^{(5)}$ for $0 \le t \le T_5$; $= \varphi_{t-T_5}^{(3)}$ for $T_5 \le t \le T_5 + T_3$; $= \varphi_{t-T_5-T_3}^{(2)}$ for $T_5 + T_3 \le t \le T_5 + T_3 + T_2$; $= \varphi_{t-T_5-T_3-T_2}^{(4)}$ for $T_5 + T_3 + T_2 \le t \le T_5 + T_3 + T_2 + T_4$ (Fig. 5)

We choose the positive δ' smaller than $\mu/4$ and the distance of y_1 from ∂D and use Theorem 1.1. For ε smaller than some ε_0 and for all $x \in \gamma$ we obtain

$$\mathsf{P}_x\{\rho_{0T}(X^\varepsilon, \varphi^x) < \delta'\} \ge \exp\{-\varepsilon^{-2}(V(O, y_0) + 0.4d + 0.05d)\}.$$

On the other hand, if a trajectory of X_t^ε passes at a distance smaller than δ' from the curve φ_t^x, then it hits the δ'-neighborhood of y_1 and intersects ∂D

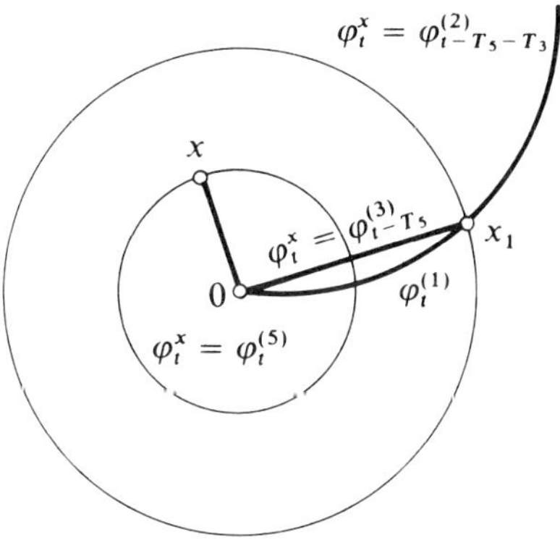

Figure 5.

on the way, not hitting γ after reaching Γ. Consequently, the probability that Z_1 belongs to ∂D is not smaller than $\exp\{-\varepsilon^{-2}(V(O, y_0) + 0.45d)\}$. □

Lemma 2.5. *For sufficiently small ε we have*

$$\mathsf{P}_x\{Z_1 \in \partial D \setminus \mathscr{E}_\delta(y_0)\} \leq \exp\{-\varepsilon^{-2}(V(O, y_0) + 0.55d)\}$$

for all $x \in \gamma$.

Proof. We recall that $Z_1 = X_{\tau_1}^\varepsilon$, where $\tau_1 = \inf\{t > \sigma_0 : X_t^2 \in \gamma \cup \partial D\}$. We introduce the notation $\tau(\gamma \cup \partial D) = \inf\{t > 0 : X_t^\varepsilon \in \gamma \cup \partial D\}$. The random variable Z_1 is nothing else but the variable $X_{\tau(\gamma \cup \partial D)}^\varepsilon$ calculated for the segment, shifted by σ_0 on the time axis, of a trajectory after time σ_0. We use the strong Markov property with respect to the Markov time σ_0. We obtain

$$\mathsf{P}_x\{Z_1 \in \partial D \setminus \mathscr{E}_\delta(y_0)\} = \mathsf{M}_x \mathsf{P} x_{\sigma_0}^\varepsilon\{X_{\tau(\gamma \cup \partial D)}^\varepsilon \in \partial D \setminus \mathscr{E}_\delta(y_0)\}.$$

Since $X_{\sigma_0}^\varepsilon \in \Gamma$, this probability does not exceed

$$\sup_{x \in \Gamma} \mathsf{P}_x\{X_{\tau(\gamma \cup \partial D)}^\varepsilon \in \partial D \setminus \mathscr{E}_\delta(y_0)\}$$

for any $x \in \gamma$. We estimate the latter probability.

By Lemma 2.2, for any $c > 0$ there exists T such that $\mathsf{P}_x\{\tau(\gamma \cup \partial D) > T\} \leq \exp\{-\varepsilon^{-2}c\}$ for all $x \in \Gamma$ and ε smaller than some ε_0. As c we take, say, $V(O, y_0) + d$. To obtain the estimate needed, it remains to estimate $\mathsf{P}_x\{\tau(\gamma \cup \partial D) \leq T, X_{\tau(\gamma \cup \partial D)}^\varepsilon \in \partial D \setminus \mathscr{E}_\delta(y_0)\}$. We obtain this estimate by means of Theorem 1.1.

We consider the closure of the $\mu/2$-neighborhood of $\partial D \setminus \mathscr{E}_\delta(y_0)$; we denote it by K. No function φ_t, $0 \le t \le T$, $\varphi_0 \in \Gamma$ such that $S_{0T}(\varphi) \le V(O, y_0) + 0.65d$ hits K. Indeed, let us assume that $\varphi_{t_1} \in K$ for some $t_1 \le T$. Then $S_{0t_1}(\varphi) = S_{0T}(\varphi) \le V(O, y_0) + 0.65d$. By Lemma 2.3, we take the functions $\varphi_t^{(1)}, 0 \le t \le T_1$, $\varphi_0^{(1)} = O$, $\varphi_{T_1}^{(1)} = \varphi_0$, with $S_{0T_1}(\varphi^{(1)}) \le 0.2d$ and $\varphi_t^{(2)}, 0 \le t \le T_2$, $\varphi_0^{(2)} = \varphi_{t_1}$, $\varphi_{T_2}^{(2)} \in \partial D \setminus \mathscr{E}_\delta(y_0)$ with $S_{0T_2}(\varphi^{(2)}) < 0.1d$ and out of pieces $\varphi^{(1)}$, φ and $\varphi^{(2)}$ we build a new function: $\hat{\varphi}(t) = \varphi_t^{(1)}$ for $0 \le t \le T_1$; $= \varphi_{t-T_1}$ for $T_1 \le t \le T_1 + t_1$; $= \varphi_{t-T_1-t_1}^{(2)}$ for $T_1 + t_1 \le t \le T_1 + t_1 + T_2$. Then $\hat{\varphi}_0 = O$, $\hat{\varphi}_{T_1+t_1+T_2} \in \partial D \setminus \mathscr{E}_\gamma(y_0)$ and $S_{0T_1+t_1+T_2}(\hat{\varphi}) \le 0.2d + V(O, y_0) + 0.1d + 0.65d$. This is smaller than the infimum of $V(O, y)$ for $y \in \partial D \setminus \mathscr{E}_\delta(y_0)$, which is impossible.

This means that all functions from $\bigcup_{x \in \Gamma} \Phi_x(V(O, y_0) + 0.65d)$ pass at a distance not smaller than $\mu/2$ from $\partial D \setminus \mathscr{E}_\delta(y_0)$. Using Theorem 1.1, we obtain for sufficiently small ε and all $x \in \Gamma$ that

$$\begin{aligned}
\mathsf{P}_x\{\tau(\gamma \cup \partial D) \le T, X^\varepsilon_{\tau(\gamma \cup \partial D)} \in \partial D \setminus \mathscr{E}_\delta(y_0)\} &\le \mathsf{P}_x\{\rho_{0T}(X^\varepsilon, \Phi_x(V(O, y_0) + 0.65d)) \ge \mu/2\} \\
&\le \exp\{-\varepsilon^{-2}(V(O, y_0) + 0.65d - 0.05d)\},
\end{aligned}$$

$$\begin{aligned}
\mathsf{P}_x\{X^\varepsilon_{\tau(\gamma \cup \partial D)} \in \partial D \setminus \mathscr{E}_\delta(y_0)\} &\le \mathsf{P}_x\{\tau(\gamma \cup \partial D) > T\} \\
&\quad + \mathsf{P}_x\{\tau(\gamma \cup \partial D) \le T, X^\varepsilon_{\tau(\gamma \cup \partial D)} \in \partial D \setminus \mathscr{E}_\delta(y_0)\} \\
&\le \exp\{-\varepsilon^{-2}(V(O, y_0) + d)\} \\
&\quad + \exp\{-\varepsilon^{-2}(V(O, y_0) + 0.6d)\} \\
&\le \exp\{-\varepsilon^{-2}(V(O, y_0) + 0.55d)\}.
\end{aligned}$$

It follows from Lemmas 2.4 and 2.5 that

$$\mathsf{P}_x\{Z_1 \in \partial D \setminus \mathscr{E}_\delta(y_0)\} \le \mathsf{P}_x\{Z_1 \in \partial D\} \exp\{-\varepsilon^{-2} \cdot 0.1d\}$$

for sufficiently small ε and all $x \in \gamma$. We denote by v the smallest n for which $Z_n \in \partial D$. Using the strong Markov property, for $x \in \gamma$ we find that

$$\begin{aligned}
\mathsf{P}_x\{|X^\varepsilon_{\tau^\varepsilon} - y_0| \ge \delta\} &= \mathsf{P}_x\{Z_v \in \partial D \setminus \mathscr{E}_\delta(y_0)\} \\
&= \sum_{n=1}^{\infty} \mathsf{P}_x\{v = n, Z_n \in \partial D \setminus \mathscr{E}_\delta(y_0)\} \\
&= \sum_{n=1}^{\infty} \mathsf{M}_x\{Z_1 \in \gamma, \ldots, Z_{n-1} \in \gamma\}; \mathsf{P}_{Z_{n-1}}\{Z_1 \in \partial D \setminus \mathscr{E}_\delta(y_0)\}\} \\
&\le \sum_{n=1}^{\infty} \mathsf{M}_x\{Z_1 \in \gamma, \ldots, Z_{n-1} \in \gamma\};
\end{aligned}$$

$$\begin{aligned}
\mathsf{P}_{Z_{n-1}}\{Z_1 \in \partial D\}\} \cdot \exp\{-\varepsilon^{-2} \cdot 0.1d\} &= \sum_{n=1}^{\infty} \mathsf{P}_x\{v = n\} \exp\{-\varepsilon^{-2} \cdot 0.1d\} \\
&= \exp\{-\varepsilon^{-2} \cdot 0.1d\} \to 0
\end{aligned}$$

as $\varepsilon \to 0$. Consequently, the theorem is proved for $x \in \gamma$.

If x is an arbitrary point in D, then

$$\mathsf{P}_x\{|X^\varepsilon_{\tau^\varepsilon} - y_0| \geq \delta\} \leq \mathsf{P}_x\{X^\varepsilon_{\tau(\gamma\cup\partial D)} \in \partial D\} + \mathsf{P}_x\{X^\varepsilon_{\tau(\gamma\cup\partial D)} \in \gamma, |X^\varepsilon_{\tau^\varepsilon} - y_0| \geq \delta\}.$$

The first probability converges to zero according to Lemma 2.1. Using the strong Markov property, we write the second one in the form

$$\mathsf{M}_x\{X^\varepsilon_{\tau(\gamma\cup\partial D)} \in \gamma; \mathsf{P}_{X^\varepsilon_{\tau(\gamma\cup\partial D)}}\{|X^\varepsilon_{\tau^\varepsilon} - y_0| \geq \delta\}\},$$

which converges to zero by what has already been proved. □

In the language of the theory of differential equations, Theorem 2.1 can be formulated in the following equivalent form.

Theorem 2.2. *Let $g(x)$ be a continuous function defined on the boundary ∂D of a domain D. Let us consider the Dirichlet problem*

$$\frac{\varepsilon^2}{2}\Delta u^\varepsilon(x) + \sum_{i=1}^r b^i(x)\frac{\partial u^\varepsilon}{\partial x^i}(x) = 0, \qquad x \in D;$$

$$u^\varepsilon(x) = g(x), \qquad x \in D,$$

in D. If the hypotheses of Theorem 2.1 *are satisfied, then* $\lim_{\varepsilon\to 0} u^\varepsilon(x) = g(y_0)$.

The proof follows easily from the formula $u^\varepsilon(x) = \mathsf{M}_x g(X^\varepsilon_{\tau^\varepsilon})$ (cf. §5, Ch. 1) if we take account of the continuity and boundedness of $g(x)$. On the other hand, we can obtain Theorem 2.1 from Theorem 2.2 by means of the same formula. □

Under additional assumptions, we can obtain more accurate information on how a trajectory of X^ε_t goes out of D for small ε.

Now it will be more convenient to use the notation $X^\varepsilon(t)$, $\varphi(t)$ instead of X^ε_t, φ_t, etc.

We have defined $V(O, y)$ as the infimum of $S_{0T}(\varphi)$ for all functions $\varphi(t)$, $0 \leq t \leq T$ going from O to y. This infimum is usually not attained (cf. examples in the next section). However, it is attained for functions defined on a semiaxis infinite from the left: there exists a function $\varphi(t)$, $-\infty \leq t \leq T$ such that $\varphi(-\infty) = O$, $\varphi(T) = y$, $S_{-\infty, T}(\varphi) = V(O, y)$. We shall not prove this but rather include it as a condition in the theorem we are going to formulate. (The assertion is contained in Wentzell and Freidlin [4] as Lemma 3.3 with the outlines of a proof.) The extremal $\varphi(t)$ is not unique: along with it, any translate $\tilde{\varphi}(t) = \varphi(t + a)$, $-\infty \leq t \leq T - a$ of it will also be an extremal.

We introduce the following definition. Let G be a neighborhood of O with smooth boundary ∂G. A curve $\varphi(t)$ leading from O to the boundary ∂D of D, necessarily intersects ∂G somewhere. Let us denote by $\theta_{\delta G}(\varphi)$ the last moment of time at which $\varphi(t)$ is on ∂G: $\theta_{\partial G}(\varphi) = \sup\{t: \varphi_t \in \partial G\}$. If for some

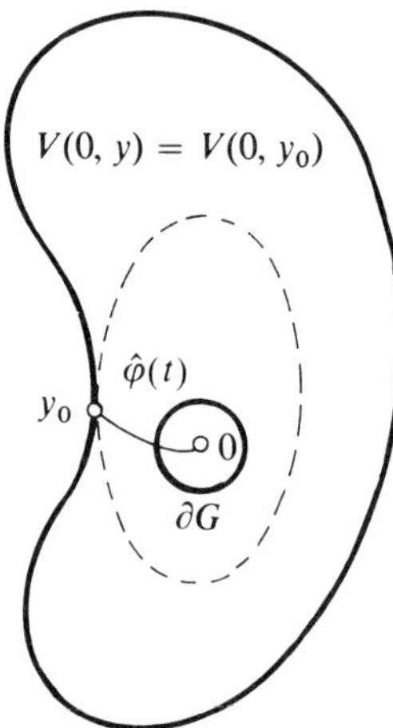

Figure 6.

$\alpha > 0$, the function $\varphi(t)$ assumes values inside G for $t \in [\theta_{\partial G}(\varphi) - \alpha, \theta_{\partial G}(\varphi)]$, then we shall say that $\varphi(t)$ leaves G in a regular manner.

Theorem 2.3. *Suppose that the hypotheses of Theorem 2.1 are satisfied and there exists an extremal $\hat{\varphi}(t)$, $-\infty < t \le T$, unique up to translations and going from O to ∂G (namely, to the point y_0, Fig. 6). Let us extend $\hat{\varphi}(t)$ by continuity to $t \ge T$ in an arbitrary way.*

Let $G \subset D$ be a neighborhood with smooth boundary of the equilibrium position and suppose that the extremal $\hat{\varphi}(t)$ leaves this neighborhood in a regular manner. Let us denote by $\theta^\varepsilon_{\partial G}$ the last moment of time at which the trajectory of $X^\varepsilon(t)$ is on ∂G until exit onto ∂D: $\theta^\varepsilon_{\partial G} = \max\{t < \tau^\varepsilon: X^\varepsilon(t) \in \partial G\}$. For any $\delta > 0$ and $x \in D$ we have

$$\lim_{\varepsilon \to 0} \mathsf{P}_x \left\{ \max_{\theta^\varepsilon_{\partial G} \le t \le \tau^\varepsilon} |X^\varepsilon(t) - \hat{\varphi}(t - \theta^\varepsilon_{\partial G} + \theta_{\partial G}(\hat{\varphi}))| < \delta \right\} = 1.$$

In other words, with probability converging to 1 as $\varepsilon \to 0$, the last portion of the Markov trajectory until exit to ∂D but after exit from G is located in a small neighborhood of the extremal $\hat{\varphi}(t)$ translated in an appropriate way.

The proof of this theorem can be carried out according to the same plan as that of Theorem 2.1. We ascertain that for functions $\varphi(t)$, $0 \le t \le T$ leading from O to ∂D and lying at a distance greater than $\delta/2$ from all translates of the extremal, we have $S_{0T}(\varphi) \ge V(O, y_0) + d$, where d is a positive constant. Then we choose $\mu \le d/5L \wedge \delta/2$ and the spheres γ and Γ. The functions $\varphi(t)$, $0 \le t \le T$ starting on Γ, lying farther than $\delta/2$ from the translates of $\hat{\varphi}(t)$ and such that $S_{0T}(\varphi) \le V(O, y_0) + 0.65d$, do not reach the

$\mu/2$-neighborhood of D. As in the proof of Theorem 2.1, from this we can conclude that

$$\mathsf{P}_x\{\tau(\gamma \cup \partial D) < T, X^\varepsilon(\tau(\gamma \cup \partial D)) \in \partial D,$$
$$\max_{\theta^\varepsilon_{\partial G} \le t \le \tau^\varepsilon} |X^\varepsilon(t) - \hat{\varphi}(t - \theta^\varepsilon_{\partial G} + \theta_{\partial G}(\hat{\varphi}))| \ge \delta\}$$
$$\le \exp\{-\varepsilon^{-2}(V(O, y_0) + 0.6d)\}$$

for all $x \in \Gamma$ and

$$\mathsf{P}_x\{X^\varepsilon(\tau(\gamma \cup \partial D) \in \partial D, \max_{\theta^\varepsilon_{\partial G} \le t \le \tau^\varepsilon} |X^\varepsilon(t) - \hat{\varphi}(t - \theta^\varepsilon_{\partial G} + \theta_{\partial G}(\hat{\varphi}))| \ge \delta\}$$
$$\le \mathsf{P}_x\{X^\varepsilon(\tau(\gamma \cup \partial D)) \in \partial D\} \exp\{-\varepsilon^{-2} \cdot 0.1d\}$$

for all $x \in \gamma$, from which we obtain the assertion of the theorem for these x and then for all $x \in D$. □

If the dynamical system (1.1) has more than one stable equilibrium position in D, then before it leaves the domain, the trajectory of X^ε_t can pass from one stable equilibrium position to another. In this case, the point of exit from the domain depends on the initial point. We consider such a more complicated structure of a dynamical system in Ch. 6. Here we discuss the case where system (1.1) has a unique limit cycle in D.

As a concrete example, let D be a planar domain homeomorphic to a ring. The boundary ∂D is assumed to be smooth, as before. We assume that system (1.1) has a unique limit cycle Π in D and D is attracted to Π, i.e., the trajectory $x_t(x)$ for $x \in D$ converges to Π with increasing t, without leaving D. On the boundary ∂D we assume that the earlier condition $(b(x), n(x)) < 0$ is satisfied.

The quasipotential of system (1.1) with respect to Π is, by definition, the function

$$V(\Pi, x) = \inf\{S_{T_1 T_2}(\varphi): \varphi_{T_1} \in \Pi, \varphi_{T_2} = x\}.$$

The following theorem can be proved in the same way as Theorem 2.1.

Theorem 2.4. *Suppose that the domain D is attracted to a stable limit cycle Π and $(b(x), n(x)) < 0$ for $x \in \partial D$. Furthermore, suppose that there exists a unique point $y_0 \in \partial D$ for which*

$$V(\Pi, y_0) = \min_{y \in \partial D} V(\Pi, y).$$

Then

$$\lim_{\varepsilon \to 0} \mathsf{P}_x\{|X^\varepsilon_{\tau^\varepsilon} - y_0| < \delta\} = 1,$$

for every $\delta > 0$ and $x \in D$.

The assertion of Theorem 2.4 remains valid in the case where not the whole domain D is attracted to Π but only its part exterior to Π, provided that the interior part does not have an exit to ∂D. Then the cycle may be stable only from the outside.

If the extremal of the functional $S(\varphi)$ from Π to ∂D is unique up to translations, then an analogue of Theorem 2.3 can be proved.

§3. Properties of the Quasipotential. Examples

In the preceding section it was shown that if we know the function $V(O, x)$, i.e., the quasipotential of the dynamical system, then we can find the point y_0 on the boundary of the domain, near which the Markov trajectories of X_t^ε starting inside the domain reach the boundary for small ε. In the next section we show that this function is important in other problems, as well. For example, the principal term of the asymptotics of the mean time $\mathsf{M}_x\tau^\varepsilon$ spent by a trajectory of X_t^ε before reaching the boundary of a domain can be expressed in terms of $V(O, x)$. The behavior of the invariant measure of X_t^ε as $\varepsilon \to 0$ can also be described in terms of $V(O, x)$.

In this section we study the problem of calculating the quasipotential $V(O, x)$, establish some properties of the extremals of the functional $S(\varphi)$ and consider examples. We shall only deal with the quasipotential of a dynamical system with respect to a stable equilibrium position; the case of the quasipotential with respect to a stable limit cycle can be considered analogously.

Theorem 3.1. *Suppose that the vector field $b(x)$ admits the decomposition*

$$b(x) = -\nabla U(x) + l(x), \tag{3.1}$$

where the function $U(x)$ is continuously differentiable in $D \cup \partial D$, $U(O) = 0$, $U(x) > 0$ and $\nabla U(x) \neq 0$ for $x \neq 0$ and $(l(x), \nabla U(x)) = 0$. Then the quasipotential $V(O, x)$ of the dynamical system (1.1) with respect to O coincides with $2U(x)$ at all points $x \in D \cup \partial D$ for which $U(x) \leq U_0 = \min_{y \in \partial D} U(y)$. If $U(x)$ is twice continuously differentiable, then the unique extremal of the functional $S(\varphi)$ on the set of functions φ_s, $-\infty \leq s \leq T$, leading from O to x is given by the equation

$$\dot{\varphi}_s = \nabla U(\varphi_s) + l(\varphi_s), \qquad s \in (-\infty, T), \qquad \varphi_T = x. \tag{3.2}$$

Proof. If the function φ_s for $s \in [T_1, T_2]$ does not exit from $D \cup \partial D$, then the relation $U(\varphi_{T_2}) - U(\varphi_{T_1}) = \int_{T_1}^{T_2} (\nabla U(\varphi_s), \dot{\varphi}_s)\, ds$ implies the inequality

$$\begin{aligned} S_{T_1 T_2}(\varphi) &= \frac{1}{2}\int_{T_1}^{T_2} |\dot{\varphi}_s - \nabla U(\varphi_s) - l(\varphi_s)|^2\, ds + 2\int_{T_1}^{T_2} (\dot{\varphi}_s, \nabla U(\varphi_s))\, ds \\ &\geq 2[U(\varphi_{T_2}) - U(\varphi_{T_1})]. \end{aligned} \tag{3.3}$$

From this we can conclude that $S_{T_1T_2}(\varphi) \geq 2U(x)$ for any curve φ_s, $\varphi_{T_1} = 0$, $\varphi_{T_2} = x$, where x is such that $U(x) \leq U_0$. Indeed, if this curve does not leave $D \cup \partial D$ over the time from T_1 to T_2, then the assertion follows from (3.3) (we recall that $U(O) = 0$). If, on the other hand, φ_s leaves $D \cup \partial D$, then it crosses the level surface $U(x) = U_0$ at some time $\tilde{T} \in (T_1, T_2)$. By the non-negativity and additivity of $S(\varphi)$, we obtain

$$S_{T_1T_2}(\varphi) \geq S_{T_1\tilde{T}}(\varphi) \geq 2U(\varphi_{\tilde{T}}) = 2U_0 \geq 2U(x).$$

On the other hand, if $\hat{\varphi}_s$ is a solution of (3.2), then $\hat{\varphi}_{-\infty} = O$. This follows from the facts that $dU(\hat{\varphi}_s)/ds = |\nabla U(\hat{\varphi}_s)|^2 > 0$ for $\hat{\varphi}_s \neq O$ and O is the only zero of $U(x)$. From this we obtain

$$S_{-\infty,T}(\hat{\varphi}) = 2\int_{-\infty}^{T} (\dot{\hat{\varphi}}_s, \nabla U(\hat{\varphi}_s))\, ds = 2[U(x) - U(O)] = 2U(x).$$

Consequently, for any curve φ_s connecting O and x we have: $S(\varphi) \geq 2U(x) = S(\hat{\varphi})$. Therefore, $V(O, x) = \inf S(\varphi) = 2U(x)$, and φ_s is an extremal.

If $U(x)$ is twice continuously differentiable, then the solution of equation (3.2) is unique, and consequently, the extremal from O to x normalized by the condition $\varphi_T = x$ is also unique. □

It follows from the theorem just proved that if $b(x)$ has a potential, i.e., $b(x) = -\nabla U(x)$, then $V(O, x)$ differs from $U(x)$ only by a multiplicative constant. This is the reason why we call V a quasipotential.

If $b(x)$ has a decomposition (3.1), then from the orthogonality condition for $\nabla U(x)$ and $l(x) = b(x) + \nabla U(x)$ we obtain the equation for the quasi-potential:

$$\tfrac{1}{2}(\nabla V(O, x), \nabla V(O, x)) + (b(x), \nabla V(O, x)) = 0, \tag{3.4}$$

which is, in essence, Jacobi's equation for the variational problem defining the quasipotential. Theorem 3.1 says that the solution of this equation, satisfying the conditions $V(O, O) = 0$, $V(O, x) > 0$ and $\nabla V(O, x) \neq 0$ for $x \neq 0$, is the quasipotential. It can be proved that conversely, if the quasi-potential $V(O, x)$ is continuously differentiable, then it satisfies equation (3.4), i.e., we have decomposition (3.1). However, it is easy to find examples showing that $V(O, x)$ may not be differentiable.

The function $V(O, x)$ cannot be arbitrarily bad: it satisfies a Lipschitz condition; this can be derived easily from Lemma 2.3.

Now we pass to the study of extremals of $S(\varphi)$ going from O to x. For them we can write Euler's equations in the usual way:

$$\ddot{\varphi}_t^k - \sum_{j=1}^{r} \left(\frac{\partial b^k}{\partial x^j}(\varphi_t) - \frac{\partial b^j}{\partial x^k}(\varphi_t)\right)\dot{\varphi}_t^j - \sum_{j=1}^{r} b^j(\varphi_t)\frac{\partial b^j}{\partial x^k}(\varphi_t) = 0, \qquad 1 \leq k \leq r.$$

If $b(x)$ has the decomposition (3.1), then we can write the simpler equations (3.2) for the extremals. From these equations we can draw a series of qualitative conclusions on the behavior of extremals. At every point of the domain, the velocity of the motion on an extremal is equal to $\nabla U(x) + l(x)$ according to (3.2), whereas $b(x) = -\nabla U(x) + l(x)$. From this we conclude that the velocity of the motion on an extremal is equal, in norm, to the velocity of the motion of trajectories of the system: $|\nabla U(x) + l(x)| = |-\nabla U(x) + l(x)| = |b(x)|$. The last assertion remains valid if we do not assume the existence of the decomposition of $b(x)$. This follows from the lemma below.

Lemma 3.1. *Let $S(\varphi) < \infty$. Let us denote by $\tilde{\varphi}$ the function obtained from φ by changing the parameter in such a way that $|\dot{\tilde{\varphi}}_s| = |b(\tilde{\varphi}_s)|$ for almost all s. We have $S(\tilde{\varphi}) \leq S(\varphi)$, where equality is attained only if $|\dot{\varphi}_s| = |b(\varphi_s)|$ for almost all s.*

Proof. For the substitution $s = s(t)$, $\tilde{\varphi}_t = \varphi_{s(t)}$ we obtain

$$\begin{aligned} S(\varphi) &= \frac{1}{2}\int_{T_1}^{T_2} |b(\varphi_s) - \dot{\varphi}_s|^2\, ds \\ &= \frac{1}{2}\int_{t(T_1)}^{t(T_2)} |b(\tilde{\varphi}_t) - \dot{\tilde{\varphi}}_t \dot{s}(t)^{-1}|^2 \dot{s}(t)\, dt \\ &= \frac{1}{2}\int_{t(T_1)}^{t(T_2)} (|b(\tilde{\varphi}_t)|^2 \dot{s}(t) + |\dot{\tilde{\varphi}}_t|^2 \dot{s}(t)^{-1})\, dt - \int_{t(T_1)}^{t(T_2)} (b(\tilde{\varphi}_t), \dot{\tilde{\varphi}}_t)\, dt \\ &\geq \int_{t(T_1)}^{t(T_2)} |b(\tilde{\varphi}_t)| \cdot |\dot{\tilde{\varphi}}_t|\, dt - \int_{t(T_1)}^{t(T_2)} (b(\tilde{\varphi}_t), \dot{\tilde{\varphi}}_t)\, dt, \end{aligned} \tag{3.5}$$

where $t(s)$ is the inverse function of $s(t)$. In (3.5) we have used the inequality $\alpha x^2 + \alpha^{-1} y^2 \geq 2xy$, true for any positive α. We define the function $s(t)$ by the equality

$$t = \int_0^{s(t)} |\dot{\varphi}_u| \, |b(\varphi_u)|^{-1}\, du.$$

This function is monotone increasing and $\tilde{\varphi}_t = \varphi_{s(t)}$ is absolutely continuous in t. For this function we have $|\dot{\tilde{\varphi}}_t| = |b(\tilde{\varphi}_t)|$ for almost all t and inequality (3.5) turns into the equality

$$S(\tilde{\varphi}) = \int_{t(T_1)}^{t(T_2)} |b(\tilde{\varphi}_t)| \cdot |\dot{\tilde{\varphi}}_t|\, dt - \int_{t(T_1)}^{t(T_2)} (b(\tilde{\varphi}_t), \dot{\tilde{\varphi}}_t)\, dt.$$

This proves the lemma. □

Now we consider some examples.

EXAMPLE 3.1. Let us consider the dynamical system $\dot{x}_t = b(x_t)$ on the real line, where $b(0) = 0$, $b(x) > 0$ for $x < 0$ and $b(x) < 0$ for $x > 0$ and D is an interval $(\alpha_1, \alpha_2) \subset R^1$ containing the point O. If we write $U(x) = -\int_0^x b(y)\,dy$, then $b(x) = -dU/dx$, so that in the one-dimensional case every field has a potential. It follows from Theorems 2.1 and 3.1 that for small ε, with probability close to one, the first exit of the process $X_t^\varepsilon = x + \int_0^t b(X_s^\varepsilon)\,ds + \varepsilon w_t$ from the interval (α_1, α_2) takes place through that endpoint α_i at which $U(x)$ assumes the smaller value. For the sake of definiteness, let $U(\alpha_1) < U(\alpha_2)$. The equation for the extremal has the form $\dot{\hat{\varphi}}_s = -b(\hat{\varphi}_s)$; we have to take that solution $\hat{\varphi}_s$, $-\infty < s \leq T$, of this equation which converges to zero as $s \to -\infty$ and for which $\hat{\varphi}_T = \alpha_1$.

In the one-dimensional case, the function $v^\varepsilon(x) = \mathsf{P}_x\{X_{\tau^\varepsilon}^\varepsilon = \alpha_1\}$ can be calculated explicitly by solving the corresponding boundary value problem. We obtain

$$v^\varepsilon(x) = \int_x^{\alpha_2} e^{2\varepsilon^{-2}U(y)}\,dy \cdot \left(\int_{\alpha_1}^{\alpha_2} e^{-2\varepsilon^{-2}U(y)}\,dy\right)^{-1}.$$

Of course it is easy to see from this formula that $\lim_{\varepsilon \to 0} \mathsf{P}_x\{X_{\tau^\varepsilon}^\varepsilon = \alpha_1\} = 1$ if $U(\alpha_1) < U(\alpha_2)$.

Using the explicit form of $v^\varepsilon(x)$, it is interesting to study how X_t exits from (α_1, α_2) if $U(\alpha_1) = U(\alpha_2)$. In this case, Theorem 2.1 is not applicable.

Using Laplace's method for the asymptotic estimation of integrals (cf., for example, Evgrafov [1]), we obtain that for $x \in (\alpha_1, \alpha_2)$,

$$\int_x^{\alpha_2} e^{2\varepsilon^{-2}U(y)}\,dy \sim \frac{\varepsilon^2}{2} \frac{\exp\{2\varepsilon^{-2}U(\alpha_2)\}}{|U'(\alpha_2)|},$$

$$\int_{\alpha_1}^{x} e^{2\varepsilon^{-2}U(y)}\,dy \sim \frac{\varepsilon^2}{2} \frac{\exp\{2_\varepsilon^{-2}U(\alpha_1)\}}{|U'(\alpha_1)|}$$

as $\varepsilon \to 0$. From this we find that if $U(\alpha_1) = U(\alpha_2)$ and $U'(\alpha_i) \neq 0$, $i = 1, 2$, then

$$\mathsf{P}_x\{X_{\tau^\varepsilon}^\varepsilon = \alpha_i\} \to \frac{|U'(\alpha_i)|^{-1}}{|U'(\alpha_1)|^{-1} + |U'(\alpha_2)|^{-1}}$$

as $\varepsilon \to 0$.

Consequently, in this case the trajectories of X_t^ε exit from the interval through both ends with positive probability as $\varepsilon \to 0$. The probability of exit through α_i is inversely proportional to $|U'(\alpha_i)| = |b(\alpha_i)|$.

EXAMPLE 3.2. Let us consider a homogeneous linear system $\dot{x}_t = Ax_t$ in R^r with constant coefficients. Assume that the matrix A is normal, i.e., $AA^* = A^*A$ and that the symmetric matrix $A + A^*$ is negative definite. In this case, the origin of coordinates O is an asymptotically stable equilibrium position. Indeed, the solution of our system can be represented in the form $x_t = e^{At}x_0$, where x_0 is the initial position. The normality of e^{At} follows from that of A. Using this observation and the relation $e^{At}e^{A^*t} = e^{(A+A^*)t}$, we obtain

$$|x_t|^2 = (e^{At}x_0, e^{At}x_0) = (e^{(A+A^*)t}x_0, x_0) \leq |x_0| \cdot |e^{(A-A^*)t}x_0|.$$

Since $A + A^*$ is negative definite, we have $|e^{(A+A^*)t}x_0| \to 0$. Consequently, $|x_t|^2 \to 0$ for any initial condition x_0.

It is easy to verify by direct differentiation that the vector field Ax admits the decomposition

$$Ax = -\nabla\left(-\frac{A + A^*}{4}x, x\right) + \frac{A - A^*}{2}x. \tag{3.6}$$

Moreover, the vector fields $\nabla(-\frac{1}{4}(A + A^*)x, x) = -\frac{1}{2}(A + A^*)x$ and $\frac{1}{2}(A - A^*)x$ are orthogonal:

$$\begin{aligned}(-\tfrac{1}{2}(A + A^*)x, \tfrac{1}{2}(A - A^*)x) &= -\tfrac{1}{4}[(Ax, Ax) - (A^*x, A^*x)] \\ &= -\tfrac{1}{4}[(A^*Ax, x) - (AA^*x, x)] = 0.\end{aligned}$$

Let the right sides of our system be perturbed by a white noise:

$$\dot{X}_t^\varepsilon = AX_t^\varepsilon + \varepsilon\dot{w}_t.$$

We are interested in how the trajectories of X_t^ε exit from a bounded domain D containing the equilibrium position O. From Theorem 3.1 and formula (3.6) we conclude that the quasipotential $V(O, x)$ of our dynamical system with respect to the equilibrium position O is equal to $-\frac{1}{2}((A + A^*)x, x)$. In order to find the point on the boundary ∂D of D near which the trajectories of X_t^ε first leave D with probability converging to 1 as $\varepsilon \to 0$, we have to find the minimum of $V(O, x)$ on ∂D. The equation for the extremals has the form

$$\dot{\varphi}_t = -\tfrac{1}{2}(A + A^*)\varphi_t + \tfrac{1}{2}(A - A^*)\varphi_t = -A^*\varphi_t.$$

If $y_0 \in \partial D$ is the unique point where $V(O, x)$ attains its smallest value on ∂D, then the last piece of a Markov trajectory is near the extremal entering this point. Up to a shift of time, the equation of this trajectory can be written in the form $\hat{\varphi}_t = e^{-A^*t}y_0$.

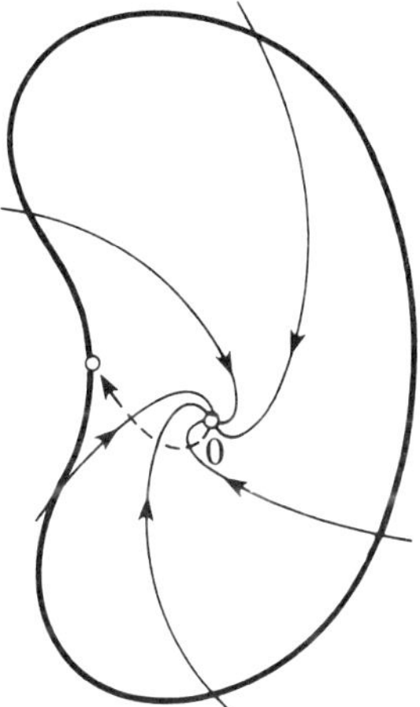

Figure 7.

For example, let a dynamical system

$$\dot{x}_t^1 = -x_t^1 - x_t^2,$$
$$\dot{x}_t^2 = x_t^1 - x_t^2,$$

be given in the plane R^2. The matrix of this system is normal and the origin of coordinates is asymptotically stable. The trajectories of the system are logarithmic spirals winding on the origin in the clockwise direction (Fig. 7). The quasipotential is equal to $[(x^1)^2 + (x^2)^2]$, so that with overwhelming probability for small ε, X_t^ε exits from D near the point $y_0 \in \partial D$ which is closest to the origin. It is easy to verify that the extremals are also logarithmic spirals but this time unwinding from O in the clockwise direction.

§4. Asymptotics of the Mean Exit Time and Invariant Measure for the Neighborhood of an Equilibrium Position

As in §2, let D be a bounded domain in R^r, $D \in O$ a stable equilibrium position of the system (1.1), τ^ε the time of first exit of a process X_t^ε from D. Under the assumption that D is attracted to O, in this section we calculate the principal term of $\ln \mathsf{M}_x \tau^\varepsilon$ as $\varepsilon \to 0$ and also that of $\ln m^\varepsilon(\bar{D})$, where $m^\varepsilon(\cdot)$ is the invariant measure of the process X_t^ε, $\bar{D} = R^r \setminus D$. Since the existence of an invariant measure and its properties depend on the behavior of $b(x)$ not only for $x \in D$, in the study of $m^\varepsilon(\bar{D})$ we have to make some assumptions concerning $b(x)$ in the whole space R^r.

Theorem 4.1. *Let O be an asymptotically stable equilibrium position of the system* (1.1) *and assume that the domain $D \subset R^r$ is attracted to O. Furthermore, assume that the boundary ∂D of D is a smooth manifold and $(b(x), n(x)) < 0$ for $x \in \partial D$, where $n(x)$ is the exterior normal of the boundary of D. Then for $x \in D$ we have*

$$\lim_{\varepsilon \to 0} \varepsilon^2 \ln \mathsf{M}_x \tau^\varepsilon = \min_{y \in \partial D} V(O, y). \tag{4.1}$$

Here the function $V(O, y)$ is the quasipotential of the dynamical system (1.1) with respect to O.

The proof of this theorem uses arguments, constructions and notation employed in the proof of Theorem 2.1.

To prove (4.1), it is sufficient to verify that for any $d > 0$, there exists an ε_0 such that for $\varepsilon < \varepsilon_0$ we have

(a) $\varepsilon^2 \ln \mathsf{M}_x \tau^\varepsilon < V_0 + d$.
(b) $\varepsilon^2 \ln \mathsf{M}_x \tau^\varepsilon > V_0 - d$.

First we prove inequality (a). We choose positive numbers μ, h, T_1, and T_2 such that the following conditions are satisfied: firstly, all trajectories of the unperturbed system, starting at points $x \in D \cup \partial D$, hit the $\mu/2$-neighborhood of the equilibrium position before time T_1 and after time T_1 they do not leave this neighborhood; secondly, for every point x lying in the ball $G = \{x \in R^r : |x - O| \leq \mu\}$, there exists a function φ_t^x such that $\varphi_0^x = x$, φ_t^x reaches the exterior of the h-neighborhood of D at a time $T(x) \leq T_2$ and φ_t^x does not hit the $\mu/2$-neighborhood of O after exit from G and $S_{0T(x)}(\varphi^x) < V_0 + d/2$.

The first of these assumptions can be satisfied because O is an asymptotically stable equilibrium position, D is attracted to O and for $x \in \partial D$, we have the inequality $(b(x), n(x)) < 0$.

The functions φ_t^x occurring in the second condition were constructed in §2.

From the definition of the action functional, we obtain for $y \in G$ that

$$\begin{aligned} \mathsf{P}_y\left\{\sup_{0 \leq t \leq T(y)} |X_t^\varepsilon - \varphi_t^y| < h\right\} &\geq \exp\left\{-\varepsilon^{-2}\left(S_{0T(y)}(\varphi^y) + \frac{d}{2}\right)\right\} \\ &\geq \exp\{-\varepsilon^{-2}(V_0 + d)\}, \end{aligned}$$

whenever ε is sufficiently small.

Since the point $\varphi_{T(y)}^y$ does not belong to the h-neighborhood of D, from the last inequality we conclude that

$$\mathsf{P}_y\{\tau^\varepsilon < T_2\} \geq \mathsf{P}_y\{\tau^\varepsilon < T(y)\} \geq \exp\{-\varepsilon^{-2}(V_0 + d)\} \tag{4.2}$$

for sufficiently small ε and $y \in G$.

We denote by σ the first entrance time of G: $\sigma = \min\{t: X_t^\varepsilon \in G\}$. Using the strong Markov property of X_t^ε, for any $x \in D$ and sufficiently small ε we obtain

$$\begin{aligned} \mathsf{P}_x\{\tau^\varepsilon < T_1 + T_2\} &\geq \mathsf{M}_x\{\sigma < T_1; \mathsf{P}_{x^\varepsilon}\{\tau^\varepsilon < T_2\}\} \\ &\geq \mathsf{P}_x\{\sigma < T_1\} \cdot \exp\{-\varepsilon^{-2}(V_0 + d)\} \\ &\geq \tfrac{1}{2}\exp\{-\varepsilon^{-2}(V_0 + d)\}. \end{aligned} \tag{4.3}$$

Here we have used inequality (4.2) for the estimation of $\mathsf{P}_{X_\sigma^\varepsilon}\{\tau^\varepsilon < T_2\}$ and the fact that the trajectories of X_t^ε converge in probability to x_t uniformly on $[0, T_1]$ as $\varepsilon \to 0$.

Then, using the Markov property of X_t^ε, from (4.3) we obtain

$$\begin{aligned} \mathsf{M}_x\tau^\varepsilon &\leq \sum_{n=0}^{\infty} (n + 1)(T_1 + T_2)\mathsf{P}_x\{n(T_1 + T_2) < \tau^\varepsilon \leq (n + 1)(T_1 + T_2)\} \\ &= (T_1 + T_2)\sum_{n=0}^{\infty} \mathsf{P}_x\{\tau^\varepsilon > n(T_1 + T_2)\} \\ &\leq (T_1 + T_2)\sum_{n=0}^{\infty}\left[1 - \min_{z\in D} \mathsf{P}_z\{\tau^\varepsilon \leq T_1 + T_2\}\right]^n \\ &= (T_1 + T_2)\left(\min_{z\in D} \mathsf{P}_z\{\tau^\varepsilon \leq T_1 + T_2\}\right)^{-1} \\ &\leq 2(T_1 + T_2)\exp\{\varepsilon^{-2}(V_0 + d)\}, \end{aligned}$$

whenever ε is sufficiently small. This implies assertion (a).

Now we prove assertion (b). We introduce the Markov times τ_k, σ_k and the Markov chain Z_n defined in the proof of Theorem 2.1. The phase space of Z_n is the set $\gamma \cup \partial D$, where $\gamma = \{x \in R^r: |x - O| = \mu/2\}$. For the one-step transition probabilities of this chain we have the estimate

$$\begin{aligned} \mathsf{P}(x, \partial D) &\leq \max_{y\in\Gamma} \mathsf{P}_y\{\tau_1 = \tau^\varepsilon\} \\ &= \max_{y\in\Gamma}\,[\mathsf{P}_y\{\tau^\varepsilon = \tau_1 < T\} + \mathsf{P}_y\{\tau^\varepsilon = \tau_1 \geq T\}]. \end{aligned} \tag{4.4}$$

As follows from Lemma 2.2, T can be chosen so large that the second probability have the estimate

$$\mathsf{P}_y\{\tau^\varepsilon = \tau_1 \geq T\} \leq \tfrac{1}{2}\exp\{-\varepsilon^{-2}(V_0 - h)\}. \tag{4.5}$$

In order to estimate the first probability, we note that the trajectories of X_t^ε, $0 \leq t \leq T$, for which $\tau^\varepsilon = \tau_1 < T$ are a positive distance from the set $\{\varphi \in \mathbf{C}_{0T}(R^r): \varphi_0 = y, S_{0T}(\varphi) < V_0 - h/2\}$ of functions provided that $h > 0$

is arbitrary and μ is sufficiently small. By virtue of the properties of the action functional, we obtain from this that

$$\mathsf{P}_y\{\tau^\varepsilon = \tau_1 < T\} \le \exp\{-\varepsilon^{-2}(V_0 - h)\}$$

for $y \in \Gamma$ and sufficiently small ε, $\mu > 0$.

It follows from the last inequality and estimates (4.4) and (4.5) that

$$\mathsf{P}(x, \partial D) \le \exp\{-\varepsilon^{-2}(V_0 - h)\}, \tag{4.6}$$

whenever ε, μ are sufficiently small.

As in Theorem 2.1, we denote by ν the smallest n for which $Z_n = X^\varepsilon_{\tau_n} \in \partial D$. It follows from (4.6) that

$$\mathsf{P}_x\{\nu > n\} \ge [1 - \exp\{-\varepsilon^{-2}(V_0 - h)\}]^{n-1}$$

for $x \in \gamma$. It is obvious that $\tau^\varepsilon = (\tau_1 - \tau_0) + (\tau_2 - \tau_1) + \cdots + (\tau_\nu - \tau_{\nu-1})$. Therefore, $\mathsf{M}_x\tau^\varepsilon = \sum_{n=1}^\infty \mathsf{M}_x\{\nu \ge n; \tau_n - \tau_{n-1}\}$.

Using the strong Markov property of X_t^ε with respect to the time σ_{n-1}, we obtain that

$$\begin{aligned}\mathsf{M}_x\{\nu \ge n; \tau_n - \tau_{n-1}\} &\ge \mathsf{M}_x\{\nu \ge n; \tau_n - \sigma_{n-1}\}\\ &\ge \mathsf{P}_x\{\nu \ge n\} \cdot \inf_{x\in\Gamma} \mathsf{M}_x\tau_1.\end{aligned}$$

The last infimum is greater than some positive constant t_1 independent of ε. This follows from the circumstance that the trajectories of the dynamical system spend some positive time going from Γ to γ.

Combining all estimates obtained so far, we obtain that

$$\begin{aligned}\mathsf{M}_x\tau^\varepsilon &> t_1 \sum_n \min_{z\in\gamma} \mathsf{P}_z\{\nu \ge n\}\\ &\ge t_1 \sum_n (1 - \exp\{-\varepsilon^{-2}(V_0 - h)\})^{n-1} = t_1 \exp\{\varepsilon^{-2}(V_0 - h)\}\end{aligned}$$

for sufficiently small ε and μ and $x \in \gamma$.

This implies assertion (b) for $x \in \gamma$. For any $x \in D$, taking into account that $\mathsf{P}_x\{\tau^\varepsilon > \tau_1\} \to 1$ as $\varepsilon \to 0$, we have

$$\begin{aligned}\mathsf{M}_x\tau^\varepsilon &= \mathsf{M}_x\{\tau^\varepsilon \le \tau_1; \tau^\varepsilon\} + \mathsf{M}_x\{\tau^\varepsilon > \tau_1; \tau^\varepsilon\}\\ &\ge \mathsf{M}_x\{\tau^\varepsilon > \tau_1; \mathsf{M}^\varepsilon_{\tau_1}\tau^\varepsilon\}\\ &> t_1 \exp\{(V_0 - h)\varepsilon^{-2}\} P_x\{\tau^\varepsilon > \tau_1\} > \frac{t_1}{2}\exp\{\varepsilon^{-2}(V_0 - h)\}.\end{aligned}$$

By the same token, we have proved assertion (b) for any $x \in D$. □

Remark. Analyzing the proof of this theorem, we can see that the assumptions that the manifold ∂D is smooth and $(b(x), n(x)) < 0$ for $x \in \partial D$ can be relaxed. It is sufficient to assume instead of them that the boundary of D and that of the closure of D coincide and for any $x \in \partial D$, the trajectory $x_t(x)$ of the dynamical system is situated in D for all $t > 0$.

We mention one more result relating to the distribution of the random variable τ^ε, the first exit time of D.

Theorem 4.2. *Suppose that the hypotheses of Theorem* 4.1 *are satisfied. Then for every* $\alpha > 0$ *and* $x \in D$ *we have*

$$\lim_{\varepsilon \to 0} \mathsf{P}_x\{e^{\varepsilon^{-2}(V_0 - \alpha)} < \tau^\varepsilon < e^{\varepsilon^{-2}(V_0 + \alpha)}\} = 1.$$

Proof. If

$$\overline{\lim_{\varepsilon \to 0}} \mathsf{P}_x\{\tau^\varepsilon > e^{\varepsilon^{-2}(V_0 + \alpha)}\} > 0$$

for some $\alpha > 0$, then

$$\lim_{\varepsilon \to 0} \varepsilon^2 \ln \mathsf{M}_x \tau^\varepsilon \geq V_0 + \alpha,$$

which contradicts Theorem 4.1. Therefore, for any $\alpha > 0$ and $x \in D$ we have

$$\lim_{\varepsilon \to 0} \mathsf{P}_x\{\tau^\varepsilon < \exp\{\varepsilon^{-2}(V_0 + \alpha)\}\} = 1. \tag{4.7}$$

Further, using the notation introduced in the proof of Theorem 2.1, we can write:

$$\begin{aligned} \mathsf{P}_x\{\tau^\varepsilon &< e^{\varepsilon^{-2}(V_0 - \alpha)}\} \\ &\leq \mathsf{M}_x\left\{\tau_1 < \tau^\varepsilon, \sum_{n=1}^{\infty} \mathsf{P}_{X^\varepsilon_{\tau_1}}\{\nu = n, \tau^\varepsilon < \exp\{\varepsilon^{-2}(V_0 - \alpha)\}\}\right\} \\ &\quad + \mathsf{P}_x\{\tau^\varepsilon = \tau_1\}. \end{aligned} \tag{4.8}$$

The last probability on the right side of (4.8) converges to zero. We estimate the remaining terms. Let $m_\varepsilon = [C \exp\{\varepsilon^{-2}(V_0 - \alpha)\}]$; we choose the constant C later. For $x \in \gamma$ we have

$$\begin{aligned} &\sum_{n=1}^{\infty} \mathsf{P}_x\{\nu = n, \tau^\varepsilon < \exp\{\varepsilon^{-2}(V_0 - \alpha)\}\} \\ &\quad \leq \mathsf{P}_x\{\nu < m_\varepsilon\} + \sum_{n=m_\varepsilon}^{\infty} \mathsf{P}_x\{\nu = n, \tau_n < \exp\{\varepsilon^{-2}(V_0 - \alpha)\}\} \\ &\quad \leq \mathsf{P}_x\{\nu < m_\varepsilon\} + \mathsf{P}_x\{\tau_{m_\varepsilon} < \exp\{\varepsilon^{-2}(V_0 - \alpha)\}\}. \end{aligned} \tag{4.9}$$

Using the inequality $\mathsf{P}_x\{\nu = 1\} < \exp\{-\varepsilon^{-2}(V_0 - h)\}$, which holds for $x \in \gamma$, $h > 0$ and sufficiently small ε, we obtain that

$$\mathsf{P}_x\{\nu < m_\varepsilon\} \le 1 - (1 - \exp\{-\varepsilon^{-2}(V_0 - h)\})^{m_\varepsilon} \to 0 \tag{4.10}$$

as $\varepsilon \to 0$, for any $C, \alpha > 0$ and h sufficiently small.

We estimate the second term on the right side of (4.9). There exists $\theta > 0$ such that $\mathsf{P}_x\{\tau_1 > \theta\} \ge 1/2$ for all $x \in \gamma$ and $\varepsilon > 0$. For the number S_m of successes in m Bernoulli trials with probability of success 1/2, we have the inequality

$$\mathsf{P}\{S_m > m/3\} > 1 - \delta$$

for $m > m_0$. Since $\tau_m = (\tau_1 - \tau_0) + (\tau_2 - \tau_1) + \cdots + (\tau_m - \tau_{m-1})$, using the strong Markov property of the process, we obtain that

$$\mathsf{P}_z\{\tau_{m_\varepsilon} < e^{\varepsilon^{-2}(V_0 - \alpha)}\} = \mathsf{P}_x\left\{\frac{\tau_{m_\varepsilon}}{m_\varepsilon} < \frac{1}{C}\right\} < \delta, \tag{4.11}$$

if $\theta/3 > 1/C$ and m_ε is sufficiently large.

Combining estimates (4.8)–(4.11), we arrive at the relation

$$\lim_{\varepsilon \to 0} \mathsf{P}_x\{\tau^\varepsilon < e^{\varepsilon^{-2}(V_0 - \alpha)}\} = 0, \qquad x \in D. \tag{4.12}$$

The assertion of the theorem follows from (4.7) and (4.12). □

Now we pass to the study of the behavior, as $\varepsilon \to 0$, of the invariant measure of X_t^ε defined by equation (1.2). For the existence of a finite invariant measure, we have to make some assumptions on the behavior of $b(x)$ in the neighborhood of infinity. If we do not make any assumptions, then the trajectories of X_t^ε may, for example, go out to infinity with probability 1; in this case no finite invariant measure exists. We shall assume that outside a sufficiently large ball with center at the origin, the projection of $b(x)$ onto the position vector $r(x)$ of the point x is negative and separated from zero, i.e., there exists a large number N such that $(b(x), r(x)) < -1/N$ for $|x| > N$.

This condition, which will be called condition A in what follows, guarantees that X_t^ε returns to the neighborhood of the origin sufficiently fast and thus there exists an invariant measure. A proof of this can be found in Khas'minskii [1]; the same book contains more general conditions guaranteeing the existence of a finite invariant measure.

If there exists an invariant measure $\mu^\varepsilon(\cdot)$ of X_t^ε, then it is absolutely continuous with respect to Lebesgue measure and the density $m^\varepsilon(x) = d\mu^\varepsilon/dx$

satisfies the stationary forward Kolmogorov equation. In our case, this equation has the form

$$\frac{\varepsilon^2}{2}\Delta m^\varepsilon(x) - \sum_1^r \frac{\partial}{\partial x^i}(b^i(x)m^\varepsilon(x)) = 0. \tag{4.13}$$

Together with the additional conditions $\int_{R^r} m^\varepsilon(x)\,dx = 1$, $m^\varepsilon(x) > 0$, this equation determines the function $m^\varepsilon(x)$ uniquely.

First we consider the case of a potential field $b(x)$: $b(x) = -\nabla U(x)$. In this case, the conditions of the existence of a finite invariant measure mean that the potential $U(x)$ increases sufficiently fast with increasing $|x|$; for example, faster than some linear function $\alpha|x| + \beta$. It turns out that if $b(x)$ has a potential, then the density of the invariant measure can be calculated explicitly. An immediate substitution into equation (4.13) shows that

$$m^\varepsilon(x) = c_\varepsilon \exp\{-2\varepsilon^{-2}U(x)\}, \tag{4.14}$$

where c_ε is a normalizing factor defined by the normalization condition $c_\varepsilon = (\int_{R^r} \exp\{-2\varepsilon^{-2}U(x)\}\,dx)^{-1}$. The convergence of the integral occurring here is a necessary and sufficient condition for the existence of a finite invariant measure in the case where a potential exists.

Let D be a domain in R^r. We have $\mu^\varepsilon(D) = c_\varepsilon \int_D \exp\{-2\varepsilon^{-2}U(x)\}\,dx$. Using this representation, we can study the limit behavior of μ^ε as $\varepsilon \to 0$. Let $U(x) \geq 0$ and assume that at some point O, the potential vanishes: $U(O) = 0$. Then it is easy to verify that

$$\varepsilon^2 \ln \mu^\varepsilon(D) = \varepsilon^2 \ln c_\varepsilon + \varepsilon^2 \ln \int_D \exp\{-2\varepsilon^{-2}U(x)\}\,dx \to \inf_{x\in D} 2U(x) \tag{4.15}$$

as $\varepsilon \to 0$. By Laplace's method, we can find a more accurate asymptotics of $\mu^\varepsilon(\cdot)$ as $\varepsilon \to 0$ (cf. Bernstein [1] and Nevel'son [1]). If $b(x)$ does not have a potential, then we cannot write an explicit expression for the density of the invariant measure in general. Nevertheless, it turns out that relation (4.15) is preserved if by $2U(x)$ we understand the quasipotential of $b(x)$.

Theorem 4.3. *Let the point O be the unique stable equilibrium position of system (1.1) and let the whole space R^r be attracted to O. Furthermore, assume that condition A is satisfied. Then the process X_t^ε has a unique invariant measure μ^ε for every $\varepsilon > 0$ and we have*

$$\lim_{\varepsilon\to 0} \varepsilon^2 \ln \mu^\varepsilon(D) = -\inf_{x\in D} V(O, x), \tag{4.16}$$

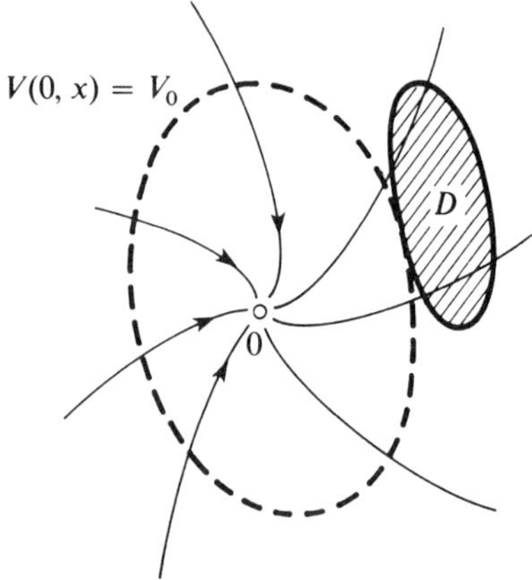

Figure 8.

for any domain $D \subset R^r$ *with compact boundary* ∂D *common for* D *and the closure of* D, *where* $V(O, x)$ *is the quasipotential of* $b(x)$ *with respect to* O:

$$V(O, x) = \inf\{S_{0T}(\varphi): \varphi \in \mathbf{C}_{0T}(R^r), \varphi_0 = O, \varphi_T = x, T > 0\}$$

(*Fig.* 8).

We outline the proof of this theorem. As we have already noted, condition A implies the existence and uniqueness of a finite invariant measure. To prove (4.16), it is sufficient to verify that for any $h > 0$ there exists $\varepsilon_0 = \varepsilon_0(h)$ such that for $\varepsilon < \varepsilon_0$ we have the inequalities

(a) $\mu^\varepsilon(D) > \exp\{-\varepsilon^{-2}(V_0 + h)\}$,

(b) $\mu^2(D) < \exp\{-\varepsilon^{-2}(V_0 - h)\}$,

where $V_0 = \inf_{x \in D} V(O, x)$.

If $V_0 = 0$, then inequalities (a) and (b) are obvious. We discuss the case $V_0 > 0$. It is clear that $V_0 > 0$ only if $\rho(O, D) = \rho_0 > 0$. For the proof of inequalities (a) and (b), we use the following representation of the invariant measure. As earlier, let γ and Γ be the spheres of radii $\mu/2$ and μ, respectively, with center at the equilibrium position O and let $\mu < \rho_0$. As in the proof of Theorem 2.1, we consider the increasing sequence of Markov times $\tau_0, \sigma_0, \tau_1, \sigma_1, \tau_2, \ldots$. Condition A implies that all these times are finite with probability 1.

The sequence $X_{\tau_1}^\varepsilon, X_{\tau_2}^\varepsilon, \ldots, X_{\tau_n}^\varepsilon, \ldots$ forms a Markov chain with compact phase space γ. The transition probabilities of this chain have positive density with respect to Lebesgue measure on γ. This implies that the chain has a unique normalized invariant measure $l^\varepsilon(dx)$. As follows, for example, from Khas'minskii [1], the normalized invariant measure $\mu^\varepsilon(\cdot)$ of X_t^ε can be expressed in terms of the invariant measure of $\{X^\varepsilon_{\tau_n}\}$ on γ in the following way:

$$\mu^\varepsilon(D) = c_\varepsilon \int_\gamma \mathsf{M}_x \int_0^{\tau_1} \chi_D(X_s^\varepsilon)\, ds\, l^\varepsilon(dx), \tag{4.17}$$

where $\chi_D(x)$ is the indicator of D and the factor c_ε is determined from the normalization condition $\mu^\varepsilon(R^r) = 1$. We set $\tau_D = \min\{t: X_t^\varepsilon \in D \cup \partial D\}$. From (4.17) we obtain

$$\begin{aligned}\mu^\varepsilon(D) &= c_\varepsilon \int_\gamma \mathsf{M}_x \int_0^{\tau_1} \chi_D(X_s^\varepsilon)\, ds l^\varepsilon(dx) \\ &\le c_\varepsilon \max_{x \in \gamma} \mathsf{P}_x\{\tau_D < \tau_1\} \cdot \max_{y \in \partial D} \mathsf{M}_y \tau_1.\end{aligned} \tag{4.18}$$

It follows from condition A and the compactness of ∂D that

$$\max_{y \in \partial D} \mathsf{M}_y \tau_1 < a_1 < \infty \tag{4.19}$$

uniformly in $\varepsilon < \varepsilon_0 < 1$. Moreover, it follows from the proof of Theorem 2.1 that for sufficiently small μ and ε and $x \in \gamma$ we have

$$\mathsf{P}_x\{\tau_D < \tau_1\} < \exp\{-\varepsilon^{-2}(V_0 - h/2)\}. \tag{4.20}$$

Taking account of the relation

$$c_\varepsilon = \left(\int_\gamma \mathsf{M}_x \tau_1 l^\varepsilon(dx)\right)^{-1},$$

it is easy to see that for sufficiently small μ and ε,

$$0 < \ln c_\varepsilon < \frac{h}{2\varepsilon^2}. \tag{4.21}$$

We conclude from estimates (4.18)–(4.21) that for sufficiently small ε we have

$$\ln \mu^\varepsilon(D) < \frac{h}{2\varepsilon^2} - \frac{V_0 - h/2}{\varepsilon^2} = -\frac{V_0 - h}{\varepsilon^2}.$$

In order to prove (b), we introduce the set $D_{-\beta} = \{x \in D: \rho(x, \partial D) > \beta\}$. For sufficiently small β, this set is nonempty and $\inf_{x \in D_{-\beta}} V(O, x) < V_0 + h/4$. If $x \in \gamma$ and $\varepsilon, \mu > 0$ are sufficiently small, then

$$\mathsf{P}_x\{\min\{t: X_t^\varepsilon \in D_{-\beta}\} < \tau_1\} > \exp\{-\varepsilon^{-2}(V_0 + h/2)\},$$

$$\inf_{x \in D_{-\beta}} \mathsf{M}_x \int_0^{\tau_1} \chi_D(X_s^\varepsilon)\, ds > a_2 > 0.$$

Combining these estimates, we obtain from (4.17) that for small ε,

$$\mu^\varepsilon(D) \geq c_\varepsilon \cdot \min_{x \in \gamma} \mathsf{P}_x\{\min\{t: X_t^\varepsilon \in D_\beta\} < \tau_1\}$$

$$\times \inf_{x \in D_{-\beta}} \mathsf{M}_x \int_0^{\tau_1} \chi_D(X_s^\varepsilon)\, ds \geq c_\varepsilon \exp\{-\varepsilon^{-2}(V_0 + h/2)\} \cdot a_2.$$

Taking account of (4.21), from this we obtain assertion (b). □

We note that if $b(x)$ has stable equilibrium positions other than zero, then the assertion of Theorem 4.3 is not true in general. The behavior of the invariant measure in the case of a field $b(x)$ of a more general form will be considered in Ch.6.

§5. Gaussian Perturbations of General Form

Let the function $b(x, y)$, $x \in R^r$, $y \in R^l$, with values in R^r be such that $|b(x_1, y_1) - b(x_2, y_2)| \leq K \cdot (|x_1 - x_2| + |y_1 - y_2|)$.

We consider a random process $X_t^\varepsilon = X_t^\varepsilon(x)$, $t \geq 0$, satisfying the differential equation

$$\dot{X}_t^\varepsilon = b(X_t^\varepsilon, \varepsilon\zeta_t), \qquad X_0^\varepsilon = x, \tag{5.1}$$

where ζ_t is a Gaussian random process in R^l.

We shall assume that ζ_t has zero mean and continuous trajectories with probability 1. As is known, for the continuity it is sufficient that the correlation matrix $a(s, t) = (a^{ij}(s, t))$, $a^{ij}(s, t) = \mathsf{M}\zeta_s^i\zeta_t^j$ have twice differentiable entries. We write $b(x) = b(x, 0)$. The process X_t^ε can be considered as a result of small perturbations of the dynamical system $x_t = x_t(x)$:

$$\dot{x}_t = b(x_t), \qquad x_0 = x.$$

It was proved in Ch. 2 that $X_t^\varepsilon \to x_t$ uniformly on every finite time interval as $\varepsilon \to 0$. In the same chapter we studied the expansion of X_t^ε in powers of the small parameter ε. In the present section we deal with large deviations of X_t^ε from the trajectories of the dynamical system x_t.

For continuous functions φ_s, $s \geq 0$, with values in R^l we define an operator $B_x(\varphi) = u$ by the formula

$$u = u_t = \int_0^t b(u_s, \varphi_s)\, ds + x.$$

In other words, $u_t = B_x(\varphi)$ is the solution of equation (5.1), where $\varepsilon\zeta_t$ is replaced by φ_t and we take the initial condition $u_0 = x$. In terms of B_x we can write: $X_t^\varepsilon = B_x(\varepsilon\zeta)$.

We shall denote by $\|\cdot\|_C$ and $\|\cdot\|_{L^2}$ the norms in $\mathbf{C}_{0T}(R^r)$ and $\mathbf{L}^2_{0T}(R^l)$, respectively.

Lemma 5.1. *Suppose that the functions $b(x, y)$ satisfy a Lipschitz condition with constant K and $u = B_x(\varphi)$, $v = B_x(\psi)$, where $\varphi, \psi \in \mathbf{C}_{0T}(R^r)$. Then*

$$\|u - v\|_C \le K\sqrt{T}\, e^{KT}\|\varphi - \psi\|_{L^2}.$$

Proof. By the definition of B_x, we have

$$\begin{aligned} |u_t - v_t| &= \left|\int_0^t [b(u_s, \varphi_s) - b(v_s, \psi_s)]\, ds\right| \\ &\le K\int_0^t |u_s - v_s|\, ds + K\int_0^t |\varphi_s - \psi_s|\, ds. \end{aligned}$$

We conclude from this by means of Lemma 1.1 of Ch. 2 that

$$\|u - v\|_C \le e^{KT}K\int_0^T |\varphi_s - \psi_s|\, ds \le K\sqrt{T}\, e^{KT}\|\varphi - \psi\|_{\mathbf{L}^2}. \quad \square$$

We denote by A the correlation operator of ζ_s. It acts in $\mathbf{L}^2_{0T}(R^l)$ by the formula

$$A\varphi_t = \int_0^T a(s, t)\varphi_s\, ds.$$

As has been shown in the preceding chapter, the action functional of the family of processes $\varepsilon\zeta_t$ in $\mathbf{L}^2_{0T}(R^l)$ as $\varepsilon \to 0$ has the form

$$S^{\zeta}_{0T}(\varphi) = \tfrac{1}{2}\|A^{-1/2}\varphi\|^2_{\mathbf{L}^2}. \tag{5.2}$$

If $A^{-1/2}\varphi$ is not defined, then we set $S^{\zeta}_{0T}(\varphi) = +\infty$.

We put

$$S(\varphi) = S^X_{0T}(\varphi) = \inf_{\psi:\, B_x(\psi) = \varphi} \tfrac{1}{2}\|A^{-1/2}\psi\|^2_{\mathbf{L}^2}.$$

Theorem 5.1. *Let X^ε_t be the random process defined by equation (5.1). The functional $S^X_{0T}(\varphi)$ is the normalized action functional for the family of processes X^ε in $\mathbf{C}_{0T}(R^r)$; the normalizing coefficient is $f(\varepsilon) = \varepsilon^{-2}$.*

Proof. Since B_x acts continuously from $\mathbf{L}^2_{0T}(R^l)$ into $\mathbf{C}_{0T}(R^r)$ and the action functional of the family ε in $\mathbf{L}^2_{0T}(R^l)$ has the form (5.2), by Theorem 3.1 in Ch. 3 the action functional of the family of processes $X^\varepsilon = B_x(\varepsilon\zeta)$ in $\mathbf{C}_{0T}(R^r)$ as $\varepsilon \to 0$ is given by the formula $S^X_{0T}(\varphi) = \inf_{\psi:\, B_x(\psi) = \varphi} \frac{1}{2}\|A^{-1/2}\psi\|^2_{\mathbf{L}^2}$. $\square$

EXAMPLE 5.1. Let $r = l = 1$ and let system (5.1) have the form $\dot{X}_t^\varepsilon = -\arctan(X_t^\varepsilon - \varepsilon\zeta_t)$, $X_t^\varepsilon = x$. In this case B_x has an inverse: $B_x^{-1}(\varphi) = \tan\dot{\varphi} + \varphi$. The action functional of the family X_t^ε can be written in the following way:

$$S_{0T}^X(\varphi) = \frac{1}{2}\int_0^T |A^{-1/2}(\tan\dot{\varphi}_s + \varphi_s)|^2\, ds,$$

where A is the correlation operator of ζ_t.

For example, if ζ_t is a Wiener process, then

$$S_{0T}^X(\varphi) = \frac{1}{2}\int_0^T \left|\frac{d}{ds}(\tan\dot{\varphi}_s + \varphi_s)\right|^2 ds = \frac{1}{2}\int_0^T \left|\frac{\ddot{\varphi}_s}{\cos^2\dot{\varphi}_s} + \dot{\varphi}_s\right|^2 ds.$$

Knowing the action functional, we can determine the rate of convergence to zero of probabilities of various events connected with the perturbed system on a finite time interval and thus obtain results analogous to Theorem 1.2 (cf. Nguen Viet Fu [1], [2]). If the perturbations are of a stationary character, we may hope to also obtain results, analogous to those of §§2, 4, concerning the most probable behavior, for small ε, of the trajectories $X_t^\varepsilon(x)$ of the perturbed system on infinite time intervals or on time intervals growing with decreasing ε.

For example, let O be an asymptotically stable equilibrium position of the system $\dot{x}_t = b(x_t)$, inside a domain D, on the boundary of which the field $b(x)$ is directed strictly inside D. We consider the time

$$\tau^\varepsilon(x) = \min\{t \geq 0; X_t^\varepsilon(x) \notin D\}$$

of first exit from D. We may try to prove the following analogue of Theorem 4.1:

$$\lim_{\varepsilon\to 0} \varepsilon^2 \ln \mathsf{M}\tau^\varepsilon(x) = V_0 = \inf\{S_{0T}(\varphi)\colon \varphi_0 = O, \varphi_T \in \partial D; T > 0\}. \quad (5.3)$$

Nevertheless, it becomes clear immediately that this is not so simple. First of all, an analysis of the presupposed plan of proof shows that the role of the limit of $\varepsilon^2 \ln \mathsf{M}\tau^\varepsilon(x)$ may also be presumably played by

$$V_0^* = \inf\{S_{-\infty,T}(\varphi)\colon \varphi_t = O, -\infty < t \leq 0, \varphi_T \in \partial D; T > 0\}$$

for the same reason. In the case of Markov perturbations, V_0 and V_0^* obviously coincide but in the non-Markov case they may not. Moreover, in the proof of Theorems 2.1, 4.1, and 4.2, we have used a construction involving cycles, dividing a trajectory of the Markov process X_t^ε into parts, the dependence among which could be accounted for and turned out to be sufficiently small.

For an arbitrary stationary perturbation $\varepsilon\zeta_t$ we do not have anything similar: we have to impose, on the stationary process ζ_t, conditions ensuring the weakening of dependence as time goes. Since we are dealing with probabilities converging to zero (probabilities of large deviations), the strong mixing property

$$\sup\{|\mathsf{P}(A \cap B) - \mathsf{P}(A)\mathsf{P}(B)|: A \in \mathscr{F}^{\zeta}_{\leq s}, B \in \mathscr{F}^{\zeta}_{\geq t}\}$$
$$\leq \alpha(t - s) \to 0 \qquad (t - s \to \infty)$$

turns out to be insufficient; we need more precise conditions. These problems are considered in Grin's works [1], [2]; in particular, for a certain class of processes X_t^ε, the infima V_0 and V_0^* coincide and (5.3) is satisfied.

Chapter 5

Perturbations Leading to Markov Processes

§1. Legendre Transformation

In this chapter we shall consider theorems on the asymptotics of probabilities of large deviations for Markov random processes. These processes can be viewed as generalizations of the scheme of summing independent random variables; the constructions used in the study of large deviations for Markov processes generalize constructions encountered in the study of sums of independent terms.

The first general limit theorems for probabilities of large deviations of sums of independent random variables are contained in Cramér's paper [1]. The basic assumption there is the finiteness of exponential moments; the results can be formulated in terms of the Legendre transforms of some convex functions connected with the exponential moments of the random variables.

The families of random processes we are going to consider are analogues of the schemes of sums of random variables with finite exponential moments, so that Legendre's transformation turns out to be essential in our case, as well. First we consider this transformation and its application to families of measures in finite-dimensional spaces.

Let $H(\alpha)$ be a function of an r-dimensional vector argument, assuming its values in $(-\infty, +\infty]$ and not identically equal to $+\infty$. Suppose that $H(\alpha)$ is convex and lower semicontinuous. (We note that the condition of semicontinuity—and even continuity—is satisfied automatically for all α with the exception of the boundary of the set $\{\alpha: H(\alpha) < \infty\}$.) To this function the Legendre transformation assigns the function defined by the formula

$$L(\beta) = \sup_{\alpha} [(\alpha, \beta) - H(\alpha)], \tag{1.1}$$

where $(\alpha, \beta) = \sum_{i=1}^{r} \alpha_i \beta^i$ is the scalar product.

It is easy to prove that L is again a function of the same class as H, i.e., it is convex, lower semicontinuous, assuming values in $(-\infty, +\infty]$ and not identically equal to $+\infty$. The following properties of Legendre's transformation can be found in Rockafellar's book [1]. The inverse of Legendre's transformation is itself:

$$H(\alpha) = \sup_{\beta} [(\alpha, \beta) - L(\beta)] \tag{1.2}$$

(Rockafellar [1], Theorem 12.2). The functions L and H coupled by relations (1.1) and (1.2) are said to be conjugate, which we shall denote in the following way: $H(\alpha) \leftrightarrow L(\beta)$. At points α_0 interior for the set $\{\alpha : H(\alpha) < \infty\}$ with respect to its affine hull, H is subdifferentiable, i.e., it has a (generally nonunique) subgradient, a vector β_0 such that for all α,

$$H(\alpha) \geq H(\alpha_0) + (\alpha - \alpha_0, \beta_0) \tag{1.3}$$

(Rockafellar [1], Theorem 23.4; geometrically speaking, a subgradient is the angular coefficient of a nonvertical supporting plane of the set of points above the graph of the function). The multi-valued mapping assigning to every point the set of subgradients of the function H at that point is the inverse of the same mapping for L, i.e., (1.3) for all α is equivalent to the inequality

$$L(\beta) \geq L(\beta_0) + (\alpha_0, \beta - \beta_0) \tag{1.4}$$

for all β (Rockafellar [1], Theorem 23.5, Corollary 23.5.1). We have $L(\beta) \to \infty$ as $|\beta| \to \infty$ if and only if $H(\alpha) < \infty$ in some neighborhood of $\alpha = 0$.

For functions H, L which are smooth inside their domains of finiteness, the determination of the conjugate function reduces to the classical Legendre transformation: we have to find the solution $\alpha = \alpha(\beta)$ of the equation $\nabla H(\alpha) = \beta$ and $L(\beta)$ is determined from the formula

$$L(\beta) = (\alpha(\beta), \beta) - H(\alpha(\beta)); \tag{1.5}$$

moreover, we have $\alpha(\beta) = \nabla L(\beta)$. If one of the functions conjugate to each other is continuously differentiable $n \geq 2$ times, and the matrix of second-order derivatives is positive definite, then the other function has the same smoothness and the matrices of the second-order derivatives at corresponding points are inverses of each other:

$$\left(\frac{\partial^2 L(\beta)}{\partial \beta^i \, \partial \beta^j}\right) = \left(\frac{\partial^2 H}{\partial \alpha_i \, \partial \alpha_j}(\alpha(\beta))\right)^{-1}.$$

EXAMPLE 1.1. Let $H(\alpha) = r(e^{\alpha} - 1) + l(e^{-\alpha} - 1)$, $\alpha \in R^1$; $r, l > 0$. Upon solving the equation $H'(\alpha) = re^{\alpha} - le^{-\alpha} = \beta$, we find that

$$\alpha(\beta) = \ln \frac{\beta + \sqrt{\beta^2 + 4rl}}{2r};$$

$$L(\beta) = \beta \ln \frac{\beta + \sqrt{\beta^2 + 4rl}}{2r} - \sqrt{\beta^2 + 4rl} + r + l.$$

It turns out that the rough asymptotics of families of probability measures in R^r can be connected with the Legendre transform of the logarithm of exponential moments. The following two theorems are borrowed from Gärtner [2], [3] with some changes.

Let μ^h be a family of probability measures in R^r and put

$$H^h(\alpha) = \ln \int_{R^r} \exp\{(\alpha, x)\}\mu^h(dx).$$

The function H^h is convex: by the Hölder inequality, for $0 < c < 1$ we have

$$\begin{aligned} H^h(c\alpha_1 + (1-c)\alpha_2) &= \ln \int_{R^r} \exp\{c(\alpha_1, x)\} \exp\{(1-c)(\alpha_2, x)\}\mu^h(dx) \\ &\leq \ln\left[\left(\int_{R^r} \exp\{(\alpha_1, x)\}\mu^h(dx)\right)^c \right. \\ &\qquad \left. \cdot \left(\int_{R^r} \exp\{(\alpha_2, x)\}\mu^h(dx)\right)^{1-c}\right] \\ &= cH^h(\alpha_1) + (1-c)H^h(\alpha_2); \end{aligned}$$

H^h is lower semicontinuous (this can be proved easily by means of Fatou's lemma), assumes values in $(-\infty, +\infty]$ and is not identically equal to $+\infty$, since $H^h(0) = 0$.

Let $\lambda(h)$ be a numerical-valued function converging to $+\infty$ as $h \downarrow 0$. We assume that the limit

$$H(\alpha) = \lim_{h \downarrow 0} \lambda(h)^{-1} H^h(\lambda(h)\alpha) \tag{1.6}$$

exists for all α. This function is also convex and $H(0) = 0$. We stipulate that it be lower semicontinuous, not assume the value $-\infty$ and be finite in some neighborhood of $\alpha = 0$. Let $L(\beta) \leftrightarrow H(\alpha)$.

Theorem 1.1. *For the family of measures μ^h and the functions λ and L, condition (II) of §3, Ch. 3 holds, i.e., for any $\delta > 0$, $\gamma > 0$, $s > 0$ there exists $h_0 > 0$ such that for all $h \leq h_0$ we have*

$$\mu^h\{y\colon \rho(y, \Phi(s)) \geq \delta\} \leq \exp\{-\lambda(h)(s - \gamma)\}, \tag{1.7}$$

where $\Phi(s) = \{\beta\colon L(\beta) \leq s\}$.

Proof. The set $\Phi(s)$ can be represented as an uncountable intersection of half-spaces:

$$\Phi(s) = \bigcap_{\alpha} \{\beta\colon (\alpha, \beta) - H(\alpha) \leq s\}.$$

This set is compact, because L is lower semicontinuous and converges to $+\infty$ at infinity. We consider the boundary

$$\partial\Phi_{+\delta}(s) = \{y: \rho(y, \Phi(s)) = \delta\}$$

of the δ-neighborhood of $\Phi(s)$. For every point y of this compact set, there exists α such that $(\alpha, y) - H(y) > s$. Hence the open half-spaces $\{y: (\alpha, y) - H(\alpha) > s\}$ cover the compact set $\partial\Phi_{+\delta}(s)$. From these α we choose a finite number $\alpha_1, \ldots, \alpha_n$. We obtain that the convex polyhedron

$$\bigcap_{i=1}^{n} \{y: (\alpha_i, y) - H(\alpha_i) \leq s\}$$

contains $\Phi(s)$ and does not intersect $\partial\Phi_{+\delta}(s)$. This implies that the polyhedron lies in the δ-neighborhood of $\Phi(s)$.

Using Chebyshev's exponential inequality, we obtain the estimate

$$\begin{aligned}
\mu^h\{y: \rho(y, \Phi(s)) \geq \delta\} &\leq \mu^h\left(\bigcup_{i=1}^{n} \{y: (\alpha_i, y) - H(\alpha_i) > s\}\right) \\
&\leq \sum_{i=1}^{n} \mu^h\{y: (\alpha_i, y) - H(\alpha_i) > s\} \\
&\leq \sum_{i=1}^{n} \int_{R^r} \exp\{\lambda(h)[(\alpha_i, y) - H(\alpha_i) - s]\}\mu^h(dy) \\
&= \sum_{i=1}^{n} \exp\{\lambda(h)[\lambda(h)^{-1}H^h(\lambda(h)\alpha_i) - H(\alpha_i)]\} \\
&\quad \times \exp\{-\lambda(h)s\}.
\end{aligned}$$

We obtain (1.7) from this by taking account of (1.6). □

We shall say that a convex function L is *strictly convex at a point* β_0 if there exists α_0 such that

$$L(\beta) > L(\beta_0) + (\alpha_0, \beta - \beta_0) \tag{1.8}$$

for all $\beta \neq \beta_0$. For a function L to be strictly convex at all points interior to the set $\{\beta: L(\beta) < \infty\}$ with respect to its affine hull (with the notation of Rockafellar's book [1], §§4, 6, at the points of the set ri(dom L)), it is sufficient that the function H conjugate to L be sufficiently smooth, i.e., that the set $\{\alpha: H(\alpha) < \infty\}$ have interior points, H be differentiable at them and if a sequence of points α_i converges to a boundary point of the set $\{\alpha: H(\alpha) < \infty\}$, then we have $|\nabla H(\alpha_i)| \to \infty$ (cf. Rockafellar [1], Theorem 26.3).

Theorem 1.2. *Let the assumptions imposed on μ^h, H^h and H earlier be satisfied. Moreover, let the function L be strictly convex at the points of a dense subset of $\{\beta: L(\beta) < \infty\}$.*

For the family of measures μ^h and the functions λ and L, condition (I) *of* §3 *of Ch.* 3 *is satisfied, i.e., for any $\delta > 0$, $\gamma > 0$ and $x \in R^r$ there exists $h_0 > 0$ such that for $h < h_0$ we have*

$$\mu^h\{y: \rho(y, x) < \delta\} \geq \exp\{-\lambda(h)[L(x) + \gamma]\}. \tag{1.9}$$

Proof. It is sufficient to prove the assertion of the theorem for points x at which L is strictly convex. Indeed, the fulfillment of the assertion of the theorem for such x is equivalent to its fulfillment for all x, the same function λ and the function $\tilde{L}(x)$ defined as $L(x)$ if L is strictly convex at x and as $+\infty$ otherwise. At the points where $L(x) < \tilde{L}(x)$, we have to use the circumstance that $L(x) = \lim_{y \to x} \tilde{L}(y)$, and the remark made in §3 of Ch. 3.

Suppose that L is strictly convex at x. We choose α_0 so that $L(\beta) > L(x) + (\alpha_0, \beta - x)$ for $\beta \neq x$. Then we have

$$H(\alpha_0) = \sup_\beta[(\alpha_0, \beta) - L(\beta)] = (\alpha_0, x) - L(x). \tag{1.10}$$

Since $H(\alpha_0)$ is finite, $H^h(\lambda(h)\alpha_0)$ is also finite for sufficiently small h. For such h we consider the probability measure μ^{h,α_0} defined by the relation

$$\mu^{h,\alpha_0}(\Gamma) = \int_\Gamma \exp\{\lambda(h)(\alpha_0, y) - H^h(\lambda(h)\alpha_0)\}\mu^h(dy).$$

We use the mutual absolute continuity of μ^h and μ^{h,α_0}:

$$\mu^h\{y: \rho(y, x) < \delta\} = \int_{\{y: \rho(y, x) < \delta\}} \exp\{-\lambda(h)(\alpha_0, y) + H^h(\lambda(h)\alpha_0)\}\mu^{h,\alpha_0}(dy). \tag{1.11}$$

We put $\delta' = \delta \wedge \gamma/3|\alpha_0|$ and estimate the integral (1.11) from below by the product of the μ^{h,α_0}-measure of the δ'-neighborhood of x with the infimum of the function under the integral sign:

$$\begin{aligned}\mu^h\{y: \rho(y, x) < \delta\} \geq \mu^{h,\alpha_0}&\{y: \rho(y, x) < \delta'\} \\ &\times \exp\{-\lambda(h)[(\alpha_0, x) - \lambda(h)^{-1}H(\lambda(h)\alpha_0)]\} \\ &\times \exp\left\{-\lambda(h)\frac{\gamma}{3}\right\}.\end{aligned}$$

By (1.6) and (1.10), the second factor here is not smaller than

$$\exp\{-\lambda(h)[L(x) + \gamma/3]$$

if h is sufficiently small. If we prove that $\mu^{h,\alpha_0}\{y: \rho(y, x) < \delta'\} \to 1$ as $h \downarrow 0$, then everything will be proved.

For this we apply Theorem 1.1 to the family of measures μ^{h, α_0}. We calculate the characteristics of this family:

$$H^{h, \alpha_0}(\alpha) = \ln \int_{R^r} \exp\{(\alpha, y)\} \mu^{h, \alpha_0}(dy)$$
$$= H^h(\alpha + \lambda(h)\alpha_0) - H^h(\lambda(h)\alpha_0);$$
$$H^{\alpha_0}(\alpha) = \lim_{h \downarrow 0} \lambda(h)^{-1} H^{h, \alpha_0}(\lambda(h)\alpha) = H(\alpha_0 + \alpha) - H(\alpha_0);$$
$$L^{\alpha_0}(\beta) = L(\beta) - [(\alpha_0, \beta) - H(\alpha_0)].$$

The function $L^{\alpha_0}(\beta)$ vanishes at $\beta = x$ and is nonnegative everywhere (since $H^{\alpha_0}(0) = 0$). H is strictly convex at x since $L(\beta)$ is. This implies that $L^{\alpha_0}(\beta)$ is strictly positive for all $\beta \neq x$ and

$$\gamma_0 = \min\{L^{\alpha_0}(\beta): \rho(\beta, x) \geq \delta'/2\} > 0.$$

We use estimate (1.7) with $\delta'/2$ instead of δ, positive $\gamma < \gamma_0$ and $s \in (\gamma, \gamma_0)$. We obtain for sufficiently small h that

$$\mu^{h, \alpha_0}\{y: \rho(y, x) \geq \delta'\} \leq \mu^{h, \alpha_0}\{y: \rho(y, \Phi^{\alpha_0}(s)) \geq \delta'/2\} \leq \exp\{-\lambda(h)(s - \gamma)\},$$

which converges to zero as $h \downarrow 0$. □

Consequently, if the hypotheses of Theorems 1.1 and 1.2 are satisfied, then $\lambda(h)L(x)$ is the action function for the family of measures μ^h as $h \downarrow 0$.

The following example shows that the requirement of strict convexity of L on a set dense in $\{\beta: L(\beta) < \infty\}$ cannot be omitted.

EXAMPLE 1.2. For the family of Poisson distributions $(\mu^h(\Gamma) = \sum_{k \in \Gamma} (h^k e^{-h}/k!))$ we have: $H^h(\alpha) = h \cdot (e^\alpha - 1)$. If we are interested in values of h converging to zero and put $\lambda(h) = -\ln h$, then we obtain

$$H(\alpha) = \lim_{h \downarrow 0} (-\ln h)^{-1} \ln H^h(-\alpha \ln h)$$

$$= \lim_{h \downarrow 0} \frac{h^{-\alpha+1} - h}{-\ln h} = \begin{cases} 0, & \alpha \leq 1, \\ +\infty, & \alpha > 1; \end{cases}$$

$$L(\beta) = \begin{cases} +\infty, & \beta < 0, \\ \beta, & \beta \geq 0. \end{cases}$$

But, the normalized action function found by us in §3, Ch. 3 is different from $+\infty$ only for nonnegative integral values of the argument.

Another example: we take a continuous finite function $S(x)$ which is not convex and is such that $S(x)/|x| \to \infty$ as $|x| \to \infty$ and $\min S(x) = 0$. As μ^h we take the probability measure with density $C(h) \exp\{-\lambda(h)S(x)\}$, where

$\lambda(h) \to \infty$ as $h \downarrow 0$. Here the normalized action function will be $S(x)$ but the Legendre transform of the function

$$H(\alpha) = \lim_{h \downarrow 0} \lambda(h)^{-1} \ln \int \exp\{\lambda(h)(\alpha, x)\}\mu^h(dx)$$

will be equal not to $S(x)$ but rather the convex hull $L(x)$ of $S(x)$. In those domains where S is not convex, L will be linear and consequently, not strictly convex.

The following examples show how the theorems proved above can be applied to obtain rough limit theorems on large deviations for sums of independent random variables. Or course, they can be derived from the precise results of Cramér [1] (at least in the one-dimensional case).

EXAMPLE 1.3. Let $\xi_1, \xi_2, \ldots, \xi_n, \ldots$ be a sequence of independent identically distributed random vectors and let

$$H_0(\alpha) = \ln \mathsf{M} e^{(\alpha, \xi_i)}$$

be finite for sufficiently small $|\alpha|$. We are interested in the rough asymptotics as $n \to \infty$ of the distribution of the arithmetic mean $(\xi_1 + \cdots + \xi_n)/n$.

We have

$$H^n(\alpha) = \ln \mathsf{M} \exp\left\{\left(\alpha, \frac{\xi_1 + \cdots + \xi_n}{n}\right)\right\} = nH_0(n^{-1}\alpha).$$

We put $\lambda(n) = n$. Then not only does $\lambda(n)^{-1}H^n(\lambda(n)\alpha)$ converge to $H_0(\alpha)$ but it also coincides with it.

The function H_0 is infinitely differentiable at interior points of the set $\{\alpha: H_0(\alpha) < \infty\}$. If it also satisfies the condition $|\nabla H_0(\alpha_i)| \to \infty$ as the points α_i converge to a boundary point of the above set, then its Legendre transform L_0 is strictly convex and the asymptotics of the distribution of the arithmetic mean is given by the action function $n \cdot L_0(x)$.

EXAMPLE 1.4. Under the hypotheses of the preceding example, we consider the distributions of the random vectors $(\xi_1 + \xi_2 + \cdots + \xi_n - nM\xi_k)/B_n$, where B_n is a sequence going to ∞ faster than $\sqrt{n}$ but slower than n. We have

$$\begin{aligned} H^n(\alpha) &= \ln \mathsf{M} \exp\left\{\left(\alpha, \frac{\xi_1 + \cdots + \xi_n - nM\xi_k}{B_n}\right)\right\} \\ &= n\left[H_0\left(\frac{\alpha}{B_n}\right) - \left(\frac{\alpha}{B_n}, \nabla H_0(0)\right)\right] \\ &= n\left[\frac{1}{2}\sum_{i,j} \frac{\partial^2 H_0}{\partial \alpha_i\, \partial \alpha_j}(0)\frac{\alpha_i \alpha_j}{B_n^2} + o\left(\frac{|\alpha|^2}{B_n^2}\right)\right]. \end{aligned}$$

If as the normalizing coefficient $\lambda(n)$ we take B_n^2/n, then we obtain

$$H(\alpha) = \lim_{n\to\infty} \lambda(n)^{-1} H^n(\lambda(n)\alpha) = \frac{1}{2}\sum_{i,j} \frac{\partial^2 H_0}{\partial\alpha_i\,\partial\alpha_j}(0)\alpha_i\alpha_j.$$

If the matrix of this quadratic form, i.e., the covariance matrix of the random vector ξ_n is nonsingular, then the Legendre transform of H has the form

$$L(\beta) = \frac{1}{2}\sum_{i,j} a_{ij}\beta^i\beta^j,$$

where $(a_{ij}) = ((\partial^2 H_0/\partial\alpha_i\,\partial\alpha_j)(0))^{-1}$. The action function for the family of the random vectors under consideration is $(B_n^2/n)L(x)$. In particular, this means that

$$\lim_{\delta\downarrow 0}\lim_{n\to\infty} \frac{n}{B_n^2}\ln \mathsf{P}\left\{\left|\frac{\xi_1 + \cdots + \xi_n - nM\xi_k}{B_n} - x\right| < \delta\right\} = -\frac{1}{2}\sum_{i,j} a_{ij}x^i x^j.$$

§2. Locally Infinitely Divisible Processes

Discontinuous Markov processes which can be considered as a result of random perturbations of dynamical systems arise in various problems. We consider an example.

Let two nonnegative functions $l(x)$ and $r(x)$ be given on the real line. For every $h > 0$ we consider the following Markov process X_t^h on those points of the real line which are multiples of h: if the process begins at a point x, then over time dt it jumps distance h to the right with probability $h^{-1}r(x)\,dt$ (up to infinitesimals of higher order as $dt \to 0$) and to the left with probability $h^{-1}l(x)\,dt$ (it jumps more than once with probability $o(dt)$). For small h, in first approximation the process can be described by the differential equation $\dot{x}_t = r(x_t) - l(x_t)$ (the exact meaning of this is as follows: under certain assumptions on r and l it can be proved that as $h \downarrow 0$, X_t^h converges in probability to the solution of the differential equation with the same initial condition).

A more concrete version of this example is as follows: in a culture medium of volume V there are bacteria whose rates c_+ and c_- of division and death depend on the concentration of bacteria in the given volume. An appropriate mathematical model of the process of the variation of concentration of bacteria with time is a Markov process X_t^h of the form described above with $h = V^{-1}$, $r(x) = x \cdot c_+(x)$ and $l(x) = x \cdot c_-(x)$.

It is natural to consider the process X_t^h as a result of a random perturbation of the differential equation $\dot{x}_t = r(x_t) - l(x_t)$ (a result of a *small* random perturbation for small h). As in the case of perturbations of the type of a white noise, we may be interested in probabilities of events of the form $\{\rho_{0T}(X^h, \varphi) < \delta\}$, etc. (probabilities of "large deviations").

As we have already mentioned, the first approximation of X_t^h for small h is the solution of the differential equation; the second approximation will be a diffusion process with drift $r(x) - l(x)$ and small local variance

$$h(r(x) + l(x)).$$

Nevertheless, this approximation does not work for large deviations: as we shall see, the probabilities of large deviations for the family of processes X_t^h can be described by means of an action functional not coinciding with the action functional of the diffusion processes.

We describe a general scheme which includes the above example.

In r-space R^r let us be given: a vector-valued function $b(x) = (b^1(x), \ldots, b^r(x))$, a matrix-valued function $(a^{ij}(x))$ (of order r, symmetric and nonnegative definite) and for every $x \in R^r$, a measure μ_x on $R^r \setminus \{0\}$ such that

$$\int_{R^r \setminus \{0\}} |\beta|^2 \mu_x(d\beta) < \infty.$$

For every $h > 0$ let (X_t^h, P_x^h) be a Markov process in R^r with right continuous trajectories and infinitesimal generator A^h defined for twice continuously differentiable functions with compact support by the formula

$$A^h f(x) = \sum_i b^i(x) \frac{\partial f(x)}{\partial x^i} + \frac{h}{2} \sum_{i,j} a^{ij}(x) \frac{\partial^2 f(x)}{\partial x^i \, \partial x^j}$$
$$+ h^{-1} \int_{R^r \setminus \{0\}} \left[f(x + h\beta) - f(x) - h \sum_i \beta^i \frac{\partial f(x)}{\partial x^i} \right] \mu_x(d\beta). \quad (2.1)$$

If $a^{ij}(x) \equiv 0$ and μ_x is finite for all x, then X_t^h moves in the following way: it jumps a finite number of times over any finite time and the density of jumps at x is $h^{-1}\mu_x(R^r \setminus \{0\})$ (i.e., if the process is near x, then over time dt it makes a jump with probability $h^{-1}\mu_x(R^r \setminus \{0\})\, dt$ up to infinitesimals of higher order as $dt \to 0$); the distribution of the length of a jump is given by the measure $\mu_x(R^r \setminus \{0\})^{-1} \cdot \mu_x(h^{-1}\, d\beta)$ (again, as $dt \to 0$); between jumps the process moves in accordance with the dynamical system $\dot{x}_t = \tilde{b}(x_t)$, where

$$\tilde{b}(x) = b(x) - \int_{R^r \setminus \{0\}} \beta \mu_x(d\beta).$$

On the other hand, if $\mu_x(R^r \setminus \{0\}) = \infty$, then the process jumps infinitely many times over a finite time.

The process under consideration above is a special case of our scheme with $r = 1$, measure μ_x concentrated at the points ± 1, $\mu_x\{1\} = r(x)$, $\mu_x\{-1\} = l(x)$ and $b(x) = r(x) - l(x)$.

If the measure μ_x is concentrated at 0 for every x, then the integral term in formula (2.1) vanishes and A^h turns into a differential operator of the second order. In this case (X_t^h, P_x) is a family of diffusion processes with a small diffusion coefficient and the corresponding trajectories are continuous with probability one. In the general case (X_t^h, P_x) combines a continuous diffusion motion and jumps.

The scheme introduced by us is a generalization of the scheme of processes with independent increments—the continuous version of the scheme of sums of independent random variables. It is known that in the study of large deviations for sums of independent random variables, an important role is played by the condition of finiteness of exponential moments (cf., for example, Cramér [1]). We introduce this condition for our scheme, as well: we shall assume that for all $\alpha = (\alpha_1, \ldots, \alpha_r)$, the expression

$$H(x, \alpha) = \sum_i b^i(x)\alpha_i + \frac{1}{2}\sum_{i,j} a^{ij}(x)\alpha_i\alpha_j + \int_{R^r\setminus\{0\}} \left(\exp\left\{\sum_i \alpha_i\beta^i\right\} - 1 - \sum_i \alpha_i\beta^i\right)\mu_x(d\beta) \tag{2.2}$$

is finite. The function H is convex and analytic in the second argument. It vanishes at zero.

The connection of H with the Markov process (X_t^h, P_x) can be described in the following way: if we apply the operator A^h defined by formula (2.1) to the function $\exp\{\sum_i \alpha_i x^i\}$, then we obtain $h^{-1}H(x, h\alpha)\exp\{\sum_i \alpha_i x^i\}$.

We denote by $L(x, \beta)$ the Legendre transform of $H(x, \alpha)$ with respect to the second variable. The equality $H(x, 0) = 0$ implies that L is nonnegative; it vanishes at $\beta = b(x)$. The function L may assume the value $+\infty$; however, inside the domain where it is finite, L is smooth.

For the example considered above we have: $H(x, \alpha) = r(x)(e^\alpha - 1) + l(x)(e^{-\alpha} - 1)$, and the function L has the form indicated in the preceding section with $r(x)$ and $l(x)$ replacing r and l.

For a function φ_t, $T_1 \le t \le T_2$, with values in R^r, we define a functional by the formula

$$S(\varphi) = S_{T_1T_2}(\varphi) = \int_{T_1}^{T_2} L(\varphi_t, \dot{\varphi}_t)\, dt, \tag{2.3}$$

if φ is absolutely continuous and the integral is convergent; otherwise we put $S_{T_1T_2}(\varphi) = +\infty$. This functional will be a normalized action functional (and the normalizing coefficient will be h^{-1}).

In particular, if the measure μ is concentrated at 0 and (a^{ij}) is the identity matrix, then, as follows from results of Ch. 4, the action functional has the form (2.3), where $L(\varphi, \dot{\varphi}) = \frac{1}{2}\cdot|\dot{\varphi} - b(\varphi)|^2$. In Wentzell and Freidlin [4] the action functional is computed for a family of diffusion processes with an arbitrary matrix (a^{ij}) (cf. the next section).

Families of infinitely divisible processes belonging to our scheme have been considered in Borovkov [1]. We return to these classes of random processes in the next section. Now we formulate a result which generalizes results of both Wentzell and Freidlin [4] and Borovkov [1].

In order that $h^{-1}S_{0T}(\varphi)$ be the action functional for the family of processes (X_t^h, P_x), it is, of course, necessary to impose some restrictions on this family. We formulate them in terms of the functions H and L.

I. There exists an everywhere finite nonnegative convex function $\overline{H}(\alpha)$ such that $\overline{H}(0) = 0$ and $H(x, \alpha) \leq \overline{H}(\alpha)$ for all x, α.
II. The function $L(x, \beta)$ is finite for all values of the arguments; for any $R > 0$ there exists positive constants M and m such that $L(x, \beta) \leq M$, $|\nabla_\beta L(x, \beta)| \leq M$, $\sum_{ij} (\partial^2 L/(\partial\beta^i\, \partial\beta^j)(x, \beta)c^i c^j \geq m \sum_i (c^i)^2$ for all $x, c \in R^r$ and all β, $|\beta| < R$.

The following requirement is between the simple continuity of H and L in the first argument, which is insufficient for our purposes, and uniform continuity, which is not satisfied even in the case of diffusion processes, i.e., for functions H and L quadratic in the second argument.

III. $$\Delta L(\delta') = \sup_{|y-y'|<\delta'} \sup_{\beta} \frac{L(y', \beta) - L(y, \beta)}{1 + L(y, \beta)} \to 0 \text{ as } \delta' \to 0.$$

Condition III implies the following continuity condition for H:

$$H(y, (1 + \Delta L(\delta'))^{-1}\alpha) - H(y', \alpha) \leq \Delta L(\delta')(1 + \Delta L(\delta'))^{-1} \tag{2.4}$$

for all α and $|y - y'| < \delta'$, where $\Delta L(\delta') \to 0$ as $\delta' \downarrow 0$.

Theorem 2.1. *Suppose that the functions H and L satisfy conditions* I–III *and the functional $S(\varphi)$ is defined by formula* (2.3). *Then $h^{-1}S(\varphi)$ is the action functional for the family of processes (X_t^h, P_x^h) in the metric*

$$\rho_{0T}(\varphi, \psi) = \sup_{0 \leq t \leq T} |\varphi_t - \psi_t|$$

uniformly in the initial point as $h \downarrow 0$.

The proof (under somewhat more relaxed conditions) is contained in Wentzell [7] (see also Wentzell [9]).

First of all we have to prove that S_{0T} is lower semicontinuous and the elements of the set $\{\varphi: S_{0T}(\varphi) \leq s\}$ are equicontinuous. We do not include this purely analytic part of the proof here; we mention that similar results can be found in the book by Ioffe and Tikhomirov [1], Ch. 9, §1.

We mention a lemma, also without proof, having to do with the properties of $S_{[0T]}$ (for a proof, see Wentzell [7], [9]).

Lemma 2.1. *Let condition* III *be satisfied. For every* $s_0 > 0$ *there exists* $\Delta t > 0$ *such that for any partition of the interval from* 0 *to* T *by points* $0 = t_0 < t_1 < \cdots < t_{n-1} < t_n = T$, $\max(t_{j+1} - t_j) \le \Delta t$, *and for any function* φ_t, $0 \le t \le T$, *for which* $S_{0T}(\varphi) \le s_0$ *we have* $S_{0T}(l) \le S_{0T}(\varphi) + \gamma$, *where* l *is a polygon with vertices at the points* (t_j, φ_{jt}).

Now, it has to be proved that for any $\delta > 0$, $\gamma > 0$ and $s_0 > 0$ there exists a positive h_0 such that for all $h < h_0$ we have

$$\mathsf{P}^h_{x0}\{\rho_{0T}(X^h, \varphi) < \delta\} \ge \exp\{-h^{-1}[S_{0T}(\varphi) + \gamma]\}, \tag{2.5}$$

where φ is an arbitrary function such that $S_{0T}(\varphi) \le s_0$, $\varphi_0 = x$ and that

$$\mathsf{P}^h_x\{\rho_{0T}(X^h, \Phi_x(s)) \ge \delta\} \le \exp\{-h^{-1}(s - \gamma)\}, \tag{2.6}$$

where $\Phi_x(s) = \{\varphi\colon \varphi_0 = x, S_{0T}(\varphi) \le s\}$ and x is any point from R^r and $s \le s_0$.

For the sake of simple notation we give the proof in the one-dimensional case. First we mention a few facts used in both parts of the proof.

Up to the term $\int_0^t b(X_s^h)\,ds$ of bounded variation, the process X_t^h is a square integrable martingale. Therefore, the stochastic integral with respect to X_t^h is meaningful. Let $f(t, \omega)$, $0 \le t \le T$ be a random function bounded in absolute value by a constant C and progressively measurable with respect to the family of σ-algebras $\mathscr{F}_{\le t} = \sigma\{X_s^h, s \le t\}$ (i.e., for any t, the function $f(s, \omega)$ for $s \le t$ is measurable in the pair (s, ω) with respect to the product of the σ-algebra of Borel subsets of the interval $[0, t]$ and the σ-algebra $\mathscr{F}_{\le t}$). Then we can define the stochastic integral

$$\int_0^T f(t - 0, \omega)\, dX_t^h \tag{2.7}$$

(cf. Kunita and Watanabe [1]). We need to integrate only random functions piecewise continuous in square mean; for them the integral (2.7) can be defined as the limit, in the sense of convergence in probability, of Riemann sums

$$\sum_{j=0}^{n-1} f(t_j, \omega)(X_{t_{j+1}}^h - X_{t_j}^h)$$

as the partition $0 = t_0 < t_1 < \cdots < t_n = T$ becomes infinitely fine. If the realizations of X_t^h have bounded variation and the realizations of $f(t, \omega)$ have one-sided limits at every point, then the integral (2.7) can be viewed as an ordinary Lebesgue–Stieltjes integral of the left-hand limit $f(t - 0, \omega)$ of the function $f(t, \omega)$.

We need the following fact:

$$\mathsf{M}_x^h \exp\left\{\int_0^T f(t-0, \omega)\, dX_t^h - h^{-1}\int_0^T H(X_t^h, hf(t, \omega))\, dt\right\} = 1 \quad (2.8)$$

(this can be derived from Ito's formula for stochastic integrals with respect to martingales, cf. Kunita and Watanabe [1]).

The proof of (2.5) is first carried out for piecewise smooth functions φ. As in the case of diffusion, we use an absolutely continuous change of measures (the idea of using such a change of measures is due to Cramér [1]).

We put $\alpha(t, x) = (\partial L/\partial \beta)(x, \dot{\varphi}_t)$. By condition II and the piecewise smoothness of φ, the function α is bounded. Then, we put

$$\tilde{\mathsf{P}}_x^h(A) = \mathsf{M}_x^h\left[A; \exp\left\{h^{-1}\left[\int_0^T \alpha(t-0, X_{t-0}^h)\, dX_t^h - \int_0^T H(X_t^h, \alpha(t, H_t^h))\, dt\right]\right\}\right]$$

for any event $A \in \mathscr{F}_{\leq t}$.

According to (2.8), P_x^h is a probability measure. With respect to this measure, X_t^h turns out to be a Markov process again (cf. Gikhman and Skorokhod [2], Ch. VII, §6) but now nonhomogeneous in time. Its infinitesimal generator is given by the formula (cf. Grigelionis [1])

$$\tilde{A}_t^h f(x) = \tilde{b}(t, x) f'(x) + \frac{h}{2} a(x) f''(x) + h^{-1}\int_{R^r\setminus\{0\}} [f(x + h\beta) - f(x) - hf'(x)\beta]\tilde{m}_{t,x}(d\beta),$$

where $ha(x)$ is the former diffusion coefficient and the drift $\tilde{b}$ and measure $\tilde{m}$ can be expressed in terms of the former ones by the formulas

$$\tilde{b}(t, x) = \frac{\partial H}{\partial \alpha}(x, \alpha(t, x)),$$

$$\tilde{m}_{t,x}(d\beta) = \exp\{h\alpha(t, x)\beta)\} m_x(d\beta).$$

By virtue of our choice of $\alpha(t, x) = (\partial L/\partial \beta)(x, \dot{\varphi}_t)$, the coefficient $\tilde{b}$ is independent of x and equal to $\dot{\varphi}_t$. This means, in particular, that with respect to $\tilde{\mathsf{P}}_x^h$, the process $X_t^h - \varphi_t$ is a martingale (with vanishing mean, since $\varphi_0 = x$) and its variance is given by the formula

$$\tilde{\mathsf{M}}_x^h(X_t^h - \varphi_t)^2 = h\tilde{\mathsf{M}}_x^h \int_0^t \frac{\partial^2 H}{\partial \alpha^2}(X_s^h, \alpha(s, X_s^h))\, ds. \quad (2.9)$$

Since the density $d\tilde{\mathsf{P}}_x^h/d\mathsf{P}_x^h$ is positive, the measure P_x^h can be expressed by integration with respect to $\tilde{\mathsf{P}}_x^h$:

$$\mathsf{P}_x^h(A) = \tilde{\mathsf{M}}_x^h\left[A; \exp\left\{h^{-1}\left[-\int_0^T \alpha(t-0, X_{t-0}^h)\,dX_t^h + \int_0^T H(X_t^h, \alpha(t, X_t^h))\,dt\right]\right\}\right]. \tag{2.10}$$

We shall use this for the event $A_\delta = \{\sup_{0 \le t \le T} |X_t^h - \varphi_t| < \delta\}$.

We put

$$D_1 = \sup_{t,y} \frac{\partial^2 H}{\partial \alpha^2}(y, \alpha(t, y)),$$

$$D_2 = \sup_{t,y} \frac{\partial^2 H}{\partial \alpha^2}(y, \alpha(t, y))\alpha(t, y)^2$$

(by condition II, these expressions are uniformly bounded for all functions φ_t for which the derivative $\dot{\varphi}_t$ is bounded by some constant). We introduce the random events

$$A_1^h = \left\{\sup_{0 \le t \le T} |X_t^h - \varphi_t| < 2h^{1/2}D_1^{1/2}\right\};$$

$$A_2^h = \left\{\left|\int_0^T \alpha(t-0, X_{t-0}^h)d(X_t^h - \varphi_t)\right| < 2h^{1/2}D_2^{1/2}\right\}.$$

In both cases here on the left side of the inequality we have a martingale (with respect to $\tilde{\mathsf{P}}_x^h$) with mean zero and on the right side a constant estimating from above the doubled square mean deviation of the martingale. By the Kolmogorov and Chebyshev inequalities we have

$$\tilde{\mathsf{P}}_x^h(A_1^h) \ge 3/4, \qquad \tilde{\mathsf{P}}_x^h(A_2^h) \ge 3/4.$$

For $h \le (4D_1)^{-1}\delta^2$, the event A_1^h implies A_δ. Therefore, $\mathsf{P}_x^h(A_\delta)$ is estimated from below by the $\tilde{\mathsf{P}}_x^h$-probability of $A_1^h \cap A_2^h$ (which is not smaller than 1/2) multiplied by the infimum of the exponential expression in (2.10) over this intersection. The sum of integrals under the exponential sign can be reduced to the form

$$-\int_0^T \alpha(t-0, X_{t-0}^h)d(X_t^h - \varphi_t) - \int_0^T [\alpha(t, X_t^h)\dot{\varphi}_t - H(X_t^h, \alpha(t, X_t^h))]\,dt.$$

The first integral here does not exceed $2h^{1/2}D_2^{1/2}$ in absolute value for $\omega \in A_1^h \cap A_2^h$ and the second one is equal to $-\int_0^T L(X_t^h, \dot{\varphi}_t)\,dt$. By virtue of condition III, this integral (without the minus sign) does not exceed

$$\int_0^T L(\varphi_t, \dot{\varphi}_t)\, dt(1 + \Delta L(2h^{1/2}D_1^{1/2})) + \Delta L(2h^{1/2}D_1^{1/2}) \cdot T.$$

Finally, for $h \le (4D_1)^{-1}\delta^2$ we obtain the estimate

$$\mathsf{P}_x^h\{\rho_{0T}(X^h, \varphi) < \delta\} \ge \tfrac{1}{2} \exp\{-h^{-1}S_{0T}(\varphi) - 2h^{-1/2}D_2^{1/2} - h^{-1}\Delta L(2h^{1/2}D_1^{1/2})[T + S_{0T}(\varphi)]\}.$$

The factor 1/2 and all terms except $-h^{-1}S_{0T}(\varphi)$ are absorbed by the term $h^{-1}\gamma$ for sufficiently small h and we obtain inequality (2.5).

In order to establish that (2.5) is satisfied uniformly for not only functions φ_t with $|\dot{\varphi}_t| \le$ const but also all functions with $S_{0T}(\varphi) \le s_0$, we use the equicontinuity of these functions and Lemma 2.1. We choose $\Delta t > 0$ such that the oscillation of each of these functions φ on any interval of length not exceeding Δt is smaller than $\delta/2$. We decrease this Δt if necessary so that for any polygon l with vertices (t_j, φ_{t_j}) such that $\max(t_{j+1} - t_j) \le \Delta t$ we have $S_{0T}(l) \le S_{0T}(\varphi) + \gamma/2$. We fix an equidistant partition of the interval from 0 to T into intervals of length not exceeding Δt. Then for all polygons l determined by functions φ with $S_{0T}(\varphi) \le s_0$ we have $|\dot{l}_t| \le \delta/2\Delta t$. For h smaller than some h_0, inequality (2.5) holds for these polygons with $\delta/2$ instead of δ and $\gamma/2$ instead of γ. We use the circumstance that $\rho_{0T}(\varphi, l) < \delta/2$. We obtain

$$\mathsf{P}_x^h\{\rho_{0T}(X^h, \varphi) < \delta\} \ge \mathsf{P}_x^h\{\rho_{0T}(X^h, l) < \delta/2\} \ge \exp\{-h^{-1}[S_{0T}(\varphi) + \gamma]\}.$$

Now we prove (2.6). Given a small positive number $\varkappa$, we choose a positive $\delta' < \delta/2$ such that the expression $\Delta L(\delta')$ in condition III does not exceed $\varkappa$. We consider a random polygon l_t^h, $0 \le t \le T$ with vertices at the points $(0, X_0^h)$, $(\Delta t, X_{\Delta t}^h)$, $(2\Delta t, X_{2\Delta t}^h)$, $\ldots$, (T, X_T^h), where the choice of $\Delta t = T/n$ (n is an integer) will be specified later. We introduce the events

$$A^h(i) = \left\{\sup_{i\Delta t \le t \le (i+1)\Delta t} |X_t^h - X_{i\Delta t}^h| < \delta'\right\},$$

$i = 0, 1, \ldots, n - 1$. If all $A^h(i)$ occurred, then we had $\rho_{0T}(X^h, l^h) < \delta$; moreover, $\rho_{0T}(X^h, \Phi_x(s)) \ge \delta$ implies $l^h \notin \Phi_x(s)$. Therefore,

$$\mathsf{P}_x^h\{\rho_{0T}(X^h, \Phi_x(s)) \ge \delta\} \le \mathsf{P}_x^h\left(\bigcup_{i=0}^{n-1} \overline{A^h(i)}\right) + \mathsf{P}_x^h\left(\bigcap_{i=0}^{n-1} A^h(i) \cap \{l^h \notin \Phi_x(s)\}\right). \tag{2.11}$$

The first probability does not exceed

$$n \cdot \sup_y \mathsf{P}_y^h\left\{\sup_{0 \le t \le \Delta t} |X_t^h - y| > \delta'\right\}.$$

By virtue of a well-known estimate (cf. Dynkin [1], Lemma 6.3), the supremum of this probability does not exceed

$$2 \sup_{t \le \Delta t} \sup_y \mathsf{P}_y^h\{|X_t^h - y| > \delta'/2\}.$$

We estimate this probability by means of the exponential Chebyshev inequality: for any $C > 0$ we have

$$\begin{aligned}\mathsf{P}_y^h\{|X_t^h - y| > \delta'/2\} \le [&\mathsf{M}_y^h \exp\{h^{-1}C \cdot (X_t^h - y)\} \\ &+ \mathsf{M}_y^h \exp\{-h^{-1}C \cdot (X_t^h - y)\}]e^{-h^{-1}C\delta'/2}. \quad (2.12)\end{aligned}$$

It follows from (2.8) that the mathematical expectation here does not exceed $\exp\{h^{-1}t\bar{H}(\pm C)\} \le \exp\{h^{-1}\Delta t\bar{H}(\pm C)\}$. Now we choose $C = 4s_0/\delta'$ and we choose Δt so that $\Delta t\bar{H}(\pm C) \le s_0$. Then the right side of (2.12) does not exceed $2\exp\{-h^{-1}s_0\}$ and the corresponding terms in (2.11) are negligible compared to $\exp\{-h^{-1}(s - \gamma)\}$.

The second term in (2.11) is estimated by means of the exponential Chebyshev inequality:

$$\begin{aligned}&\mathsf{P}_x^h\left(\bigcap_{i=0}^{n-1} A^h(i) \cap \{l^h \notin \Phi_x(s)\}\right) \\ &\quad = \mathsf{P}_x^h\left(\bigcap_{i=0}^{n-1} A^h(i) \cap \{S_{0T}(l^h) > s\}\right) \\ &\quad \le \mathsf{M}_x^h\left[\bigcap_{i=0}^{n-1} A^h(i); \exp\{h^{-1}(1 + \varkappa)^{-2}S_{0T}(l^h)\}\right] \\ &\quad\quad \times \exp\{-h^{-1}(1 + \varkappa)^{-2}s\}. \quad (2.13)\end{aligned}$$

Using the choice of δ', we obtain

$$\begin{aligned}S_{0T}(l^h) &= \sum_{i=0}^{n-1} \int_{i\Delta t}^{(i+1)\Delta t} L\left(l_t^h, \frac{X_{(i+1)\Delta t}^h - X_{i\Delta t}^h}{\Delta t}\right) dt \\ &\le \sum_{i=0}^{n-1} \Delta t\left[L\left(X_{i\Delta t}^h, \frac{X_{(i+1)\Delta t}^h - X_{i\Delta t}^h}{\Delta t}\right) \cdot (1 + \varkappa) + \varkappa\right] \\ &= (1 + \varkappa)\sum_{i=0}^{n-1} \Delta t \cdot L\left(X_{i\Delta t}^h, \frac{X_{(i+1)\Delta t}^h - X_{i\Delta t}^h}{\Delta t}\right) + \varkappa T\end{aligned}$$

for $\omega \in \bigcap_{i=0}^{n-1} A^h(i)$. Taking account of this, we obtain from (2.13) that

$$\mathsf{P}_x^h\left(\bigcap_{i=0}^{n-1} A^h(i) \cap \{l^h \notin \Phi_x(s)\}\right)$$
$$\leq \mathsf{M}_x^h\left[\bigcap_{i=0}^{n-1} A^h(i); \prod_{i=0}^{n-1} \exp\left\{h^{-1}(1+\varkappa)^{-1}\Delta t L\left(X_{i\Delta t}^h, \frac{X_{(i+1)\Delta t}^h - X_{i\Delta t}^h}{\Delta t}\right)\right\}\right]$$
$$\times \exp\{h^{-1}[-(1+\varkappa)^{-2}s + (1+\varkappa)^{-2}\varkappa T]\}. \tag{2.14}$$

Using the Markov property with respect to the times $\Delta t, 2\Delta t, \ldots, (n-1)\Delta t$, we obtain that the mathematical expectation in this formula does not exceed

$$\left[\sup_y \mathsf{M}_y^h\left[A^h(0); \exp\left\{h^{-1}(1+\varkappa)^{-1}\Delta t L\left(y, \frac{X_{\Delta t}^h - y}{\Delta t}\right)\right\}\right]\right]^n. \tag{2.15}$$

Here the second argument of L does not exceed $\delta'/\Delta t$. For fixed y, we approximate the function $L(y, \beta)$, convex on the interval $|\beta| < \delta'/\Delta t$, by means of a polygon $L'(y, \beta)$ circumscribed from below to an accuracy of $\varkappa$. This can be done by means of a polygon with the number N of links not depending on y, for example, in the following way: we denote by β_{-i}, β_{+i} $(i = 0, 1, 2, \ldots)$ the points at which $L(y, \beta) = i\varkappa$ (for $i \neq 0$, there are two such points) and choose the polygon formed by the tangent lines to the graph of $L(y, \beta)$ at these points, i.e., we put

$$L'(y, \beta) = \max_{-i_0 \leq i \leq i_0}\left[L(y, \beta_i) + \frac{\partial L}{\partial \beta}(y, \beta_i)(\beta - \beta_i)\right]. \tag{2.16}$$

Here the number N of points is equal to $2i_0 + 1$, where i_0 is the integral part of $\varkappa^{-1} \sup_y \sup_{|\beta| \leq \delta'/\Delta t} L(y, \beta)$. It is easy to see that $L(y, \beta) - L'(y, \beta) \leq \varkappa$ for $|\beta| < \delta'/\Delta t$ (Fig. 9).

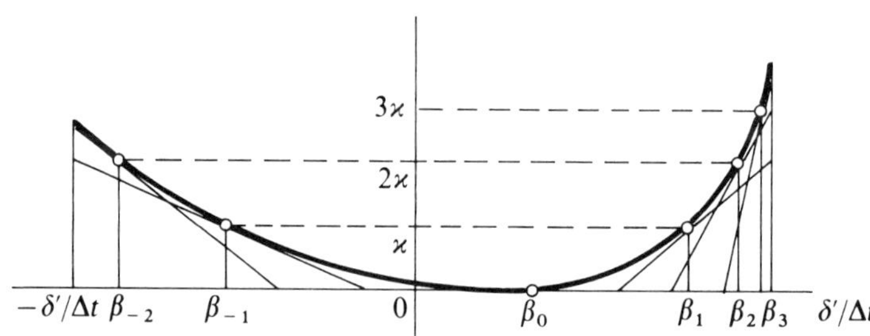

Figure 9.

Using the definition of L, the expression (2.16) can be rewritten in the form

$$L'(y, \beta) = \max_{-i_0 \le i \le i_0} [\alpha_i \beta - H(y, \alpha_i)], \tag{2.17}$$

where the $\alpha_i = \partial L(y, \beta_i)/\partial\beta$ depend on y. Taking account of this, we can see that the mathematical expectation in formula (2.15) does not exceed

$$\begin{aligned} &\mathsf{M}_y^h[A^h(0); \exp\{h^{-1}(1+\varkappa)^{-1} \max_{-i_0 \le i \le i_0} [\alpha_i(X_{\Delta t}^h - y) \\ &\qquad - \Delta t H(y, \alpha_i)] + h^{-1}(1+\varkappa)^{-1}\varkappa\Delta t\}] \\ &\quad \le \exp\{h^{-1}\varkappa(1+\varkappa)^{-1}\Delta t\} \\ &\quad \cdot \sum_{i=-i_0}^{i_0} \mathsf{M}_y^h[A^h(0); \exp\{h^{-1}[(1+\varkappa)^{-1}\alpha_i(X_{\Delta t}^h - y) \\ &\qquad - \Delta t(1+\varkappa)^{-1}H(y, \alpha_i)]\}]. \end{aligned} \tag{2.18}$$

By (2.8) we have

$$\mathsf{M}_y^h \exp\left\{h^{-1}\left[(1+\varkappa)^{-1}\alpha_i(X_{\Delta t}^h - y) - \int_0^{\Delta t} H(X_t^h, (1+\varkappa)^{-1}\alpha_i)\,dt\right]\right\} = 1.$$

In order to use this, we subtract and add $h^{-1}\int_0^{\Delta t} H(X_t^h, (1+\varkappa)^{-1}\alpha_i)\,dt$ under the exponential sign on the right side of (2.18). Using the choice of δ' and inequality (2.4), we obtain that the ith mathematical expectation on the right side of (2.18) does not exceed $\exp\{h^{-1}\Delta t \cdot \varkappa(1+\varkappa)^{-1}\}$.

We combine the estimates obtained so far:

$$\begin{aligned} \mathsf{P}_x^h\{\rho_{0T}(X^h, \Phi_x(s)) \ge \delta\} &\le 4n \exp\{-h^{-1}s\} + (2i_0+1)^n \\ &\times \exp\{h^{-1}[-(1+\varkappa)^{-2}s + \varkappa(1+\varkappa)^{-2}T + 2\varkappa(1+\varkappa)^{-1}T]\}. \end{aligned} \tag{2.19}$$

Since $\varkappa > 0$ can be chosen arbitrarily small, for h smaller than or equal to some h_0 we obtain estimate (2.6) for all $s \le s_0$.

§3. Special Cases. Generalizations

In this section we derive theorems on large deviations for certain families of Markov processes. They either follow from Theorem 2.1 or are close to it.

First of all, the hypotheses of Theorem 2.1 are satisfied for the families of diffusion processes in R^r, given by the stochastic equations

$$\dot{X}_t^\varepsilon = b(X_t^\varepsilon) + \varepsilon\sigma(X_t^\varepsilon)\dot{w}_t, \tag{3.1}$$

as the parameter ε converges to zero, provided that the drift coefficients $b^i(x)$ and diffusion coefficients $a^{ij}(x) = \sum_{k=1}^r \sigma_k^i(x)\sigma_k^i(x)$, $i, j = 1, \ldots, r$ are bounded and uniformly continuous in x and the diffusion matrix is uniformly non-degenerate: $\sum_{ij} a^{ij}(x)c_i c_j \geq \mu \sum_i c_i^2$, $\mu > 0$. We calculate the characteristics of the family of processes $(X_t^\varepsilon, \mathsf{P}_x^\varepsilon)$: the infinitesimal generator for functions $f \in \mathbf{C}^{(2)}$ has the form

$$A^\varepsilon f(x) = \sum_i b^i(x) \frac{\partial f(x)}{\partial x^i} + \frac{\varepsilon^2}{2} \sum_{ij} a^{ij}(x) \frac{\partial^2 f(x)}{\partial x^i \partial x^j}. \tag{3.2}$$

The role of the parameter h of the preceding paragraph is played by ε^2 and the normalizing coefficient is ε^{-2}. Moreover, we have

$$H(x, \alpha) = \sum b^i(x)\alpha_i + \frac{1}{2} \sum a^{ij}(x)\alpha_i \alpha_j; \tag{3.3}$$

the Legendre transform of this function is

$$L(x, \beta) = \frac{1}{2} \sum_{ij} a_{ij}(x)(\beta^i - b^i(x))(\beta^j - b^j(x)), \tag{3.4}$$

where $(a_{ij}(x)) = (a^{ij}(x))^{-1}$; the normalized action functional has the form (for absolutely continuous φ_t, $T_1 \leq t \leq T_2$)

$$S_{T_1 T_2}(\varphi) = \frac{1}{2} \int_{T_1}^{T_2} \sum_{ij} a_{ij}(\varphi_t)(\dot{\varphi}_t^i - b^i(\varphi_t))(\dot{\varphi}_t^j - b^j(\varphi_t))\, dt. \tag{3.5}$$

We verify conditions I–III of Theorem 2.1. As the function $\bar{H}$ we may choose

$$\bar{H}(\alpha) = B|\alpha| + \tfrac{1}{2}A|\alpha|^2,$$

where B and A are constants majorizing $|b(x)|$ and the largest eigenvalue of the matrix $(a^{ij}(x))$. As the constant $M = M(R)$ of condition II we may choose $\mu^{-1}(R + B + 1)^2$ and as m we may take A^{-1}. Condition III is also satisfied: the function $\Delta L(\delta')$ can be expressed in terms of the moduli of continuity of the functions $b^i(x)$, $a^{ij}(x)$ and the constants μ, A and B.

The following theorem deals with a slight generalization of this scheme and in this case the proof of Theorem 2.1 needs only minor changes.

Theorem 3.1. *Let the functions $b^i(x)$ and $a^{ij}(x)$ be bounded and uniformly continuous in R^r, let the matrix $(a^{ij}(x))$ be symmetric for any x and let*

$$\sum_{ij} a^{ij}(x)c_i c_j \geq \mu \sum_i c_i^2,$$

where μ is a positive constant. Furthermore, suppose that $b^{1\varepsilon}(x), \ldots, b^{r\varepsilon}(x)$ uniformly converge to $b^1(x), \ldots, b^r(x)$, respectively, as $\varepsilon \to 0$, $(X_t^\varepsilon, \mathsf{P}_x^\varepsilon)$ is a diffusion process in R^r with drift $b^\varepsilon(x) = (b^{1\varepsilon}(x), \ldots, b^{r\varepsilon}(x))$ and diffusion matrix $\varepsilon^2(a^{ij}(x))$, and the functional S is given by formula (3.5). Then $\varepsilon^{-2}S_{0T}(\varphi)$ is the action functional for the family of processes $(X_t^\varepsilon, \mathsf{P}_x^\varepsilon)$ in the sense of the metric $\rho_{0T}(\varphi, \psi) = \sup_{0 \le t \le T} |\varphi_t - \psi_t|$, uniformly with respect to the initial point as $\varepsilon \to 0$.

Moreover, this result can be carried over to diffusion processes on a differentiable manifold.

Theorem 3.2. *Let M be an r-dimensional manifold of class $\mathbf{C}^{(3)}$ with metric ρ. Suppose that there exists a positive constant λ such that in the λ-neighborhood of every point $x_0 \in M$ we can introduce a common coordinate system K_{x_0} and the distance ρ differs from the corresponding Euclidean distance only by a numerical factor uniformly bounded for all K_{x_0}. For every $\varepsilon > 0$ let $(X_t^\varepsilon, \mathsf{P}_x^\varepsilon)$ be a diffusion process on M (cf. McKean [1]) and suppose that in the coordinate system K_{x_0}, its infinitesimal generator can be written in the form*

$$\sum_i b^{i\varepsilon}(x) \frac{\partial}{\partial x^i} + \frac{\varepsilon^2}{2} \sum_{ij} a^{ij}(x) \frac{\partial^2}{\partial x^i \partial x^j}. \tag{3.6}$$

(In passage to another coordinate system, the $a^{ij}(x)$ are transformed as a tensor and in the $b^{i\varepsilon}(x)$ there also occur terms of order ε^2 containing derivatives of $a^{ij}(x)$). *Suppose that $b^{i\varepsilon}(x) \to b^i(x)$ uniformly in all x such that $\rho(x_0, x) < \lambda$ and all coordinate systems K_{x_0} as $\varepsilon \to 0$, the functions $b^i(x)$, $a^{ij}(x)$ are bounded and uniformly continuous in all x, $\rho(x_0, x) < \lambda$, and all K_{x_0}, and*

$$\sum_{ij} a^{ij}(x) c_i c_j \ge \mu \sum_i c_i^2$$

for all $c_1, \ldots, c_r$, where $\mu = \text{const} > 0$. Let the functional S be given by formula (3.5).

Then $\varepsilon^{-2}S_{0T}(\varphi)$ is the action functional for the family of processes $(X_t^\varepsilon, \mathsf{P}_x^\varepsilon)$ with respect to the metric $\rho_{0T}(\varphi, \psi) = \sup_{0 \le t \le T} \rho(\varphi_t, \psi_t)$, uniformly with respect to the initial point as $\varepsilon \to 0$.

The proof of Theorem 3.2 (and of Theorem 3.1) is contained in §1 of Wentzell and Freidlin [4].

We consider some more families of Markov processes for which the action functional can be written out easily by using formulas of the preceding section, but the fact that the expression thus obtained is indeed the action functional does not follow from Theorem 2.1.

Let $\pi_t^{1h^{-1}}, \ldots, \pi_t^{rh^{-1}}$ be independent Poisson processes with parameters $h^{-1}\lambda^1, \ldots, h^{-1}\lambda^r$, respectively ($\lambda^1, \ldots, \lambda^r$ are positive constants). We consider the process $\xi_t^h = (\xi_1^{1h}, \ldots, \xi_t^{rh})$, where $\xi_t^{ih} = h\pi_t^{ih^{-1}}$. The infinitesimal generator of this Markov process is

$$A^h f(x^1, \ldots, x^r) = h^{-1}\lambda^1[f(x^1 + h, x^2, \ldots, x^r) - f(x^1, x^2, \ldots, x^r)] + \cdots$$
$$+ h^{-1}\lambda^r[f(x^1, \ldots, x^r + h) - f(x^1, \ldots, x^{r-1}, x^r)], \tag{3.7}$$

$$H(x, \alpha) \equiv H(\alpha) = \sum_{i=1}^r \lambda^i(e^{\alpha_i} - 1), \tag{3.8}$$

$$L(x, \beta) \equiv L(\beta) = \begin{cases} \sum_{i=1}^r \left[\beta^i \ln \dfrac{\beta^i}{\lambda^i} - \beta^i + \lambda^i\right] & \text{for } \beta^1 \geq 0, \ldots, \beta^r \geq 0, \\ +\infty, \quad \text{if at least one of the } \beta^i < 0; \end{cases} \tag{3.9}$$

the functional $S(\varphi)$ is defined as the integral

$$S_{T_1T_2}(\varphi) = \int_{T_1}^{T_2} L(\dot{\varphi}_t)\, dt \tag{3.10}$$

for absolutely continuous $\varphi_t = (\varphi_t^1, \ldots, \varphi_t^r)$ and as $+\infty$ for all remaining φ (the integral is equal to $+\infty$ for all φ_t for which at least one coordinate is not a nondecreasing function).

The hypotheses of Theorem 2.1 are not applicable, since L equals $+\infty$ outside the set $\{\beta^1 \geq 0, \ldots, \beta^r \geq 0\}$. Nevertheless, the following theorem holds.

Theorem 3.3. *The functional given by formulas* (3.10) *and* (3.9) *is the normalized action functional for the family of processes ξ_t^h, uniformly with respect to the initial point as $h \downarrow 0$* (*as normalizing coefficient we take, of course, h^{-1}*).

This is a special case of Borovkov's results [1] (in fact, these results are concerned formally only with the one-dimensional case; for $r > 1$ we can refer to later work: Mogul'skii [1] and Wentzell [7]).

The preceding results were related to the case where with the increase of the parameter, the jumps of the Markov process occur as many times more often as many times smaller they become. We formulate a result concerning the case where this is not so.

As above, let $\pi_t^{1\alpha}, \ldots, \pi_t^{r\alpha}$ be independent Poisson processes with parameters $\alpha\lambda^1, \ldots, \alpha\lambda^r$. For $\alpha > 0$ and $h > 0$ we put $\xi_t^{i\alpha h} = h(\pi_t^{i\alpha} - \alpha\lambda^i t)$, $i = 1, \ldots, r$.

Theorem 3.4. *Put*

$$S_{T_1T_2}(\varphi) = \frac{1}{2}\int_{T_1}^{T_2} \sum_{i=1}^r (\lambda^i)^{-1}(\dot{\varphi}_t^i)^2\, dt \tag{3.11}$$

for absolutely continuous φ_t, $T_1 \le t \le T_2$ *and* $S_{T_1T_2}(\varphi) = +\infty$ *for the remaining* φ_t. *Then* $(h^2\alpha)^{-1}S_{0T}(\varphi)$ *is the action functional for the family of processes* $\xi^{\alpha h} = (\xi_t^{1\alpha h}, \ldots, \xi_t^{r\alpha h})$, $0 \le t \le T$, *uniformly with respect to the initial point as* $h\alpha \to \infty$, $h^2\alpha \to 0$.

This is a special case of one of the theorems of the same article [1] by Borovkov (the multidimensional case can be found in Mogul'skii [1]).

§4. Consequences. Generalization of Results of Chapter 4

Let us see whether the results obtained by us in §§2, 3 and 4 of Ch. 4 for small perturbations of the type of a "white noise" of a dynamical system can be carried over to small jump-like perturbations (or to perturbations of the type of diffusion with varying diffusion).

Let $\dot{x}_t = b(x_t)$ be a dynamical system with one stable equilibrium position O and let (X_t^h, P_x^h) be a family of Markov processes of the form described in §2. For this family, we may pose problems on the limit behavior as $h \downarrow 0$ of the invariant measure μ^h, of the distribution of the point $X_{\tau^h}^h$ of exit from a domain and of the mean time $\mathsf{M}_x^h \tau^h$ of exit from a domain. We may conjecture that the solutions will be connected with the function $V(O, x) = \inf\{S_{T_1T_2}(\varphi): \varphi_{T_1} = O, \varphi_{T_2} = x\}$. Namely, the invariant measure μ^h must be described by the action function $h^{-1}V(O, x)$, the distribution of $X_{\tau^h}^h$ as $h \downarrow 0$ must to be concentrated near those points of the boundary at which $\min_{y \in \partial D} V(O, y)$ is attained, $\mathsf{M}_x^h \tau^h$ must be logarithmically equivalent to

$$\exp\left\{h^{-1}\min_{y \in \partial D} V(O, y)\right\};$$

and for small h, the exit from a domain must take place with overwhelming probability along an extremal of $S(\varphi)$ leading from O to the boundary, etc.

The proofs in §§2, 4 of Ch. 4 have to be changed in the following way. Instead of the small spheres γ and Γ about the equilibrium position, we have to take a small ball γ containing the equilibrium position and the exterior Γ of a sphere of a somewhat larger radius (a jump like process may simply jump over a sphere). Instead of the chain Z_n on the set $\gamma \cup \partial D$, we consider a chain on the sum of γ and the complement of D. A trajectory of X_t^h beginning at a point of γ is not necessarily on a sphere of small radius (the boundary of Γ) at the first entrance time of Γ. Nevertheless, the probability that at this time the process will be at a distance larger than some $\delta > 0$ from this sphere converges to zero faster than any exponential $\exp\{-Kh^{-1}\}$ as $h \downarrow 0$. Theorems 2.1, 2.3, 2.4, 4.1 and 4.2 of Ch. 4 remain true for families (X_t^h, P_x^h) satisfying the hypotheses of Theorem 2.1.

Of course, these theorems are also true for families of diffusion processes satisfying the hypotheses of Theorems 3.1 and 3.2. We give the corresponding formulation in the language of differential equations, i.e., the generalization of Theorem 2.2 of Ch. 4.

Theorem 4.1. *Let O be a stable equilibrium position of the dynamical system $\dot{x}_t = b(x_t)$ on a manifold M and let D be a domain in M with compact closure and smooth boundary ∂D. Suppose that the trajectories of the dynamical system beginning at any point of $D \cup \partial D$ are attracted to O as $t \to \infty$ and the vector $b(x)$ is directed strictly inside D at every boundary point. Furthermore, let $(X_t^\varepsilon, \mathsf{P}_x^\varepsilon)$ be a family of diffusion processes on M with drift $b^\varepsilon(x)$ (in local coordinates) converging to $b(x)$ as $\varepsilon \to 0$ and diffusion matrix $\varepsilon^2(a^{ij}(x))$ and let the hypotheses of Theorem 3.2 be satisfied. For every $\varepsilon > 0$ let $u^\varepsilon(x)$ be the solution of Dirichlet's problem*

$$\frac{\varepsilon^2}{2} \sum_{ij} a^{ij}(x) \frac{\partial^2 u^\varepsilon(x)}{\partial x^i \, \partial x^j} + \sum_i b^{i\varepsilon}(x) \frac{\partial u^\varepsilon(x)}{\partial x^i} = 0, \qquad x \in D,$$

$$u^\varepsilon(x) = g(x), \qquad x \in \partial D,$$

with continuous boundary function g. Then $\lim_{\varepsilon \to 0} u^\varepsilon(x) = g(y_0)$, where y_0 is a point on the boundary at which $\min_{y \in \partial D} V(O, y)$ is attained (it is assumed that this point is unique).

The proof of the theorem below is entirely analogous to that of Theorem 4.3 of Ch. 4.

Theorem 4.2. *Let the dynamical system $\dot{x}_t = b(x_t)$ in R^r have a unique equilibrium position O which is stable and attracts the trajectories beginning at any point of R^r. For sufficiently large $|x|$ let the inequality $(x, b(x)) \leq -c|x|$, where c is a positive constant, be satisfied. Suppose that the family of diffusion processes $(X_t^\varepsilon, \mathsf{P}_x^\varepsilon)$ with drift $b^\varepsilon(x) \to b(x)$ and diffusion matrix $\varepsilon^2(a^{ij}(x))$ satisfies the hypotheses of Theorem 3.1. Then the normalized invariant measure μ^ε of $(X_t^\varepsilon, \mathsf{P}_x^\varepsilon)$ as $\varepsilon \to 0$ is described by the action function $\varepsilon^{-2} V(O, x)$.*

The formulation and proof of the corresponding theorem for diffusion processes on a manifold is postponed until §4 of Ch. 6. The matter is that this theorem will be quite simple for compact manifolds but the trajectories of a dynamical system on such a manifold cannot be attracted to one stable equilibrium position. For families of jump-like processes, the corresponding theorem will also be more complicated than Theorem 4.3 of Ch. 4 or Theorem 4.2 of this chapter.

We indicate the changes in Theorem 3.1 of Ch. 4 enabling us to determine the quasipotential.

Let A be a subset of the domain D with boundary ∂D. The problem $\mathbf{R}_A$ for a first-order differential equation in D is, by definition, the problem of finding a function U continuous in $D \cup \partial D$, vanishing on A and positive outside A, continuously differentiable and satisfying the equation in question in $(D \cup \partial D) \setminus A$ and such that $\nabla U(x) \neq 0$ for $x \in (D \cup \partial D) \setminus A$.

Theorem 4.3. *Let $H(x, \alpha) \leftrightarrow L(x, \beta)$ be strictly convex functions smooth in the second argument and coupled by the Legendre transformation. Let $S_{T_1T_2}(\varphi) = \int_{T_1}^{T_2} L(\varphi_t, \dot{\varphi}_t)\, dt$ (for absolutely continuous functions φ and $+\infty$ otherwise). Let A be a compact subset of D. Let us put*

$$V(A, x) = \inf\{S_{T_1T_2}(\varphi): \varphi_{T_1} \in A, \varphi_{T_2} = x; -\infty \leq T_1 < T_2 \leq \infty\}. \tag{4.1}$$

Suppose that U is a solution of problem $\mathbf{R}_A$ for the equation

$$H(x, \nabla U(x)) = 0 \tag{4.2}$$

in $D \cup \partial D$. Then $U(x) = V(A, x)$ for all x for which $U(x) \leq \min\{U(y): y \in \partial D\}$. The infimum in (4.1) is attained at the solution of the equation

$$\dot{\varphi}_t = \nabla_\alpha H(\varphi_t, \nabla U(\varphi_t)) \tag{4.3}$$

with the final condition $\varphi_{T_2} = x$ for any $T_2 < \infty$ (then the value of φ automatically belongs to A for some T_1, $-\infty \leq T_1 < T_2$).

Theorem 3.1 of Ch. 4 is a special case of this with $H(x, \alpha) = (b(x), \alpha) + |\alpha|^2/2$, $A = \{0\}$.

Proof. Let φ_t, $T_1 \leq t \leq T_2$, be any curve connecting A with x and lying entirely in $D \cup \partial D$. We use the inequality $L(x, \beta) \geq \sum_i \alpha_i \beta^i - H(x, \alpha)$, following from definition (1.1). Substituting φ_t, $\dot{\varphi}_t$ and $\nabla U(\varphi_t)$ for x, β and α, we obtain

$$S_{T_1T_2}(\varphi) = \int_{T_1}^{T_2} L(\dot{\varphi}_t, \varphi_t)\, dt \geq \int_{T_1}^{T_2} \sum_I \frac{\partial U}{\partial x^i}(\varphi_t)\dot{\varphi}_t^i\, dt - \int_{T_1}^{T_2} H(\varphi_t, \nabla U(\varphi_t))\, dt.$$

The second integral is equal to zero by (4.2). The first one is $U(\varphi_{T_2}) - U(\varphi_{T_1}) = U(x)$. Hence $S_{T_1T_2}(\varphi) \geq U(x)$. If $U(x) \leq \min\{U(y): y \in \partial D\}$, then the consideration of curves going through the boundary before hitting x does not yield any advantage and $V(A, x) \geq U(x)$.

Now we show that the value $U(x)$ can be attained. For this we construct an extremal. We shall solve equation (4.3) for $t \leq T_2$ with the condition $\varphi_{T_2} = x$. A solution exists as long as φ_t does not go out of $(D \cup \partial D) \setminus A$, since the right side is continuous (the solution may not be unique). We have $(d/dt)U(\varphi_t) = \sum_i (dU/\partial x^i)(\varphi_t) \cdot \dot{\varphi}_t^i = L(\varphi_t, \dot{\varphi}_t) + H(\varphi_t, \nabla U(\varphi_t))$. The second term is equal to zero. The first term is positive, since $L(x, \beta) \geq 0$ and vanishes only for $\beta = b(x) = \nabla_\alpha H(x, 0)$, which is not equal to

$\nabla_\alpha H(x, \nabla U(x))$ for $x \notin A$ (because $\nabla U(x) \neq 0$). Hence $U(\varphi_t)$ decreases with decreasing t. Consequently, φ_t does not go out of $D \cup \partial D$. We prove that for some $T_1 < T_2$ (maybe for $T_1 = -\infty$) the curve φ_t enters A. If this were not so, then the solution φ_t would be defined for all $t < T_2$ and the limit $\lim_{t \to -\infty} U(\varphi_t) > 0$ would exist. However, this would mean that φ_t does not hit some neighborhood of the compactum A. Outside this neighborhood we have $(d/dt)U(\varphi_t) = L(\varphi_t, \dot{\varphi}_t) \geq \text{const} > 0$, which contradicts the existence of a finite limit of $U(\varphi_t)$ as $t \to -\infty$.

Now we determine the value of S at the function we have just obtained:

$$S_{T_1 T_2}(\varphi) = \int_{T_1}^{T_2} L(\varphi_t, \dot{\varphi}_t)\, dt = \int_{T_1}^{T_2} \frac{d}{dt} U(\varphi_t)\, dt = U(\varphi_{T_2}) - U(\varphi_{T_1}) = U(x).$$

The theorem is proved. □

The hypotheses of this theorem can be satisfied only in the case where all trajectories of the dynamical system $\dot{x}_t = b(x_t)$ beginning at points of $D \cup \partial D$ are attracted to A as $t \to \infty$ (or they enter A). The same is true for Theorem 3.1 of Ch. 4.

As an example, we consider the family of one-dimensional jump-like processes introduced at the beginning of §2. If $l(x) > r(x)$ for $x > x_0$ and $l(x) < r(x)$ for $x < x_0$, then the point x_0 is a stable equilibrium position for the equation $\dot{x}_t = r(x_t) - l(x_t)$. We solve problem $\mathbf{R}_{x_0}$ for the equation $H(x, U'(x)) = 0$, where $H(x, \alpha) = r(x)(e^{\alpha} - 1) + l(x)(e^{-\alpha} - 1)$. Given x, the function $H(x, \alpha)$ vanishes at $\alpha = 0$ and at $\alpha = \ln(l(x)/r(x))$. For $x \neq x_0$ we have to take the second value, because it is required that $U'(x) \neq 0$ for $x \neq x_0$. We find that $U(x) = \int_{x_0}^{x} \ln(l(y)/r(y))\, dy$ (this function is positive for both $x > x_0$ and $x < x_0$). It is in fact the quasipotential. An extremal of the normalized action functional is nothing else but a solution of the differential equation $\dot{\varphi}_t = (\partial H/\partial\alpha)(\varphi_t, U'(\varphi_t)) = l(\varphi_t) - r(\varphi_t)$.

Now we obtain immediately the solution of a series of problems connected with the family of processes (X_t^h, P_x^h): the probability that the process leaves an interval $(x_1, x_2) \ni x_0$ through the left endpoint converges to 1 as $h \downarrow 0$ if $U(x_1) < U(x_2)$ and to 0 if the opposite inequality holds (for $U(x_1) = U(x_2)$ the problem remains open); the mathematical expectation of the first exit time is logarithmically equivalent to $\exp\{h^{-1} \min[U(x_1), U(x_2)]\}$ (this result has been obtained by another method by Labkovskii [1]) and so on. The asymptotics of the invariant measure of (X_t^h, P_x^h) as $h \downarrow 0$ is given by the action function $h^{-1}U(x)$ (this does not follow from the results formulated by us; however, in this case as well as in the potential case, considered in §4, of Ch. 4, the invariant measure can easily be calculated explicitly).

Chapter 6

Markov Perturbations on Large Time Intervals

§1. Auxiliary Results. Equivalence Relation

In this chapter we consider families of diffusion processes $(X_t^\varepsilon, \mathsf{P}_x^\varepsilon)$ on a connected manifold M. We shall assume that these families satisfy the hypotheses of Theorem 3.2 of Ch. 5 and the behavior of probabilities of large deviations from the "most probable" trajectory—the trajectory of the dynamical system $\dot{x}_t = b(x_t)$—can be described as $\varepsilon \to 0$, by the action functional $\varepsilon^{-2}S(\varphi) = \varepsilon^{-2}S_{T_1T_2}(\varphi)$, where

$$S(\varphi) = \frac{1}{2}\int_{T_1}^{T_2} \sum_{ij} a_{ij}(\varphi_t)(\dot{\varphi}_t^i - b^i(\varphi_t))(\dot{\varphi}_t^j - b^j(\varphi_t))\, dt.$$

We set forth results of Wentzell and Freidlin [4], [5] and some generalizations thereof.

In problems connected with the behavior of X_t^ε in a domain $D \subseteq M$ on large time intervals (problems of exit from a domain and for $D = M$ the problem of invariant measure; concerning special cases, cf. §§2, 4 of Ch. 4 and §4 of Ch. 5), an essential role is played by the function

$$V_D(x, y) = \inf\{S_{0T}(\varphi)\colon \varphi_0 = x, \varphi_T = y, \varphi_t \in D \cup \partial D \text{ for } t \in [0, T]\},$$

where the infimum is taken over intervals $[0, T]$ of arbitrary length. For $D = M$ we shall use the notation $V_M(x, y) = V(x, y)$.

For small ε, the function $V_D(x, y)$ characterizes the difficulty of passage from x to a small neighborhood of y, without leaving D (or $D \cup \partial D$), within a "reasonable" time. For $D = M$ we can give this, for example, the following exact meaning: it can be proved that

$$V(x, y) = \lim_{T\to\infty} \lim_{\delta\to 0} \lim_{\varepsilon\to 0}[-\varepsilon^2 \ln \mathsf{P}_x^\varepsilon\{\tau_\delta \le T\}],$$

where τ_δ is the first entrance time of the δ-neighborhood of y for the process X_t^ε.

It is easy to see that V_D is everywhere finite. Let us study its more complicated properties.

Lemma 1.1. *There exists a constant $L > 0$ such that for any $x, y \in M$ with $\rho(x, y) < \lambda$, there exists a function φ_t, $\varphi_0 = x$, $\varphi_T = y$, $T = \rho(x, y)$, for which $S_{0T}(\varphi) \leq L \cdot \rho(x, y)$.*

This is Lemma 2.3 of Ch. 4, adapted to manifolds. Here $\lambda > 0$ is the radius of the neighborhoods in which the coordinate systems K_{x_0} are defined (cf. §3, Ch. 5).

This lemma implies that $V_D(x, y)$ is continuous for $x, y \in D$ and if the boundary is smooth, for $x, y \in D \cup \partial D$.

Lemma 1.2. *For any $\gamma > 0$ and any compactum $K \subseteq D$ (in the case of a smooth boundary, for $K \subseteq D \cup \partial D$) there exists T_0 such that for any $x, y \in K$ there exists a function φ_t, $0 \leq t \leq T$, $\varphi_0 = x$, $\varphi_T = y$, $T \leq T_0$ such that $S_{0T}(\varphi) \leq V_D(x, y) + \gamma$.*

For the proof, we choose a sufficiently dense finite δ-net $\{x_i\}$ of points in K; we connect them with curves at which the action functional assumes values differing from the infimum by less than $\gamma/2$ and complete them with end sections by using Lemma 1.1: from x to a point x_i near x, then from x_i to a point x_j near y, and from x_j to y.

Let φ_t, $0 \leq t \leq T$, be a continuous curve in M and let $0 = t_0 < t_1 < \cdots < t_n = T$ be a partition of the interval from 0 to T, such that all values φ_t for $t \in [t_i, t_{i+1}]$ lie in one coordinate neighborhood K_{x_i}. We can define a "polygon" l inscribed in φ by defining l_t for $t \in [t_i, t_{i+1}]$ by means of coordinates in K_{x_i}: $l_t^j = \varphi_{t_i}^j[(t_{i+1} - t)/(t_{i+1} - t_i)] + \varphi_{t_{i+1}}^j[(t - t_i)/(t_{i+1} - t_i)]$.

Lemma 1.3. *For any $K > 0$ and any $\gamma > 0$ there exists a positive h such that for any function φ_t, $0 \leq t \leq T$, $T + S_{0T}(\varphi) \leq K$, and any partition of the interval from 0 to T with $\max(t_{i+1} - t_i) \leq h$ we have*

$$S_{0T}(l) \leq S_{0T}(\varphi) + \gamma.$$

The first step of the proof is application of the equicontinuity of all such φ in order to establish that for sufficiently small h, the values φ_t, $t_i \leq t \leq t_{i+1}$, lie in one coordinate neighborhood K_{x_i} and the number of different coordinate systems K_{x_i} is bounded from above by a number depending only on K. This is the assertion of Lemma 2.1 of Ch. 5 concerning which we referred the reader to Wentzell [7] (however, for a function $L(x, \beta)$ quadratic in β, the proof is simpler).

We have already seen in the proof of Theorem 2.1 of Ch. 4 that in our arguments we have to go out to a small distance from the boundary of D or, on the other hand, we have to consider only points lying inside D at a positive distance from the boundary. Let $\rho(x,y)$ be a Riemannian distance corresponding to a smooth tensor $g_{ij}(x)$: $(d\rho)^2 = \sum_{ij} g_{ij}(x)\,dx^i\,dx^j$. We denote by $D_{+\delta}$ the δ-neighborhood of D and by $D_{-\delta}$ the set of points of D at a distance greater than δ from the boundary. In the case of a compact twice continuously differentiable boundary ∂D, the boundaries $\partial D_{+\delta}$ and $\partial D_{-\delta}$ are also twice continuously differentiable for sufficiently small δ. For points x lying between $\partial D_{-\delta}$ and $\partial D_{+\delta}$ or on them, the following points are defined uniquely: $(x)_0$ is the point on ∂D closest to x; $(x)_{-\delta}$ is the point on $\partial D_{-\delta}$ closest to x; $(x)_{+\delta}$ is the point on $\partial D_{+\delta}$ closest to x. The mappings $x \to (x)_0$, $x \to (x)_{-\delta}$ and $x \to (x)_{+\delta}$ are twice continuously differentiable.

Lemma 1.4. *Let ∂D be compact and twice continuously differentiable. For any $K > 0$ and any $\gamma > 0$ there exists $\delta_0 > 0$ such that for any positive $\delta \le \delta_0$ and any function φ_t, $0 \le t \le T$, $T + S_{0T}(\varphi) \le K$, assuming its values in $D \cup \partial D$, there exists a function $\tilde{\varphi}_t$, $0 \le t \le \tilde{T}$, having the following properties: The value $\tilde{\varphi}_0$ coincides with φ_0 if $\varphi_0 \in D_{-\delta} \cup \partial D_{-\delta}$ and is equal to $(\varphi_0)_{-\delta}$ otherwise; the same is true for $\tilde{\varphi}_T$, φ_T and $S_{0\tilde{T}}(\tilde{\varphi}) \le S_{0\tilde{T}}(\varphi) + \gamma$.*

The same is true for $D_{+\delta}$ and D replacing D and $D_{-\delta}$.

Proof. First of all we locally "straighten" the boundary, possibly decreasing λ and in neighborhoods covering the boundary we choose new local coordinates $x^1, x^2, \ldots, x^r$ in which ∂D becomes the hyperplane $\{x^1 = 0\}$, for points x in a small neighborhood of the boundary outside D the coordinate x^1 is equal to $\rho(x, \partial D)$ and for points near ∂D inside D, $x^1 = -\rho(x, \partial D)$. In these coordinate systems, the coefficients $a^{ij}(x)$, $a_{ij}(x)$ will differ from those in the original standard systems K_{x_0} but will be uniformly bounded and continuous for all these systems (with constants and moduli of continuity depending on the curvature of ∂D).

The fact that $S(\varphi)$ is quadratic is insignificant now. We only use the circumstance that it can be represented in the form

$$S_{0T}(\varphi) = \int_0^T L(\varphi_t, \dot{\varphi}_t)\,dt,$$

where $L(x, \beta)$ is convex and satisfies conditions II and III of Theorem 2.1 of Ch. 5 in every local coordinate system.

We choose a positive λ', not exceeding $\lambda/2$ or $\gamma/3L$ (L is the constant in Lemma 1.1). Using the equicontinuity of all functions φ under consideration, we choose $h_0 > 0$ such that the values of any of the φ on any interval of length not greater than h_0 lie in the λ'-neighborhood of the left endpoint. We decrease h_0 so that for the "polygon" l inscribed in φ with time step not exceeding h_0, we have $S_{0T}(l) \le S_{0T}(\varphi) + \gamma/3$ (cf. Lemma 1.3). Let us put $n = [K/h_0] + 1$.

For any function φ_t defined on an interval of length $T > h_0$, we take the partition of the interval from 0 to T into n equal parts and consider the corresponding polygon $l_t, 0 \le t \le T$. In each of the local coordinate systems chosen by us, the modulus of the derivative of l_t does not exceed $R = n\lambda'/h_0$ at any point. We choose a positive $\delta_0 \le \lambda'$ such that $\Delta L(\delta_0) \le \gamma/(3K + \gamma)$ (cf. condition III of Theorem 2.1, Ch. 5) and $\delta_0 \cdot n \sup_{x \in M, |\beta| \le R} |(\partial L/\partial \beta')(x, \beta)| \le \gamma/3$ (cf. condition II).

For every $\delta \le \delta_0$ and each of the functions φ_t under consideration we define a function $\tilde{\varphi}_t, 0 \le t \le \tilde{T} = T$:

$$\tilde{\varphi}_t = \begin{cases} l_t & \text{if } l_t \in D_{-\delta} \cup \partial D_{-\delta} \\ (l_t)_{-\delta} & \text{otherwise} \end{cases}$$

This function satisfies the conditions concerning the initial and terminal points. We estimate $S_{0T}(\tilde{\varphi})$. We have

$$S_{0T}(\tilde{\varphi}) - S_{0T}(l) = \int_0^T [L(\tilde{\varphi}_t, \dot{\tilde{\varphi}}_t) - L(\tilde{\varphi}_t, \dot{l}_t) \, dt + \int_0^T [L(\tilde{\varphi}_t, \dot{l}_t) - L(l_t, \dot{l}_t)] \, dt. \tag{1.1}$$

(The expression $L(\tilde{\varphi}_t, \dot{l}_t)$ is meaningful, since because of the choice $\delta_0 \le \lambda' \le \lambda/2$, on the whole interval from iT/n to $(i + 1)T/n$ the points $\tilde{\varphi}_t$ and l_t are in a neighborhood where one of the local coordinate systems chosen by us acts.)

The value $\tilde{\varphi}_t$ is different from l_t only when l_t is in the δ-strip along the boundary, i.e., on not more than n little intervals $[t_i, t_{i+1}] \subseteq [iT/n, (i + 1)T/n]$. Moreover, only the first coordinates may be different. In the first integral in (1.1) we have $\dot{\tilde{\varphi}}_t$ and $\dot{l}_t$; the first of these derivatives, just as the second one, does not exceed R in modulus (because it either coincides with l_t or differs from l_t by a vanishing first coordinate). We have

$$\int_{iT/n}^{(i+1)T/n} [L(\tilde{\varphi}_t, \dot{\tilde{\varphi}}_t) - L(\tilde{\varphi}_t, \dot{l}_t)] \, dt = \int_{t_i}^{t_{i+1}} \frac{\partial L}{\partial \beta^1}(\tilde{\varphi}_t, \beta(t)) \cdot (\dot{\tilde{\varphi}}_t^1 - \dot{l}_t^1) \, dt. \tag{1.2}$$

Here $\beta(t)$ is a point of the interval connecting $\dot{\tilde{\varphi}}_t$ with $\dot{l}_t$ (which varies with t in general), $\dot{\tilde{\varphi}}_t^1 - \dot{l}_t^1$ does not depend on t and is equal to

$$(\tilde{\varphi}_{t_{i+1}}^1 - l_{t_{i+1}}^1 - \tilde{\varphi}_{t_i}^1 + l_{t_i}^1)(t_{i+1} - t_i).$$

We pull this constant out of the integral. We use the facts that $|\partial L/\partial \beta^1| \le \gamma/3n\delta_0$ and that $0 \le l_t^1 - \tilde{\varphi}_t^1 \le \delta$ for all t. We obtain that the integral (1.2) does not exceed $\gamma/3n$ and the first integral in (1.1) is not greater than $\gamma/3$.

The second integral in (1.1) does not exceed

$$\Delta L(\delta) \cdot \int_0^T (1 + L(l_t, \dot{l}_t)) \, dt \le \frac{\gamma}{3K + \gamma} \cdot (T + S_{0T}(l)) \le \frac{\gamma}{3}.$$

Finally,

$$S_{0T}(\tilde{\varphi}) \le S_{0T}(l) + 2\gamma/3 \le S_{0T}(\varphi) + \gamma.$$

On the other hand, if $T \le h_0$, then $\rho(\varphi_0, \varphi_T) \le \lambda'$; the distances of φ_0 and φ_T from the points $\tilde{\varphi}_0$ and $\tilde{\varphi}_{\tilde{T}}$ do not exceed $\delta \le \delta_0 \le \lambda'$, so that $\rho(\tilde{\varphi}_0, \tilde{\varphi}_{\tilde{T}}) \le 3\lambda' \le \gamma/L$. Using Lemma 1.1, we obtain that these points are connected by the curve $\tilde{\varphi}_t$, $0 \le t \le \tilde{T} = \rho(\tilde{\varphi}_0, \tilde{\varphi}_T)$, $S_{0\tilde{T}}(\tilde{\varphi}) \le \gamma \le S_{0T}(\varphi) + \gamma$; this curve does not leave $D_{-\delta} \cup \partial D_{-\delta}$.

The proof can be preserved for curves φ_t in $D_{+\delta} \cup \partial D_{+\delta}$ and $\tilde{\varphi}_t$ in $D \cup \partial D$. □

It follows from Lemma 1.4 that $V_D(x, y)$ changes little as D is changed to $D_{+\delta}$ with small δ, uniformly in x and y, varying within the boundaries of some compactum in $D \cup \partial D$. It also changes little as D is changed to $D_{-\delta}$. In this case, if x or y lies outside $D_{-\delta} \cup \partial D_{-\delta}$, then they have to be replaced by $(x)_{-\delta}$ or $(y)_{-\delta}$, respectively.

With $V_D(x, y)$ there is associated an equivalence relation $\sim_D$ between points of $D \cup \partial D$: $x \sim_D y$ if $V_D(x, y) = V_D(y, x) = 0$. This relation depends only on the dynamical system and does not change if the matrix $(a^{ij}(x))$ is changed to another matrix for which the maximal and minimal eigenvalues are different from each other by a bounded factor. For $D = M$, we simply write $x \sim y$ in place of $x \sim_M y$. It is clear that $x \sim_D y$ implies $x \sim y$.

It follows from Lemma 1.1 that in the case of a smooth boundary ∂D, the points equivalent to each other form a closed set.

Lemma 1.5. *Let $x \sim_D y$, $y \neq x$. The trajectory $x_t(x)$ of the dynamical system $\dot{x}_t = b(x_t)$ beginning at x lies in the set of points $z \sim_D x$.*

Proof. There exists a sequence of functions $\varphi_t^{(n)}$, $0 \le t \le T_n$, $\varphi_0^{(n)} = x$, $\varphi_{T_n}^{(n)} = y$, lying entirely in $D \cup \partial D$ and such that $S_{0T_n}(\varphi^{(n)}) \to 0$. The T_n are bounded from below by a positive constant, say T. For the functions $\varphi_t^{(n)}$ on the interval from 0 to T, the values of S_{0T} converge to 0. Therefore, some subsequence of these functions converges, uniformly on $[0, T]$, to a function φ_t with $S_{0T}(\varphi) = 0$, i.e., to the trajectory $x_t(x)$ of the dynamical system. The points $x_t(x)$, $0 \le t \le T$, are equivalent to x and y; this follows from the fact that $V_D(x, \varphi_t^{(n)})$ and $V_D(\varphi_t^{(n)}, y)$ do not exceed $S_{0T_n}(\varphi^{(n)}) \to 0$ as $n \to \infty$.

We continue with $x_T(x)$. We choose that one of the points x and y which is farther from $x_T(x)$. If it is, say, x, then we have $\rho(x_T(x), x) \ge \frac{1}{2}\rho(x, y)$. In the same way as earlier, we obtain that over some time interval, $x_t(x)$ goes within the set of points equivalent to x. These intervals are bounded from below by a positive constant and thus we obtain our assertion for all $t > 0$ by successive application of the above argument. It is easy to see that our assertion is also true for $t < 0$. □

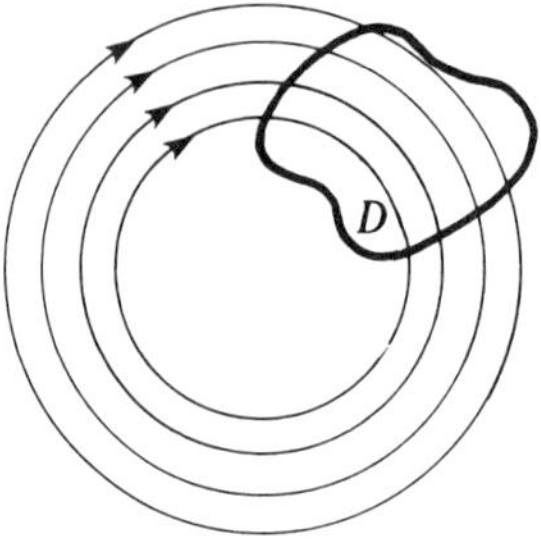

Figure 10.

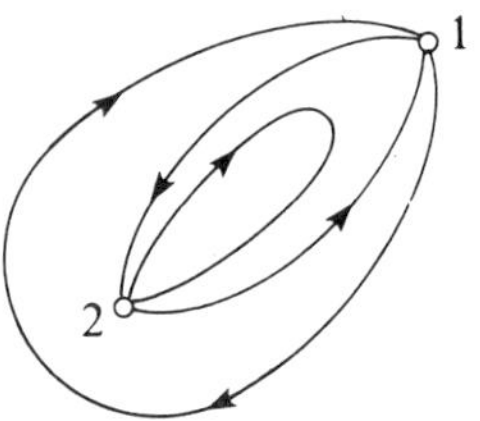

Figure 11.

Any ω-limit set, i.e., any set of partial limits of a trajectory of the dynamical system as $t \to \infty$, consists of equivalent points; this follows from Lemma 1.1. A maximal set of points equivalent to each other, containing some ω-limit set may consist of one ω-limit set, it may be the sum of a finite or infinite number of ω-limit sets, or it may contain further points not belonging to any ω-limit set. Moreover, such a set contains, with every point, a whole trajectory of the dynamical system.

We give some examples. Let the trajectories of a system be concentric circles (Fig. 10). In this case, for any two points x, y of a closed trajectory we have $x \sim y$. However, $x \nsim_D y$ for $x \neq y$ if D has the form indicated in the figure.

In Fig. 11, if the trajectories not drawn approach the trajectories drawn, there are six different ω-limit sets: the point 1, the point 2, the exterior curve (including the point 1), the union of the exterior curve and the intermediate curve, the union of the intermediate curve and the interior curve, the interior curve; the union of these ω-limit sets forms a maximal set of points equivalent to each other. In the case where the trajectories not drawn move away from those drawn, the set of equivalent points remains the same but there remain only two ω-limit sets: the points 1 and 2.

Lemma 1.6. *Let all points of a compactum $K \subseteq D \cup \partial D$ be equivalent to each other but not equivalent to any other point in $D \cup \partial D$. For any $\gamma > 0$, $\delta > 0$ and $x, y \in K$ there exists a function φ_t, $0 \le t \le T$, $\varphi_0 = x$, $\varphi_T = y$, entirely in the intersection of $D \cup \partial D$ with the δ-neighborhood of K and such that $S_{0T}(\varphi) < \gamma$.*

Proof. We connect the points x and y with curves

$$\varphi_t^{(n)} \in D \cup \partial D, \qquad 0 \le t \le T_n, \qquad \gamma > S_{0T_n}(\varphi^{(n)}) \to 0.$$

If all curves $\varphi_t^{(n)}$ left the δ-neighborhood of K, then they would have a limit point outside this δ-neighborhood, equivalent to x and y. □

Lemma 1.7. *Let all points of a compactum K be equivalent to each other and let $K \neq M$. Let us denote by τ_G the time of first exit of the process X_t^ε from the*

δ-neighborhood G of K. For any $\gamma > 0$ there exists $\delta > 0$ such that for all sufficiently small ε, $x \in G$ we have

$$\mathsf{M}_x^\varepsilon \tau_G < e^{\gamma\varepsilon^{-2}}. \tag{1.3}$$

Proof. We choose a point z outside K such that $\rho(z, K) < \gamma/3L \wedge \lambda$ where L and λ are the constants from Lemma 1.1. We put $\delta = \rho(z, K)/2$ and consider the δ-neighborhood $G \supset K$. We denote by x' the point of K closest to x (or any of the closest points if there are several such points) and by y the point closest to z. According to Lemma 1.2, for every pair of points x', $y \in K$ there exists a function φ_t, $0 \le t \le T$, $\varphi_0 = x'$, $\varphi_T = y$, such that $S_{0T}(\varphi) < \gamma/3$, and T is bounded by some constant independent of the initial and terminal points. We complete the curve φ_t at the beginning and at the end with little segments leading from x to x' and from y to z, with the values of S not exceeding $\gamma/6$ and $\gamma/3$, respectively. Then the length of the time interval on which each of the functions $\tilde{\varphi}_t$ is defined is uniformly bounded for $x \in G$ by a constant T_0 and the value of S at each of these functions does not exceed $5\gamma/6$. We extend the definition of each of these functions up to the end of the interval $[0, T_0]$ as the solution of $\dot{x}_t = b(x_t)$; this does not increase the value of S.

Now we use Theorem 3.2 of Ch. 5: for $x \in G$ we have

$$\mathsf{P}_x^\varepsilon\{\tau_G < T_0\} \ge \mathsf{P}_x^\varepsilon\{\rho_{0T_0}(X^\varepsilon, \tilde{\varphi}) < \delta\} \ge e^{-0.9\gamma\cdot\varepsilon^{-2}}.$$

Using the Markov property, we obtain that

$$\mathsf{P}_x^\varepsilon\{\tau_G \ge nT_0\} \le [1 - e^{-0.9\gamma\cdot\varepsilon^{-2}}]^n.$$

This yields

$$\mathsf{M}_x^\varepsilon \tau_G < T_0 \sum_{n=0}^{\infty} [1 - e^{-0.9\gamma\cdot\varepsilon^{-2}}]^n = T_0 e^{0.9\gamma\cdot\varepsilon^{-2}}.$$

Sacrificing 0.1γ in order to get rid of T_0, we obtain the required estimate. □

Lemma 1.8. *Let K be an arbitrary compactum and let G be a neighborhood of K. For any $\gamma > 0$ there exists $\delta > 0$ such that for all sufficiently small ε and x belonging to the closed δ-neighborhood $g \cup \partial g$ of K we have*

$$\mathsf{M}_x^\varepsilon \int_0^{\tau_G} \chi_g(X_t^\varepsilon)\, dt > e^{-\gamma\varepsilon^{-2}}. \tag{1.4}$$

Proof. We connect the point $x \in g \cup \partial g$ and the closest point x' of K with a curve φ_t with the value of S not greater than $\gamma/3$ (this can be done for sufficiently small δ). We continue this curve by a solution of $\dot{x}_t = b(x_t)$ until exit from $g \cup \partial g$ in such a way that the interval on which the curve is defined has length not greater than some $T < \infty$ independent of x. The trajectory of X_t^ε is at a distance not greater than $\delta/2$ from φ_t with probability not less than $e^{-2\gamma\varepsilon^{-2}/3}$ (for small ε); moreover, it is in g until exit from G, it spends a time bounded from below by a constant t_0 and the mathematical expectation in (1.4) is greater than $t_0 e^{-2\gamma\varepsilon^{-2}/3}$. □

Now we prove a lemma which is a generalization of Lemma 2.2 of Ch. 4.

Lemma 1.9. *Let K be a compact subset of M not containing any ω-limit set entirely. There exist positive constants c and T_0 such that for all sufficiently small ε and any $T > T_0$ and $x \in K$ we have*

$$\mathsf{P}_x^\varepsilon\{\tau_K > T\} \le e^{-\varepsilon^{-2}\cdot c(T-T_0)}, \tag{1.5}$$

where τ_K is the time of first exit of X_t^ε from K.

Proof. Using the continuous dependence of a solution on the initial conditions, it is easy to see that for sufficiently small δ, the closed δ-neighborhood $K_{+\delta}$ of K does not contain any ω-limit set entirely, either. For $x \in K_{+\delta}$ we denote by $\tau(x)$ the time of first exit of the solution $x_t(x)$ from $K_{+\delta}$. We have $\tau(x) < \infty$ for all $x \in K_{+\delta}$. The function $\tau(x)$ is upper semicontinuous, and consequently, it attains its largest value $\max_{x\in K_{+\delta}} \tau(x) = T_1 < \infty$.

We put $T_0 = T_1 + 1$ and consider all functions φ_t defined for $0 \le t \le T_0$ and assuming values only in $K_{+\delta}$. The set of these functions is closed in the sense of uniform convergence, and consequently, S_{0T} attains its minimum A on this set. The minimum is positive, since there are no trajectories of the dynamical system among the functions under consideration.

Then, using Theorem 3.2 of Ch. 5, in the same way as in the proof of Lemma 2.2 of Ch. 4, we obtain:

$$\mathsf{P}_x^\varepsilon\{\tau_K > T_0\} \le \exp\{-\varepsilon^{-2}(A-\gamma)\},$$

$$\mathsf{P}_x^\varepsilon\{\tau_K > T\} \le \exp\left\{-\varepsilon^{-2}\left(\frac{T}{T_0} - 1\right)(A-\gamma)\right\}. \quad \square$$

Corollary. *It follows from Lemma 1.9 that for ε smaller than some ε_0 and for all $x \in K$ we have*

$$\mathsf{M}_x^\varepsilon \tau_K \le T_0 + \varepsilon^2/c < T_0' = T_0 + \varepsilon_0^2/c.$$

§2. Markov Chains Connected with the Process $(X_t^\varepsilon, \mathsf{P}_x^\varepsilon)$

In this section we shall assume that D is a domain with smooth boundary and compact closure on a manifold M. We impose the following restrictions on the structure of the dynamical system in $D \cup \partial D$:

(**A**) in D there exist a finite number of compacta $K_1, K_2, \ldots, K_l$ such that:

(1) for any two points x, y belonging to the same compactum we have $x \sim_D y$;
(2) if $x \in K_i$ and $y \notin K_i$, then $x \nsim_D y$;
(3) every ω-limit set of the dynamical system $\dot{x}_t = b(x_t)$, lying entirely in $D \cup \partial D$, is contained in one of the K_i.

We have seen (§§2, 4, Ch. 4 and §4, Ch. 5) that in the case of a dynamical system with one (stable) equilibrium position O, for the study of the behavior of the process X_t^ε on large time intervals for small ε, an essential role is played by the Markov chain Z_n on the set $\gamma \cup \partial D$, where γ is the boundary of a small neighborhood of O. We have also seen that the asymptotics of the transition probabilities of this chain are almost independent of the initial point $x \in \gamma$, so that for small ε, the chain Z_n behaves as a simple Markov chain with a finite number of states. The asymptotics of the transition probabilities of the chain were determined by the quantities $V_0 = \min\{V(O, y): y \in \partial D\}$, $\min\{V(O, y): y \in \partial D \setminus \mathscr{E}_\delta(y_0)\}$. For example, for small ε, the transition probability $P(x, \partial D)$ for $x \in \gamma$ is between $\exp\{-\varepsilon^{-2}(V_0 \pm \gamma)\}$, where $\gamma > 0$ is small. We use an analogous construction in the case of systems satisfying condition (**A**).

We introduce the following notation:

$$\tilde{V}_D(K_i, K_j) = \inf\Big\{S_{0T}(\varphi): \varphi_0 \in K_i, \varphi_T \in K_j, \varphi_t \in (D \cup \partial D) \setminus \bigcup_{s \neq i, j} K_s \text{ for } 0 < t < T\Big\}$$

(if there are no such functions, we set $\tilde{V}_D(K_i, K_j) = +\infty$). For $x, y \in D \cup \partial D$ we set

$$\tilde{V}_D(x, K_j) = \inf\Big\{S_{0T}(\varphi): \varphi_0 = x, \varphi_T \in K_j, \varphi_t \in (D \cup \partial D) \setminus \bigcup_{s \neq j} K_s \text{ for } 0 < t < T\Big\};$$

$$\tilde{V}_D(K_i, y) = \inf\Big\{S_{0T}(\varphi): \varphi_0 \in K_i, \varphi_T = y, \varphi_t \in (D \cup \partial D) \setminus \bigcup_{s \neq i} K_s \text{ for } 0 < t < T\Big\};$$

$$\tilde{V}_D(x, y) = \inf\Big\{S_{0T}(\varphi): \varphi_0 = x, \varphi_T = y, \varphi_t \in (D \cup \partial D) \setminus \bigcup_{s} K_s \text{ for } 0 < t < T\Big\}.$$

Finally,

$$\tilde{V}_D(K_i, \partial D) = \min_{y \in \partial D} \tilde{V}_D(K_i, y);$$

$$\tilde{V}_D(x, \partial D) = \min_{y \in \partial D} \tilde{V}_D(x, y).$$

It is these quantities which determine the asymptotics of transition probabilities of the Markov chain connected with the process X_t^ε.

We consider an example. Let $M = R^2$ and let the trajectories of the dynamical system have the form depicted in Fig. 12. Then there are four compacta consisting of equivalent points and containing the ω-limit sets.

It is easy to see that $\tilde{V}_D(K_1, y) = 0$ for all y inside the cycle K_2 and in particular, on K_2 we have $\tilde{V}_D(K_1, K_2) = 0$; $\tilde{V}_D(K_1, y) = \infty$ for all y outside K_2, whence $\tilde{V}_D(K_1, K_3) = \tilde{V}_D(K_1, K_4) = \tilde{V}_D(K_1, \partial D) = \infty$. In the same way, $\tilde{V}_D(K_3, K_1) = \tilde{V}_D(K_4, K_1) = \infty$. Further, the values $\tilde{V}_D(K_2, y)$ are finite and positive if $y \notin K_2$, etc. The following matrix of $\tilde{V}_D(K_i, K_j)$ is consistent with the structure, depicted in Fig. 12, of trajectories of the dynamical system:

$$\begin{pmatrix} 0 & 0 & \infty & \infty \\ 1 & 0 & 9 & 9 \\ \infty & 6 & 0 & 6 \\ \infty & 0 & 0 & 0 \end{pmatrix} \tag{2.1}$$

(That $\tilde{V}_D(K_2, K_4)$ is equal to $\tilde{V}_D(K_2, K_3)$ and $\tilde{V}_D(K_4, K_2) = \tilde{V}_D(K_4, K_3)$ can be proved easily.)

Knowing $\tilde{V}_D(K_i, K_j)$ for all i, j, it is easy to determine

$$V_D(K_i, K_j) = V_D(x, y)|_{x \in K_i, y \in K_j}.$$

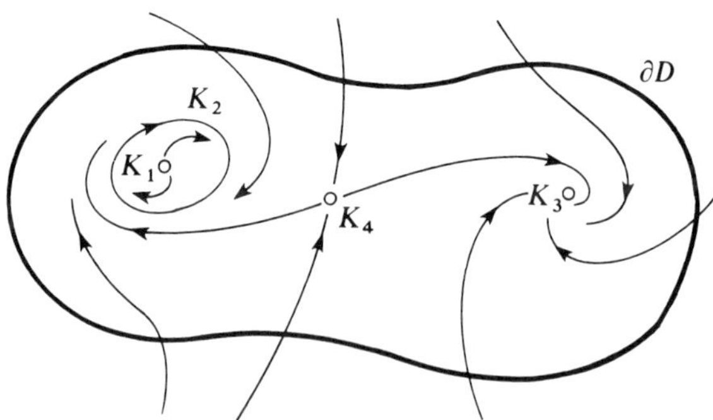

Figure 12.

Namely,

$$V_D(K_i, K_j) = \tilde{V}_D(K_i, K_j) \wedge \min_s[\tilde{V}_D(K_i, K_s) + \tilde{V}_D(K_s, K_j)]$$
$$\wedge \min_{s_1, s_2}[\tilde{V}_D(K_i, K_{s_1}) + \tilde{V}_D(K_{s_1}, K_{s_2}) + \tilde{V}_D(K_{s_2}, K_j)]$$
$$\wedge \cdots \wedge \min_{s_1, s_2, \ldots, s_{l-2}}[\tilde{V}_D(K_i, K_{s_1}) + \cdots + \tilde{V}_D(K_{s_{l-2}}, K_j)]$$

We can express $V_D(K_i, y)$, $V_D(x, K_j)$, $V_D(x, y)$ similarly. This can be proved by using Lemma 1.6. In the example considered by us, the $V_D(K_i, K_j)$ form the following matrix:

$$\begin{pmatrix} 0 & 0 & 9 & 9 \\ 1 & 0 & 9 & 9 \\ 7 & 6 & 0 & 6 \\ 1 & 0 & 0 & 0 \end{pmatrix} \tag{2.2}$$

Let ρ_0 be a positive number smaller than half of the minimum of the distances between K_i, K_j and between K_i, ∂D. Let ρ_1 be a positive number smaller than ρ_0. We denote by C the set $D \cup \partial D$ from which we delete the ρ_0-neighborhoods of K_i, $i = 1, \ldots, l$, by Γ_i the boundaries of the ρ_0-neighborhoods of the K_i, by g_i the ρ_1-neighborhoods of the K_i, and by g the union of the g_i. We introduce the random times $\tau_0 = 0$, $\sigma_n = \inf\{t \geq \tau_n\colon X_t^\varepsilon \in C\}$, $\tau_n = \inf\{t \geq \sigma_{n-1}\colon X_t^\varepsilon \in \partial g \cup \partial D\}$ and consider the Markov chain $Z_n = X_{\tau_n}^\varepsilon$. From $n = 1$ on, Z_n belongs to $\partial g \cup \partial D$. As far as the times σ_n are concerned, $X_{\sigma_0}^\varepsilon$ can be any point of C; all the following $X_{\sigma_n}^\varepsilon$ until the time of exit of X_t^ε to ∂D belong to one of the surfaces Γ_i and after exit to the boundary we have $\tau_n = \sigma_n = \tau_{n+1} = \sigma_{n+1} = \cdots$ and the chain Z_n stops.

The estimates of the transition probabilities of the Z_n provide the following two lemmas.

Lemma 2.1. *For any $\gamma > 0$ there exists $\rho_0 > 0$ (which can be chosen arbitrarily small) such that for any ρ_2, $0 < \rho_2 < \rho_0$, there exists ρ_1, $0 < \rho_1 < \rho_2$ such that for any δ_0 smaller than ρ_0 and sufficiently small ε, for all x in the ρ_2-neighborhood G_i of the compactum K_i $(i = 1, \ldots, 1)$ the one-step transition probabilities of Z_n satisfy the inequalities*

$$\exp\{-\varepsilon^{-2}(\tilde{V}_D(K_i, K_j) + \gamma)\} \leq P(x, \partial g_j) \leq \exp\{-\varepsilon^{-2}(\tilde{V}_D(K_i, K_j) - \gamma)\}; \tag{2.3}$$

$$\exp\{-\varepsilon^{-2}(\tilde{V}_D(K_i, \partial D) + \gamma)\} \leq P(x, \partial D) \leq \exp\{-\varepsilon^{-2}(\tilde{V}_D(K_i, \partial D) - \gamma)\}; \tag{2.4}$$

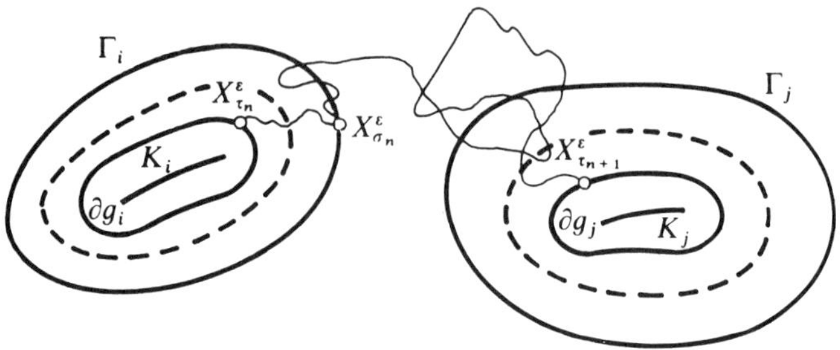

Figure 13.

for all $y \in \partial D$ *we have*

$$\exp\{-\varepsilon^{-2}(\tilde{V}_D(K_i, y) + \gamma)\} \leq P(x, \partial D \cap \mathscr{E}_{\delta_0}(y)) \leq \exp\{-\varepsilon^{-2}(\tilde{V}_D(K_i, y) - \gamma)\}. \tag{2.5}$$

In particular, if D coincides with the whole manifold M (and M is compact), then $\partial D = \varnothing$, the chain Z_n has ∂g as its space of states, and (2.3) implies that for $x \in \partial g_i$, $P(x, \partial g_j)$ lies between $\exp\{-\varepsilon^{-2}(\tilde{V}(K_i, K_j) \pm \gamma)\}$.

We provide a figure for this and the next lemma (Fig. 13).

Proof. First of all, $\tilde{V}_D(K_i, K_j) = +\infty$ (or $\tilde{V}_D(K_i, \partial D) = +\infty$, $\tilde{V}_D(K_i, y) = +\infty$) means that there is no smooth curve connecting K_i with K_j in $D \cup \partial D$ and not touching the other compacta (or connecting K_i with ∂D or y on the boundary, respectively). From this it is easy to derive that they cannot be connected even with a continuous curve not touching the indicated compacta. This implies that for $\tilde{V}_D(K_i, K_j) = \infty$ (or $\tilde{V}_D(K_i, \partial D) = \infty$, $\tilde{V}_D(K_i, y) = \infty$), the transition probabilities in (2.3) (or (2.4), (2.5), respectively) are equal to zero. As far as the finite $\tilde{V}_D(K_i, K_j)$, $\tilde{V}_D(K_i, y)$ are concerned, they are bounded by some $V_0 < \infty$.

We note that it is sufficient to prove (2.3) and (2.5); estimate (2.5) will imply (2.4), since ∂D can be covered by a finite number of δ_0-neighborhoods.

We choose a positive ρ_0 smaller than $\gamma/10L$, $\lambda/2$ (L and λ are constants from Lemma 1.1) and a third of the minimum distance between K_i, K_j and ∂D. Let a ρ_2, $0 < \rho_2 < \rho_0$ be chosen. By Lemma 1.4 there exists a positive $\delta \leq \rho_0/2$ such that for all $i, j = 1, \ldots, 1$ and $y \in \partial D$ we have

$$\tilde{V}_{D+\delta}(K_i, K_j) \geq \tilde{V}_D(K_i, K_j) - 0.1\gamma,$$

$$\tilde{V}_{D-\delta}(K_i, K_j) \leq \tilde{V}_D(K_i, K_j) + 0.1\gamma,$$

$$\tilde{V}_{D+\delta}(K_i, y) \geq \tilde{V}_D(K_i, y) - 0.1\gamma,$$

$$\tilde{V}_{D-\delta}(K_i, (y)_{-\delta}) \leq \tilde{V}_D(K_i, y) + 0.1\gamma.$$

For every pair K_i, K_j for which $\tilde{V}_D(K_i, K_j) < \infty$ we choose a function $\varphi_t^{K_i, K_j}$, $0 \le t \le T = T(K_i, K_j)$, such that $\varphi_0^{K_i, K_j} \in K_i$, $\varphi_T^{K_i, K_j} \in K_j$, $\varphi_t^{K_i, K_j}$ does not touch $\bigcup_{s \ne i, j} K_s$, does not leave $D_{-\delta} \cup \partial D_{-\delta}$ for $0 \le t \le T$, and for which

$$S_{0T}(\varphi^{K_i, K_j}) \le \tilde{V}_D(K_i, K_j) + 0.2\gamma.$$

Further, on ∂D we choose a ρ_2-net $y_1, \ldots, y_N$. For every pair K_i, y_k for which $\tilde{V}_D(K_i, y_k) < \infty$, we choose a function $\varphi_t^{K_i, y_k}$, $0 \le t \le T = T(K_i, y_k)$, $\varphi_0^{K_i, y_k} \in K_i$, $\varphi_T^{K_i, y_k} = (y_k)_{-\delta}$, which does not touch $\bigcup_{s \ne i} K_s$, does not leave $D_{-\delta} \cup \partial D_{-\delta}$, and for which

$$S_{0T}(\varphi^{K_i, y_k}) \le \tilde{V}_D(K_i, y_k) + 0.2\gamma.$$

We fix a positive ρ_1 smaller than ρ_2, $\rho_0/2$,

$$\tfrac{1}{2} \min\left\{\rho\left(\varphi_t^{K_i, K_j}, \bigcup_{s \ne i, j} K_s\right) : 0 \le t \le T(K_i, K_j), i, j = 1, \ldots, l\right\}$$

and

$$\tfrac{1}{2} \min\left\{\rho\left(\varphi_t^{K_i, y_k}, \bigcup_{s \ne i} K_s\right) : 0 \le t \le T(K_i, y_k), i = 1, \ldots, l; k = 1, \ldots, N\right\}.$$

Let an arbitrary positive $\delta_0 \le \rho_0$ be chosen. We derive estimates (2.3) and (2.5).

We choose a positive δ' not exceeding δ, ρ_1 or $\rho_0 - \rho_2$ and such that the δ'-neighborhood of the segment of the normal passing through any point $y \in \partial D$ intersects the boundary in $\mathscr{E}_{\delta_0}(y)$. First we derive the lower estimates.

Let $\tilde{V}_D(K_i, K_j) < \infty$. In accordance with Lemma 1.1, for any $x \in G_i$ we take a curve connecting x with a point $x' \in K_i$ for which the value of S does not exceed 0.1γ; the distance between this curve and the set C is not smaller than δ'. Then, according to Lemma 1.6, we find a curve in G_i which connects x' with $\varphi_0^{K_i, K_j} \in K_i$, with the value of S not greater than 0.1γ again. We combine these curves, complete them with the curve $\varphi_t^{K_i, K_j}$ and obtain a function φ_t, $0 \le t \le T$ (φ_t and T depend on $x \in G_i$ and j), $\varphi_0 = x$, $\varphi_T \in K_j$ such that $S_{0T}(\varphi) \le \tilde{V}_D(K_i, K_j) + 0.4\gamma$. For $j = i$ we define φ_t so that it connects $x \in G_i$ with a point x'' at distance $\rho_0 + \delta'$ from K_i and then with the closest point of K_i; then $S_{0T}(\varphi) \le 0.6\gamma = \tilde{V}_D(K_i, K_i) + 0.6\gamma$. The lengths of the intervals of definition of the functions φ_t constructed for all possible compacta K_i, K_j and points $x \in G_i$ can be bounded from above by a constant $T_0 < \infty$ (cf. Lemmas 1.1 and 1.2). We extend all functions φ_t to the intervals from T to T_0 to be a solution of $\dot{x}_t = b(x_t)$ so that $S_{0T_0}(\varphi) = S_{0T}(\varphi)$.

If a trajectory of X_t^ε passes at a distance from φ_t smaller than δ' for $0 \le t \le T_0$, then the trajectory intersects Γ_i and reaches the δ'-neighborhood of

K_j without getting closer than $\rho_2 + \delta'$ to any of the other compacta; moreover, $X^\varepsilon_{\tau_1} \in \partial g_j$. Using Theorem 3.2 of Ch. 5, we obtain for ε not exceeding some ε_0 depending only on γ, V_0, T_0, and δ':

$$\begin{aligned} P(x, \partial g_j) &\geq \mathsf{P}^\varepsilon_x\{\rho_{0T_0}(X^\varepsilon, \varphi) < \delta'\} \\ &\geq \exp\{-\varepsilon^{-2}(S_{0T_0}(\varphi) + 0.1\gamma)\} \\ &> \exp\{-\varepsilon^{-2}(\tilde{V}_D(K_i, K_j) + \gamma)\}. \end{aligned}$$

The lower estimate in (2.3) has been obtained; we pass to estimate (2.5). Let $x \in G_i$, $y \in \partial D$, $\tilde{V}_D(K_i, y) < \infty$. We choose a point y_k from our ρ_0-net such that $\rho(y_k, y) < \rho_0$; then $\rho((y_k)_{-\delta}, (y)_{-\delta}) < 2\rho_0$. We connect x and $x' \in K_i$ with a curve with the value of S not greater than 0.1γ and x' and $\varphi_0^{K_i, y_k} \in K_i$ with the "value" of S not greater than 0.1γ. We complete the curve thus obtained with $\varphi_t^{K_i, y_k}$. Then we reach $(y_k)_{-\delta}$ and the total value of S is not greater than $\tilde{V}_D(K_i, y) + 0.5\gamma$. We connect $(y_k)_{-\delta}$ with $(y)_{+\delta}$ through $(y)_{-\delta}$, increasing the value of S by no more than 0.3γ. Finally, we extend all functions constructed so far to the same interval $[0, T_0]$ to be a solution of $\dot{x}_t = b(x_t)$ in such a way that $S_{0T_0}(\varphi) \leq \tilde{V}_D(K_i, y) + 0.8\gamma$.

Using Theorem 3.2 of Ch. 5 again, we obtain the lower estimate in (2.5).

Now we obtain the upper estimates.

It is sufficient to prove estimates (2.3) and (2.5) for $x \in \Gamma_i$ (this follows from the strong Markov property). By virtue of the choice of ρ_0 and δ', for any curve φ_t, $0 \leq t \leq T$ beginning on Γ_i, touching the δ'-neighborhood of ∂g_j (the $(\delta_0 + \delta')$-neighborhood of $y \in \partial D$), not touching the compacta K_s, $s \neq i, j$ and not leaving $D_{+\delta} \cup \partial D_{+\delta}$, we have: $S_{0T}(\varphi) \geq \tilde{V}_D(K_i, K_j) - 0.3\gamma (S_{0T}(\varphi) \geq \tilde{V}_D(K_i, y) - 0.4\gamma)$. Using Lemma 1.9, we choose T_1 such that for all sufficiently small $\varepsilon > 0$ and $x \in (D \cup \partial D)\backslash g$ we have $\mathsf{P}^\varepsilon_x\{\tau_1 > T_1\} \leq \exp\{-\varepsilon^{-2}V_0\}$.

Any trajectory of X^ε_t beginning at a point $x \in \Gamma_i$ and being in ∂g_j (in $\partial D \cap \mathscr{E}_{\delta_0}(y)$) at time τ_1 either spends time T_1 without touching $\partial g \cup \partial D$ or reaches ∂g_j ($\partial D \cap \mathscr{E}_{\delta_0}(y)$) over time T_1; in this case

$$\rho_{0T_1}(X^\varepsilon, \Phi_x(\tilde{V}_D(K_i, K_j) - 0.3\gamma)) \geq \delta'$$
$$(\rho_{0T_1}(X^\varepsilon, \Phi_x(\tilde{V}_D(K_i, y) - 0.4\gamma)) \geq \delta').$$

Therefore, for any $x \in \Gamma_i$ we have

$$\begin{aligned} \mathsf{P}^\varepsilon_x\{X^\varepsilon_{\tau_1} \in \partial g_j\} &\leq \mathsf{P}^\varepsilon_x\{\tau_1 > T_1\} \\ &\quad + \mathsf{P}^\varepsilon_x\{\rho_{0T_1}(X^\varepsilon, \Phi_x(\tilde{V}_D(K_i, K_j) - 0.3\gamma)) \geq \delta'\}; \end{aligned} \tag{2.6}$$

$$\begin{aligned} \mathsf{P}^\varepsilon_x\{X^\varepsilon_{\tau_1} \in \partial D \cup \mathscr{E}_{\delta_0}(y)\} &\leq \mathsf{P}^\varepsilon_x\{\tau_1 > T_1\} \\ &\quad + \mathsf{P}^\varepsilon_x\{\rho_{0T_1}(X^\varepsilon, \Phi_x(\tilde{V}_D(K_i, y) - 0.4\gamma)) \geq \delta'\}. \end{aligned} \tag{2.7}$$

For small ε the first probability on the right side of (2.6) and (2.7) does not exceed $\exp\{-\varepsilon^{-2}V_0\}$ and by virtue of Theorem 3.2 of Ch. 5 the second one is smaller than $\exp\{-\varepsilon^{-2}(\tilde{V}_D(K_i, K_j) - 0.5\gamma)\}$ ($\exp\{-\varepsilon^{-2}(\tilde{V}_D(K_i, y) - 0.5\gamma)\}$) for all $x \in \Gamma_i$. From this we obtain the upper estimates in (2.3) and (2.5). □

Lemma 2.2. *For any $\gamma > 0$ there exists $\rho_0 > 0$ (which can be chosen arbitrarily small) such that for any ρ_2, $0 < \rho_2 < \rho_0$ there exists ρ_1, $0 < \rho_1 < \rho_2$ such that for any δ_0, $0 < \delta_0 \le \rho_0$, sufficiently small ε and all x outside the ρ_2-neighborhood of the compacta K_i and the boundary ∂D, the one-step transition probabilities of the chain Z_n satisfy the inequalities*

$$\exp\{-\varepsilon^{-2}(\tilde{V}_D(x, K_j) + \gamma)\} \le P(x, \partial g_j) \le \exp\{-\varepsilon^{-2}(\tilde{V}_D(x, K_j) - \gamma)\}; \tag{2.8}$$

$$\exp\{-\varepsilon^{-2}(\tilde{V}_D(x, \partial D) + \gamma)\} \le P(x, \partial D) \le \exp\{-\varepsilon^{-2}(\tilde{V}_D(x, \partial D) - \gamma)\}; \tag{2.9}$$

$$\exp\{-\varepsilon^{-2}(\tilde{V}_D(x, y) + \gamma)\} \le P(x, \partial D \cap \mathscr{E}_{\delta_0}(y)) \le \exp\{-\varepsilon^{-2}(\tilde{V}_D(x, y) - \gamma)\}. \tag{2.10}$$

Proof. We choose a ρ_0 as in the proof of the preceding lemma. Let ρ_2, $0 < \rho_2 < \rho_0$, be given. Let us denote by C_{ρ_2} the compactum consisting of the points of $D \cup \partial D$ except the ρ_2-neighborhoods of K_i and ∂D. We choose δ, $0 \le \delta \le \rho_0/2$, such that for all $x \in C_{\rho_2}$, $j = 1, \dots, 1$ and $y \in \partial D$ we have

$$\tilde{V}_{D+\delta}(x, K_j) \ge \tilde{V}_D(x, K_j) - 0.1\gamma,$$

$$\tilde{V}_{D-\delta}(x, K_j) \le \tilde{V}_D(x, K_j) + 0.1\gamma,$$

$$\tilde{V}_{D+\delta}(x, y) \ge \tilde{V}_D(x, y) - 0.1\gamma,$$

$$\tilde{V}_{D-\delta}(x, (y)_{-\delta}) \le \tilde{V}_D(x, y) + 0.1\gamma.$$

We choose ρ_2-nets $x_1, \dots, x_M$ in C_{ρ_2} and $y_1, \dots, y_N$ on ∂D. For every pair x_i, K_j for which $\tilde{V}_D(x_i, K_j) < \infty$ we select a function $\varphi_t^{x_i, K_j}$, $0 \le t \le T = T(x_i, K_j)$, $\varphi_0^{x_i, K_j} = x_i$, $\varphi_T^{x_i, K_j} \in K_j$, which does not touch $\bigcup_{s \ne j} K_s$, does not leave $D_{-\delta} \cup \partial D_{-\delta}$ and for which $S_{0T}(\varphi^{x_i, K_j}) \le \tilde{V}_D(x_i, K_j) + 0.2\gamma$. For every pair x_i, y_k, $\tilde{V}_D(x_i, y_k) < \infty$, we select a function $\varphi_t^{x_i, y_k}$, $0 \le t \le T = T(x_i, y_k)$, $\varphi_0^{x_i, y_k} = x_i$, $\varphi_T^{x_i, y_k} = (y_k)_{-\delta}$, not touching $\bigcup_s K_s$, not leaving $D_{-\delta} \cup \partial D_{-\delta}$ and such that $S_{0T}(\varphi^{x_i, y_k}) \le \tilde{V}_D(x_i, y_k) + 0.2\gamma$.

Further, we pick a positive ρ_1 less than ρ_2, $\rho_0/2$ and half the minimum distance of the curves constructed in the proof from those compacta which they must not approach and we pick a δ satisfying the same conditions as in the proof of Lemma 2.1 and not exceeding $\rho_2 - \rho_1$; the proof can be carried out analogously. □

Lemma 2.2 has a very simple special case where there are no compacta K_i, i.e., there are no ω-limit sets of the dynamical system in $D \cup \partial D$. In this case we obtain immediately the following theorem.

Theorem 2.1. *Suppose that the family $(X_t^\varepsilon, \mathsf{P}_x^\varepsilon)$ of diffusion processes satisfies the hypotheses of Theorem 3.2 of Ch. 5 and the drift $b(x)$ satisfies a Lipschitz condition. Let D be a domain with compact closure and smooth boundary and assume that none of the ω-limit sets of the system $\dot{x}_t = b(x_t)$ lies entirely in $D \cup \partial D$. Then the asymptotics as $\varepsilon \to 0$ of the distribution, at the time of exit to the boundary, of the process beginning at x can be described by the action function $\varepsilon^{-2} V_D(x, y)$, $y \in \partial D$, uniformly in the initial point strictly inside D, i.e., we have*

$$\lim_{\delta_0 \to 0} \lim_{\varepsilon \to 0} \varepsilon^2 \ln \mathsf{P}_x^\varepsilon\{X_{\tau_\varepsilon}^\varepsilon \in \mathscr{E}_{\delta_0}(y)\} = -V_D(x, y),$$

uniformly in x belonging to any compact subset of D and in $y \in \partial D$, where $\tau^\varepsilon = \inf\{t: X_t^\varepsilon \notin D\}$.

In particular, with probability converging to 1 as $\varepsilon \to 0$, the exit to the boundary takes place in a small neighborhood of the set of the points at which the trajectory $x_t(x)$, $t \geq 0$, touches the boundary until exit from $D \cup \partial D$. (Of course, this simple result can be obtained more easily.)

The formulation of Theorem 2.1 in the language of differential equations reads as follows:

Theorem 2.2. *Suppose that $u^\varepsilon(x)$ is the solution of the Dirichlet problem $L^\varepsilon u^\varepsilon(x) = 0$ in D, $u^\varepsilon(x) = \exp\{\varepsilon^{-2} F(x)\}$ on ∂D, where in local coordinates we have $L^\varepsilon = \sum_i b^{i\varepsilon}(x)(\partial/\partial x^i) + \varepsilon^2/2 \sum_{ij} a^{ij}(x)(\partial^2/\partial x^i \partial x^j)$, $b^{i\varepsilon}(x) \to b^i(x)$ uniformly in x and the local coordinate systems K_{x_0} as $\varepsilon \to 0$ and F is a continuous function. If none of the ω-limit sets of the system $\dot{x}_t = b(x_t)$ lies entirely in $D \cup \partial D$, then $u^\varepsilon(x) \asymp \exp\{\varepsilon^{-2} \max_{y \in \partial D} (F(y) - V_D(x, y))\}$ as $\varepsilon \to 0$, uniformly in x belonging to any compact subset of D, where $V_D(x, y)$ is defined by the coefficients $b^i(x)$, $a^{ij}(x)$ in the same way as in §1.*

In the case where there are ω-limit sets in $D \cup \partial D$ (and namely, in D) and condition **A** is satisfied, for the study of problems connected with the behavior of $(X_t^\varepsilon, \mathsf{P}_x^\varepsilon)$ we first have to study the limit behavior of Markov chains with exponential asymptotics of the transition probabilities.

§3. Lemmas on Markov Chains

For finite Markov chains an invariant measure, the distribution at the time of exit from a set, the mean exit time, etc., can be expressed explicitly as the ratio of some determinants, i.e., sums of products consisting of transition probabilities (since the values of the invariant measure and of the other quantities involved are positive, these sums only contain terms with a plus sign). Which products appear in the various sums, can be described conveniently by means of graphs on the set of states of the chain.

For chains on an infinite phase space divided into a finite number of parts for which there are upper and lower estimates (of the type of those obtained in the preceding section) of the transition probabilities, in this section we obtain estimates of values of an invariant measure, probabilities of exit to a set sooner than to another (others), and so on. In the application of these results to a chain Z_n with estimates of the transition probabilities obtained in the preceding section, in each sum we select one or several terms decreasing more slowly than the remaining ones. The constant characterizing the rate of decrease of the sum of products is, of course, determined as the minimum of the sums of constants characterizing the rate of decrease of each of the transition probabilities (cf. §§4, 5).

Let L be a finite set, whose elements will be denoted by the letters i, j, k, m, n, etc. and let a subset W be selected in L. A graph consisting of arrows $m \to n$ ($m \in L \setminus W$, $n \in L$, $n \neq m$) is called a W-graph if it satisfies the following conditions:

(1) every point $m \in L \setminus W$ is the initial point of exactly one arrow;
(2) there are no closed cycles in the graph.

We note that condition (2) can be replaced by the following condition:

(2′) for any point $m \in L \setminus W$ there exists a sequence of arrows leading from it to some point $n \in W$.

We denote by $G(W)$ the set of W-graphs; we shall use the letter g to denote graphs. If p_{ij} ($i, j \in L, j \neq i$) are numbers, then $\prod_{(m \to n) \in g} p_{mn}$ will be denoted by $\pi(g)$.

Lemma 3.1. *Let us consider a Markov chain with set of states L and transition probabilities p_{ij} and assume that every state can be reached from any other state in a finite number of steps. Then the stationary distribution of the chain is $\{(\sum_{i \in L} Q_i)^{-1} Q_i, i \in L\}$, where*

$$Q_i = \sum_{g \in G\{i\}} \pi(g). \tag{3.1}$$

Proof. The numbers Q_i are positive. It is sufficient to prove that they satisfy the system of equations

$$Q_i = \sum_{j \in L} Q_j p_{ji} \qquad (i \in L),$$

since it is well known that the stationary distribution is the unique (up to a multiplicative constant) solution of this system. In the ith equation we carry the ith term from the right side to the left side; we obtain that we have to verify the equality

$$Q_i \sum_{k \neq i} p_{ik} = \sum_{j \neq i} Q_j p_{ji}. \tag{3.2}$$

It is easy to see that if we substitute the numbers defined by formulas (3.1) in (3.2), then on both sides we obtain the sum $\pi(g)$ over all graphs g satisfying the following conditions:

(1) Every point $m \in L$ is the initial point of exactly one arrow $m \to n$ $(n \neq m, n \in L)$;
(2) in the graph there is exactly one closed cycle and this cycle contains the point i. □

Lemma 3.2. *Let us be given a Markov chain on a phase space X divided into disjoint sets X_i, where i runs over a finite set L. Suppose that there exist non-negative numbers p_{ij} $(j \neq i, i, j \in L)$ and a number $a > 1$ such that*

$$a^{-1} p_{ij} \leq P(x, X_j) \leq a p_{ij} \qquad (x \in X_i, i \neq j).$$

for the transition probabilities of our chain. Furthermore, suppose that every set X_j can be reached from any state x sooner or later (for this it is necessary and sufficient that for any j there exist a $\{j\}$-graph g such that $\pi(g) > 0$). *Then*

$$a^{2-2l}\left(\sum_{i \in L} Q_i\right)^{-1} Q_i \leq \nu(X_i) \leq a^{2l-2}\left(\sum_{i \in L} Q_i\right)^{-1} Q_i$$

for any normalized invariant measure ν of our chain, where l is the number of elements in L and the Q_i are defined by formula (3.1).

Proof. For any pair i, j there exists a number s of steps such that the transition probabilities $P^{(s)}(x, X_j)$ for $x \in X_i$ can be estimated from below by a positive constant. It follows from this that all $\nu(X_j)$ are positive. Let us consider a Markov chain with transition probabilities $p_{ij} = (1/\nu(X_i)) \int_{X_i} \nu(dx) P(x, X_j)$. The stationary distribution of this chain is $\{\nu(X_i),\ i \in L\}$, which can be estimated by means of the expression given for it in Lemma 3.1. □

Now we formulate an assertion which we shall use in the study of exit to the boundary.

For $i \in L \setminus W$, $j \in W$ we denote by $G_{ij}(W)$ the set of W-graphs in which the sequence of arrows leading from i into W (cf. condition (2′)) ends at the point j.

Lemma 3.3. *Let us be given a Markov chain on a phase space* $X = \bigcup_{i \in L} X_i$, $X_i \cap X_j = \varnothing$ $(i \neq j)$, *and assume that the transition probabilities of the chain satisfy the inequalities*

$$a^{-1} p_{ij} \leq P(x, X_j) \leq a p_{ij} \qquad (x \in X_i, j \neq i), \tag{3.3}$$

where a is a number greater than one. For $x \in X$ and $B \subseteq \bigcup_{k \in W} X_k$ we denote by $q_W(x, B)$ the probability that at the first entrance time of $\bigcup_{k \in W} X_k$, the particle performing a random walk in accordance with our chain hits B provided that it starts from x.

If the number of points in $L \setminus W$ is equal to r, then

$$a^{-4^r} \frac{\sum_{g \in G_{ij}(W)} \pi(g)}{\sum_{g \in G(W)} \pi(g)} \leq q_W(x, X_j) \leq a^{4^r} \frac{\sum_{g \in G_{ij}(W)} \pi(g)}{\sum_{g \in G(W)} \pi(g)} \tag{3.4}$$

$$(x \in X_i, i \in L \setminus W, j \in W),$$

provided that the denominator is positive.

We shall prove this by induction on r. For $r = 1$ we have $W = L \setminus \{i\}$ Using the Markov property, we obtain for $x \in X_i$ that

$$q_{L \setminus \{i\}}(x, X_j) = P(x, X_j) + \int_{X_i} P(x, dy) P(y, X_j)$$
$$+ \int_{X_i} P(x, dy_1) \int_{X_i} P(y_1, dy_2) P(y_2, X_j) + \cdots.$$

Let us denote by $\underline{A}_{ij}$ the infimum of $P(x, X_j)$ over $x \in X_i$; let $\underline{B}_i = \inf_{x \in X_i} P(x, X_i)$; let us denote by $\bar{A}_{ij}$, $\bar{B}_i$ the corresponding suprema. By assumption, $\sum_{k \neq i} p_{ik} > 0$. This means that $\underline{B}_i$, $\bar{B}_i < 1$. We have

$$\underline{A}_{ij} + \underline{A}_{ij} \underline{B}_i + \underline{A}_{ij} \underline{B}_i^2 + \cdots \leq q_{L \setminus \{i\}}(x, X_j)$$
$$\leq \bar{A}_{ij} + \bar{A}_{ij} \bar{B}_i + \bar{A}_{ij} \bar{B}_i^2 + \cdots.$$

i.e.,

$$\frac{\underline{A}_{ij}}{1 - \underline{B}_i} \leq q_{L \setminus \{i\}}(x, X_j) \leq \frac{\bar{A}_{ij}}{1 - \bar{B}_i}.$$

But, by assumption,

$$a^{-1}p_{ij} \leq \underline{A}_{ij} \leq \bar{A}_{ij} \leq ap_{ij};\ 1 - \underline{B}_i = \sup_{\alpha \in X_i} P\left(x, \bigcup_{k \neq i} X_k\right) \leq a \sum_{k \neq i} p_{ik}.$$

(Analogously, $1 - \bar{B}_i \geq a^{-1} \sum_{k \neq i} p_{ik}$). We obtain from this that (3.4) is satisfied even with a^2 instead of a^4.

Now let (3.4) hold for all W such that $L \setminus W$ contains r elements, for all $i \in L \setminus W$ and for all $j \in W$. We prove inequalities (3.4) for a set W such that there are $r + 1$ points in $L \setminus W$. Let $i \in L \setminus W$, $j \in W$ and put

$$F = \bigcup_{\substack{k \in L \setminus W \\ k \neq i}} X_k.$$

We may hit X_j immediately after exit from X_i; we may hit F first and then X_j; we may hit F first, then return to X_i and then hit X_j without calling on F, etc. In accordance with this, using the strong Markov property, we obtain

$$\begin{aligned} q_W(x, X_j) = q_{L \setminus \{i\}}(x, X_j) &+ \int_F q_{L \setminus \{i\}}(x, dy) q_{W \cup \{i\}}(y, X_j) \\ &+ \int_F q_{L \setminus \{i\}}(x, dy) \int_{X_i} q_{W \cup \{i\}}(y, dx_1) q_{L \setminus \{i\}}(x_1, X_j) \\ &+ \int_F q_{L \setminus \{i\}}(x, dy_1) \int_{X_i} q_{W \cup \{i\}}(y_1, dx_1) \int_F q_{L \setminus \{i\}}(x_1, dy_2) \\ &\times q_{W \cup \{i\}}(y_2, X_j) + \int_F q_{L \setminus \{i\}}(x, dy_1) \int_{X_i} q_{W \cup \{i\}}(y_1, dx_1) \\ &\times \int_F q_{L \setminus \{i\}}(x_1, dy_2) \int_{X_i} q_{W \cup \{i\}}(y_2, dx_2) q_{L \setminus \{i\}}(x_2, X_j) + \cdots \end{aligned}$$

We introduce the notation

$$\begin{aligned} \underline{C}_{ij} &= \inf_{x \in X_i} q_{L \setminus \{i\}}(x, X_j), \\ \underline{D}_{ij} &= \inf_{x \in X_i} \int_F q_{L \setminus \{i\}}(x, dy) q_{W \cup \{i\}}(y, X_j), \\ \underline{E}_i &= \inf_{x \in X_i} \int_F q_{L \setminus \{i\}}(x, dy) q_{W \cup \{i\}}(y, X_i) \end{aligned}$$

and denote the corresponding suprema by $\bar{C}_{ij}$, $\bar{D}_{ij}$ and $\bar{E}_i$ (as $\bar{B}_i$, $\bar{E}_i$ is also smaller than 1). Using this notation, we can write

$$\begin{aligned} \underline{C}_{ij} + \underline{D}_{ij} + \underline{C}_{ij}\underline{E}_i + \underline{D}_{ij}\underline{E}_i + \underline{C}_{ij}\underline{E}_i^2 + \underline{D}_{ij}\underline{E}_i^2 + \cdots \leq q_W(x, X_j) \\ \leq \bar{C}_{ij} + \bar{D}_{ij} + \bar{C}_{ij}\bar{E}_i + \bar{D}_{ij}\bar{E}_i + \bar{C}_{ij}\bar{E}_i^2 + \bar{D}_{ij}\bar{E}_i^2 + \cdots, \end{aligned}$$

i.e.,

$$\frac{\underline{C}_{ij} + \underline{D}_{ij}}{1 - \underline{E}_i} \le q_W(x, X_j) \le \frac{\bar{C}_{ij} + \bar{D}_{ij}}{1 - \bar{E}_i}.$$

In order to make the formulas half as bulky, we shall only consider the upper estimate. Since (3.4) is proved for $r = 1$, we have $\bar{C}_{ij} \le a^4(p_{ij}/\sum_{k \ne i} p_{ik})$. By the induction hypothesis, we have

$$\bar{D}_{ij} \le \sum_{\substack{k \in L \setminus W \\ k \ne i}} a^4 \frac{p_{ik}}{\sum_{k \ne i} p_{ik}} a^{4^r} \frac{\sum_{g \in G_{kj}(W \cup \{i\})} \pi(g)}{\sum_{g \in G(W \cup \{i\})} \pi(g)} = a^{4+4^r} \frac{H_{ij}}{\sum_{k \ne i} p_{ik} \cdot K_i}.$$

The H_{ij} here is the sum of the products $\pi(g)$ over those graphs in $G_{ij}(W)$ in which the arrow beginning at i does not lead immediately to j and K_i is the sum of the same products over all $(W \cup \{i\})$-graphs. From this we obtain

$$\bar{C}_{ij} + \bar{D}_{ij} \le a^{4+4^r} \frac{p_{ij} K_i + H_{ij}}{\sum_{k \ne i} p_{ik} \cdot K_i},$$

where in the numerator we now have the sum of the $\pi(g)$ over all graphs belonging to $G_{ij}(W)$.

Now we estimate the denominator $1 - \bar{E}_i$. From the condition that the denominator in (3.4) does not vanish we obtain that if the chain begins at an arbitrary point, it will, with probability 1, hit $\bigcup_{k \in W} X_k$, and consequently, $\bigcup_{k \in W \cup \{i\}} X_k$. Therefore, $q_{W \cup \{i\}}(y, X_i) = 1 - q_{W \cup \{i\}}(y, \bigcup_{k \in W} X_k)$ and

$$\begin{aligned} 1 - \bar{E}_i &= \inf_{x \in X_i} \left\{ 1 - \int_F q_{L \setminus \{i\}}(x, dy) \left[1 - q_{W \cup \{i\}}\left(y, \bigcup_{k \in W} X_k \right) \right] \right\} \\ &= \inf_{x \in X_i} \left\{ q_{L \setminus \{i\}}\left(x, \bigcup_{k \in W} X_k \right) + \int_F q_{L \setminus \{i\}}(x, dy) q_{W \cup \{i\}}\left(y, \bigcup_{k \in W} X_k \right) \right\}. \end{aligned}$$

The first term is not less than $a^{-4}(\sum_{k \in W} p_{ik}/\sum_{k \ne i} p_{ik})$; and the second one is not less than

$$\sum_{\substack{t \in L \setminus W \\ k \ne i}} a^{-4} \frac{p_{it}}{\sum_{k \ne i} p_{ik}} \sum_{k \in W} a^{-4^r} \frac{\sum_{g \in G_{tk}(W \cup \{i\})} \pi(g)}{\sum_{g \in G(W \cup \{i\})^-} \pi(g)} = a^{-4-4^r} \frac{L_i}{\sum_{k \ne i} p_{ik} \cdot K_i}.$$

The L_i here is the sum of the products $\pi(g)$ over those graphs in $G(W)$ in which the arrow beginning at i leads to a point belonging to $L \setminus W$. Bringing the estimates of the first and second terms to a common denominator, we obtain that

$$1 - \bar{E}_i \geq a^{-4-4^r} \frac{\sum_{k \in W} p_{ik} \cdot K_i + L_i}{\sum_{k \neq i} p^{ik} \cdot K_i}.$$

In the numerator here we have the sum of the $\pi(g)$ over all graphs $g \in G(W)$. Finally, we obtain

$$\frac{\bar{C}_{ij} + \bar{D}_{ij}}{1 - \bar{E}_i} \leq a^{8+2\cdot 4^r} \frac{p_{ij} K_i + H_{ij}}{\sum_{k \in W} p_{ik} \cdot K_i + L_i} \leq a^{4^{r+1}} \frac{p_{ij} K_i + H_{ij}}{\sum_{k \in W} p_{ik} \cdot K_i + L_i},$$

which gives the upper estimate in (3.4). Performing analogous calculations for $(\underline{C}_{ij} + \underline{D}_{ij})/(1 - \underline{E}_i)$, we obtain that (3.4) is proved for the case where the number of elements in $L \setminus W$ is equal to $r + 1$.

The lemma is proved. □

Lemma 3.4. *Let us be given a Markov chain on a phase space*

$$X = \bigcup_{i \in L} X_i, \qquad X_i \cap X_j = \varnothing \ (i \neq j),$$

with the estimates (3.3) *for the transition probabilities. We denote by $m_W(x)$ the mathematical expectation of the number of steps until the first entrance of $\bigcup_{k \in W} X_k$, calculated under the assumption that the initial state is x. If the number of points in $L \setminus W$ is equal to r, then for $x \in X_i$, $i \in L \setminus W$ we have*

$$\begin{aligned} \alpha^{-4^r} \frac{\sum_{g \in G(W \cup \{i\})} \pi(g) + \sum_{i \in L \setminus W, j \neq i} \sum_{g \in G_{ij}(W \cup \{j\})} \pi(g)}{\sum_{g \in G(W)} \pi(g)} &\leq m_W(x) \\ \leq a^{4^r} \frac{\sum_{g \in G(W \cup \{i\})} \pi(g) + \sum_{j \in L \setminus W, j \neq i} \sum_{g \in G_{ij}(W, \cup \{j\})} \pi(g)}{\sum_{g \in G(W)} \pi(g)}&. \end{aligned} \tag{3.5}$$

If $L \setminus W$ consists of only one point i, then in the sum in the numerator we have only one graph—the empty one; the product $\pi(g)$ is, of course, taken to be equal to 1. If $L \setminus W$ consists of more than one point, then the graphs over which the sum is taken in the numerator can be described as follows: they are the graphs without cycles, consisting of $(r - 1)$ arrows $m \to n$, $m \in L \setminus W$, $n \in L$, $m \neq n$, and not containing chains of arrows leading from i into W. We shall denote by $G(i \nrightarrow W)$ the set of these graphs.

The proof will be carried out by induction again. First let $r = 1$, i.e., we consider the first exit from X_i. The smallest number of steps until exit is equal to 1; we have to add one to this if we hit X_i again in the first step; we have to add one more 1 if the same happens in the second step, etc. Using the Markov property, we obtain

$$m_W(x) = m_{L\setminus\{i\}}(x) = 1 + P(x, X_i) + \int_{X_i} P(x, dx_1)P(x_1, X_i)$$
$$+ \int_{X_i} P(x, dx_1) \int_{X_i} P(x_1, dx_2)P(x_2, X_i) + \cdots.$$

This expression is between $1/(1 - \underline{B}_i)$ and $1/(1 - \bar{B}_i)$, where $\underline{B}_i$ and $\bar{B}_i$ are introduced in the proof of Lemma 3.3. We obtain (3.5) for $r = 1$ from the estimates of $1 - \underline{B}_i$ and $1 - \bar{B}_i$.

Now let (3.5) hold for all sets $L\setminus W$ with r elements and for all $i \in L\setminus W$. We prove (3.5) for $L\setminus W$ consisting of $r + 1$ points. As in the proof of Lemma 3.3, we put $F = \bigcup_{k\in L\setminus W;\, k\neq i} X_k$. The smallest value of the first entrance time of $\bigcup_{k\in W} X_k$ is the first exit time of X_i; if at this time we hit F, then we have to add the time spent in F; if after exit from F we hit X_i again, then we also have to add the time spent in X_i at this time, etc. Using the strong Markov property, we can write this in terms of the functions $m_{L\setminus\{i\}}(x)$, $m_{W\cup\{i\}}(x)$ and the measures $q_{L\setminus\{i\}}(x, \cdot)$, $q_{W\cup\{i\}}(x_1, \cdot)$:

$$m_W(x) = m_{L\setminus\{i\}}(x) + \int_F q_{L\setminus\{i\}}(x, dy)m_{W\cup\{i\}}(y)$$
$$+ \int_F q_{L\setminus\{i\}}(x, dy) \int_{X_i} q_{W\cup\{i\}}(y, dx_1)m_{L\setminus\{i\}}(x_1)$$
$$+ \int_F q_{L\setminus\{i\}}(x, dy_1) \int_{X_i} q_{W\cup\{i\}}(y_1, dx_1) \int_F q_{L\setminus\{i\}}(x_1, dy)$$
$$\times m_{W\cup\{i\}}(y_2) + \int_F q_{L\setminus\{i\}}(x, dy_1) \int_{X_i} q_{W\cup\{i\}}(y_1, dx_1)$$
$$\times \int_F q_{L\setminus\{i\}}(x_1, dy_2) \int_{X_i} q_{W\cup\{i\}}(y_2, dx_2)m_{L\setminus\{i\}}(x_2) + \cdots.$$

We introduce the notation

$$\underline{M}_i = \inf_{x\in X_i} m_{L\setminus\{i\}}(x),$$

$$\underline{N}_i = \inf_{x\in X_i} \int_F q_{L\setminus\{i\}}(x, dy)m_{W\cup\{i\}}(y).$$

The symbols $\overline{M}_i$ and $\overline{N}_i$ denote the corresponding suprema. Using this notation and $\underline{E}_i$, $\overline{E}_i$ (introduced earlier), we obtain for $x \in X_i$ that

$$\frac{\underline{M}_i + \underline{N}_i}{1 - \underline{E}_i} \geq m_W(x) \leq \frac{\overline{M}_i + \overline{N}_i}{1 - \overline{E}_i}.$$

We have already estimated the denominators in this formula in the proof of the preceding lemma and we have already proved that $\underline{M}_i$ and $\overline{M}_i$ are between $a^{\mp 4^r}/\sum_{k \neq i} p_{ik}$. To estimate $\underline{N}_i$ and $\overline{N}_i$, we use the estimates of $q_{L \setminus \{i\}}(x, X_k)$ and the estimates of $m_{W \cup \{i\}}(y)$ for $y \in X_k$, which hold by the induction hypothesis. We obtain that N_i and $\overline{N}_i$ are between

$$\sum_{\substack{k \in L \setminus W \\ k \neq i}} \frac{a^{\mp k} p_{ik}}{\sum_{k \neq i} p_{ik}} a^{\mp 4^r} \frac{Q_{ik} + R_{ik}}{K_i},$$

where K_i has the same meaning as in the proof of the preceding lemma, Q_{ik} is the sum of $\pi(g)$ over the graphs belonging to $G(W \cup \{i\} \cup \{k\})$ and R_{ik} is the sum of $\pi(g)$ over those graphs without cycles consisting of $(r-1)$ arrows $m \to n$, $m \in L \setminus W$, $m \neq i$, $n \in L$, $n \neq m$, in which the chain of arrows beginning at the point k does not lead to W. Bringing the fractions to a common denominator and making the necessary simplifications, we obtain that for $x \in X_i$, $m_W(x)$ is between

$$a^{\mp 4^{r+1}} \frac{K_i + \sum_{\substack{k \in L \setminus W \\ k \neq i}} p_{ik} Q_{ik} + \sum_{\substack{k \in L \setminus W \\ k \neq i}} p_{ik} R_{ik}}{\sum_{k \in W} p_{ik} K_i + L_i} \tag{3.6}$$

(the notation L_i was introduced in Lemma 3.3). Here in the numerator we have the sum of products $\pi(g)$ which in fact has to appear in the numerator of formula (3.5): namely, K_i is the sum of $\pi(g)$ over the $(W \cup \{i\})$-graphs, the second term is the sum over those graphs without cycles consisting of $(r-1)$ arrows $m \to n$, $m \in L \setminus W$, $n \in W$, $n \neq m$, in which an arrow $i \to k$, $k \notin W$, goes from i and there is no arrow going from k, and the third term is the sum of $\pi(g)$ over the same graphs in which a chain begins at i from more than one arrow and ends outside W. This, together with the denominator already computed, gives the assertion of the lemma.

In order not to get confused, with the empty graph, we have to consider separately the passage from $r = 1$ to $r = 2$. In this case, $L \setminus W$ consists of two elements i and k and $K_i = p_{ki} + \sum_{j \in W} p_{kj}$, $Q_{ik} + R_{ik} = 1$. In the numerator in (3.6) we have $p_{ki} + \sum_{j \in W} p_{kj} + p_{ik}$, so that the assertion of the lemma is true in this case, as well. $\square$

§4. The Problem of the Invariant Measure

By means of the results of the preceding two paragraphs, here we solve the problem of rough asymptotics of the invariant measure of a diffusion process with small diffusion on a compact manifold. We shall use the following formula, which expresses, up to a factor, the invariant measure μ^ε of a diffusion process $(X_t^\varepsilon, \mathsf{P}_x^\varepsilon)$ in terms of the invariant measure ν^ε of the chain Z_n on ∂g (we recall that $D = m$, $\partial D = \varnothing$):

$$\mu^\varepsilon(B) = \int_{\partial g} \nu^\varepsilon(dy)\mathsf{M}_y^\varepsilon \int_0^{\tau_1} \chi_B(X_t^\varepsilon)\, dt \tag{4.1}$$

(cf. Khas'minskii [1]). It is clear that for the determination of the asymptotics of sums of products $\pi(g)$ consisting of the numbers $\exp\{-\varepsilon^{-2}\tilde{V}(K_i, K_j)\}$ we need the quantities

$$W(K_i) = \min_{g \in G\{i\}} \sum_{(m \to n) \in g} \tilde{V}(K_m, K_n) \tag{4.2}$$

(here $\tilde{V}(K_m, K_n) = \tilde{V}_M(K_m, K_n)$ is the infimum of the values of the normalized action functional on curves connecting the mth compactum with the nth one without touching the others).

Lemma 4.1. *The minimum* (4.2) *can also be written in the form*

$$W(K_i) = \min_{g \in G\{i\}} \sum_{(m \to n) \in g} V(K_m, K_n). \tag{4.3}$$

Proof. It is clear that the minimum in (4.3) is not greater than that in (4.2) (because the inequality holds for the corresponding terms). It remains to prove the reverse inequality. Let the minimum in (4.3) be attained for the graph g.

If $m \to n$ is an arrow from g and $V(K_m, K_n) = \tilde{V}(K_m, K_n)$, we leave $m \to n$ as it is. If, on the other hand,

$$V(K_m, K_n) = \tilde{V}(K_m, K_{i_1}) + \cdots + \tilde{V}(K_{i_s}, K_n)$$

(cf. §2), we replace it by the arrows $m \to i_1, \ldots, i_s \to n$. Then the sum does not change but the $\{i\}$-graph ceases to be an $\{i\}$-graph. First of all, it may turn out that i coincides with one of the intermediate points i_j, $j = 1, \ldots, s$ and in the new graph there is an arrow $i \to i_{j+1}$ or $i \to n$; in this case we omit it. Then in the graph there may be a cycle containing the point n:

$$n \to i_j \to i_{j+1} \to \cdots \to i_s \to n.$$

This cycle did not appear in g; therefore, one of the points $i_j, i_{j+1}, \ldots, i_s$ was the origin of an arrow leading to some other point; in this case we open the cycle by omitting the new arrow ($i_j \to i_{j+1}$ or $i_{j+1} \to i_{j+2} \cdots$ or $i_s \to n$). Finally, there may still be points i_j from which two arrows are issued. In this case we omit the old arrows; then the sum does not increase and the graph thus constructed will be an $\{i\}$-graph. Looking over all arrows in this way, we arrive at a graph $\tilde{g}$ for which

$$\sum_{(m \to n) \in \tilde{g}} \tilde{V}(K_m, K_n) \leq \sum_{(m \to n) \in g} V(K_m, K_n). \quad \square$$

Theorem 4.1. *Suppose that the system* $\dot{x}_t = b(x_t)$ *on a compact manifold* M *satisfies condition* **A** (*cf.* §2). *Let* μ^ε *be the normalized invariant measure of the diffusion process* $(X_t^\varepsilon, \mathsf{P}_x^\varepsilon)$ *and assume that the family of these processes satisfies the hypotheses of Theorem* 3.2 *of Ch.* 5. *Then for any* $\gamma > 0$ *there exists* $\rho_1 > 0$ (*which can be chosen arbitrarily small*) *such that the* μ^ε*-measure of the* ρ_1*-neighborhood* g_i *of the compactum* K_i *is between*

$$\exp\{-\varepsilon^{-2}(W(K_i) - \min_i W(K_i) \pm \gamma)\}$$

for sufficient small ε, *where the* $W(K_i)$ *are constants defined by formulas* (4.2) *and* (4.3).

Proof. In accordance with Lemmas 2.1, 1.7 and 1.8, we choose small positive $\rho_1 < \rho_2 < \rho_0$ such that estimates (2.3), (1.3) and (1.4) are satisfied for small ε with $\gamma/4l$ replacing γ. By Lemma 3.2, the values of the normalized invariant measure ν^ε of the chain Z_n lie between

$$\exp\left\{-\varepsilon^{-2}\left(W(K_i) - \min_i W(K_i) \pm \frac{l-1}{2l}\right)\right\}.$$

For the estimation of $\mu^\varepsilon(g_i)$ we use formula (4.1):

$$\mu^\varepsilon(g_i) = \int_{\partial g} \nu^\varepsilon(dy) \mathsf{M}_y^\varepsilon \int_0^{\tau_1} \chi_{g_i}(X_t^\varepsilon)\, dt = \int_{\partial g_i} \nu^\varepsilon(dy) \mathsf{M}_y^\varepsilon \int_0^{\sigma_0} \chi_{g_i}(X_t^\varepsilon)\, dt.$$

For ε small, this does not exceed $\exp\{-\varepsilon^{-2}(W(K_i) - \min_i W(K_i) - [(2l-1)/4l]\gamma)\}$ and is not less than $\exp\{-\varepsilon^{-2}(W(K_i) - \min_i W(K_i) + [(2l-1)/4l]\gamma)\}$. The sum of these numbers is not less than

$$\exp\left\{-\varepsilon^{-2}\left[\frac{2l-1}{l}\right]\gamma\right\},$$

from which we obtain

$$\mu^\varepsilon(M) \geq \exp\left\{-\varepsilon^{-2}\left[\frac{2l-l}{4l}\right]\gamma\right\}.$$

In order to estimate $\mu^\varepsilon(M)$ from above, we use formula (4.1) again:

$$\mu^\varepsilon(M) = \int_{\partial g} \nu^\varepsilon(dy)\mathsf{M}_y^\varepsilon \tau_1 = \int_{\partial g} \nu^\varepsilon(dy)[\mathsf{M}_y^\varepsilon \sigma_0 + \mathsf{M}_y^\varepsilon \mathsf{M}_X^\varepsilon \tau_1]$$
$$\leq \sup_{y \in \partial g} \mathsf{M}_y^\varepsilon \sigma_0 + \sup_{x \in C} \mathsf{M}_x^\varepsilon \tau_1.$$

The first mean does not exceed $\exp\{\varepsilon^{-2}(\gamma/4l)\}$ by virtue of (1.3) and the second is not greater than some constant by virtue of the corollary to Lemma 1.9.

Normalizing μ^ε by dividing by $\mu^\varepsilon(M)$, we obtain the assertion of the theorem. □

If $b(x)$ has a potential, i.e., it can be represented in the form $b(x) = -\nabla U(x)$, where ∇ is the operator of taking gradient in the metric $ds^2 = \sum a_{ij}(x)dx^i\, dx^j$, then we can write the following explicit formula for the density of the invariant measure: $m^\varepsilon(x) = C_\varepsilon \exp\{-2\varepsilon^{-2}U(x)\}$, where C_ε is a normalizing factor. This can be verified by substituting in the forward Kolmogorov equation; in the case where (a_{ij}) is the identity matrix, cf. formula (4.14), Ch. 4. This representation of the invariant measure reduces the study of the limit behavior of the invariant measure to the study of the asymptotics of a Laplace type integral (cf. Kolmogorov [1]). Theorem 4.1 gives us an opportunity to study the limit behavior of the invariant measure when no potential exists, and therefore, the explicit form of the solution of the Kolmogorov equation cannot be used. However, the results of this theorem are, of course, less sharp.

It follows from Theorem 4.1, in particular, that as $\varepsilon \to 0$, the measure μ^ε is concentrated in a small neighborhood of the union of the K_i for which $\min W(K_i)$ is attained. This result was obtained in Wentzell and Freidlin [2], [4]. In some cases, the character of the limit behavior of μ^ε can be given more accurately.

Theorem 4.2. *Suppose that the hypotheses of the preceding theorem are satisfied, $\min_i W(K_i)$ is attained at a unique K_{i_0}, and there exists only one normalized invariant measure μ_0 concentrated in K_{i_0} of the dynamical system $\dot{x}_t = b(x_t)$. Then μ^ε converges weakly to μ_0 as $\varepsilon \to 0$.*

The proof is standard: of the facts related to our concrete family of processes $(X_t^\varepsilon, \mathsf{P}_x^\varepsilon)$, we have to use that $\mathsf{M}_x^\varepsilon f(X_t^\varepsilon) \to f(x_t(x))$ uniform in x as $\varepsilon \to 0$ for any continuous function f (the corollary to Theorem 1.2 of Ch. 2).

Results concerning the limit behavior of μ^ε in the case where more than one invariant measure of the unperturbed dynamical system is concentrated in K_{i_0} have been obtained in two distinct situations. Let K_{i_0} be a smooth submanifold of M.

Kifer [2] considered the case where the dynamical system on K_{i_0} is a transitive Anosov system. (The class of Anosov systems is characterized by the condition that the tangent fibering can be represented as a sum of three invariant fiberings; in the first one the tangent vectors undergo an exponential expansion as translated on trajectories of the system, in the second one they undergo an exponential contraction, and the third one is a one-dimensional fibering induced by the vector $b(x)$ at every point. These systems form a sufficiently large set in the space of all dynamical systems.) This case is close to that of a unique invariant measure. More precisely, in the case of Anosov systems, from the infinite set of normalized invariant measures we can select one, say μ_*, connected with the smooth structure on the manifold under consideration in a certain manner; this measure has the property that the following condition of exponentially fast mixing is satisfied for smooth functions:

$$\left| \int \varphi_1(x_{t_1}(x))\varphi_2(x_{t_2}(x))\mu_*(dx) - \int \varphi_1(x)\mu_*(dx) \int \varphi_2(x)\mu_*(dx) \right| \leq \text{const} \cdot \|\varphi_1\|_1 \|\varphi_2\|_1 e^{-k|t_1 - t_2|},$$

where $\| \ \|_1$ is the norm in the space $\mathbf{C}^{(1)}$ of continuously differentiable functions. The measure μ_* appears in various limit problems concerning dynamical systems, connected with smoothness. It also turns out to be the limit of the invariant measures μ^ε of the perturbed system, independently of the concrete characteristics of the perturbations.

The second class of examples considered relates to the case where the manifold K_{i_0} can be fibered into invariant manifolds on each of which the invariant measure of the unperturbed system is unique. The limit behavior of μ^ε-depends on the structure of $b(x)$ outside K_{i_0} and the concrete form of perturbations; it can be determined if in the process X_t^ε we select a "fast" and a "slow" motion and use a technique connected with the averaging principle (cf. §9, Ch 7). In Khas'minskii's paper [3] the limit of μ^ε is found in the following special case: K_{i_0} coincides with the whole manifold M, which is the two-dimensional torus, and in natural coordinates, the system has the form $\dot{x}_t^1 = b^1(x)$, $\dot{x}_t^2 = \gamma b^1(x)$, $b^1(x) > 0$. If γ is irrational, then the invariant measure of the dynamical system is unique: it can be given by the density $C \cdot b^1(x)^{-1}$. If γ is rational, then the torus can be fibered into invariant circles with an invariant measure on each of them. The density of the limit measure can be calculated and it does not coincide with $C \cdot b^1(x)^{-1}$ in general.

We introduce a definition helping to understand the character of the distribution of μ^ε among the neighborhoods of the K_i. We shall say that a set $N \subset M$ is stable if for any $x \in N$, $y \notin N$ we have $V(x, y) > 0$. Similarly to the equivalence of points, the property of stability depends only on the structure of the system $\dot{x}_t = b(x_t)$. The example illustrated in Fig. 11 shows that there may exist a stable compact set not containing any stable ω-limit set (i.e., such that any trajectory of the dynamical system beginning near this set does not leave a small neighborhood of the set). In the example in Fig. 12, K_2 and K_4 are stable and K_1 and K_3 are unstable.

Lemma 4.2. *If a compactum K_i is unstable, then there exists a stable compactum K_j such that $V(K_i, K_j) = 0$.*

Proof. There exists $x \notin K_i$ such that $V(K_i, x) = 0$. We issue a trajectory $x_t(x)$, $t \geq 0$, from x. It leads us to its ω-limit set contained in one of the compacta K_j; furthermore, $V(K_i, K_j) = V(x, K_j) = 0$. The compactum K_j does not coincide with K_i, otherwise x would have to be in K_i; if K_j is unstable, in the same way we pass from it to another compactum, etc. Finally we arrive at a stable compactum. □

Lemma 4.3.

(a) *Among the $\{i\}$-graphs for which the minimum* (4.3) *is attained there is one in which from the index m, $m \neq i$ of each unstable compactum an arrow $m \to j$ is issued with $V(K_m, K_j) = 0$ and with K_j stable.*

(b) *For a stable compactum K_i, the value $W(K_i)$ can be calculated according to* (4.3), *considering graphs on the set of indices of only stable compacta.*

(c) *If K_j is an unstable compactum, then*

$$W(K_j) = \min[W(K_i) + V(K_i, K_j)], \tag{4.4}$$

where the minimum is taken over all stable compacta K_i.

Proof. (a) For an $\{i\}$-graph for which the minimum (4.3) is attained we consider all m for which assertion (a) is not satisfied. Among these there are ones not containing any arrow from the index of an unstable compactum.

If no arrow ends in m, we replace $m \to n$ by $m \to j$ with $V(K_m, K_j) = 0$. Then the $\{i\}$-graph remains an $\{i\}$-graph and the sum of the values of V corresponding to the arrows decreases.

If the arrows ending in m are $s_1 \to m, \ldots, s_t \to m$, and $K_{s_1}, \ldots, K_{s_t}$ are stable, we also replace $m \to n$ by $m \to j$, $V(K_m, K_j) = 0$. If no cycle is formed, then we obtain an $\{i\}$-graph with a smaller value of the sum. However, a cycle $m \to j \to \cdots \to s_k \to m$ may be formed. Then we replace $s_k \to m$ by $s_k \to n$; have $V(K_m, K_j) + V(K_{s_k}, K_n) = V(K_{s_k}, K_n) \leq V(K_{s_k}, K_m) + V(K_m, K_n)$, so that the sum of the values of V corresponding to the arrows does not increase.

Repeating this operation, we get rid of all "bad" arrows.

(b) That the minimum over $\{i\}$-graphs on the set of indices of stable compacta is not greater than the former minimum follows from (a). The reverse inequality follows from the fact that every $\{i\}$-graph on the set of indices of stable compacta can be completed to an $\{i\}$-graph on the whole set $\{1, \ldots, l\}$ by adding arrows with vanishing V, beginning at indices of unstable compacta.

(c) For any $i \neq j$ we have $W(K_j) \leq W(K_i) + V(K_i, K_j)$. Indeed, on the right side we have the minimum of $\sum_{(m \to n) \in g} V(K_m, K_n)$ over graphs in which every point is the initial point of exactly one arrow and there is exactly one cycle $i \to j \to \cdots \to i$. Removing the arrow beginning at j, we obtain an $\{i\}$-graph without increasing the sum.

If K_j is an unstable compactum, then we choose a stable compactum K_s such that $V(K_j, K_s) = 0$ and a graph g for which $\min_{g \in G(j)} \sum_{(m \to n) \in g} V(K_m, K_n)$ is attained and in which indices of unstable compacta are initial points of only arrows with $V(K_m, K_n) = 0$. We add the arrow $j \to s$ to this graph. Then a cycle $j \to s \to \cdots \to j$ is formed. In this cycle we choose the last index i of a stable compactum before j. To the arrow $i \to k$ issued from i (k may be equal to j) there corresponds $V(K_i, K_k) = V(K_i, K_j)$. We throw this arrow out and obtain an $\{i\}$-graph g' for which $\sum_{(m \to n) \in g} V(K_m, K_n) = W(K_j) - V(K_i, K_j)$. This implies that $W(K_j) \geq \min[W(K_i) + V(K_i, K_j)]$ over all stable compacta K_i. □

Formula (4.4) implies, among other things, that the minimum of $W(K_i)$ may be attained only at stable compacta.

We consider the example of a dynamical system on a sphere whose trajectories, illustrated in the plane, have the form depicted in Fig. 12. Of course, the system has another singular point, not in the figure, which is unstable; we have to introduce the compactum K_5 consisting of this point. If the values $\tilde{V}(K_i, K_j)$, $1 \leq i, j \leq 4$ are given by the matrix (2.1), then the values $W(K_j)$ for stable compacta are $W(K_2) = 6$ and $W(K_3) = 9$. We obtain that as $\varepsilon \to 0$, the invariant measure μ^ε is concentrated in a small neighborhood of the limit cycle K_2 and converges weakly to the unique invariant measure, concentrated in K_2, of the system $\dot{x}_t = b(x_t)$ (it is given by a density, with respect to arc length, inversely proportional to the length of the vector $b(x)$).

Theorem 4.3. *For $x \in M$ let us set*

$$W(x) = \min[W(K_i) + V(K_i, x)], \tag{4.5}$$

where the minimum can be taken over either all compacta or only stable compacta. Let γ be an arbitrary positive number. For any sufficiently small neighborhood $\mathscr{E}_\rho(x)$ of x there exists $\varepsilon_0 > 0$ such that for $\varepsilon \leq \varepsilon_0$ we have

$$\exp\left\{-\varepsilon^{-2}\left(W(x) - \min_i W(K_i) + \gamma\right)\right\}$$
$$\leq \mu^\varepsilon(\mathscr{E}_\rho(x)) \leq \exp\left\{-\varepsilon^{-2}\left(W(x) - \min_i W(K_i) - \gamma\right)\right\}.$$

Proof. For a point x not belonging to any of the compacta K_i, we use the following device: to the compact $K_1, \ldots, K_l$ we add another compactum $\{x\}$. The system of disjoint compacta thus obtained continues to satisfy condition **A** of §2 and we can apply Theorem 4.1. The compactum $\{x\}$ is unstable. Therefore, the minimum of the values of W is attained at a compactum other than $\{x\}$ and $W(\{x\})$ can be calculated according to (4.5).

For a point x belonging to some K_i, we have $W(x) = W(K_i)$. The value of $\mu^\varepsilon(\mathscr{E}_\rho(x))$ can be estimated from above by $\mu^\varepsilon(g_i)$ and a lower estimate can be obtained by means of Lemma 1.8. □

Theorem 4.3 means that the asymptotics as $\varepsilon \to 0$ of the invariant measure μ^ε is given by the action function

$$\varepsilon^{-2}\left(W(x) - \min_i W(K_i)\right).$$

We consider the one-dimensional case, where everything can be calculated to the end. Let the manifold M be the interval from 0 to 6, closed into a circle. Let us consider a family of diffusion processes on it with infinitesimal generators $b(x)(d/dx) + (\varepsilon^2/2)(d^2/dx^2)$, where $b(x) = -U'(x)$ and the graph of $U(x)$ is given in Fig. 14. The function $U(x)$ has local extrema at the points 0, 1, 2, 3, 4, 5, 6 and its values at these points are 7, 1, 5, 0, 10, 2, 11, respectively. (This is not the case, considered in §3, Ch. 4, of a potential field $b(x)$, since U is not continuous on the circle M.) There are six compacta containing ω-limit sets of $\dot{x}_t = b(x_t)$: the point 0 (which is the same as 6), 1, 2, 3, 4 and 5; the points 1, 3 and 5 are stable.

We can determine the values of $V(1, x)$ for $0 \le x \le 2$ by solving problem $\mathbf{R}_1$ for the equation $b(x)V'_x(1, x) + \frac{1}{2}(V'_x(1, x))^2 = 0$. We obtain $V(1, x) = 2[U(x) - U(1)]$. Analogously, $V(3, x) = 2[U(x) - U(3)]$ for $2 \le x \le 4$ and $V(5, x) = 2[U(x) - U(5)]$ for $4 \le x \le 6$ (in all three cases it can be verified separately that for curves leading to a point x on a route different

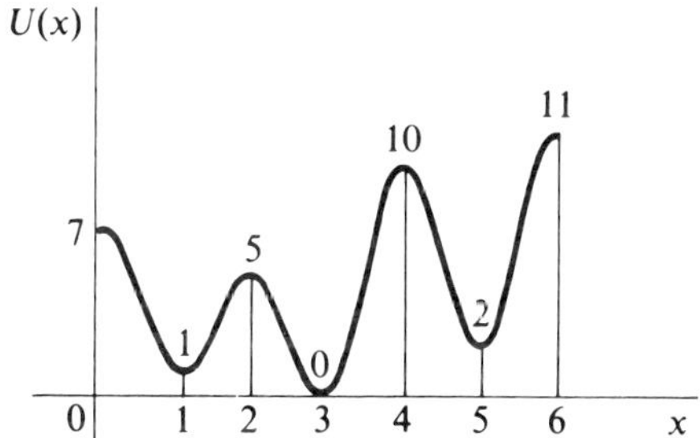

Figure 14.

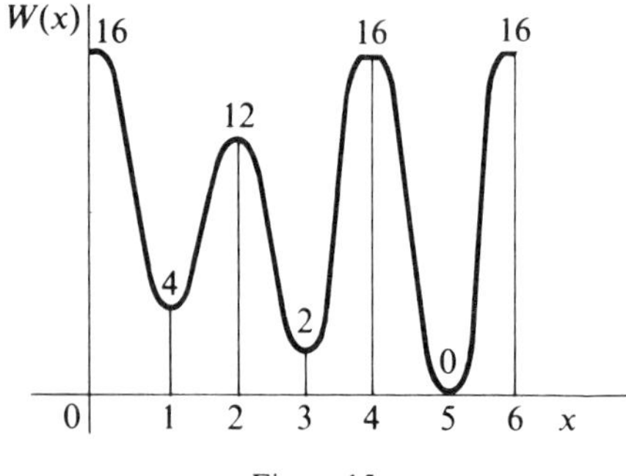

Figure 15.

from the shortest, the value of the functional is greater). Moreover, we find that

$$V(1, 3) = V(1, 2) = 8,\ V(1, 5) = V(1, 0) = 12,\ V(3, 1) = V(3, 2) = 10,$$
$$V(3, 5) = V(3, 4) = 20,\ V(5, 1) = V(5, 0) = V(5, 6) = 18,$$
$$V(5, 3) = V(5, 4) = 16.$$

On the set $\{1, 3, 5\}$ we consider $\{i\}$-graphs and from them we select those which minimize the sums (4.3). For $i = 1$ this turns out to be the graph $5 \to 3,\ 3 \to 1$; consequently $W(1) = 26$. For $i = 3$ the sum is minimized by the graph $1 \to 3,\ 5 \to 3$ and $W(3) = 24$. The value $W(5) = 22$ is attained for the graph $3 \to 1,\ 1 \to 5$. The function $W(x)$ can be expressed in the following way:

$$W(x) = \begin{cases} 24 + 2U(x) & \text{for } 0 \le x \le 3, \\ (24 + 2U(x)) \wedge 38 & \text{for } 3 \le x \le 4, \\ 18 + 2U(x) & \text{for } 4 \le x \le 5, \\ (18 + 2U(x)) \wedge 38 & \text{for } 5 \le x \le 6. \end{cases}$$

Subtracting its minimum $W(5) = 22$ from W, we obtain the normalized action function for the invariant measure μ^ε as $\varepsilon \to 0$; its graph is given in Fig. 15 (we recall that the normalizing coefficient here is ε^{-2}).

The reader may conjecture that as $\varepsilon \to 0$, the invariant measure of a one-dimensional diffusion process with a small diffusion is concentrated at the bottom of the potential well with the highest walls; this conjecture is wrong.

§5. The Problem of Exit from a Domain

In this section we do not assume any more that the manifold M is compact. Instead, we assume that a domain D is given on it with smooth boundary and compact closure and condition **A** of §2 is satisfied.

We consider graphs on the set of symbols $\{K_1, \ldots, K_l, x, y, \partial D\}$. For $x \in D$, $y \in \partial D$ we put

$$W_D(x, y) = \min_{g \in G_{xy}\{y, \partial D\}} \sum_{(\alpha \to \beta) \in g} \tilde{V}_A(\alpha, \beta). \tag{5.1}$$

(We recall that $G_{xy}\{y, \partial D\}$ is the set of all graphs consisting of $(l + 1)$ arrows emanating from the points $K_1, \ldots, K_l, x$ and such that for each of these points there exists a chain of arrows leading from the point to y or ∂D and for the initial point x this chain ends at y.)

Lemma 5.1. *The minimum* (5.1) *can also be written in the form*

$$W_D(x, y) = \min_{g \in G_{xy}\{y, \partial D\}} \sum_{(\alpha \to \beta) \in g} V_D(\alpha, \beta). \tag{5.2}$$

The minimum value of $W_D(x, y)$ over all $y \in \partial D$ does not depend on x and is equal to

$$W_D = \min_{g \in G\{\partial D\}} \sum_{(\alpha \to \beta) \in g} V_D(\alpha, \beta), \tag{5.3}$$

where either we consider $\{\partial D\}$-graphs on the set $\{K_1, \ldots, K_1, \partial D\}$ or from this set we delete the symbols K_i denoting unstable compacta.

The minima (5.1), (5.2) *can also be written in the form*

$$W_D(x, y) = [\tilde{V}_D(x, y) + W_D] \wedge \min_i[\tilde{V}_D(x, K_i) + W_D(K_i, y)] = [V_D(x, y) + W_D] \wedge \min_i[V_D(x, K_i) + W_D(K_i, y)], \tag{5.4}$$

where $W_D(K_i, y)$ is defined as one of the following minima, in which there occur graphs on the set $\{K_1, \ldots, K_l, y, \partial D\}$:

$$W_D(K_i, y) = \min_{g \in G_{K_i y}\{y, \partial D\}} \sum_{(\alpha \to \beta) \in g} \tilde{V}_D(\alpha, \beta) = \min_{g \in G_{K_i y}\{y, \partial D\}} \sum_{(\alpha \to \beta) \in g} V_D(\alpha, \beta). \tag{5.5}$$

In formula (5.3), *$V_D(\alpha, \beta)$ can also be replaced by $V(\alpha, \beta)$ (here $\alpha = K_1, \ldots, K_l$ and β is one of the K_i or ∂D).*

The proof is analogous to those of Lemmas 4.1 and 4.3. Lemma 4.2 is replaced by the following lemma, which can be proved in the same way:

Lemma 5.2. *If α is an unstable compactum K_i or a point $x \in D \setminus \bigcup_i K_i$, then either there exists a stable compactum K_j such that $V_D(\alpha, K_j) = 0$ or $V_D(\alpha, \partial D) = 0$.*

Theorem 5.1. *Let τ^ε be the time of first exit of the process X_t^ε from D. For any compact subset F of D, any $\gamma > 0$ and any $\delta > 0$ there exist δ_0, $0 < \delta_0 \le \delta$ and $\varepsilon_0 > 0$ such that for all $\varepsilon \le \varepsilon_0$, $x \in F$ and $y \in \partial D$ we have*

$$\exp\{-\varepsilon^{-2}(W_D(x, y) - W_D + \gamma)\} \le \mathsf{P}_x^\varepsilon\{X_\tau^\varepsilon \in \mathscr{E}_{\delta_0}(y)\} \le \exp\{-\varepsilon^{-2}(W_D(x, y) - W_D - \gamma)\}, \tag{5.6}$$

where W_D is defined by formula (5.3) and $W_D(x, y)$ by formulas (5.1), (5.2) or (5.5)

In other words, for a process beginning at the point x, the asymptotics of the distribution as $\varepsilon \to 0$ at the time of exit to the boundary is given by the action function $\varepsilon^{-2}(W_D(x, y) - W_D)$, uniformly for all initial points strictly inside D.

Proof. We put $\gamma' = \gamma \cdot 4^{-l-1}$ and choose a corresponding ρ_0 according to Lemmas 2.1 and 2.2. We choose a positive ρ_2 smaller than ρ_0 and the distance between F and ∂D. According to the same lemmas, we choose a positive $\rho_1 < \rho_2$ and use the construction described in §2, involving Z_n.

For $x \in \bigcup_i G_i$ we use Lemma 3.3 with $L = \{K_1, \ldots, K_l, y, \partial D\}$, $W = \{y, \partial D\}$ and the following sets: X_α, $\alpha \in L$: $G_1, \ldots, G_l, \partial D \cup \mathscr{E}_{\delta_0}(y), \partial D \setminus \mathscr{E}_{\delta_0}(y)$ (beginning with $n = 1$, $Z_n \in G_i$ implies $Z_n \in \partial g_i$). The estimates for the probabilities $P(x, X_\beta)$ for $x \in G_i$ are given by formulas (2.3)–(2.5); we have $p_{\alpha\beta} = \exp\{-\varepsilon^{-2}\tilde{V}_D(\alpha, \beta)\}$ and $a = \exp\{\varepsilon^{-2}\gamma'\}$. The sum $\sum_{g \in G_{K_i y}\{y, \partial D\}} \pi(g)$ is equivalent to a positive constant N multiplied by $\exp\{-\varepsilon^{-2}W_D(K_i, y)\}$. ($N$ is equal to the number of graphs $g \in G_{K_i y}\{y, \partial D\}$ at which the minimum of $\sum_{(\alpha \to \beta) \in g} \tilde{V}_D(\alpha, \beta)$ is attained. The denominator in (3.3) is equivalent to a positive constant multiplied by $\exp\{-\varepsilon^{-2}W_D\}$.) Taking into account that for $x \in G_i$, $W_D(x, y)$ differs from $W_D(K_i, y)$ by not more than γ', this implies the assertion of the theorem for $x \in \bigcup_i G_i$.

If $x \in F \setminus \bigcup_i G_i$, we use the strong Markov property with respect to the Markov time τ_1:

$$\mathsf{P}_x^\varepsilon\{X_{\tau^\varepsilon}^\varepsilon \in \mathscr{E}_{\delta_0}(y)\} = \mathsf{P}_x^\varepsilon\{X_{\tau_1}^\varepsilon \in \mathscr{E}_{\delta_0}(y)\} + \sum_{i=1}^{l} \mathsf{M}_x^\varepsilon\{X_{\tau_1}^\varepsilon \in \partial g_i; \mathsf{P}_{X_{\tau_1}^\varepsilon}^\varepsilon\{X_{\tau^\varepsilon}^\varepsilon \in \mathscr{E}_{\delta_0}(y)\}\}. \tag{5.7}$$

According to (2.10), the first probability is between $\exp\{-\varepsilon^{-2}(\tilde{V}_D(x, y) \pm \gamma')\}$; and according to what has already been proved, the probability under the sign of mathematical expectation is between $\exp\{-\varepsilon^{-2}(W_D(K_i, y) - W_D \pm (4^l + 1)\gamma')\}$. Using estimate (2.8), we obtain that the ith mathematical expectation in (5.7) falls between $\exp\{-\varepsilon^{-2}(\tilde{V}_D(x, K_i) + W_D(K_i, y) + W_D \pm (4^l + 1)\gamma')\}$, and the whole sum (5.7) is between $\exp\{-\varepsilon^{-2}(W_D(x, y) - W_D \pm (4^l + 2)\gamma')\}$, where $W_D(x, y)$ is given by the first of formulas (5.4). This proves the theorem. □

Theorem 5.1 enables us to establish, in particular, the most probable place of exit of X_t^ε to the boundary for small ε.

Theorem 5.2 (Wentzell and Freidlin [3], [4]). *For every $j = 1, \ldots, l$ let Y_i be the set of points $y \in \partial D$ at which the minimum of $V_D(K_i, y)$ is attained. Let the point x be such that the trajectory $x_t(x)$, $t \geq 0$, of the dynamical system, issued from x, does not go out of D and is attracted to K_i. From the $\{\partial D\}$-graphs on the set $\{K_1, \ldots, K_l, \partial D\}$ we select those at which the minimum (5.3) is attained. In each of these graphs we consider the chain of arrows leading from K_i into ∂D; let the last arrow in the chain be $K_j \to \partial D$. We denote by $M(i)$ the set of all such j in all selected graphs.*

Then with probability converging to 1 as $\varepsilon \to 0$, the first exit to the boundary of the trajectory of X_t^ε, beginning at x, takes place in a small neighborhood of the set $\bigcup_{j \in M(i)} Y_j$.

The assertion remains valid if all the V_D, including those in formula (5.3), are replaced by $\tilde{V}_D$.

We return to the example illustrated in Fig. 12. We reproduce this figure here (Fig. 16). Besides $\tilde{V}_D(K_i, K_j)$, let $\tilde{V}_D(K_2, \partial D) = 8$, $\tilde{V}_D(K_3, \partial D) = 2$, $\tilde{V}_D(K_4, \partial D) = 1$ be given; $\tilde{V}_D(K_1, \partial D)$ is necessarily equal to $+\infty$. On ∂D we single out the sets Y_2, Y_3 and Y_4. Now there will be two $\{\partial D\}$-graphs minimizing $\sum_{(\alpha \to \beta) \in g} \tilde{V}_D(\alpha, \beta)$: the first one consists of the arrows $K_1 \to K_2$, $K_2 \to \partial D$, $K_3 \to \partial D$ and $K_4 \to K_2$ and the second one is the same with $K_4 \to K_3$ replacing $K_4 \to K_2$. Consequently, $M(1) = M(2) = \{2\}$, $M(3) = \{3\}$, and $M(4) = \{2, 3\}$.

The trajectories of the dynamical system emanating from a point x in the left-hand part of D (to the left of the separatrices ending at the point K_4) are attracted to the cycle K_2 with the exception of the unstable equilibrium position K_1. The points of the right-hand part are attracted to K_3 and the points on the separating line to K_4. Consequently, for small ε, from points of the left half of D, the process X_t^ε goes out to ∂D in a small neighborhood of

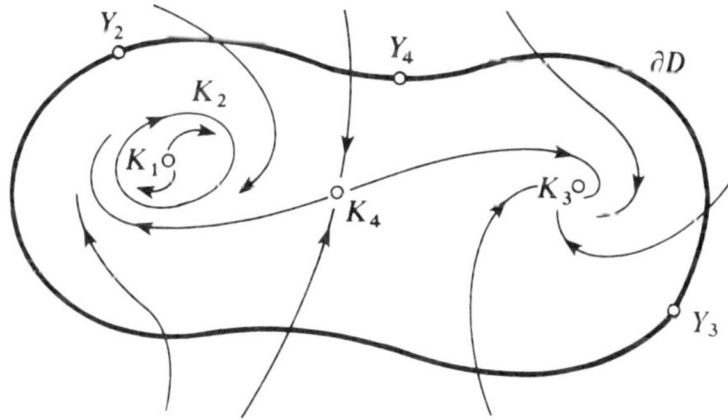

Figure 16.

Y_2, from points of the right half, it hits ∂D near Y_3 and from points of the separating line near Y_2 or Y_3.

If we increase $\tilde{V}_D(K_i, \partial D)$ so that $\tilde{V}_D(K_2, \partial D) = 16$, $\tilde{V}_D(K_3, \partial D) = 10$, $\tilde{V}_D(K_4, \partial D) = 9$, then again there are two $\{\partial D\}$-graphs minimizing the sum (5.3): $K_1 \to K_2$, $K_2 \to K_3$, $K_3 \to \partial D$, $K_4 \to K_2$ or $K_4 \to K_3$; for all i we have $M(i) = \{3\}$. Consequently, for small ε, the exit to the boundary from all points of the domain will take place near Y_3.

Now we turn to the problem of the time spent by the process X_t^ε in D until exit to the boundary.

We consider graphs on the set $L = \{K_1, \ldots, K_l, x, \partial D\}$. We put

$$M_D(x) = \min_{g \in G(x \nrightarrow \{\partial D\})} \sum_{(\alpha \to \beta) \in g} \tilde{V}_D(\alpha, \beta). \tag{5.8}$$

The notation $G(\alpha \nrightarrow W)$ was introduced in §3 after formulating Lemma 3.4.

Lemma 5.3. *The minimum* (5.8) *can also be written as*

$$M_D(x) = \min_{g \in G(x \nrightarrow \{\partial D\})} \sum_{(\alpha \to \beta) \in g} V_D(\alpha, \beta); \tag{5.9}$$

$$M_D(x) = W_D \wedge \min_i [V_D(x, K_i) + M_D(K_i)], \tag{5.10}$$

where $M_D(K_i)$ is defined by the equality

$$M_D(K_i) = \min_{g \in G(K_i \nrightarrow \{\partial D\})} \sum_{(\alpha \to \beta) \in g} V_D(\alpha, \beta), \tag{5.11}$$

where in the minimum we have graphs on the set $\{K_1, \ldots, K_l, \partial D\}$ (and W_D is defined by formula (5.3)*).*

In determining the minima (5.9) *or* (5.11), (5.10), *one can omit all unstable compacta K_i.*

The proof is analogous to that of Lemmas 4.1 and 4.3 again.

Theorem 5.3. *We have*

$$\lim_{\varepsilon \to 0} \varepsilon^2 \ln \mathbf{M}_x^\varepsilon \tau^\varepsilon = W_D - M_D(x) \tag{5.12}$$

uniformly in x belonging to any compact subset F of D.

Proof. We choose γ', ρ_0, ρ_1, ρ_2 as in the proof of the preceding theorem, but with the additional condition that the mean exit time from the ρ_0-neighborhood of the K_i does not exceed $\exp\{\varepsilon^{-2}\gamma'\}$ (cf. Lemma 1.7). We consider Markov times τ_0 $(=0)$, $\tau_1, \tau_2, \ldots$ and the chain $Z_n = X_{\tau_n}^\varepsilon$. We denote by

ν the index of the step in which Z_n first goes out to ∂D, i.e., the smallest n for which $\tau^\varepsilon = \tau_n$. Using the strong Markov property, we can write:

$$\mathsf{M}_x^\varepsilon \tau^\varepsilon = \sum_{n=0}^{\infty} \mathsf{M}_x^\varepsilon\{Z_n \notin \partial D; \mathsf{M}_{Z_n}^\varepsilon \tau_1\}.$$

Lemmas 1.7, 1.8 and 1.9 yield that in this sum $\mathsf{M}_{Z_n}^\varepsilon \tau_1$ does not exceed $2\exp\{\varepsilon^{-2}\gamma'\}$ but is greater than $\exp\{-\varepsilon^{-2}\cdot\gamma'\}$ (for small ε). Hence up to a factor between $[2\exp\{\varepsilon^{-2}\gamma'\}]^{\mp 1}$, $\mathsf{M}_x^\varepsilon \tau^\varepsilon$ coincides with $\sum_{n=0}^\infty \mathsf{P}_x^\varepsilon\{Z_n \notin \partial D\} = \mathsf{M}_x^\varepsilon \nu$. This mathematical expectation can be estimated by means of Lemma 3.4.

First for $x \in \bigcup_i G_i$ we obtain, using estimates (2.3)–(2.5), that

$$\begin{aligned} &\exp\{\varepsilon^{-2}(W_D - M_D(K_i) - (4^l + 1)\gamma')\} \\ &\quad \le \mathsf{M}_x^\varepsilon \tau_1 \le 2\exp\{\varepsilon^{-2}(W_D - M_D(K_i) + (4^l + 1)\gamma')\}, \end{aligned} \tag{5.13}$$

for small ε. Then for an initial point $x \in F \setminus \bigcup_i G_i$ we obtain

$$\mathsf{M}_x^\varepsilon \tau^\varepsilon = \mathsf{M}_x^\varepsilon \tau_1 + \sum_{i=1}^{l} \mathsf{M}_x^\varepsilon\{Z_1 \in \partial g_i; \mathsf{M}_{Z_1}^\varepsilon \tau^\varepsilon\}.$$

Taking account of the inequality $M_x^\varepsilon \tau_1 < 2\exp\{\varepsilon^{-2}\gamma'\}$ and estimates (2.8) (2.10) and (5.13), we obtain that $\mathsf{M}_x^\varepsilon \tau^\varepsilon$ is between $\exp\{\varepsilon^{-2}(W_D - M_D(x) \mp \gamma)\}$. This is true for $x \in \bigcup_i G_i$, as well. Since $\gamma > 0$ was arbitrarily small, we obtain the assertion of the theorem. □

We return to the example considered above (with $\tilde{V}_D(K_i, K_j)$ given by the matrix (2.1) and $\tilde{V}_D(K_2, \partial D) = 8$, $\tilde{V}_D(K_3, \partial D) = 2$, $\tilde{V}_D(K_4, \partial D) = 1$). We calculate the asymptotics of the mathematical expectation of the time τ^ε of exit to the boundary for trajectories beginning at the stable equilibrium position K_3. We find that $W_D = 10$ (the minimum (5.3) is attained at the two graphs $K_1 \to K_2$, $K_2 \to \partial D$, $K_3 \to \partial D$, $K_4 \to K_2$ or $K_4 \to K_3$) and $M_D(K_3) = 6$ (the minimum (5.11) is attained at the graphs $K_1 \to K_2$, $K_3 \to K_4$, $K_4 \to K_2$; $K_1 \to K_2$, $K_3 \to K_2$, $K_4 \to K_2$; $K_1 \to K_2$, $K_3 \to K_2$, $K_4 \to K_3$). Hence the mathematical expectation of the exit time is logarithmically equivalent to $\exp\{\varepsilon^{-2}(W_D - M_D(K_3))\} = \exp\{4\varepsilon^{-2}\}$.

Considering Z_n, we can understand why we obtain this average exit time. Beginning in K_3, the chain Z_n makes a number of order

$$\exp\{\varepsilon^{-2} V_D(K_3, \partial D)\} = \exp\{2\varepsilon^{-2}\}$$

of steps on ∂g_3, spending an amount of time of the same order with probability close to 1 for small ε. After this, with probability close to 1, it goes out to ∂D and with probability of order

$$\exp\{-\varepsilon^{-2}(V_D(K_3, K_2) - V_D(K_3, \partial D))\} = \exp\{-4\varepsilon^{-2}\},$$

it passes to the stable cycle K_2 (it may be delayed for a relatively small number of steps near the unstable equilibrium position K_4). After this has taken place, over a time of order $\exp\{\varepsilon^{-2}V_D(K_2, \partial D)\} = \exp\{8\varepsilon^{-2}\}$ the chain Z_n performs transitions within the limits of ∂g_2 and (approximately $\exp\{\varepsilon^{-2} \cdot V_D(K_2, K_1)\} = \exp\{\varepsilon^{-2}\}$ many times less often) ∂g_1 with overwhelming probability. After this it goes out to the boundary. Hence a mathematical expectation of order $\exp\{4\varepsilon^{-2}\}$ arises due to the less likely—of probability of order $\exp\{-4\varepsilon^{-2}\}$—values of order $\exp\{8\varepsilon^{-2}\}$.

We recall that in the case of a domain attracted to one stable equilibrium position, the average exit time $\mathsf{M}_x^\varepsilon \tau^\varepsilon$ has the same order as the boundaries of the range of the most probable values of τ^ε (Theorem 4.2 of Ch. 4). In particular, any quantile of the distribution of τ is logarithmically equivalent to the average. Our example shows that this is not so in general in the case where there are several compacta K_i containing ω-limit sets in the domain D; the mathematical expectation may tend to infinity essentially faster than the median and quantiles.

This restricts seriously the value of the above theorem as a result characterizing the limit behavior of the distribution of τ^ε.

We mention the corresponding result formulated in the language of differential equations.

Theorem 5.4. *Let $g(x)$ be a positive continuous function on $D \cup \partial D$ and let $v^\varepsilon(x)$ be the solution of the equation $L^\varepsilon v^\varepsilon(x) = -g(x)$ in D with vanishing boundary conditions on ∂D. We have*

$$v^\varepsilon(x) \asymp \exp\{\varepsilon^{-2}(W_D - M_D(x))\}$$

uniformly in x belonging to any compact subset as $\varepsilon \to 0$.

§6. Decomposition into Cycles. Sublimit Distributions

In the problems to which this chapter is devoted there are two large parameters: ε^{-2} and t, the time over which the perturbed dynamical system is considered. It is natural to study what happens when the convergence of these parameters to infinity is coordinated in one way or another. We shall be interested in the limit behavior of the measures $\mathsf{P}_x^\varepsilon\{X_t^\varepsilon \in \Gamma\}$; we restrict ourselves to the case of a compact manifold (as in §4).

The simplest case is where first ε^{-2} goes to infinity and then t does. Then all is determined by the behavior of the unperturbed dynamical system. It is clear that $\lim_{t\to\infty} \lim_{\varepsilon\to 0} \mathsf{P}_x^\varepsilon\{X_t^\varepsilon \in \Gamma\} = 1$, if the open set Γ contains the whole

ω-limit set of the trajectory $x_t(x)$ beginning at the point $x_0(x) = x$. This limit is equal to zero if Γ is at a positive distance from the ω-limit set.

In §4 we considered the case where first t goes to infinity and then ε^{-2} does. Theorem 4.1 gives an opportunity to establish that

$$\lim_{\varepsilon \to 0} \lim_{t \to \infty} \mathsf{P}_x^\varepsilon\{X_t^\varepsilon \in \Gamma\} = 1$$

for open sets Γ containing all compacta K_i at which the minimum of $W(K_i)$ is attained. In the case of general position this minimum is attained at one compactum. If on this compactum there is concentrated a unique normalized invariant measure μ_0 of the dynamical system, then

$$\lim_{\varepsilon \to 0} \lim_{t \to \infty} \mathsf{P}_x^\varepsilon\{X_t^\varepsilon \in \Gamma\} = \mu_0(\Gamma)$$

for all Γ with boundary of μ_0-measure zero.

We study the behavior of X_t^ε on time intervals of length $t(\varepsilon^{-2})$ where $t(\varepsilon^{-2})$ is a function monotone increasing with increasing ε^{-2}. It is clear that if $t(\varepsilon^{-2})$ increases sufficiently slowly, then over time $t(\varepsilon^{-2})$ the trajectory of X_t^ε cannot move far from that stable compactum in whose domain of attraction the initial point is. Over larger time intervals there are passages from the neighborhood of this compactum to neighborhoods of others; first to the "closest" compactum (in the sense of the action functional) and then to more and more "far away" ones. First of all we establish in which order X_t^ε enters the neighborhoods of the compacta K_i.

Theorem 6.1. *Let $L = \{1, 2, \ldots, l\}$ and let Q be a subset of L. For the process X_t^ε let us consider the first entrance time τ_Q of the boundaries ∂g_j of the ρ-neighborhoods g_j of the K_j with indices in $L \setminus Q$. Let the process begin in $g_i \cup \partial g_i$, $i \in Q$. Then for sufficiently small ρ, with probability converging to 1 as $\varepsilon \to 0$, X_τ^ε belongs to one of the sets ∂g_j such that in one of the $(L \setminus Q)$-graphs g at which the minimum*

$$A(Q) = \min_{g \in G(L \setminus Q)} \sum_{(m \to n) \in g} \tilde{V}(K_m, K_n) \tag{6.1}$$

is attained, the chain of arrows beginning at i leads to $j \in L \setminus Q$.

The proof can be carried out easily by means of Lemmas 2.1 and 3.3.

In this theorem $\tilde{V}(K_m, K_n)$ can be replaced by $V(K_m, K_n)$ and in this case we can also omit all unstable compacta and consider only passages from one stable compactum to another.

We consider an example. Let K_i, $i = 1, 2, 3, 4, 5$ be stable compacta containing ω-limit sets and let the values $V(K_i, K_j)$ be given by the matrix

$$\begin{pmatrix} 0 & 4 & 9 & 13 & 12 \\ 7 & 0 & 5 & 10 & 11 \\ 6 & 8 & 0 & 17 & 15 \\ 3 & 6 & 8 & 0 & 2 \\ 5 & 7 & 10 & 3 & 0 \end{pmatrix}$$

Let the process begin near K_1. We determine the $\{2, 3, 4, 5\}$-graph minimizing the sum of values of V. This graph consists of the only arrow $1 \to 2$. Therefore, the first of the compacta approached by the process will be K_2.

Further, we see where we go from the neighborhood of K_2. We put $Q = \{2\}$. We find that the $\{1, 3, 4, 5\}$-graph $2 \to 3$ minimizes the sum (6.1). Consequently, the next compactum approached by the process will be K_3, with overwhelming probability. Then, the graph $3 \to 1$ shows that the process returns to K_1. Passages from K_1 to K_2, from K_2 to K_3 and from K_3 back to K_1 take place many times, but ultimately the process comes to one of the compacta K_4, K_5. In order to see to which one exactly, we use Theorem 6.1 for $Q = \{1, 2, 3\}$. We find that the minimum (6.1) is attained for the $\{4, 5\}$-graph $3 \to 1$, $1 \to 2$, $2 \to 4$. Hence with probability close to 1 for small ε, if the process begins near K_1, K_2 or K_3, then it reaches the neighborhood of K_4 sooner than that of K_5. (We are not saying that until this, the process performs passages between the compacta K_1, K_2 and K_3 only in the most probable order $K_1 \to K_2 \to K_3 \to K_1$. What is more, in our case it can be proved that with probability close to 1, passages will take place in the reverse order before reaching the neighborhood of K_4.)

Afterwards the process goes from K_4 to K_5 and backwards (this is shown by the graphs $4 \to 5$ and $5 \to 4$). Then it returns to the neighborhoods of the compacta K_1, K_2 and K_3; most probably to K_1 first (as is shown by the graph $5 \to 4, 4 \to 1$). On large time intervals passages will take place between the "cycles" $K_1 \to K_2 \to K_3 \to K_1$ and $K_4 \to K_5 \to K_4$; they take place the most often in the way described above. Consequently, we obtain a "cycle of cycles"—a cycle of rank two.

The passages between the K_i can be described by means of the hierarchy of cycles in the general case, as well (Freidlin [9], [10]). Let $K_1, \ldots, K_{l_0}$ be stable compacta and let Q be a subset of $L = \{1, 2, \ldots, l_0\}$. We assume that there exists a unique $(L \setminus Q)$-graph g^* for which the minimum (6.1) is attained. We define $R_Q(i)$, $i \in Q$ as that element of $L \setminus Q$ which is the terminal point of the chain of arrows going from i to $L \setminus Q$ in the graph g^*.

Now we describe the decomposition of L into hierarchy of cycles. We begin with cycles of rank one. For every $i_0 \in L$ we consider the sequence $i_0, i_1, i_2, \ldots, i_n, \ldots$ in which $i_n = R_{\{i_{n-1}\}}(i_{n-1})$. Let n be the smallest index for which there is a repetition: $i_n = i_m$, $0 \le m < n$ and for k smaller than n all i_k are different. Then the cycles of rank one generated by the element

$i_0 \in L$ are, by definite, the groups $\{i_0\}, \{i_1\}, \ldots, \{i_{m-1}\}, \{i_m \to i_{m+1} \to \cdots \to i_{n-1} \to i_m\}$, where the last group is considered with the indicated cyclic order. Cycles generated by distinct initial points $i_0 \in L$ either do not intersect or coincide; in the latter case the cyclic order on them is one and the same. Hence the cycles of rank one have been selected (some of them consist of only one point).

We continue the definition by recurrence. Let the cycles of rank $(k-1)$ (briefly $(k-1)$-cycles) $\pi_1^{k-1}, \pi_2^{k-1}, \ldots, \pi_{n_{k-1}}^{k-1}$ be already defined. They are sets of $(k-2)$-cycles equipped with a cyclic order. Ultimately, every cycle consists of points—elements of L; we shall denote the set of point which constitute a cycle by the same symbol as the cycle itself. We shall say that a cycle π_j^{k-1} is a successor of π_i^{k-1} and write $\pi_i^{k-1} \to \pi_j^{k-1}$ if $R_{\pi_i}^{k-1}(m) \in \pi_j^{k-1}$ for $m \in \pi_i^{k-1}$. It can be proved that the function $R_{\pi_i^{k-1}}(m)$ assumes the same value for all $m \in \pi_i^{k-1}$, so that the above definition is unambiguous.

Now we consider a cycle $\pi_{i_0}^{k-1}$ and a sequence of cycles $\pi_{i_0}^{k-1} \to \pi_{i_1}^{k-1} \to \cdots \to \pi_{i_m}^{k-1} \to \cdots$ beginning with it. In this sequence repetition begins from a certain index. Let n be the smallest such index: $\pi_{i_n}^{k-1} = \pi_{i_m}^{k-1}$, $0 \le m < n$. We shall say that the cycle $\pi_{i_0}^{k-1}$ generates the cycles of rank k $\{\pi_{i_0}^{k-1}\}$, $\{\pi_{i_1}^{k-1}\}, \ldots, \{\pi_{i_{m-1}}^{k-1}\}$ (m cycles of rank k, each consisting of one cycle of the preceding rank) and $\{\pi_{i_m}^{k-1} \to \pi_{i_{m-1}}^{k-1} \to \cdots \to \pi_{i_{n-1}}^{k-1} \to \pi_{i_m}^{k-1}\}$. Taking all initial $(k-1)$-cycles $\pi_{i_0}^{k-1}$, we decompose all cycles of rank $(k-1)$ into k-cycles.

The cycles of rank zero are the points of L; for some k all $(k-1)$-cycles participate in one k-cycle, which exhausts the whole set L.

For small ε, the decomposition into cycles completely determines the most probable order of traversing the neighborhoods of the stable compacta by trajectories of X_t^ε (of course, all this concerns the case of "general position," where every minimum (6.1) is attained only for one graph). Now we turn our attention to the time spent by the process in one cycle or another.

Theorem 6.2. *Let π be a cycle. Let us put*

$$C(\pi) = A(\pi) - \min_{i \in \pi} \min_{g \in G_\pi\{i\}} \sum_{(m \to n) \in g} V(K_m, K_n), \tag{6.2}$$

where $A(\pi)$ is defined by formula (6.1) and $G_\pi\{i\}$ is the set of $\{i\}$-graphs over the set π. Then for sufficiently small $\rho > 0$ we have

$$\lim_{\varepsilon \to 0} \varepsilon^2 \ln \mathsf{M}_x^\varepsilon \tau_\pi = C(\pi) \tag{6.3}$$

uniformly in x belonging to the ρ-neighborhood of $\cup_{i \in \pi} K_i$ and for any $\gamma > 0$ we have

$$\lim_{\varepsilon \to 0} \mathsf{P}_x^\varepsilon\{e^{\varepsilon^{-2}(C(\pi)-\gamma)} < \tau_\pi < e^{\varepsilon^{-2}(C(\pi)+\gamma)}\} = 1 \tag{6.4}$$

uniformly in all indicated x.

Proof. Relation (6.3) can be derived from Theorem 5.3. We recall that for any set $Q \subset L$ (not a cycle), $\lim_{\varepsilon \to 0} \varepsilon^2 \ln \mathsf{M}_x^\varepsilon \tau_Q$ depends on the choice of the point $x \in \cup_{i \in Q} g_i$ in general (cf. §5). The proof of assertion (6.4) is analogous to that of Theorem 4.2 of Ch. 4 and we omit it. □

The decomposition into cycles and Theorem 6.2 enable us to answer the problem of behavior of X_t^ε on time intervals of length $t(\varepsilon^{-2})$, where $t(\varepsilon^{-2})$ has exponential asymptotics as $\varepsilon^{-2} \to \infty$: $t(\varepsilon^{-2}) \asymp e^{C\varepsilon^{-2}}$. Let $\pi, \pi', \ldots, \pi^{(s)}$ be cycles of next to the last rank, unified into the last cycle, which exhausts L. If the constant C is greater than $C(\pi), C(\pi'), \ldots, C(\pi^{(s)})$, then over time of order $\exp\{C\varepsilon^{-2}\}$, the process can traverse all these cycles many times (and all cycles of smaller rank inside them) and for it the limit distribution found in §4 can be established. In terms of the hierarchy of cycles, this limit distribution is concentrated on that one of the cycles $\pi, \pi', \ldots, \pi^{(s)}$ for which the corresponding constant $C(\pi), C(\pi'), \ldots$ or $C(\pi^{(s)})$ is the greatest; within this cycle, it is concentrated on that one of the subcycles for which the corresponding constant defined by (6.2) is the greatest possible, and so on up to points i and the compacta K_i corresponding to them. The same is true if we do not have $t(\varepsilon^{-2}) \asymp \exp\{C\varepsilon^{-2}\}$ but only

$$\varliminf_{\varepsilon \to 0} \varepsilon^2 \ln t(\varepsilon^{-2}) > \max\{C(\pi), C(\pi'), \ldots, C(\pi^{(s)})\}.$$

Over a time $t(\varepsilon^{-2})$ having smaller order, the process cannot traverse all cycles and no limit distribution common for all initial points sets. Nevertheless, on cycles of not maximum rank, which can be traversed by the process, sublimit distributions depending on the initial point set.

We return to the example considered after Theorem 6.1. We illustrate the hierarchy of cycles in Fig. 17: the cycles of rank zero (the points) are unified in the cycles $\{1 \to 2 \to 3 \to 1\}$ and $\{4 \to 5 \to 4\}$ of rank one and they are unified in the only cycle of rank two. On the arrow beginning at each cycle π we indicated the value of the constant $C(\pi)$. If $\lim_{\varepsilon \to 0} \varepsilon^2 \ln t(\varepsilon^{-2})$ is between 0 and 2, then over time $t(\varepsilon^{-2})$, the process does not succeed in moving away from that compactum K_i near which it began and the sublimit

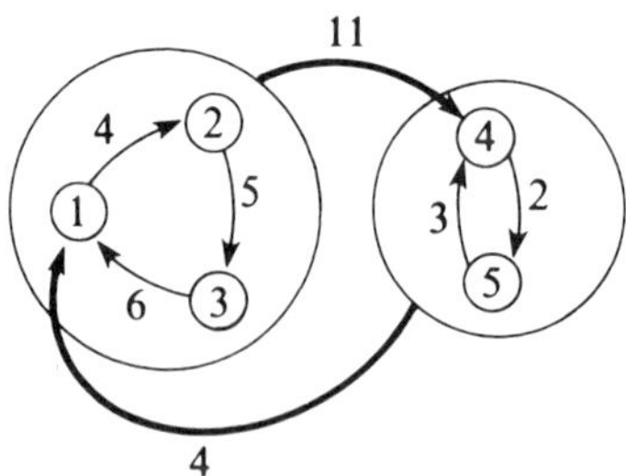

Figure 17.

distribution corresponding to the initial point x is concentrated on that K_i to whose domain of attraction x belongs. Over time $t(\varepsilon^{-2})$ for which $\lim_{\varepsilon\to 0} \varepsilon^2 \ln t(\varepsilon^{-2})$ is between 2 and 3, a moving away from the 0-cycle $\{4\}$ takes place; i.e., the only thing what happens is that if the process was near K_4, then it passes to K_5. Then the sublimit distribution is concentrated on K_5, for initial points in the domain of attraction of K_4 and in the domain of attraction of K_5. If $\lim_{\varepsilon\to 0} \varepsilon^2 \ln t(\varepsilon^{-2})$ is between 3 and 4, then over time $t(\varepsilon^{-2})$ a limit distribution is established on the cycle $\{4 \to 5 \to 4\}$, but nothing else takes place. Since this distribution is concentrated at the point 5 corresponding to the compactum K_5, the result is the same as for

$$\lim_{\varepsilon\to 0} \varepsilon^2 \ln t(\varepsilon^{-2})$$

between 2 and 3.

If $\lim_{\varepsilon\to 0} \varepsilon^2 \ln t(\varepsilon^{-2})$ is between 4 and 5, a moving away from the cycle $\{4 \to 5 \to 4\}$ takes place, the process hits the neighborhood of K_1 and passes from there to K_2, but not farther; the sublimit distribution is concentrated on K_2 for initial points attracted to any compactum except K_3 (and for points attracted to K_3 on K_3). Finally, if $\lim_{\varepsilon\to 0} \varepsilon^2 \ln t(\varepsilon^{-2}) > 5$, then a limit distribution is established (although a moving away from the cycle $\{1 \to 2 \to 3 \to 1\}$ and even from the cycle $\{3\}$ may not have taken place).

Now we can determine the sublimit distributions for initial points belonging to the domain of attraction of any stable compactum. For example, for x attracted to K_4 we have $\lim_{\varepsilon\to 0} \mathsf{P}_x^\varepsilon\{X_{t(\varepsilon^{-2})}^\varepsilon \in \Gamma\} = 1$ for open $\Gamma \supset K_4$ provided that $0 < \underline{\lim}_{\varepsilon\to 0} \varepsilon^2 \ln t(\varepsilon^{-2}) \le \overline{\lim}_{\varepsilon\to 0} \varepsilon^2 \ln t(\varepsilon^{-2}) < 2$; $\lim_{\varepsilon\to 0} \mathsf{P}_x^\varepsilon\{X_{t(\varepsilon^{-2})}^\varepsilon \in \Gamma\} = 1$ for open $\Gamma \supset K_5$ provided that $2 < \underline{\lim}_{\varepsilon\to 0} \varepsilon^2 \ln t(\varepsilon^{-2}) \le \overline{\lim}_{\varepsilon\to 0} \varepsilon^2 \ln t(\varepsilon^{-2}) < 4$. If these lower and upper limits are greater than 4 and smaller than 5, then $\lim_{\varepsilon\to 0} \mathsf{P}_x^\varepsilon\{X_{t(\varepsilon^{-2})}^\varepsilon \in \Gamma\} = 1$ for open sets Γ containing K_2. Finally, if $\underline{\lim}_{\varepsilon\to 0} \varepsilon^2 \ln t(\varepsilon^{-2}) > 5$, then the distribution of $X_{t(\varepsilon^{-2})}^\varepsilon$ is attracted to K_3, independently of the initial point x.

A general formulation of the result and a scheme of proof can be found in Freidlin [10]. Sublimit distributions over an exponential time can also be established in the case of a process X_t^ε stopping after reaching the boundary of the domain (the situation of §5).

We note that the problem of limit distribution of $X_{t(\varepsilon^{-2})}^\varepsilon$ is closely connected with the stabilization problem of solutions of parabolic differential equations with a small parameter (cf. the same article).

§7. Eigenvalue Problems

Let L be an elliptic differential operator in a bounded domain D with smooth boundary ∂D. As is known, the smallest eigenvalue λ_1 of the operator $-L$ with zero boundary conditions is real, positive, and simple. It admits the following probability theoretic characterization. Let (X_t, P_x) be the diffusion process with generator L and let τ be the time of first exit of X_t from D. Then λ_1 forms the division between those λ for which $\mathsf{M}_x e^{\lambda\tau} < \infty$ and those for which $\mathsf{M}_x e^{\lambda\tau} = \infty$ (cf. Khas'minskii [2]).

The results concerning the action functional for the family of processes $(X_t^\varepsilon, \mathsf{P}_x^\varepsilon)$ corresponding to the operators

$$L^\varepsilon = \frac{\varepsilon^2}{2} \sum a^{ij}(x) \frac{\partial^2}{\partial x^i \partial x^j} + \sum b^i(x) \frac{\partial}{\partial x^i} \tag{7.1}$$

can be applied to determine the asymptotics of eigenvalues of $-L^\varepsilon$ as $\varepsilon \to 0$. Two qualitatively different cases arise according as all trajectories of the dynamical system $\dot{x}_t = b(x_t)$ leave $D \cup \partial D$ or there are stable ω-limit sets of the system in D.

The first, simpler, case was considered in Wentzell [6]. As has been established in §1 (Lemma 1.9), the probability that the trajectory of X_t^ε spends more than time T in D can be estimated from above by the expression $\exp\{-\varepsilon^{-2}c(T - T_0)\}$, $c > 0$. This implies that $\mathsf{M}_x^\varepsilon e^{\lambda\tau^\varepsilon} < \infty$ for $\lambda < \varepsilon^{-2}c$, and therefore, $\lambda_1^\varepsilon \geq \varepsilon^{-2}c$. The rate of convergence of λ_1^ε to infinity is given more accurately by the following theorem.

Theorem 7.1. *As $\varepsilon \to 0$ we have: $\lambda_1^\varepsilon = (c_1 + o(1))\varepsilon^{-2}$, where*

$$c_1 = \lim_{T \to \infty} T^{-1} \min\{S_{0T}(\varphi): \varphi_t \in D \cup \partial D, 0 \leq t \leq T\} \tag{7.2}$$

and S_{0T} is the normalized action functional for the family of diffusion processes $(X_t^\varepsilon, \mathsf{P}_x^\varepsilon)$.

Proof. First of all we establish the existence of the limit (7.2). We denote the minimum appearing in (7.2) by $a(T)$. Firstly, it is clear that $a(T/n) \leq a(T)/n$ for any natural number n (because the value of $S(\varphi)$ is less than or equal to $S_{0T}(\varphi)/n$ for at least one of the intervals $[kT/n, (k + 1)T/n]$). From this we obtain that for any $T, \tilde{T} > 0$,

$$a(T) \geq [T/\tilde{T}] \cdot \alpha(\tilde{T}). \tag{7.3}$$

In order to obtain the opposite estimation, we use the fact that there exist positive T_0 and A such that any two points x and y of $D \cup \partial D$ can be connected with a curve $\varphi_t(x, y)$, $0 \leq t \leq T(x, y) \leq T_0$, such that

$$S_{0T(x,y)}(\varphi(x, y)) \leq A.$$

For some large $\tilde{T}$ let the minimum $a(\tilde{T})$ be attained for a function $\tilde{\varphi}$. We put $x = \tilde{\varphi}_{\tilde{T}}$, $y = \tilde{\varphi}_0$ and construct φ_t from pieces: $\varphi_t = \tilde{\varphi}_t$ for $0 \le t \le \tilde{T}$; $\varphi_t = \varphi_{t-\tilde{T}}(x, y)$ for $\tilde{T} \le t \le \tilde{T} + T(x, y)$; we extend φ periodically with period $\tilde{T} + T(x, y)$. For any positive T we obtain

$$S_{0T}(\varphi) \le ([T/(\tilde{T} + T(x, y))] + 1)(S_{0\tilde{T}}(\tilde{\varphi}) + S_{0T(x,y)}(\varphi(x, y))), \quad (7.4)$$

$$a(T) \le ([T/\tilde{T}] + 1)(a(\tilde{T}) + A). \quad (7.5)$$

Dividing (7.3) and (7.5) by T and passing to the limit as $T \to \infty$, we obtain

$$\tilde{T}^{-1}a(\tilde{T}) \le \underline{\lim}_{T\to\infty} T^{-1}a(T) \le \overline{\lim}_{T\to\infty} T^{-1}a(T) \le \tilde{T}^{-1}(a(\tilde{T}) + A). \quad (7.6)$$

Passing to the limit as $\tilde{T} \to \infty$, we obtain that the limit (7.2) exists (and is finite).

Now let τ^ε be the time of first exit of X_t^ε from D. We prove that $\mathsf{M}_x^\varepsilon \exp\{\varepsilon^{-2}b\tau^\varepsilon\} = \infty$ for sufficiently small ε if $b > c_1$. It is easy to see that

$$\lim_{T\to\infty} T^{-1} \sup_{x,y\in D} \min\{S_{0T}(\varphi): \varphi_0 = x, \varphi_T = y, \varphi_t \in D \cup \partial D, 0 \le t \le T\}$$

is also equal to c_1. We choose a positive $\varkappa$ smaller than $(b - c_1)/3$, and T such that

$$T^{-1} \sup_{x,y\in D} \min\{S_{0T}(\varphi): \varphi_0 = x, \varphi_T = y, \varphi_t \in D \cup \partial D, 0 \le t \le T\} < c_1 + \varkappa. \quad (7.7)$$

Now we choose $\delta > 0$ such that

$$\min\{S_{0T}(\varphi): \varphi_0 = x, \varphi_T = y, \varphi_t \in D_{-\delta} \cup \partial D_{-\delta}, 0 \le t \le T\} < T(c_1 + 2\varkappa) \quad (7.8)$$

for all $x, y \in D_{-\delta}$, where $D_{-\delta}$ is the set of points of D at a distance greater than δ from the boundary. This can be done according to Lemma 1.4 ($\tilde{T} = T$, because the functions φ_t are defined on the same interval $[0, T]$; cf. the proof of Lemma 1.4). For arbitrary $x \in D_{-\delta}$, $y \in D_{-2\delta}$ we choose a curve φ_t, $0 \le t \le T$, connecting them inside $D_{-\delta} \cup \partial D_{-\delta}$ and such that

$S_{0T}(\varphi) \leq T(c_1 + 2\varkappa)$. By Theorem 3.2 of Ch. 5 we obtain that for sufficiently small ε and all $x \in D_{-\delta}$,

$$\begin{aligned} \mathsf{P}_x^\varepsilon\{\tau^\varepsilon > T, X_T^\varepsilon \in D_{-\delta}\} &\geq \mathsf{P}_x^\varepsilon\{\rho_{0T}(X^\varepsilon, \varphi) < \delta\} \\ &\geq \exp\{-\varepsilon^{-2}[S_{0T}(\varphi) + \varkappa T]\} \\ &\geq \exp\{-\varepsilon^{-2}T(c_1 + 3\varkappa)\}. \end{aligned} \tag{7.9}$$

Successive application of the Markov property $(n - 1)$ times gives

$$\mathsf{P}_x^\varepsilon\{\tau^\varepsilon > nT\} \geq \exp\{-n\varepsilon^{-2}T(c_1 + 3\varkappa)\}.$$

From this we obtain for small ε and all n that

$$\begin{aligned} \mathsf{M}_x^\varepsilon \exp\{\varepsilon^{-2}b\tau^\varepsilon\} &\geq \exp\{\varepsilon^{-2}bnT\}\mathsf{P}_x^\varepsilon\{\tau^\varepsilon > nT\} \\ &\geq \exp\{n\varepsilon^{-2}T(b - c_1 - 3\varkappa)\}, \end{aligned}$$

which tends to infinity as $n \to \infty$. Therefore,

$$\mathsf{M}_x^\varepsilon \exp\{\varepsilon^{-2}b\tau^\varepsilon\} = \infty.$$

Now we prove that if $b < c_1$, then $\mathsf{M}_x \exp\{\varepsilon^{-2}b\tau^\varepsilon\} < \infty$ for small ε. Again we choose $0 < \varkappa < (c_1 - b)/3$ and T and $\delta > 0$ such that

$$\min\{S_{0T}(\varphi) \colon \varphi_t \in D_{+\delta} \cup \partial D_{+\delta}, 0 \leq t \leq T\} > T(c_1 - 2\varkappa) \tag{7.10}$$

(we have applied Lemma 1.4 once more). The distance between the set of functions ψ_t entirely in D for $0 \leq t \leq T$ and any of the sets

$$\Phi_x(T(c_1 - 2\varkappa)) = \{\varphi \colon \varphi_0 = x, S_{0T}(\varphi) \leq T(c_1 - 2\varkappa)\}$$

is not less than δ. Therefore, by the upper estimate in Theorem 3.2 of Ch. 5, for sufficiently small ε and all $x \in D$ we have

$$\begin{aligned} \mathsf{P}_x^\varepsilon\{\tau^\varepsilon > T\} &\leq \mathsf{P}_x^\varepsilon\{\rho_{0T}(X^\varepsilon, \Phi_x(T(c_1 - 2\varkappa))) \geq \delta\} \\ &\leq \exp\{-\varepsilon^{-2}T(c_1 - 3\varkappa)\}. \end{aligned} \tag{7.11}$$

The Markov property gives us $\mathsf{P}_x^\varepsilon\{\tau^\varepsilon > nT\} \leq \exp\{-n\varepsilon^{-2}T(c_1 - 3\varkappa)\}$, and

$$\begin{aligned} \mathsf{M}_x^\varepsilon \exp\{\varepsilon^{-2}b\tau^\varepsilon\} &\leq \sum_{n=0}^{\infty} \exp\{\varepsilon^{-2}b(n + 1)T\}\mathsf{P}_x^\varepsilon\{nT < \tau^\varepsilon \leq (n + 1)T\} \\ &\leq \sum_{n=0}^{\infty} \exp\{\varepsilon^{-2}b(n + 1)T\}\mathsf{P}_x^\varepsilon\{\tau\varepsilon > nT\} \\ &\leq \exp\{\varepsilon^{-2}bT\} \sum_{n=0}^{\infty} \exp\{n\varepsilon^{-2}T(b - c_1 + 3\varkappa)\} < \infty. \end{aligned}$$

The theorem is proved. □

The following theorem helps us to obtain a large class of examples in which the limit (7.2) can be calculated.

Theorem 7.2. *Let the field $b(x)$ have a potential with respect to the Riemannian metric connected with the diffusion matrix, i.e., let there exist a function $U(x)$, smooth in $D \cup \partial D$ and such that*

$$b^i(x) = -\sum_j a^{ij}(x) \frac{\partial U(x)}{\partial x^j}. \tag{7.12}$$

Then c_1 is equal to

$$\min_{x \in D \cup \partial D} \left\{ \frac{1}{2} \sum_{ij} a_{ij}(x) b^i(x) b^j(x) \right\}, \tag{7.13}$$

where $(a_{ij}(x)) = (a^{ij}(x))^{-1}$.

Proof. We denote by c_1^* the minimum (7.13). Let the minimum $a(T)$ be attained for the function φ:

$$a(T) = \int_0^T \frac{1}{2} \sum a_{ij}(\varphi_t)(\dot{\varphi}_t^i - b^i(\varphi_t))(\dot{\varphi}_t^j - b^j(\varphi_t))\, dt. \tag{7.14}$$

We transform the integral in the following way:

$$\begin{aligned} a(T) = {} & \int_0^T \frac{1}{2} \sum a_{ij}(\varphi_t)\dot{\varphi}_t^i \dot{\varphi}_t^j \, dt \\ & - \int^T \sum a_{ij}(\varphi_t) b^i(\varphi_t) \dot{\varphi}_t^j \, dt + \int_0^T \frac{1}{2} \sum a_{ij}(\varphi_t) b^i(\varphi_t) b^j(\varphi_t)\, dt. \end{aligned}$$

The first integral here is nonnegative, the second is equal to $U(\varphi_0) - U(\varphi_t)$ by (7.12), and the third is not smaller than c_1^* multiplied by T. This implies that $T^{-1}a(T) \geq c_1^* - T^{-1}C$, where C is the maximum of $U(x) - U(y)$ for all $x, y \in D \cup \partial D$. Passing to the limit as $T \to \infty$, we obtain that $c_1 \geq c_1^*$. The inequality $c_1 \leq c_1^*$ follows from the fact that for the function $\varphi_t \equiv x_0$, where x_0 is a point where the minimum (7.13) is attained, we have $T^{-1}S_{0T}(\varphi) = c_1^*$. □

Now we consider the opposite case—the case where D contains ω-limit sets of the system $\dot{x}_t = b(x_t)$. We assume that condition **A** of §2 is satisfied (there exist a finite number of compacta $K_1, \ldots, K_l$ consisting of equivalent points and containing all ω-limit sets). As we have already said, in this case our process with a small diffusion can be approximated well by a Markov chain with $(l + 1)$ states corresponding to the compacta K_i and the boundary ∂D, having transition probabilities of order

$$\begin{aligned} &\exp\{-\varepsilon^{-2}\tilde{V}_D(K_i, K_j)\}, \\ &\exp\{-\varepsilon^{-2}\tilde{V}_D(K_i, \partial D)\} \end{aligned} \tag{7.15}$$

(the transition probabilities from ∂D to K_i are assumed to be equal to 0 and the diagonal elements of the matrix of transition probabilities are such that the sums in rows are equal to 1).

It seems plausible that the condition of finiteness of $\mathsf{M}_x^\varepsilon e^{\lambda\tau^\varepsilon}$ is close to the condition that the mathematical expectation of $e^{\lambda\nu^\varepsilon}$ is finite, where ν^ε is the number of steps in the chain with the indicated transition probabilities until the first entrance of ∂D. The negative of the logarithm of the largest eigenvalue of the matrix of transition probabilities after the eigenvalue 1 turns out to be the divisor between those λ for which the above mathematical expectation is finite and those for which it is infinite.

The asymptotics of eigenvalues of a matrix with entries of order (7.15) can be found in Wentzell's note [3]. The justification of passage from the chain to the diffusion process X_t^ε is discussed in Wentzell [2].

Theorem 7.3. *Let $\lambda_1^\varepsilon, \lambda_2^\varepsilon, \ldots, \lambda_l^\varepsilon$ be the eigenvalues, except the eigenvalue* 1, *indexed in decreasing order of their absolute values, of a stochastic matrix with entries logarithmically equivalent to the expressions* (7.15) *as $\varepsilon \to 0$. We define constants $V^{(k)}$, $k = 1, \ldots, l, l + 1$ by the formula*

$$V^{(k)} = \min_{g \in G(k)} \sum_{(\alpha \to \beta) \in g} \tilde{V}_A(\alpha, \beta), \tag{7.16}$$

where $G^{(k)}$ is the set of W-graphs over the set $L = \{K_1, \ldots, K_l, \partial D\}$ such that W contains k elements.

(We note that for $k = l + 1$, we have $W = L$; in this case there is exactly one W-graph, which is the empty graph and the sum in (7.16), just as $V^{(l+1)}$, is equal to zero.)

Then for $\varepsilon \to 0$ we have

$$\operatorname{Re}(1 - \lambda_k^\varepsilon) \asymp \exp\{-(V^{(k)} - V^{(k+1)})\varepsilon^{-2}\}. \tag{7.17}$$

The proof is analogous to those of Lemma 3.1 and other lemmas of §3: the coefficients of the characteristic polynomial of the matrix of transition probabilities from which the identity matrix is subtracted can be expressed as sums of products $\pi(g)$ over W-graphs. From this we obtain the asymptotics of the coefficients, which in turn yields the asymptotics of the roots.

We note that the assertion of the theorem implies that $V^{(k-1)} - V^{(k)} \le V^{(k)} - V^{(k+1)} \le V^{(k+1)} - V^{(k+2)}$. In the case of "general position," where these inequalities are strict, the eigenvalue λ_k^ε turns out to be real and simple for small ε (the principal eigenvalue λ_1 is always real and simple).

As far as the justification of the passage from the asymptotic problem for the Markov chain to the problem for the diffusion process is concerned, it is divided into two parts: a construction, independent of the presence of the small parameter ε, which connects diffusion processes with certain discrete chains and estimates connected with the Markov chains Z_n introduced in §2.

For the first part, the following lemma turns out to be essential.

Suppose that (X_t, P_x) is a diffusion process with generator L in a bounded domain D with smooth boundary, ∂g is a surface inside D, Γ is the surface of a neighborhood of ∂g, separating ∂g from ∂D. Let us introduce the random times

$$\sigma_0 = \min\{t; X_t \in \Gamma\}, \tau_1 = \min\{t \ge \sigma_0 \colon X_t \in \partial g \cup \partial D\}.$$

For $a > 0$ we introduce the operators q_a acting on bounded functions defined on ∂g:

$$q_a f(x) = \mathsf{M}_x\{X_{\tau_1} \in \partial g; e^{a\tau_1} f(X_{\tau_1})\}, \qquad x \in \partial g. \tag{7.18}$$

Lemma 7.1. *The smallest eigenvalue λ_1 of $-L$ with zero boundary conditions on ∂D is the division between those a for which the largest eigenvalue of q_a is smaller than 1 and those for which it is larger than 1.*

The proof of this assertion and a generalization of it to other eigenvalues can be found in Wentzell [5].

In order to apply this to diffusion processes with small diffusion, we choose $\partial g = \bigcup_i \partial g_i$ and $\Gamma = \bigcup_i \Gamma_i$ in the same was as in §2 and for the corresponding operators q_a^ε we prove the following estimates.

Lemma 7.2. *For any $\gamma > 0$ one can choose the radii of the neighborhoods of K_i so small for all $a \le e^{-\gamma\varepsilon^{-2}}$ and sufficiently small ε we have*

$$1 + ae^{-\gamma\varepsilon^{-2}} \le \mathsf{M}_x^\varepsilon e^{a\tau_1} \le 1 + ae^{\gamma\varepsilon^{-2}}, \qquad x \in \partial g; \tag{7.19}$$

$$\begin{aligned} \exp\{-(\tilde{V}_D(K_i, K_j) + \gamma)\varepsilon^{-2}\} &\le q_a^\varepsilon \chi_{\partial g_j}(x) \\ = \mathsf{M}_x^\varepsilon\{X_{\tau_1}^\varepsilon \in \partial g_j; e^{a\tau_1}\} &\le \exp\{-(\tilde{V}_D(K_i, K_j) - \gamma)\varepsilon^{-2}\}, \qquad x \in \partial g_i; \end{aligned} \tag{7.20}$$

$$\begin{aligned} \exp\{-(\tilde{V}_D(K_i, \partial D) + \gamma)\varepsilon^{-2}\} &\le \mathsf{M}_x^\varepsilon\{X_{\tau_1}^\varepsilon \in \partial D; e^{a\tau_1}\} \\ &\le \exp\{-(\tilde{V}_D(K_i, \partial D) - \gamma)\varepsilon^{-2}\}, \qquad x \in \partial g_i. \end{aligned} \tag{7.21}$$

The proof is analogous to that of Lemmas 2.1 and 1.7; the upper estimates are obtained with somewhat more work.

In conclusion, we obtain the following result.

Theorem 7.4. *The smallest eigenvalue λ_1^ε of $-L^\varepsilon$ is logarithmically equivalent to*

$$\exp\{-\varepsilon^{-2}(V^{(1)} - V^{(2)}\} \tag{7.22}$$

as $\varepsilon \to 0$, where $V^{(1)}$, $V^{(2)}$ are defined by formula (7.16).

In particular, if D contains a unique asymptotically stable equilibrium position O, then

$$\lambda_1^\varepsilon \asymp \exp\left\{-\varepsilon^{-2}\min_{y\in\partial D} V(O, y)\right\}, \tag{7.23}$$

where V is a quasipotential.

Theorems 7.1 and 7.4 do not give anything interesting in the case where there are only unstable sets inside D or the stable limit sets are on the boundary of the domain. Some results concerning this case are contained in the paper [1] by Devinatz, Ellis, and Friedman.

The results contained in Theorems 7.3 and 7.4 are related to results of §6. In particular, the constants which are to be crossed by $\lim_{\varepsilon\to 0} \varepsilon^2 \ln t(\varepsilon^{-2})$ as a sublimit distribution changes to another one are nothing else but $V^{(k)} - V^{(k+1)}$.

We clarify the mechanisms of this connection. Let $(X_t^\varepsilon, \mathsf{P}_x^\varepsilon)$ be a diffusion process with small diffusion on a compact manifold M or a process in a closed domain $D \cup \partial D$, stopped after exit to the boundary and let $f(x)$ be a continuous function vanishing on ∂D. We assume that the function $u^\varepsilon(t, x) = \mathsf{M}_x^\varepsilon f(X_t^\varepsilon)$—the solution of the corresponding problem for the equation $\partial u^\varepsilon/\partial t = L^\varepsilon u^\varepsilon$—can be expanded in a series in the eigenfunctions $e_i^\varepsilon(x)$ of $-L^\varepsilon$:

$$u^\varepsilon(t, x) = \sum_i c_i^\varepsilon e^{-\lambda_i^\varepsilon t} e_i^\varepsilon(x) \tag{7.24}$$

with coefficient c_i^ε obtained by integrating f multiplied by the corresponding eigenfunction $\tilde{e}_i^\varepsilon$ of the adjoint operator. (Since L^ε is not self-adjoint, the existence of such an expansion is not automatic.) We assume that the first l_0 eigenvalues are real, nonnegative and decrease exponentially fast:

$$\lambda_i^\varepsilon \asymp \exp\{-C_i \varepsilon^{-2}\}, \qquad +\infty \geq C_1 > C_2 > \cdots > C_{l_0}.$$

Let the function $t(\varepsilon^{-2})$ be such that $C_{j+1} < \underline{\lim}_{\varepsilon\to 0} \varepsilon^2 \ln t(\varepsilon^{-2}) \leq \overline{\lim}_{\varepsilon\to 0} \varepsilon^2 \ln t(\varepsilon^{-2}) < C_j$. We have $\lim_{\varepsilon\to 0} e^{-\lambda_i t(\varepsilon^{-2})} = 1$ for $1 \leq i \leq j$ and this limit is equal to zero for $i > j$. If the infinite remainder of the sum (7.24) does not prevent it, we obtain

$$\lim_{\varepsilon \to 0} u^\varepsilon(t(\varepsilon^{-2}), x) = \sum_{i=1}^{j} \lim_{\varepsilon \to 0} \int f(y) \tilde{e}_i^\varepsilon(y)\, dy \cdot \lim_{\varepsilon \to 0} e_i^\varepsilon(x). \tag{7.25}$$

In other words, the character of the limit behavior of $u^\varepsilon(t(\varepsilon^{-2}), x)$ changes as $\lim_{\varepsilon \to 0} \varepsilon^2 \ln t(\varepsilon^{-2})$ crosses C_j.

This outlines the route following which we can derive conclusions concerning the behavior of a process on exponentially increasing time intervals from results involving exponentially small eigenvalues and the corresponding eigenfunctions. The same method can be used for the more difficult inverse problem.

Chapter 7

The Averaging Principle. Fluctuations in Dynamical Systems with Averaging

§1. The Averaging Principle in the Theory of Ordinary Differential Equations

Let us consider the system

$$\dot{Z}_t^\varepsilon = \varepsilon b(Z_t^\varepsilon, \xi_t), \qquad Z_0^\varepsilon = x \tag{1.1}$$

of ordinary differential equations in R^r, where ξ_t, $t \geq 0$, is a function assuming values in R^l, ε is a small numerical parameter and

$$b(x, y) = (b^1(x, y), \ldots, b^r(x, y)).$$

If the functions $b^i(x, y)$ do not increase too fast, then the solution of equation (1.1) converges to $Z_t^0 \equiv x$ as $\varepsilon \to 0$, uniformly on every finite time interval $[0, T]$. However, the behavior of Z_t^ε on time intervals of order ε^{-1} or of larger order is usually of main interest since it is only times of order ε^{-1} over which significant changes—such as exit from the neighborhood of an equilibrium position or of a periodic trajectory—take place in system (1.1). In the study of the system on intervals of the form $[0, T\varepsilon^{-1}]$, it is convenient to pass to new coordinates in order that the time interval do not depend on ε. We set $X_t^\varepsilon = Z_{t/\varepsilon}^\varepsilon$. Then the equation for X_t^ε assumes the form

$$\dot{X}_t^\varepsilon = b(X_t^\varepsilon, \xi_{t/\varepsilon}), \qquad X_0^\varepsilon = x. \tag{1.2}$$

The study of this system on a finite time interval is equivalent to the study of system (1.1) on time intervals of order ε^{-1}.

Let the function $b(x, y)$ be bounded, continuous in x and y, and let it satisfy a Lipschitz condition in x with a constant independent of y:

$$|b(x_1, y) - b(x_2, y)| \leq K|x_1 - x_2|.$$

We assume that

$$\lim_{T \to \infty} \frac{1}{T} \int_0^T b(x, \xi_s)\, ds = \bar{b}(x) \tag{1.3}$$

for $x \in R^r$. If this limit exists, then the function $\bar{b}(x)$ is obviously bounded and satisfies a Lipschitz condition with the same constant K. Condition (1.3) is satisfied, for example, if ξ_t is periodic or is a sum of periodic functions.

The displacement of the trajectory X_t^ε over a small time Δ can be written in the form

$$\begin{aligned} X_\Delta^\varepsilon - x &= \int_0^\Delta b(X_s^\varepsilon, \xi_{s/\varepsilon})\, ds \\ &= \int_0^\Delta b(x, \xi_{s/\varepsilon})\, ds + \int_0^\Delta [b(X_s^\varepsilon, \xi_{s/\varepsilon}) - b(x, \xi_{s/\varepsilon})]\, ds \\ &= \Delta\left(\frac{\varepsilon}{\Delta}\int_0^{\Delta/\varepsilon} b(x, \xi_s)\, ds\right) + \rho_\varepsilon(\Delta). \end{aligned}$$

The coefficient of Δ in the first term of the last side converges to $\bar{b}(x)$ as $\varepsilon/\Delta \to 0$, according to (1.3). The second term satisfies the inequality $|\rho_\varepsilon(\Delta)| < K\Delta^2$. Consequently, the displacement of the trajectory X_t^ε over a small time differs from the displacement of the trajectory $\bar{x}_t$ of the differential equation

$$\dot{\bar{x}}_t = \bar{b}(\bar{x}_t), \qquad \bar{x}_0 = x. \tag{1.4}$$

only by an infinitely small quantity compared to Δ as $\Delta \to 0$, $\varepsilon/\Delta \to 0$. If we assume that the limit in (1.3) is uniform in x, then from this we obtain a proof of the fact that the trajectory X_t^ε converges to the solution of equation (1.4), uniformly on every finite time interval as $\varepsilon \to 0$.

The assertion that the trajectory X_t^ε is close to $\bar{x}_t$ is called the averaging principle.

An analogous principle can also be formulated in a more general situation We consider, for example, the system

$$\begin{aligned} \dot{X}_t^\varepsilon &= b_1(X_t^\varepsilon, \xi_t^\varepsilon), \qquad & X_0^\varepsilon &= x, \\ \dot{\xi}_t^\varepsilon &= \varepsilon^{-1} b_2(X_t^\varepsilon, \xi_t^\varepsilon), \qquad & \xi_0^\varepsilon &= y, \end{aligned} \tag{1.5}$$

where $x \in R^r$, $\xi \in R^l$, and b_1 and b_2 are bounded, continuously differentiable functions on $R^r \otimes R^l$ with values in R^r and R^l, respectively. The velocity of the motion of the variables ξ has order ε^{-1} as $\varepsilon \to 0$. Therefore, the ξ are called fast variables, the space R^l the space of fast motion and the x slow variables.

We consider the fast motion $\xi_t(x)$ for fixed slow variables $x \in R^r$:

$$\dot{\xi}_t(x) = b_2(x, \xi_t(x)), \qquad \xi_0(x) = y,$$

and assume that the limit

$$\lim_{T\to\infty} \frac{1}{T}\int_0^T b_1(x, \xi_s(x))\, ds = \bar{b}_1(x). \tag{1.6}$$

exists. For the sake of simplicity, let this limit be independent of the initial point y of the trajectory $\xi_s(x)$. The averaging principle for system (1.5) is the assertion that under certain assumptions, displacement in the space of slow motion can be approximated by the trajectory of the averaged system

$$\dot{\bar{x}}_t = \bar{b}_1(\bar{x}_t), \qquad \bar{x}_0 = x.$$

For equation (1.2), the role of fast motion is played by $\xi_t^\varepsilon = \xi_{t/\varepsilon}$. In this case the velocity of fast motion does not depend on the slow variables.

Although the averaging principle has long been applied to problems of celestial mechanics, oscillation theory and radiophysics, no mathematically rigorous justification of it had existed for a long time. The first general result in this area was obtained by N. N. Bogolyubov (cf. Bogolyubov and Mitropol'skii [1]). He proved that if the limit (1.3) exists uniformly in x, then the solution X_t^ε of equation (1.2) converges to the solution of the average system (1.4), uniformly on every interval. Under certain assumptions, the rate of convergence was also estimated and an asymptotic expansion in powers of the small parameter was constructed. In Bogolyubov and Zubarev [1] (cf. also Bogolyubov and Mitropol'skii [1]), these results were extended to some cases of system (1.5), namely to systems in which the fast motion is one-dimensional and the equation for ξ^ε has the form $\dot{\xi}_t^\varepsilon = \varepsilon^{-1} b_2(X_t^\varepsilon)$ and to some more general systems. V. M. Volosov obtained a series of results concerning the general case of system (1.5) (cf. Volosov [1]). Nevertheless, in the case of multidimensional fast motions, the requirement of uniform convergence to the limit in (1.6), which is usually imposed, excludes a series of interesting problems, for example, problems arising in perturbations of Hamiltonian systems. In a sufficiently general situation it can be proved that for every $T > 0$ and $\rho > 0$, the Lebesgue measure of the set F_ρ^ε of those initial conditions in problem (1.5) for which $\sup_{0 \le t \le T} |X_t^\varepsilon - \bar{x}_t| > \rho$ converges to zero with ε. This result, obtained in Anosov [1], was later sharpened for systems of a special form (cf. Neishtadt [1]).

Consequently, if a system of differential equations is reduced to the form (1.2) or (1.5), then it is clear, at least formally, what the equation of zeroth approximation looks like. In some cases a procedure can be found for the determination of higher approximations. In the study of concrete problems first we have to choose the variables in such a way that the fast and slow motions be separated.

As an example, we consider the equation

$$\ddot{x}_t + \omega^2 x = \varepsilon(x, \dot{x}, t), \qquad x \in R^1 \tag{1.7}$$

If $f(x, \dot{x}, t) = (1 - x^2)\dot{x}$, then this equation turns into the so-called van der Pol equation describing oscillations in a lamp generator. For $\varepsilon = 0$ we obtain the equation of a harmonic oscillator. In the phase plane $(x, \dot{x})$ the solutions of this equation are the ellipses $x = r\cos(\omega t + \theta)$, $\dot{x} = -r\omega\sin(\omega t + \theta)$, on which the phase point rotates with constant angular velocity ω. Without perturbations ($\varepsilon = 0$), the amplitude r is determined by the initial conditions and does not change with time and for the phase we have $\varphi_t = \omega t + \theta$. If ε is now different from zero, but small, then the amplitude r and the difference $\varphi_t - \omega t$ are not constant in general. Nevertheless, one may expect that the rate of change of them is small provided that ε is small. Indeed, passing from the variables $(x, \dot{x})$ to the van der Pol variables (r, θ), we obtain

$$\begin{aligned} \frac{dr^\varepsilon}{dt} &= \varepsilon F_1(\omega t + \theta^\varepsilon, r^\varepsilon, t), & r_0^\varepsilon &= r_0, \\ \frac{d\theta^\varepsilon}{dt} &= \varepsilon F_2(\omega t + \theta^\varepsilon, r^\varepsilon, t), & \theta_0^\varepsilon &= \theta_0, \end{aligned} \tag{1.8}$$

where the functions $F_1(s, r, t)$ and $F_2(s, r, t)$ are given by the equalities

$$F_1(s, r, t) = -\frac{1}{\omega} f(r \cos s, -r \sin s, t) \sin s,$$

$$F_2(s, r, t) = -\frac{1}{r\omega} f(r \cos s, -r \sin s, t) \cos s.$$

Consequently, in the van der Pol variables (r, θ), the equation can be written in the form (1.1). If $f(x, \dot{x}, t)$ does not depend explicitly on t, then $F_1(\omega t + \theta, r)$ and $F_2(\omega t + \theta, r)$ are periodic in t and condition (1.3) is satisfied. Therefore, the averaging principle is applicable to system (1.8). In the case where $f(x, \dot{x}, t)$ depends periodically on the last argument, condition (1.3) is also satisfied. If the frequency ω and the frequency ν of the function f in t are incommensurable, then the averaged right sides of system (1.8) have the form (cf., for example, Bogolyubov and Mitropol'skii [1])

$$\begin{aligned} \bar{F}_1(r) &= -\frac{1}{4\pi^2\omega}\int_0^{2\pi}\int_0^{2\pi} f(r\cos\psi, -r\sin\psi, \nu t)\sin\psi\, d\psi\, dt, \\ \bar{F}_2(r) &= -\frac{1}{4\pi^2 r\omega}\int_0^{2\pi}\int_0^{2\pi} f(r\cos\psi, -r\sin\psi, \nu t)\cos\psi\, d\psi\, dt. \end{aligned} \tag{1.9}$$

It is easy to calculate the averaged right sides of system (1.8) for commensurable ω and ν, as well. Relying on the averaging principle, we can derive that as $\varepsilon \to 0$, the trajectory $(r_t^\varepsilon, \theta_t^\varepsilon)$ can be approximated by the trajectory $(\bar{r}_t^\varepsilon, \bar{\theta}_t^\varepsilon)$ of the averaged system

$$\dot{\bar{r}}_t = \varepsilon \bar{F}_1(\bar{r}_t), \qquad \dot{\bar{\theta}}_t = \bar{F}_2(\bar{r}_t), \qquad \bar{r}_0 = r_0, \qquad \bar{\theta}_0 = \theta_0,$$

uniformly on the interval $[0, T\varepsilon^{-1}]$. Using this approximation, we can derive a number of interesting conclusions on the behavior of solutions of equation (1.7). For example, let $\bar{F}_1(r_0) = 0$ and let the function $\bar{F}_1(r)$ be positive to the left of r_0 and negative to the right of it, i.e., let r_0 be the amplitude of an asymptotically stable periodic solution of the averaged system. It can be shown by means of the averaging principle that oscillations with amplitude close to r_0 and frequency close to ω occur in the system described by equation (1.7) with arbitrary initial conditions, provided that ε is sufficiently small. In the case where $\bar{F}_1(r)$ has several zeros at which it changes from positive to negative, the amplitude of oscillations over a long time depends on the initial state. We shall return to the van der Pol equation in §8.

Equation (1.7) describes a system which is obtained as a result of small perturbations of the equation of an oscillator. Perturbations of a mechanical system of a more general form can also be considered. Let the system be conservative, let us denote by $H(p, q)$ its Hamiltonian, and let the equations of the perturbed system have the form

$$\frac{dp_i^\varepsilon}{dt} = \frac{\partial H}{\partial q_i}(p^\varepsilon, q^\varepsilon) + \varepsilon f_p^i(p^\varepsilon, q^\varepsilon),$$

$$\frac{dq_i^\varepsilon}{dt} = -\frac{\partial H}{\partial p_i}(p^\varepsilon, q^\varepsilon) + \varepsilon f_q^i(p^\varepsilon, q^\varepsilon), \qquad i = 1, 2, \ldots, n.$$

If $n = 1$, $p = x$, $q = \dot{x}$ and $H(p, q) = q^2 + \omega^2 p^2$, $f_p^1 = 0$, $f_q^1 = f(p, q)$, then this system is equivalent to equation (1.7). In the case of one degree of freedom ($n = 1$), if the level sets $H(p, q) = C =$ const are smooth curves homeomorphic to the circle, new variables can also be introduced in such a way that the fast and slow motions be separated. As such variables, we can take the value H of the Hamiltonian $H(p, q)$ and the angular coordinate φ on a level set. Since $\dot{H}(p, q) = 0$ for the unperturbed system, $H(p_t^\varepsilon, q_t^\varepsilon)$ will change slowly.

It is clear that if we choose (H', φ) as variables, where H' is a smooth function of H, then H' also changes slowly. For example, in the van der Pol variables, the slow variable is $r = \sqrt{H}$. In order that the system preserve the Hamiltonian form, in place of the variables (H, φ), one often considers the so-called action-angle variables (I, φ) (cf., for example, Arnold [1]), where $I = H/\omega$. In the multidimensional case the slowly varying variables will be first integrals of the unperturbed system. In some cases (cf. Arnold [1], Ch. 10), in Hamiltonian systems with n degrees of freedom we can also introduce action-angle variables $(\mathbf{I}, \boldsymbol{\varphi})$; in these variables the fast and slow motions are separated and the system of equations preserves the Hamiltonian form.

§2. The Averaging Principle when the Fast Motion is a Random Process

As has already been mentioned, condition (1.3), which implies the averaging principle, is satisfied if the function ξ_t is periodic. On the other hand, this condition can be viewed as a certain kind of law of large numbers: (1.3) is satisfied if the values assumed by ξ_t at moments of time far away from each other are "almost independent".

In what follows we shall assume that the role of fast variables is played by a random process. First we discuss the simpler case of (1.2), where the fast motion does not depend on the slow variables. Hence let ξ_t be a random process with values in R^l. We shall assume that the function $b(x, y)$ satisfies a Lipschitz condition: $|b(x_1, y_1) - b(x_2, y_2)| \leq K(|x_1 - x_2| + |y_1 - y_2|)$. Concerning the process ξ_t, we assume that its trajectories are continuous with probability one or on every finite time interval they have a finite number of discontinuities of the first kind and there are no discontinuities of the second kind. Under these assumptions, the solution of equation (1.2) exists with probability 1 for any $x \in R^r$ and it is defined uniquely for all $t \geq 0$.

If condition (1.3) is satisfied with probability 1 uniformly in $x \in R^r$, then the ordinary averaging principle implies that with probability 1, the trajectory of X_t^ε converges to the solution of equation (1.4), uniformly on every finite interval ($\bar{b}(x)$ and $\bar{x}_t$ may depend on ω in general).

Less stringent assumptions can be imposed concerning the type of convergence in (1.3). Then we obtain a weaker result in general.

We assume that there exists a vector field $\bar{b}(x)$ in R^r such that for any $\delta > 0$ and $x \in R^r$ we have

$$\lim_{T \to \infty} \mathsf{P}\left\{\left|\frac{1}{T}\int_t^{t+T} b(x, \xi_s)\, ds - \bar{b}(x)\right| > \delta\right\} = 0, \tag{2.1}$$

uniformly in $t > 0$. It follows from (2.1) that $\bar{b}(x)$ satisfies a Lipschitz condition (with the same constant as $b(x, y)$). Therefore, there exists a unique solution of the problem

$$\dot{\bar{x}}_t = \bar{b}(\bar{x}_t), \qquad \bar{x}_0 = x. \tag{2.2}$$

The random process X_t^ε can be considered as a result of random perturbations of the dynamical system (2.2), small on the average. Relation (2.1) is an assumption on the average in time smallness of random perturbations.

Theorem 2.1. *Suppose that condition* (2.1) *is satisfied and* $\sup_t \mathsf{M}|b(x, \xi_t)|^2 < \infty$. *Then for any* $T > 0$ *and* $\delta > 0$ *we have*

$$\lim_{\varepsilon \to 0} \mathsf{P}\left\{\sup_{0 \leq t \leq T} |X_t^\varepsilon - \bar{x}_t| > \delta\right\} = 0.$$

The assertion of this theorem follows easily from Theorem 1.3 of Ch. 2. For this we need to put $b(\varepsilon, s, x, \omega) = b(x, \xi_{s/\varepsilon}(\omega))$ and note that condition

(2.1) can be written in the following form: for any T, $\delta > 0$ and $x \in R^r$ we have

$$\lim_{\varepsilon \to 0} \mathsf{P}\left\{\left| \int_t^{t+T} b(\varepsilon, s, x, \omega)\, ds - T\bar{b}(x) \right| > \delta\right\} = 0,$$

uniformly in t. This is exactly the condition in Theorem 1.3 of Ch. 2.

We note that our arguments repeat, in essence, the proof of the averaging principle in the deterministic case, which is contained in Gikhman [1] and Krasnosel'skii and Krein [1]. A similar result is contained in Khas'minskii [4].

Condition (2.1), which is assumed in Theorem 2.1, is satisfied under quite relaxed assumptions on the process $\eta_s^x = b(x, \xi_s)$ ($x \in R^r$ is a parameter). For example, if η_s^x is stationary in the wide sense, then it is sufficient that the diagonal entries of its correlation matrix $(R^{ij}(\tau))$ converge to zero as $\tau \to \infty$; in this case, $\bar{b}(x) = \mathsf{M}b(x, \xi_s)$. In the nonstationary case it is sufficient that there exist a function $r(x, \tau)$ such that $\lim_{\tau \to \infty} r(x, \tau) = 0$ and

$$|\mathsf{M}(b(x, \xi_s) - \bar{b}(x), b(x, \xi_{s+\tau}) - \bar{b}(x))| < r(x, \tau).$$

We postpone examples until §8 and now study the difference $X_t^\varepsilon - \bar{x}_t$ in more detail. In the deterministic case, where, for example, ξ_t is a periodic function, this difference is of order ε and we can write down the other terms of the asymptotic expansion in integral powers of ε. In the study of probability theoretical problems we apparently have to consider typical the situation where the random process ξ_t satisfies some condition of weak dependence, i.e., a condition that the dependence between the variables ξ_t and $\xi_{t+\tau}$ becomes weaker in some sense with increasing τ. It turns out that in this case the deviation of X_t^ε from $\bar{x}_t$ is of another character. The difference $X_t^\varepsilon - \bar{x}_t$ is of order $\sqrt{\varepsilon}$, but no other terms of the asymptotic expansion can be written down: as $\varepsilon \to 0$, the expression $(1/\sqrt{\varepsilon})(X_t^\varepsilon - \bar{x}_t)$ does not converge to any limit in general and only has a limit distribution. In other words, whereas the averaging principle itself—the assertion of Theorem 2.1—can be considered as a result of the type of laws of large numbers, the behavior of the standardized difference $(1/\sqrt{\varepsilon})(X_t^\varepsilon - \bar{x}_t)$ can be described by an assertion of the type of the central limit theorem. In order to clarify this, we consider the simplest system $\dot{X}_t^\varepsilon = b(\xi_{t/\varepsilon})$, in which the right sides do not depend on x. If ξ_t satisfies the strong mixing condition of (cf. below), then under weak additional restrictions, the distribution of the normalized difference $\zeta_t^\varepsilon = (1/\sqrt{\varepsilon})(X_t^\varepsilon - \bar{x}_t)$ converges to a normal distribution as $\varepsilon \to 0$ (cf., for example, Ibragimov and Linnik [1]). In the next section we show that under some additional assumptions, the distributions of the normalized differences converge to Gaussian distributions in the case of systems of general form, as well. What is more, following Khas'minskii [4], we show that not only do the distributions of the variables ζ_t^ε converge to Gaussian distributions for every fixed t, but as $\varepsilon \to 0$, the processes ζ_t^ε also converge, in the sense of weak convergence, to a Gaussian Markov process and we also determine the characteristics of the limit process. In the remaining sections we also study large deviations of order 1 of X_t^ε from $\bar{x}_t$ and large deviations of order $\varepsilon^\varkappa$, where $\varkappa \in (0, \frac{1}{2})$.

§3. Normal Deviations from an Averaged System

We pass to the study of the difference between the solution X_t^ε of system (1.2) and the solution $\bar{x}_t$ of the averaged system. It has been shown in the preceding section that with probability close to 1 for ε small, the trajectory of X_t^ε is situated in a small neighborhood of the function $\bar{x}_t$ for $t \in [0, T]$, $T < \infty$. Therefore, if we take the smoothness of the field $b(x, y)$ into account, we may hope that the difference $X_t^\varepsilon - \bar{x}_t$ can be approximated with the deviation, from $\bar{x}_t$, of the solution of the system obtained from (1.2) by linearization in the neighborhood of the trajectory of the averaged system. Consequently, the study of the normalized difference $\zeta_t^\varepsilon = \varepsilon^{-1/2}(X_t^\varepsilon - \bar{x}_t)$ may be carried out according to the following plan: firstly, we study the normalized deviation in the case of a linearized system; secondly, we verify that the trajectory of the original system differs from that of the linearized one by a quantity infinitely small compared to $\sqrt{\varepsilon}$ as $\varepsilon \to 0$. In implementing the first part of our plan, we have to introduce notions and carry out arguments very similar to those usually employed in the proof of the central limit theorem for random processes. Moreover, since we would like to prove the weak convergence of the processes rather than only convergence of the finite-dimensional distributions, we also need to verify the weak compactness of the family of processes ζ_t^ε.

We note that the study of large deviations of order $\varepsilon^\varkappa$, where $\varkappa \in (0, \frac{1}{2})$, can also be reduced to the study of deviations of the same order for the linearized system. As to the probabilities of deviations of order 1 of X_t^ε from $\bar{x}_t$ for system (1.2) and the linearized system, they have essentially different asymptotics.

We pass to the implementation of the above program of the study of deviations of order $\sqrt{\varepsilon}$. For this we recall, the notion of strong mixing and some properties of random processes satisfying the strong mixing condition.

In a probability space $\{\Omega, \mathscr{F}, \mathsf{P}\}$ let us be given an increasing family of σ-algebras $\mathscr{F}_s^t$: $\mathscr{F}_{s_1}^{t_1} \subseteq \mathscr{F}_{s_2}^{t_2}$ for $0 < s_2 \le s_1 \le t_1 \le t_2 \le \infty$. We say that this family satisfies the strong mixing condition with coefficient $\alpha(\tau)$ if

$$\sup_t \sup_{\xi, \eta} |\mathsf{M}\xi\eta - \mathsf{M}\xi\mathsf{M}\eta| = \alpha(\tau) \downarrow 0 \tag{3.1}$$

as $\tau \to \infty$, where the supremum is taken over all $\mathscr{F}_0^t$-measurable ξ, $|\xi| \le 1$ and $\mathscr{F}_{t+\tau}^\infty$-measurable variables η, $|\eta| \le 1$.

It can be proved (cf. Rozanov [1]) that if ξ is an $\mathscr{F}^t_0$-measurable and η is an $\mathscr{F}^\infty_{t+\tau}$-measurable random variable and $\mathsf{M}|\xi|^{2+\delta} < \infty$, $\mathsf{M}|\eta|^{2+\delta} < \infty$, then

$$|\mathsf{M}\xi\eta - \mathsf{M}\xi\mathsf{M}\eta| \leq (\mathsf{M}|\xi|^{2+\delta})^{1/(2+\delta)}(\mathsf{M}|\eta|^{2+\delta})^{1/(2+\delta)} \cdot 7[\alpha(\tau)]^{\delta/(4+\delta)}. \quad (3.2)$$

If $0 \leq s_1 \leq t_1 \leq s_2 \leq t_2 \leq \cdots \leq s_m \leq t_m$ are arbitrary numbers, $\Delta = \min_{2 \leq k \leq m}(s_k - t_{k-1})$ and the random variables $\eta_1, \ldots, \eta_m$ are such that the η_k are $\mathscr{F}^{t_k}_{s_k}$-measurable and $|\eta_k| \leq 1$, then

$$\left| \mathsf{M} \prod_{k=1}^m \eta_k - \prod_{k=1}^m \mathsf{M}\eta_k \right| \leq (m-1)\alpha(\Delta), \quad (3.3)$$

where $\alpha(\Delta)$ is the mixing coefficient for the σ-algebras $\mathscr{F}^t_s$.

We say that a random process ξ_t, $t \geq 0$, satisfies the condition of strong mixing with mixing coefficient $\alpha(\tau)$ if the σ-algebras $\mathscr{F}^t_s$ generated by the values of the processes ξ_u for $u \in [s, t]$ satisfy the condition (3.1) of strong mixing.

Let $\eta_t = (\eta^1(t), \ldots, \eta^{2k}(t))$ be a random process satisfying the condition of strong mixing with coefficient $\alpha(\tau)$. Suppose that for some $m > 2$ we have

$$\mathsf{M}|\eta^i(t)|^{m(2k-1)} < C,$$

$$\int_0^\infty \tau^{n-1}[\alpha(\tau)]^{(m-2)/m}\, d\tau = A_n \quad \text{for } n = 1, 2, \ldots, k;$$

$$\int_{t_0}^{t_0+T} |\mathsf{M}\eta^i(t)|\, dt < B\sqrt{T}$$

for $i = 1, 2, \ldots, 2k$, where C, B, and $A_1, \ldots, A_k$ are positive constants. Then there exists a constant $C_{(2k)}$, determined only by the constants $C, B, A_1, \ldots, A_k$, such that

$$\left| \mathsf{M} \int \cdots \int_{D_{2k}} \eta^1(s_1)\eta^2(s_2) \cdot \ldots \cdot \eta^{2k}(s_{2k})\, ds_1 \cdots ds_{2k} \right| < C_{(2k)}T^k, \quad (3.4)$$

where $D_{2k} = \{(s_1, \ldots, s_{2k}): s_i \in [t_0, t_0 + T] \text{ for } i = 1, \ldots, 2k\}$. If D is the direct product of two-dimensional convex domains $D^{(1)}, \ldots, D^{(k)}$ such that each of them can be enclosed in a square with side T and for all $s \in [t_0, t_0 + T]$ we have the inequality $|\int_{t_0}^s \mathsf{M}\eta^i(t)\, dt| < C$, then

$$\left| \mathsf{M} \int \cdots \int_{D_{2k}} \eta^1(c_1)\eta^2(s_2) \cdot \ldots \cdot \eta^{2k}(s_{2k})\, ds_1 \cdots ds_{2k} \right| < C'_{(2k)}T^k. \quad (3.5)$$

The proof of estimates of the type of (3.4) takes its origin in the first publications on limit theorems for weakly dependent random variables (cf. Bernstein

[2]). In the form presented here, estimates (3.4) and (3.5) are proved in Khas'minskii [4].

Now we formulate the fundamental result of this section.

Theorem 3.1. *Let the functions $b^i(x, y)$, $x \in R^r$, $y \in R^l$, $i = 1, \ldots, r$, have bounded continuous first and second partial derivatives on the whole space. Suppose that the random process ξ_t with values in R^l has piecewise continuous trajectories with probability 1 and satisfies the condition of strong mixing with coefficient $\alpha(\tau)$ such that $\int_0^\infty \tau[\alpha(\tau)]^{1/5}\, d\tau < \infty$ and $\sup_{x,t} \mathsf{M}|b(x, \xi_t)|^3 < N < \infty$. Moreover, let the following conditions be satisfied:*

1. *The limits*

$$\lim_{T \to \infty} \frac{1}{T} \int_{t_0}^{t_0+T} \mathsf{M}b(x, \xi_s)\, ds = \bar{b}(x),$$

$$\lim_{T \to \infty} \frac{1}{T} \int_{t_0}^{t_0+T} \int_{t_0}^{t_0+T} A^{ki}(x, s, t)\, ds\, dt = A^{ki}(x),$$

exist uniformly in $x \in R^r$, $t_0 \geq 0$, where

$$A^{ki}(x, s, t) = \mathsf{M}[b^k(x, \xi_s) - \mathsf{M}b^k(x, \xi_s)][b^i(x, \xi_t) - \mathsf{M}b^i(x, \xi_t)].$$

2. *For some $C < \infty$ we have*

$$\left| \int_0^\tau [\mathsf{M}b(\bar{x}_s, \xi_{s/\varepsilon}) - \bar{b}(\bar{x}_s)]\, ds \right| < C\varepsilon,$$

$$\max_{k,i} \left| \int_0^\tau \left[\mathsf{M} \frac{\partial b^k}{\partial x^i}(\bar{x}_s, \xi_{s/\varepsilon}) - \frac{\partial \bar{b}^k}{\partial x^i}(\bar{x}_s) \right] ds \right| < C\varepsilon$$

for all $\tau \in [0, T_0]$.

Then as $\varepsilon \to 0$, the process

$$\zeta_t^\varepsilon = \frac{1}{\sqrt{\varepsilon}}(X_t^\varepsilon - \bar{x}_t)$$

converges weakly on the interval $[0, T_0]$ to a Gaussian Markov process ζ_t^0 satisfying the system of linear differential equations

$$\dot{\zeta}_t^0 = \dot{w}_t^0 + B(\bar{x}_t)\zeta_t^0, \qquad \zeta_0^0 = 0, \tag{3.6}$$

where w_t^0 is a Gaussian process with independent increments, vanishing mathematical expectation and correlation matrix $(R^{ki}(t))$, $R^{ki}(t) = \mathsf{M}w_t^{0,k}w_t^{0,i} = \int_0^t A^{ki}(\bar{x}_s)\, ds$ and $B(x) = (B_j^i(x)) = (\partial \bar{b}^i/\partial x^j(x))$.

Proof. We introduce the notation

$$B(x, y) = (B^i_j(x, y)) = \left(\frac{\partial b^i}{\partial x^j}(x, y)\right), \qquad \lambda^\varepsilon_t = \frac{1}{\sqrt{\varepsilon}} \int_0^t [b(\bar{x}_s, \xi_{s/\varepsilon}) - \bar{b}(\bar{x}_s)]\, ds,$$

$$\Phi(s, \varepsilon, \omega) = B(\bar{x}_s, \xi_{s/\varepsilon}) - B(\bar{x}_s),$$

$$\Psi(s, \varepsilon, \omega) = \frac{1}{\sqrt{\varepsilon}} [b(\bar{x}_s + \sqrt{\varepsilon}\zeta^\varepsilon_s, \xi_{s/\varepsilon}) - b(\bar{x}_s, \xi_{s/\varepsilon}) - B(\bar{x}_s, \xi_{s/\varepsilon})\zeta^\varepsilon_s \sqrt{\varepsilon}].$$

It follows from the definition of X^ε_t and $\bar{x}_t$ that we have the following relation for the normalized difference ζ^ε_t:

$$\begin{aligned} \zeta^\varepsilon_t &= \frac{1}{\sqrt{\varepsilon}} \int_0^t [b(X^\varepsilon_s, \xi_{s/\varepsilon}) - \bar{b}(\bar{x}_s)]\, ds \\ &= \lambda^\varepsilon_t + \int_0^t B(\bar{x}_s)\zeta^\varepsilon_s\, ds + \int_0^t \Phi(s, \varepsilon, \omega)\zeta^\varepsilon_s\, ds + \int_0^t \Psi(s, \varepsilon, \omega)\, ds. \end{aligned} \tag{3.7}$$

It is easy to see (it follows from the existence of bounded second derivatives of $b^i(x, y)$) that $\Psi(s, \varepsilon, \omega)$ is of order $\sqrt{\varepsilon}(\zeta^\varepsilon_s)^2$. In general, the linearized equation must contain a term corresponding to the third term on the right side of (3.7). Nevertheless, as will follow from the discussion below, this term has a vanishing effect as $\varepsilon \to 0$, and therefore, we do not include it in the linearized system from the very beginning.

Hence we consider the simplified linearized equation

$$Z^\varepsilon_t = \lambda^\varepsilon_t + \int_0^t B(\bar{x}_s) Z^\varepsilon_s\, ds. \tag{3.8}$$

In accordance with our plan, we first have to prove that as $\varepsilon \to 0$, Z^ε_t converges to a process ζ^0_t satisfying equation (3.6). For this we need the following lemma.

Lemma 3.1. *The process λ^ε_t converges weakly to the process w^0_t defined in the formulation of Theorem* 3.1 *as* $\varepsilon \to 0$.

We divide the proof of the lemma into several steps.

1. First of all, we prove that the family of processes λ^ε_t is weakly compact in $\mathbf{C}_{0T_0}$. For this it is sufficient to show (cf. Prokhorov [1]) that for any $s, s + h \in [0, T_0]$ we have

$$\mathsf{M}|\lambda^\varepsilon_{s+h} - \lambda^\varepsilon_s|^4 \le Ch^2, \tag{3.9}$$

where C is a constant independent of ε. It is obviously sufficient to establish an analogous estimation for every component of the process

$$\lambda^\varepsilon = (\lambda^{\varepsilon,1}(s), \ldots, \lambda^{\varepsilon,r}(s)).$$

From the assumption concerning the mixing coefficient of ξ_t and condition 2 it follows that estimate (3.4) is applicable to the process $\lambda^{\varepsilon,k}(s)$:

$$\begin{aligned}\mathsf{M}|\lambda^{\varepsilon,k}(s+h) - \lambda^{\varepsilon,k}(s)|^4 &= \varepsilon^2 \mathsf{M} \int_{s/\varepsilon}^{(s+h)/\varepsilon} \cdots \int_{s/\varepsilon}^{(s+h)/\varepsilon} \\ &\quad \times \prod_{i=1}^{i} [b^k(\bar{x}_{\varepsilon s_i}, \xi_{s_i}) - \bar{b}^k(\bar{x}_{\varepsilon s_i})]\, ds_1\, ds_2\, ds_3\, ds_4 \\ &\le \varepsilon^2 C_1 \frac{h^2}{\varepsilon^2} = C_1 h^2.\end{aligned}$$

This implies (3.9) and weak compactness.

2. The hypotheses of Theorem 3.1 imply the relations

$$\lim_{\varepsilon\to 0} \mathsf{M}\lambda_t^\varepsilon = 0; \qquad \lim_{\varepsilon\to 0} \mathsf{M}\lambda^{\varepsilon,k}(t)\lambda^{\varepsilon,l}(t) = \int_0^t A^{kl}(\bar{x}_s)\, ds. \tag{3.10}$$

The first of these equalities follows from condition 2 of Theorem 3.1. From the same condition we obtain that

$$\mathsf{M}\lambda^{\varepsilon,k}(t)\lambda^{\varepsilon,l}(t) = \frac{1}{\varepsilon} \int_0^t \int_0^t g^{kl}(u, s, \varepsilon)\, ds\, du + O_\varepsilon(1), \tag{3.11}$$

where we have used the notation $g^{kl}(u, s, \varepsilon) = \mathsf{M}[b^k(\bar{x}_u, \xi_{u/\varepsilon}) - \mathsf{M}b^k(\bar{x}_u, \xi_{u/\varepsilon})] \times [b^l(\bar{x}_s, \xi_{s/\varepsilon}) - \mathsf{M}b^l(\bar{x}_s, \xi_{s/\varepsilon})]$. In order to derive the second equality in (3.10) from this, we put $H = \{(s, u): 0 \le s \le t, 0 \le u \le t\}$, $\Delta = t/n$, $A_i = \{(s, u): i\Delta \le s \le (i+1)\Delta, i\Delta \le u \le (i+1)\Delta\}$, where n is an integer; $A = \bigcup_{i=0}^{n-1} A_i$ and $B = H \setminus A$. From (3.2) with $\delta = 1$ we obtain the estimate

$$|g^{kl}(u, s, \varepsilon)| \le C_2 \left[\alpha\left(\frac{u-s}{\varepsilon}\right)\right]^{1/5}.$$

From this we obtain

$$\begin{aligned}\int_B g^{kl}(u, s, \varepsilon)\, du\, ds &\le 2C_2 \varepsilon^2 \sum_{i=0}^{n-1} \int_0^{i\Delta/\varepsilon} du \int_{i\Delta/\varepsilon}^{(i+1)\Delta/\varepsilon} ds[\alpha(s-u)]^{1/5} \\ &\le 4C_2 \varepsilon^2 n \int_0^\infty u|\alpha(u)|^{1/5}\, du.\end{aligned} \tag{3.12}$$

Taking account of the boundedness of the derivatives of $b^k(x, \xi)$ and condition 1 of Theorem 3.1, we see that

$$\begin{aligned}\int_A g^{kl}(u, s, \varepsilon)\, du\, ds &= \varepsilon^2 \sum_{i=0}^{n-1} \int_{i\Delta/\varepsilon}^{(i+1)\Delta/\varepsilon} ds \int_{i\Delta/\varepsilon}^{(i+1)\Delta/\varepsilon} du \\ &\quad \times \mathsf{M}[b^k(\bar{x}_{i\Delta}, \xi_s) - \mathsf{M}b^k(\bar{x}_{i\Delta}, \xi_s)] \\ &\quad \times [b^l(\bar{x}_{i\Delta}, \xi_u) - \mathsf{M}b^l(\bar{x}_{i\Delta}, \xi_u)] + O(n\Delta^3) \\ &= \varepsilon\left[\sum_{i=0}^{n-1} A^{kl}(\bar{x}_{i\Delta}) \cdot \Delta + O_\varepsilon(1)\right] + O(1/n^2). \qquad (3.13)\end{aligned}$$

This equality holds as $\varepsilon n \to 0$ and $n \to \infty$. From the boundedness of the derivatives of $b(x, y)$, condition 1 of Theorem 3.1 and the condition of strong mixing it follows that the functions $A^{kl}(x)$ are continuous. Taking account of this continuity and relations (3.11)–(3.13), we find that

$$\mathsf{M}\lambda^{\varepsilon, k}(t)\lambda^{\varepsilon, l}(t) = \int_0^t A^{kl}(\bar{x}_s)\, ds + \gamma_{\varepsilon, k},$$

where $\gamma_{\varepsilon, n} \to 0$ as $\varepsilon n \to 0$ and $\varepsilon n^2 \to \infty$.

3. In conclusion, we show that λ_t^ε converges weakly to the process w_t^0 defined in the formulation of Theorem 3.1. From the weak compactness of the family of measures corresponding to the processes λ_t^ε in the space $\mathbf{C}_{0T_0}$ it follows that every sequence of such processes contains a subsequence converging to some process $\tilde{\lambda}_t$. If we show that the distribution of the limit process $\tilde{\lambda}_t$ does not depend on the choice of the subsequence, then weak convergence will be proved.

It follows from step 2 that $\mathsf{M}\tilde{\lambda}_t = 0$ and the entries of the covariance matrix of the process $\tilde{\lambda}_t = (\tilde{\lambda}^1(t), \ldots, \tilde{\lambda}^r(t))$ have the form $\mathsf{M}\tilde{\lambda}^k(t)\tilde{\lambda}^l(t) = \int_0^t A^{kl}(\bar{x}_s)\, ds$. Moreover, $\tilde{\lambda}_t$ is a process with independent increments. Indeed, let $s_1 \le t_1 \le s_2 \le t_2 \le \cdots \le s_m \le t_m$ be arbitrary nonnegative numbers and write $\Delta = \min_{2 \le k \le m}(s_k - t_{k-1})$. We apply inequality (3.3) to the variables $\eta_l^\varepsilon = \exp\{i(z, \lambda_{t_l}^\varepsilon - \lambda_{s_l}^\varepsilon)\}$, $z \in R^r$:

$$\left| \mathsf{M} \exp\left\{ i\left(z, \sum_{l=1}^m (\lambda_{t_l}^\varepsilon - \lambda_{s_l}^\varepsilon) \right) \right\} - \prod_{l=1}^m \mathsf{M} \exp\{i(z, \lambda_{t_l}^\varepsilon - \lambda_{s_l}^\varepsilon)\} \right| \le (m-1)\alpha(\Delta/\varepsilon).$$

Taking into account that $\lim_{\tau \to \infty} \alpha(\tau) = 0$, we conclude from this that for the limit process $\tilde{\lambda}_t$, the multivariate characteristic function of the vector $(\tilde{\lambda}_{t_1} - \tilde{\lambda}_{s_1}, \tilde{\lambda}_{t_2} - \tilde{\lambda}_{s_2}, \ldots, \tilde{\lambda}_{t_m} - \tilde{\lambda}_{s_m})$ is equal to the product of the characteristic functions of the separate increments. Consequently, the limit process $\tilde{\lambda}_t$ has independent increments.

Therefore, the limit process $\tilde{\lambda}_t$ has continuous trajectories, independent increments, mean zero and the given covariance matrix $(\mathsf{M}\tilde{\lambda}^k(t)\tilde{\lambda}^l(t))$. As is known, these properties determine $\tilde{\lambda}_t$ uniquely and $\tilde{\lambda}_t$ is necessarily Gaussian (cf. Skorokhod [1]) and coincides with w_t^0 (in the sense of distributions). The weak compactness of the family of measures corresponding to the processes λ_t^ε and the fact that this family has a unique limit point imply the weak convergence of λ_t^ε to w_t^0. Lemma 3.1 is proved. □

Now it is very easy to prove the weak convergence of the measure corresponding to the process Z_t^ε to the measure corresponding to ζ_t^0. Indeed, equation (3.8) defines a continuous mapping $G: \lambda^\varepsilon \to Z^\varepsilon$ of $\mathbf{C}_{0T_0}$ into itself. It is clear that if the measure corresponding to λ^ε converges weakly to the measure corresponding to w^0, then the measure corresponding to $Z^\varepsilon = G(\lambda^\varepsilon)$ converges weakly to the measure corresponding to $G(w^0) = \zeta^0$.

Hence we have carried out the first part of our plan. Now we estimate the difference $\zeta_t^\varepsilon - Z_t^\varepsilon = U_t^\varepsilon$. From (3.7) and (3.8), for U_t^ε we obtain the relation

$$U_t^\varepsilon - \int_0^t B(\bar{x}_s, \zeta_{s/\varepsilon})U_s^\varepsilon\, ds = \int_0^t \Phi(s, \varepsilon, \omega)Z_s^\varepsilon\, ds + \int_0^t \Psi(s, \varepsilon, \omega)\, ds. \quad (3.14)$$

Since the entries of the matrix $B(x, y)$ are bounded, relying on Lemma 1.1 of Ch. 2, we conclude from the last equality that

$$|U_t^\varepsilon| \le e^{ct}\left[\left|\int_0^t \Phi(s, \varepsilon, \omega)Z_s^\varepsilon\, ds\right| + \left|\int_0^t \Psi(s, \varepsilon, \omega)\, ds\right|\right], \quad (3.15)$$

where c is a constant. If we show that the family of measures induced by the processes ζ_t^ε in $\mathbf{C}_{0T_0}$ is weakly compact and the right side of the last inequality converges to zero in probability, then the proof of the theorem will be completed.

First we prove that the right side of (3.15) converges to zero in probability as $\varepsilon \to 0$. It follows from the boundedness of the partial derivatives of the functions $b^i(x, y)$ that for some C_1 we have

$$|\Psi(s, \varepsilon, \omega)| \le C_1\sqrt{\varepsilon}|\zeta_s^\varepsilon|^2. \quad (3.16)$$

Taking account of the boundedness of $\mathsf{M}|\lambda_t^\varepsilon|^2$, it is easy to derive from (3.7) that $\mathsf{M}|\zeta_s^\varepsilon|^2 \le C_2 < \infty$ for $s \in [0, T_0]$. This and (3.16) imply the estimate

$$\mathsf{M}\left|\int_0^t \Psi(s, \varepsilon, \omega)\, ds\right| \le C_3 t\sqrt{\varepsilon}. \quad (3.17)$$

Now we estimate the first term on the right side of (3.15). We use the following representation of the solution of problem (3.8) in terms of the Green's function $K(t, s) = (K_j^i(t, s))$:

$$Z_t^\varepsilon = \lambda_t^\varepsilon + \int_0^t K(t, s)\lambda_s^\varepsilon\, ds. \quad (3.18)$$

As is known, $K(t, s)$ is continuously differentiable in the triangle

$$\{(t, s): 0 \le s \le t \le T_0\}.$$

For the norm of the matrix $K(t, s)$ we have the estimate

$$\|K(t, s)\| \le \|B(\bar{x}_s)\| \exp\{\|B(\bar{x}_s)\| \, |t - s|\} < C_4.$$

Using the representation (3.18), we find that

$$\tfrac{1}{2}\mathsf{M}\left|\int_0^t \Phi(s, \varepsilon, \omega) Z_s^\varepsilon \, ds\right|^2 \le \mathsf{M}\left|\int_0^t [B(\bar{x}_s, \xi_{s/\varepsilon}) - B(\bar{x}_s)]\lambda_s^\varepsilon \, ds\right|^2$$
$$+ \mathsf{M}\left|\int_0^t ds[B(\bar{x}_s, \xi_{s/\varepsilon}) - B(\bar{x}_s)] \int_0^s K(s, u)\lambda_u^\varepsilon \, du\right|^2. \tag{3.19}$$

Let us denote by I_1 and I_2 the first and second terms on the right side of (3.19), respectively. Let us put

$$\varphi^k(s/\varepsilon, \varepsilon, \omega) = b^k(\bar{x}_s, \xi_{s/\varepsilon}) - \bar{b}^k(\bar{x}_s),$$

$$\varphi_j^k(s/\varepsilon, \varepsilon, \omega) = \frac{\partial b^k}{\partial x^j}(\bar{x}_s, \xi_{s/\varepsilon}) - \frac{\partial \bar{b}^k}{\partial x^j}(\bar{x}_s).$$

Taking account of the definition of λ_t^ε and applying estimate (3.5), we obtain

$$I_1 \le C(r) \max_{k, j} \mathsf{M}\left|\frac{1}{\sqrt{\varepsilon}} \int_0^t ds \varphi_j^k(s/\varepsilon, \varepsilon, \omega) \int_0^s \varphi^j(u/\varepsilon, \varepsilon, \omega) \, du\right|^2$$
$$= \varepsilon^3 C(r) \max_{k, j} \int_0^{t/\varepsilon} ds_1 \int_0^{t/\varepsilon} ds_2 \int_0^{s_1} du_1 \int_0^{s_2} du_2 \, \mathsf{M}[\varphi_j^k(s_1, \varepsilon, \omega)$$
$$\times \varphi_j^k(s_2, \varepsilon, \omega)\varphi^j(u_1, \varepsilon, \omega)\varphi^j(u_2, \varepsilon, \omega)] \le C(r)C_5\varepsilon^3(t/\varepsilon)^2 = C_6\varepsilon t^2. \tag{3.20}$$

Here $C(r)$ is a constant depending on the dimension of the space.

In order to estimate I_2, we note that the differentiability of the entries $K_j^i(t, s)$ of the matrix-valued Green's function and condition 2 of the theorem imply the estimate

$$\left|\int_{u/\varepsilon}^{t/\varepsilon} \mathsf{M}\varphi_j^i(s, \varepsilon, \omega) K_k^j(\varepsilon s, u) \, ds\right| \le C_7 < \infty$$

for $0 \le u \le t \le T_0$. Using this estimate and inequality (3.5), we obtain

$$\mathsf{M}\left[\frac{1}{\sqrt{\varepsilon}}\int_0^t \varphi_j^i(s/\varepsilon, \varepsilon, \omega)\,ds \int_0^s K_k^j(s, u)\,du \int_0^u \varphi^k(v/\varepsilon, \varepsilon, \omega)\,dv\right]^2$$
$$\le \varepsilon^3 \int_0^t du_1 \int_0^t du_2 \,\mathsf{M} \int_{u_1/\varepsilon}^{t/\varepsilon} ds_1 \int_{u_2/\varepsilon}^{t/\varepsilon} ds_2 \int_0^{u_1/\varepsilon} dv_1 \int_0^{u_2/\varepsilon} dv_2\, \tilde{\varphi}(\varepsilon, s_1, u_1, \omega)$$
$$\times\, \tilde{\varphi}(\varepsilon, s_2, u_2, \omega)\varphi^k(v_1, \varepsilon, \omega)\varphi^k(v_2, \varepsilon, \omega) \le C_8 \varepsilon t^4,$$

where $\tilde{\varphi}(\varepsilon, s, u, \omega) = \varphi_j^i(s, \varepsilon, \omega)K_k^j(\varepsilon s, u)$. These inequalities imply the estimate $I_2 \le C_9\varepsilon$. Taking account of (3.19) and (3.20), we arrive at the inequality

$$\mathsf{M}\left|\int_0^t \Phi(s, \varepsilon, \omega)Z_s^\varepsilon\,ds\right|^2 \le C_{10}\varepsilon, \tag{3.21}$$

which holds for $t \in [0, T_0]$. It follows from (3.15), (3.17) and (3.21) that $\mathsf{M}|U_t^\varepsilon| \to 0$ as $\varepsilon \to 0$.

In order to prove the weak compactness of the family of measures corresponding to the ζ_t^ε, we note that ζ_t^ε and λ_t^ε are connected with the relation

$$\zeta_t^\varepsilon = \lambda_t^\varepsilon + \frac{1}{\sqrt{\varepsilon}}\int_0^t [b(X_s^\varepsilon, \xi_{s/\varepsilon}) - b(\bar{x}_s, \xi_{s/\varepsilon})]\,ds.$$

Taking account of estimate (3.9) and the boundedness of $b(x, y)$, we can easily obtain an estimate for ζ_t^ε, analogous to (3.9):

$$\mathsf{M}|\zeta_{t+h}^\varepsilon - \zeta_t^\varepsilon|^4 \le \tilde{C}h^2.$$

This estimate guarantees the weak compactness of the family of the processes ζ_t^ε, $t \in [0, T_0]$. The weak compactness and the convergence of the finite-dimensional distributions imply the weak convergence of ζ_t^ε to ζ_t^0. This completes the proof of Theorem 3.1. □

In §8 we shall consider some examples of the application of this theorem and now we only make one remark. According to Theorem 3.1 we have

$$\lim_{\varepsilon\to 0} \mathsf{M}F(\zeta^\varepsilon) = \mathsf{M}F(\zeta^0) \tag{3.22}$$

if the functional $F(\varphi)$ is bounded and continuous on $\mathbf{C}_{0T_0}$. For discontinuous functionals, this passage to the limt is impossible in general. Nevertheless, if for the limit process ζ_t^0, the set of points of discontinuity of F has probability zero, then, as is easy to prove, relation (3.22) is preserved. For example, let us be given a domain $D \subset R^r$ with smooth boundary ∂D and let $F(\varphi) = 1$ if $\tau(\varphi) = \inf\{t: \varphi_t \in \partial D\} < T$, and $F(\varphi) = 0$ for the remaining functions in $\mathbf{C}_{0T}(R^r)$. This functional is discontinuous at those φ which reach ∂D but do not leave $D \cup \partial D$ until time T and at those φ for which $\tau(\varphi) = T$. If the matrix of the $A^{ij}(\bar{x}_s)$ is nonsingular, then for the limit process ζ_t^0, the set of trajectories reaching ∂D until time T but not leaving D has probability zero. This follows from the strong Markov property of the process and from the fact that a nondegenerate diffusion process beginning at a point $x \in \partial D$ hits both sides of the smooth surface ∂D before any time $t > 0$ with probability 1. The vanishing of the probability $\mathsf{P}\{\tau(\zeta^0) = T\}$ follows from the existence of the transition probability density of ζ_t^0. Consequently, $\mathsf{P}\{\tau(\zeta^\varepsilon) < T\}$ converges to $\mathsf{P}\{\tau(\zeta^0) < T\}$ as $\varepsilon \to 0$. In particular, choosing the ball of radius δ with center at the point 0 as D, we obtain

$$\lim_{\varepsilon \to 0} \mathsf{P}\left\{\sup_{0 \le t \le T} |X_t^\varepsilon - \bar{x}_t| > \delta\sqrt{\varepsilon}\right\} = \mathsf{P}\{\tau(\zeta^0) > T\}.$$

The last probability can be calculated by solving the corresponding Kolmogorov equation.

We mention another situation in which we encounter approximation by diffusion processes for deviations from trajectories of the averaged system. If we return to the "slow" time in which equation (1.1) is written, then the averaging principle contained in Theorem 2.1 can be formulated in the following way: If condition (2.1) is satisfied, then for any $\delta > 0$ we have

$$\lim_{\varepsilon \to 0} \mathsf{P}\left\{\sup_{0 \le t \le T/\varepsilon} |Z_t^\varepsilon - \bar{x}_{\varepsilon t}| > \delta\right\} = 0,$$

where Z_t^ε is the solution of equation (1.1) and $\bar{x}_t$ is the solution of the averaged system (1.4). In the case $\bar{b}(x) \equiv 0$ this theorem implies that in the time interval $[0, T/\varepsilon]$ the process Z_t^ε does not move away noticeably from the initial position. It turns out that in this case, displacements of order 1 take place over time intervals of order ε^{-2}. Apparently, it was Stratonovich [1], who first called attention to this fact. On the level of the rigor of physics, he established (cf. the same publication) that under certain conditions, the family of processes $Z_{t/\varepsilon^2}^\varepsilon$ converges to a diffusion process and computed the characteristics of the limit process. A mathematically rigorous proof of this result was given by Khas'minskii [5]. A proof is given in Borodin [1] under essentially less stringent assumptions.

Without precisely formulating the conditions, which, in addition to the equality $\bar{b}(x) \equiv 0$, contain some assumptions concerning the boundedness of the derivatives of the $b(x, y)$ and also the sufficiently good mixing of ξ_t and the existence of moments, we include the result here.

Let us introduce the notation

$$a^{ik}(x, s, t) = \mathsf{M}b^i(x, \xi_s)b^k(x, \xi_t),$$

$$B(x, y) = (B^i_j(x, y)) = \left(\frac{\partial b^i(x, y)}{\partial x^j}\right),$$

$$K^i(x, s, t) = \sum_{j=1}^{r} \mathsf{M}B^i_j(x, \xi_s)b^j(x, \xi_t).$$

Suppose that the limits

$$\bar{a}^{ik}(x) = \lim_{T\to\infty} \frac{1}{T} \int_{t_0}^{t_0+T} \int_{t_0}^{t_0+T} a^{ik}(x, s, t)\, ds\, dt,$$

$$\bar{K}^i(x) = \lim_{T\to\infty} \frac{1}{T} \int_{t_0}^{t_0+T} \int_{t_0}^{t_0+T} K^i(x, s, t)\, ds\, dt$$

exist, uniformly in $t_0 > 0$ and $x \in R^r$.

Then on the interval $[0, T]$, the process $\eta_t^\varepsilon = Z^\varepsilon_{t/\varepsilon^2}$ converges weakly to the diffusion process with generating operator

$$L = \frac{1}{2} \sum_{i,k=1}^{r} \bar{a}^{ik}(x) \frac{\partial^2}{\partial x^i \partial x^k} + \sum_{i=1}^{r} \bar{K}^i(x) \frac{\partial}{\partial x^i}.$$

as $\varepsilon \to 0$. A precise formulation and proof can be found in Khas'minskii [5] and Borodin [1].

A natural generalization of the last result is the limit theorem describing the evolution of first integrals of a dynamical system. Consider equation (1.2), and assume that the corresponding averaged system (1.4) has a first integral $H(x) : H(\bar{X}_t) = H(x) =$ constant for $t \geq 0$. Let $H(x)$ be a smooth function with compact connected level sets $C(y) = \{x \in R^r : H(x) = y\}$. Since $X_t^\varepsilon \to \bar{X}_t$ as $\varepsilon \downarrow 0$ uniformly on any finite time interval $[0, T]$ in probability, $\max_{0 \leq t \leq T} |H(X_t^\varepsilon) - H(x)| \to 0$ as $\varepsilon \downarrow 0$ in probability. To observe the evolution of $H(X_t^\varepsilon)$, one should consider time intervals growing together with ε^{-1}. Let us rescale time: $\hat{X}_t^\varepsilon = X^\varepsilon_{t/\varepsilon}$. The functions $\hat{X}_t^\varepsilon$ satisfy the equation

$$\dot{\hat{X}}_t^\varepsilon = \frac{1}{\varepsilon} b(\hat{X}_t^\varepsilon, \xi_{t/\varepsilon^2}), \qquad \hat{X}_0^\varepsilon = x \in R^r. \tag{3.23}$$

We have:

$$H(\hat{X}_t^\varepsilon) - H(x) = \frac{1}{\varepsilon} \int_0^t (\nabla H(\hat{X}_s^\varepsilon), b(\hat{X}_s^\varepsilon, \xi_{s/\varepsilon^2}))\, ds$$

$$= \frac{1}{\varepsilon} \int_0^t (\nabla H(\hat{X}_s^\varepsilon), (b(\hat{X}_s^\varepsilon, \xi_{s/\varepsilon^2}) - \bar{b}(\hat{X}_s^\varepsilon)))\, ds.$$

Here we used the equality $(\nabla H(x), \bar{b}(x)) = 0$, $x \in R^r$, which holds since $H(x)$ is a first integral of system (1.4).

Let ξ_t be a stationary stochastic process with good enough mixing properties. Then the limit in (1.3) is equal to $\bar{b}(x) = \mathsf{M}b(x, \xi_t)$, and, due to the central limit theorem,

$$\frac{1}{\varepsilon}\int_0^t (\nabla H(x), b(x, \xi_{s/\varepsilon^2}) - \bar{b}(x))\, ds$$

converges as $\varepsilon \downarrow 0$ in the distribution to a Gaussian random variable. Of course, the characteristics of this Gaussian variable depend on $x \in R^r$.

Taking into account that the rates of change for $\hat{X}_t^\varepsilon$ and ξ_{t/ε^2} have different order, and that X_t^ε converges to $\overline{X}_t$ as $\varepsilon \downarrow 0$, one can expect that, if the dynamical system $\overline{X}_t$ has some ergodic properties on the level sets $C(y)$, the characteristics of the limit of $H(\hat{X}_t^\varepsilon)$ as $\varepsilon \downarrow 0$ depend only on $H(\hat{X}_t^\varepsilon)$ but not on the position of $\hat{X}_t^\varepsilon$ on $C(H(\hat{X}_t^\varepsilon))$. This means that the process $H(\hat{X}_t^\varepsilon)$, $0 \le t \le T$, should converge to a diffusion process Y_t:

$$dY_t = \sigma(Y_t)\, dW_t + B(Y_t)\, dt, \qquad Y_0 = H(x),$$

where W_t is a standard one-dimensional Wiener process. Thus, the convergence to a diffusion process is the result not just of averaging with respect to the fast oscillating process ξ_{t/ε^2}, but also of the ergodic behavior of the averaged dynamical system $\overline{X}_t$ on the level sets of the first integral.

The result of Stratonovich, Khasminskii, and Borodin is a special case of the limit theorem for first integrals: If $\bar{b}(x) \equiv 0$, then each coordinate x^i is a first integral of the averaged system, and the evolution of $\hat{X}_t^\varepsilon$ is, actually, the evolution of the first integrals. No assumptions concerning the ergodicity on the level sets is needed here, since each level set consists of one point.

To formulate the exact result, we restrict ourselves to the two-dimensional case. Consider equation (3.23) for $r = 2$. Let ξ_t, $-\infty < t < \infty$, be a stationary process with the values in R^m. Assume that the trajectories ξ_t have at most a finite number of simple discontinuities on each finite time interval with probability 1. Equation (1.2) is satisfied at the points where ξ_{t/ε^2} is continuous. The solution is assumed to be continuous for all $t \ge 0$. We assume also that the vector field $b(x, u)$, $x \in R^2$, $u \in R^m$, is Lipschitz continuous and $|b(x, u)|$ is bounded or grows slowly; the function $\bar{b}(x) = \mathsf{M}b(x, \xi_t)$ then is also Lipschitz continuous.

Introduce the following conditions.

1. Let $H(x)$, $x \in R^2$, be a first integral for averaged system (1.4). Assume that $H(x)$ is three times differentiable and the sets $C(y) = \{x \in R^2 : H(x) = y\}$ are closed connected curves in R^2 without self-intersections. This means, in particular, that $H(x)$ has a unique extremum point. We can assume without loss of generality, that this point is the origin

$O \in R^2$, and that $H(O) = 0$, $H(x) > 0$ for $x \neq 0$. The trajectory $\overline{X}_t$ performs periodic motion along $C(y)$, $y = H(\overline{X}_0)$ with a period $T(y)$. Assume that $T(y) \leq c(1 + y^2)$ for $y \geq 0$ and some $c > 0$.

2. Let

$$|H(x)| \leq c(|x|^2 + 1), \qquad |\nabla H(x)| \leq c(|x| + 1),$$
$$\left|\frac{\partial^2 H(x)}{\partial x^i \partial x^j}\right| \leq c, \qquad \left|\frac{\partial^3 H(x)}{\partial x^i \partial x^j \partial x^k}\right| \leq c$$

for some $c > 0$ and for $i, j, k \in \{1, 2\}$, $x \in R^2$. Let

$$|x|^\mu \leq c(H(x) + 1)$$

for some $1 < \mu \leq 2$.

3. Put $g(x, z) = b(x, z) - \bar{b}(x)$. Suppose a positive function $q(z)$ and a constant $c > 0$ exist such that

$$|\bar{b}(x)| \leq c(|x| + 1), \qquad |\nabla \bar{b}(x)| \leq c,$$
$$\left|\frac{\partial^2 \bar{b}(x)}{\partial x^i \partial x^j}\right| \leq c, \qquad |g(x, z)| \leq cq(z),$$
$$\left|\frac{\partial g(x, z)}{\partial x^i}\right| \leq \frac{c}{1 + |x|} q(z), \qquad i, j \in \{1, 2\}, x \in R^2, z \in R^m.$$

Let, for some $p > (4(\mu + 1))/(\mu - 1)$,

$$\mathsf{M}|q(\xi_t)|^p \leq c.$$

The next assumption concerns the mixing properties of the process $\xi_t(\omega)$, $\omega \in \Omega$, $-\infty < t < \infty$. Let $\mathscr{F}_s^t$ be the σ-field generated by the process ξ_v when $-\infty \leq s \leq v \leq t \leq \infty$. Define

$$\beta(\tau) = \sup_{B \in \mathscr{F}_{-\infty}^0 \times \mathscr{F}_\tau^\infty} |(P_{0,\tau}(B) - (P_0 \times P_\tau)(B))|,$$

where the measures $P_{0,\tau}$ and $P_0 \times P_\tau$ on the space $\Omega \times \Omega$ are defined by the relations:

$$P_{0,\tau}(A_1 \times A_2) = P(A_1 A_2),$$
$$(P_0 \times P_\tau)(A_1 \times A_2) = P(A_1)P(A_2),$$

for $A_1 \in F_{-\infty}^0$, $A_2 \in F_\tau^\infty$.

We say that the process ξ_t satisfies the absolute regularity mixing condition with the coefficient $\beta(\tau)$, if $\beta(\tau) \downarrow 0$ as $\tau \to \infty$.

It is known that the absolute regularity mixing condition is stronger than mixing condition (3.1) (strong mixing condition). Note that since ξ_t is a stationary process, the supremum in t in (3.1) can be omitted. Some sufficient conditions for these types of mixing properties and bounds for the coefficients $\alpha(\tau)$ and $\beta(\tau)$ can be found in Ibragimov and Rozanov [1].

4. Assume that the stationary process ξ_t satisfies the absolute regularity mixing conditions with the coefficient $\beta(\tau)$ such that

$$\int_0^\infty \tau^3 \beta^{(p-8)/p}(\tau)\, d\tau < \infty, \qquad \beta^{(p-8)/p}(\tau) < c \cdot \min(1, \tau^{-4}) \tag{3.24}$$

for some $c > 0$ and the constant p defined in condition 3.

If

$$b(x,z) = \sum_{k=1}^{n} u_k(x) v_k(z), \qquad n < \infty,$$

one can replace the absolute regularity condition by the strong mixing condition with the coefficient $\alpha(\tau)$ satisfying (3.24) with $\beta(\tau)$ replaced by $\alpha(\tau)$.

Define

$$\begin{aligned} g(x,z) &= b(x,z) - \bar{b}(x), \\ F(x,z) &= (\nabla H(x), g(x,z)), \\ D(x,s) &= \mathsf{M} F(x,\xi_s) F(x,\xi_0), \\ Q(x,s) &= \mathsf{M}(\nabla F(x,\xi_s), g(x,\xi_0)), \end{aligned}$$

where ∇ is the gradient in $x \in R^2$. Let

$$\begin{aligned} D(x) &= 2\int_0^\infty D(x,s)\, ds, \\ Q(x) &= 2\int_0^\infty Q(x,s)\, ds, \\ \sigma^2(y) &= \frac{1}{T(y)} \oint_{C(y)} \frac{D(x)\, dl}{|b(x)|}, \\ B(y) &= \frac{1}{T(y)} \oint_{C(y)} \frac{Q(x)\, dl}{|b(x)|}, \end{aligned}$$

where dl is the length element on $C(y) = \{x \in R^2 : H(x) = y\}$ and $T(y) = \oint_{C(y)} (dl/|b(x)|)$ is the period of the rotation along $C(y)$.

5. Assume that the functions $\sigma^2(y)$ and $B(y)$ are Lipschitz continuous.

Theorem 3.2 (Borodin and Freidlin [1]). *Let $Y_t^\varepsilon = H(\hat{X}_t^\varepsilon)$, where $\hat{X}_t^\varepsilon$ is the solution of* (3.22) *and $H(x)$ is the first integral of averaged equation* (1.4). *Suppose that conditions* 1–5 *hold. Then for any $0 < T < \infty$, the processes Y_t^ε, $0 \le t \le T$, converge weakly in the space of continuous functions $\phi : [0, T] \to R^1$ to the diffusion process Y_t defined by the equation*

$$dY_t = \sigma(Y_t)\, dW_t + B(Y_t)\, dt, \qquad Y_0 = H(x).$$

The proof of this theorem and some of its modifications and generalizations can be found in Borodin and Freidlin [1]. Some examples related to Theorem 3.2 are considered in §8.

We assumed in Theorem 3.2 that the first integral $H(x)$ has just one critical point. This assumption is essential. If for some y the level sets $C(y)$ consist of more than one connected component, the processes $Y_t^\varepsilon = H(\hat{X}_t^\varepsilon)$, will not, in general, converge to a Markov process. To have in the limit a Markov process, one should extend the phase space. We consider such questions in the next chapter for white noise perturbations of dynamical systems. The situation is similar in the case of fast oscillating perturbations.

§4. Large Deviations from an Averaged System

We have established that for small ε, over the time $[0, T]$ the process X_t^ε is near the trajectory $\bar{x}_t$ of the averaged system with overwhelming probability and the normalized deviations $(1/\sqrt{\varepsilon})(X_t^\varepsilon - \bar{x}_t)$ form a random process, which converges weakly to a Gaussian Markov process as $\varepsilon \to 0$. If the averaged dynamical system has an asymptotically stable equilibrium position or limit cycle and the initial point $X_0^\varepsilon = x$ is situated in the domain of attraction of this equilibrium position or cycle, then it follows from the above results that with probability close to 1 for ε small, the trajectory of X_t^ε hits the neighborhood of the equilibrium position or the limit cycle and spends an arbitrarily long time T in it provided that ε is sufficiently small. By means of Theorem 3.1 we can estimate the probability that over a fixed time $[0, T]$ the trajectory of X_t^ε does not leave a neighborhood D^ε of the equilibrium position if this neighborhood has linear dimensions of order $\sqrt{\varepsilon}$. However, the above results do not enable us to estimate, in a precise way, the probability that over the time $[0, T]$ the process X_t^ε leaves a given neighborhood, independent of ε, of the equilibrium position. We can only say that this probability converges to zero. Theorems 2.1 and 3.1 do not enable us to study events determined by the behavior of X_t^ε on time intervals increasing with ε^{-1}. For example, by means of these theorems we cannot estimate the time spent by X_t^ε in a neighborhood D of an asymptotically stable equilibrium position until the first exit time of D. Over a sufficiently long time, X_t^ε goes from the neighborhood of one equilibrium position of the averaged system to neighborhoods of others. These passages take place "in spite of" the averaged motion, due to prolonged deviations of ξ_t from its "typical" behavior. In one word, the situation here is completely analogous to that confronted in Chs. 4–6: in order to study all these questions, we need theorems on large deviations for the family of processes X_t^ε. For this family we are going to introduce an action functional and by means of it we shall study probabilities of events having small probability for $\varepsilon \ll 1$ and also the behavior of the process on time intervals increasing with decreasing ε. These results were obtained in Freidlin [9], [11].

In what follows we assume for the sake of simplicity that not only are the partial derivatives of the $b^i(x, y)$, $x \in R^r$, $y \in R^r$, bounded but so are the $b^i(x, y)$ themselves:

$$\sup_{i, j, x \in R^r, y \in R^l} \left(|b^i(x, y)| + \left| \frac{\partial b^i}{\partial x^j}(x, y) \right| + \left| \frac{\partial b^i}{\partial y^j}(x, y) \right| \right) < K < \infty.$$

The assumption that $|b(x, y)|$ is bounded could be replaced by an assumption on the finiteness of some exponential moments of $|b(x, \xi_t)|$ but this would lengthen the proofs.

We shall say that condition F is satisfied if there exists a numerical-valued function $H(x, \alpha)$, $x \in R^r$, $\alpha \in R^r$, such that

$$\lim_{\varepsilon \to 0} \varepsilon \ln \mathsf{M} \exp\left\{ \frac{1}{\varepsilon} \int_0^T (\alpha_s, b(\varphi_s, \xi_{s/\varepsilon}))\, ds \right\} = \int_0^T H(\varphi_s, \alpha_s)\, ds, \tag{4.1}$$

for any step functions φ_s, α_s on the interval $[0, T]$ with values in R^r.

If as φ_s and α_s we choose constants $\varphi, \alpha \in R^r$, then we obtain from (4.1) that

$$\lim_{T \to \infty} \frac{1}{T} \ln \mathsf{M} \exp\left\{ \int_0^T (\alpha, b(\varphi, \xi_s))\, ds \right\} = H(\varphi, \alpha). \tag{4.2}$$

Lemma 4.1. *The function $H(x, \alpha)$ is jointly continuous in its variables and convex in the second argument.*

Indeed, it follows from (4.2) that

$$|H(x + \Delta, \alpha + \delta) - H(x, \alpha)| \le K|\delta| + K|\alpha|\,|\Delta|$$

and therefore, continuity is proved. The convexity in α also follows from (4.2) if we take account of the convexity of the exponential function and the montonicity and concavity of the logarithmic function.

We define the function $L(x, \beta)$ as the Legendre transform of $H(x, \alpha)$ in the variables α:

$$L(x, \beta) = \sup_{\alpha}[(\alpha, \beta) - H(x, \alpha)].$$

The function $L(x, \beta)$ is convex in β and jointly lower semicontinuous in all variables; it assumes nonnegative values including $+\infty$. It follows from the boundedness of $b(x, y)$ that $L(x, \beta) = +\infty$ outside some bounded set in the space of the variables β.

The function $L(x, \beta)$ is jointly lower semicontinuous in all variables. Indeed, it follows from the definition of $L(x, \beta)$ that for any $x, \beta \in R^r$ and $n > 0$ there exists $\alpha_n = \alpha_n(x, \beta)$ such that

$$L(x, \beta) < (\alpha_n, \beta) - H(x, \alpha_n) + 1/n.$$

Taking account of the continuity of $H(x, \alpha)$, from this we obtain that for some $\delta_n = \delta_n(x, \beta, \alpha_n)$ and $|x - x'| < \delta_n$, $|\beta - \beta'| < \delta_n$ we have

$$(\alpha_n, \beta) - H(x, \alpha_n) < (\alpha_n, \beta') + H(x', \alpha_n) + 1/n \leq L(x', \beta') + 1/n.$$

Consequently, $L(x, \beta) < L(x', \beta') + 2/n$ if $|x - x'| < \delta_n$ and $|\beta - \beta'| < \delta_n$, i.e., $L(x, \beta)$ is jointly lower semicontinuous in all variables.

Remark 1. Condition F is equivalent to the assumption that the limit (4.1) exists for every continuous φ_s, α_s.

Remark 2. In general, the variables (x, α) and (x, β) vary in different spaces. If equation (1.2) is considered on a manifold G, then x is a point in the manifold, α is an element of the cotangent space at x, (x, α) is a point of the cotangent bundle and (x, β) is a point of the tangent bundle.

On $\mathbf{C}_{0T}(R^r)$ we introduce a functional $S_{0T}(\varphi)$:

$$S_{0T}(\varphi) = \int_0^T L(\varphi_s, \dot{\varphi}_s)\, ds$$

if φ_s is absolutely continuous; we put $S_{0T}(\varphi) = +\infty$ for the remaining elements φ of $\mathbf{C}_{0T}(R^r)$.

Lemma 4.2. *For any compactum $F_0 \subset R^r$ and any $s < \infty$, the set $\Phi_{F_0}(s) = \{\varphi \in \mathbf{C}_{0T}(R^r): \varphi_0 \in F_0, S_{0T}(\varphi) \leq s\}$ is compact in $\mathbf{C}_{0T}(R^r)$. The functional $S_{0T}(\varphi)$ is lower semicontinuous in $\mathbf{C}_{0T}(R^r)$.*

Proof. Since $L(x, \beta)$ is equal to $+\infty$ outside a bounded set in the space of the variables β, the set $\Phi_{F_0}(s)$ may only contain functions whose derivatives are uniformly bounded. Taking account of the compactness of F_0, it follows from this that all functions in $\Phi_{F_0}(s)$ are uniformly bounded and equicontinuous. Consequently, for the proof of the compactness of $\Phi_{F_0}(s)$ we only have to show that $\Phi_{F_0}(s)$ is closed. The closedness of $\Phi_{F_0}(s)$ obviously follows from the lower semicontinuity of S_{0T}. As in §2, Ch. 5, concerning semicontinuity we refer to the book [1] by Ioffe and Tichomirov. □

Theorem 4.1 (Freidlin [9], [11]). *Let condition F be satisfied and let $H(x, \alpha)$ be differentiable with respect to α. The functional $S_{0T}(\varphi)$ is the normalized action functional in $\mathbf{C}_{0T}(R^r)$ for the family of processes X_t^ε as $\varepsilon \to 0$, the normalizing coefficient being $f(\varepsilon) = \varepsilon^{-1}$, i.e., the set*

$$\Phi_x(s) = \{\varphi \equiv \mathbf{C}_{0T}(R^r): \varphi_0 = x, S_{0T}(\varphi) \leq s\}$$

is compact in $\mathbf{C}_{0T}(R^r)$ and for any $s, \delta, \gamma > 0$ and $\varphi \in \mathbf{C}_{0T}(R^r)$, $\varphi_0 = x$, there exists $\varepsilon_0 > 0$ such that for $\varepsilon < \varepsilon_0$ we have

$$\mathsf{P}\{\rho_{0T}(X^\varepsilon, \varphi) < \delta\} \geq \exp\{-\varepsilon^{-1}(S_{0T}(\varphi) + \gamma)\}, \tag{4.3}$$

$$\mathsf{P}\{\rho_{0T}(X^\varepsilon, \Phi_x(s)) > \delta\} \leq \exp\{-\varepsilon^{-1}(s - \gamma)\}, \tag{4.4}$$

where X_t^ε is the solution of equation (1.2) *with the initial condition $X_0^\varepsilon = x$.*

We postpone the proof of this theorem until the next section and now discuss some consequences of it and the verification of the conditions of the theorem for some classes of processes. First of all we note that for any set $A \subseteq \mathbf{C}_{0T}^x(R^r) = \{\varphi \in \mathbf{C}_{0T}(R^r): \varphi_0 = x\}$ we have

$$\begin{aligned} -\inf_{\varphi \in (A)} S(\varphi) &\leq \varliminf_{\varepsilon \to 0} \varepsilon \ln \mathsf{P}\{X^\varepsilon \in A\} \\ &\leq \varlimsup_{\varepsilon \to 0} \varepsilon \ln \mathsf{P}[X^\varepsilon \in A] \leq -\inf_{\varphi \in [A]} S(\varphi), \end{aligned} \tag{4.5}$$

where $[A]$ is the closure of A and (A) is the interior of A in $\mathbf{C}_{0T}^x(R^r)$. Estimates (4.5) follow from general properties of an action functional (cf. Ch. 3). If the infima in (4.5) over $[A]$ and (A) coincide, then (4.5) implies the relation

$$\lim_{\varepsilon \to 0} \varepsilon \ln \mathsf{P}\{X^\varepsilon \in A\} = -\inf_{\varphi \in A} S_{0T}(\varphi). \tag{4.6}$$

We would like to mention that because of the boundedness of $|b(x, \xi_s)|$ and the possible degeneracy of the random variables $b(x, \xi_s)$, the condition of coincidence of the infima of $S_{0T}(\varphi)$ over the sets $[A]$ and (A) is more stringent than, say, in the case of a functional corresponding to an additive perturbation of the type of a white noise (cf. Ch. 4).

It follows from the compactness of $\Phi_x(s)$ that there exists $\varepsilon_0 > 0$ such that estimate (4.3) holds for $\varepsilon < \varepsilon_0$ for any function $\varphi \in \Phi_x(s)$. Since $b(x, y)$ satisfies a Lipschitz condition, for any $\varphi^x \in \mathbf{C}_{0T}^x(R^r)$, $\varphi^y \in \mathbf{C}_{0T}^y(R^r)$ and $\delta > 0$ we have

$$\{\omega: \rho_{0T}(X^\varepsilon, \varphi^x) < \delta\} \subseteq \{\omega: \rho_{0T}(Y^\varepsilon, \varphi^y) < (e^{KT} + 2)\delta\},$$

if $\rho_{0T}(\varphi^x, \varphi^y) < \delta$, where $X^\varepsilon = X_t^\varepsilon$ and $Y^\varepsilon = Y_t^\varepsilon$ are solutions of equation (1.2) with initial conditions x and y, respectively and K is the Lipschitz constant. Relying on estimates (4.3) and (4.5), from this we obtain that if $\rho(\varphi^x, \varphi^y) < \delta$, then

$$\inf_{\psi \in C_{0T}^y(R^r),\, \rho(\psi, \varphi^y) < \delta'} S_{0T}(\psi) \leq S_{0T}(\varphi^x),$$

where $\delta' = (e^{KT} + 2)\delta$. Taking account of the last inequality, one can easily see that

$$\{\rho(X^\varepsilon, \Phi_x(s)) < \delta\} \subseteq \{\rho(Y^\varepsilon, \Phi_y(s)) < \delta'\}$$

for $|x - y| \leq \delta$. With use of these estimates and inclusions, it is easy to derive the following "uniform" version of Theorem 4.1′ from Theorem 4.1.

Theorem 4.1′. *Suppose that condition F is satisfied and $H(x, \alpha)$ is differentiable with respect to α. The functional $\varepsilon^{-1}S_{0T}(\varphi)$ is the action functional for the family of processes X_t^ε, uniformly with respect to the initial point x in any compactum $Q \subset R^r$, as $\varepsilon \to 0$. This means that the assertions of Theorem 4.1 hold and for any s, δ, $\gamma > 0$ and any compactum $Q \subset R^r$ there exists $\varepsilon_0 > 0$ such that inequalities (4.3) and (4.4) hold for every initial point $x \in Q$ and every $\varphi \in \Phi_x(s)$.*

For a large class of important events, the infimum in (4.6) can be expressed in terms of the function $u_x(t, z) = \inf\{S_{0T}(\varphi): \varphi_0 = x, \varphi_t = z\}$. As a rule, the initial point $\varphi_0 = x$ is assumed to be fixed, and therefore, we shall omit the subscript x in $u_x(t, z)$. As is known, the function $u(t, z)$ satisfies the Hamilton–Jacobi equation. Since the Legendre transformation is involutive, the Hamilton–Jacobi equation for $u(t, z)$ has the form (cf., for example, Gel'fand and Fomin [1])

$$\frac{\partial u(t, z)}{\partial t} - H\left(x; \frac{\partial u}{\partial x^1}, \ldots, \frac{\partial u}{\partial x^n}\right) = 0. \tag{4.7}$$

The functional $S_{0T}(\varphi)$ vanishes for trajectories of the averaged system. Indeed, from the concavity of $\ln x$ it follows that

$$\begin{aligned} H(x, \alpha) &= \lim_{T\to\infty} \frac{1}{T} \ln \mathsf{M} \exp\left\{\int_0^T (\alpha, b(x, \xi_s))\, ds\right\} \\ &\geq \lim_{T\to\infty} \frac{1}{T} \mathsf{M} \int_0^T (\alpha, b(x, \xi_s))\, ds \\ &= \left(\alpha, \lim_{T\to\infty} \frac{1}{T} \int_0^T \mathsf{M} b(x, \xi_s)\, ds\right) = (\alpha, \bar{b}(x)), \end{aligned}$$

and consequently, $L(\varphi, \dot{\varphi}) = \sup_\alpha[(\dot{\varphi}, \alpha) - H(\varphi, \alpha)] = 0$ for $\dot{\varphi} = \bar{b}(\varphi)$. It follows from the differentiability of $H(x, \alpha)$ with respect to α that the action functional vanishes only at trajectories of the averaged system. We take for the set $A = \{\varphi \in \mathbf{C}_{0T}^x(R^r): \sup_{0\leq t\leq T}|\varphi_t - \bar{x}_t| > \delta\}$. Then we conclude from (4.5) that for any $\delta > 0$ we have

$$\mathsf{P}\left\{\sup_{0\leq t\leq T} |X_t^\varepsilon - \bar{x}_t| > \delta\right\} \leq \exp\{-c\varepsilon^{-1}\}$$

for sufficiently small ε, where c is an arbitrary number less than $\inf_{\varphi\in[A]} S(\varphi)$. This infimum is positive, since $S(\varphi) > 0$ for $\varphi \in [A]$ and S is lower semi-continuous. Consequently, if the hypotheses of Theorem 4.1 are satisfied, then the probability of deviations of order 1 from the trajectory of the averaged system is exponentially small.

In §8 we consider some examples of the application of Theorem 4.1 and now discuss the problem of fulfilment of the hypotheses of Theorem 4.1 in the case where ξ_t is a Markov process.

Lemma 4.3. *Suppose that ξ_t is a homogeneous Markov process with values in $D \subseteq R^r$ and for any $x, \alpha \in R^r$ let*

$$\lim_{T\to\infty} \frac{1}{T} \ln \mathsf{M}_y \exp\left\{\int_0^T (\alpha, b(x, \xi_s))\, ds\right\} = H(x, \alpha) \tag{4.8}$$

uniformly in $y \in D$. Then condition F is satisfied.

Proof. Let α_s and z_s be step functions and let α_k and z_k be their values on $[t_{k-1}, t_k)$; $0 \leq t_0 < t_1 < t_2 < \cdots < t_n = T$, respectively. Using the Markov property, we can write

$$\begin{aligned}
\lim_{\varepsilon\to 0} \varepsilon \ln \mathsf{M}_y \exp\left\{\frac{1}{\varepsilon}\int_0^T (\alpha_s, b(z_s, \xi_{s/\varepsilon}))\, ds\right\}
&= \lim_{\varepsilon\to 0} \varepsilon \ln \mathsf{M}_y \exp\left\{\frac{1}{\varepsilon}\int_0^{t_1} (\alpha_1, b(z_1, \xi_{s/\varepsilon}))\, ds\right\} \\
&\quad \times \mathsf{M}_{\xi_{t_1/\varepsilon}} \exp\left\{\frac{1}{\varepsilon}\int_0^{t_2-t_1} (\alpha_2, b(z_2, \xi_{s/\varepsilon}))\, ds\right\} \cdots \\
&\quad \times \mathsf{M}_{\xi_{t_{n-1}/\varepsilon}} \exp\left\{\frac{1}{\varepsilon}\int_0^{t_n-t_{n-1}} (\alpha_n, b(z_n, \xi_{s/\varepsilon}))\, ds\right\}.
\end{aligned}$$

From (4.8) it follows that

$$\left|\varepsilon \ln \mathsf{M}_y \exp\left\{\frac{1}{\varepsilon}\int_0^{t_k-t_{k-1}} (\alpha_k, b(z_k, \xi_{s/\varepsilon}))\, ds\right\} - H(z_k, \alpha_k)(t_k - t_{k-1})\right| < \delta_k, \tag{4.9}$$

where $\delta_k \to 0$ uniformly in $y \in D$ as $\varepsilon \to 0$. Repeating this estimation on every interval $[t_{k-1}, t_k)$, we obtain from (4.9) that

$$\left|\varepsilon \ln \mathsf{M} \exp\left\{\frac{1}{\varepsilon}\int_0^T (\alpha_s, b(z_s, \xi_{s/\varepsilon}))\, ds - \sum_{k-1}^{n} H(z_k, \alpha_k)(t_k - t_{k-1})\right\}\right| < \sum_{k=1}^{n} \delta_k.$$

This relation implies condition F. □

For Markov processes we can formulate general conditions of Feller type and on the positivity of transitional probabilities which guarantee the validity of condition F and differentiability of $H(x, \alpha)$ with respect to α. We shall not discuss the general case but rather the case where ξ_t is a Markov process with a finite number of states. In §9 we consider the case of a diffusion process ξ_t.

Let ξ_t, $t \geq 0$, be a homogeneous stochastically continuous Markov process with N states $\{1, 2, \ldots, N\}$, let $p_{ij}(t)$ be the probability of passage from i to j over time t, and let $P(t) = (p_{ij}(t))$. We denote by $Q = (q_{ij})$ the matrix consisting of the derivatives $dp_{ij}(t)/dt$ at $t = 0$; as in known, these derivatives exist.

Theorem 4.2. *Suppose that all entries of Q are different from zero. Let us denote by $Q^{\alpha,x} = (q_{ij}^{\alpha,x})$ the matrix whose entries are given by the equalities $q_{ij}^{\alpha,x} = q_{ij} + \delta_{ij}\cdot(\alpha, b(x, i))$, where $\delta_{ij} = 1$ for $i = j$ and $\delta_{ij} = 0$ for $i \neq j$. Then $Q^{\alpha,x}$*

has a simple real eigenvalue $\lambda = \lambda(x, \alpha)$ exceeding the real parts of all other eigenvalues. This eigenvalue is differentiable with respect to α. Condition F is satisfied and $H(x, \alpha) = \lambda(x, \alpha)$.

Proof. If ξ_t is a Markov process, then the family of operators T_t, $t \geq 0$, acting on the set of bounded measurable functions on the phase space of ξ_t according to the formula

$$T_t f(z) = \mathsf{M}_z f(\xi_t) \exp\left\{\int_0^t (\alpha, b(x, \xi_s))\, ds\right\},$$

forms a positive semigroup. In our case the phase space consists of a finite number of points and the semigroup is a semigroup of matrices acting in the N-dimensional space of vectors $f = (f(1), \ldots, f(N))$. It is easy to calculate the infinitesimal generator A of this semigroup:

$$A = \lim_{t \downarrow 0} \frac{T_t - E}{t} = (q_{ij} + \delta_{ij} \cdot (\alpha, b(x, i))) = Q^{\alpha, x}.$$

By means of the infinitesimal generator the semigroup T_t can be represented in the form $T_t = \exp\{tQ^{\alpha, x}\}$. Since by assumption $q_{ij} \neq 0$, the entries of the matrix T_t are positive if $t > 0$. By Frobenius' theorem (Gantmakher [1]), the eigenvalue with largest absolute value $\mu = \mu(t, x, \alpha)$ of such a matrix is real, positive and simple. To it there corresponds an eigenvector $\mathbf{e}(t, x, \alpha) = (e_1, \ldots, e_N)$, $\sum_{k=1}^N e_k = 1$, all of whose components are positive.

It is easy to derive from the semigroup property of the operators T_t that $\mathbf{e}(t, x, \alpha)$ does not actually depend on t and is an eigenvector of the matrix $Q^{x, \alpha}$, i.e., $Q^{x, \alpha}\mathbf{e}(x, \alpha) = \lambda(x, \alpha)\mathbf{e}(x, \alpha)$. The corresponding eigenvalue $\lambda(x, \alpha)$ is real, simple, exceeds the real parts of all other eigenvalues of $Q^{x, \alpha}$ and $\mu(t, x, \alpha) = \exp\{t \cdot \lambda(x, \alpha)\}$. The differentiability of $\lambda(x, \alpha)$ with respect to α follows from the differentiability of the entries of $Q^{x, \alpha}$ and the simplicity of the eigenvalue $\lambda(x, \alpha)$ (cf. Kato [1]).

By Lemma 4.3, in order to complete the proof of the theorem it is sufficient to show that

$$\lim_{T \to \infty} \frac{1}{T} \ln \mathsf{M}_i \exp\left\{\int_0^T (\alpha, b(x, \xi_s))\, ds\right\} = \lambda(x, \alpha)$$

for $i = 1, 2, \ldots, N$. This equality can obviously be rewritten in the following equivalent form:

$$\lim_{T \to \infty} \frac{1}{T} \ln(T_t \mathbf{1})(i) = \lambda(x, \alpha), \tag{4.10}$$

where $\mathbf{1}$ is the vector with components equal to one.

To prove (4.10) we use the fact that all components of the eigenvector $\mathbf{e} = \mathbf{e}(x, \alpha) = (e_1, \ldots, e_N), \sum_{k=1}^N e_k = 1$, are positive: $0 < c < \min_{1 \le k \le N} e_k \le \max_{1 \le k \le N} e_k \le 1$. Using the positivity of the semigroup T_t, we can conclude that

$$c(T_t \mathbf{1})(i) < (T_t \mathbf{e})(i) = e^{\lambda(x, \alpha)t} e_i \le (T_t \mathbf{1})(i)$$

for $i = 1, 2, \ldots, N$. We take the logarithm of this relation and divide by t:

$$\frac{1}{t} \ln c + \frac{1}{t} \ln(T_t \mathbf{1})(i) < \frac{1}{t} \ln(T_t \mathbf{e})(i)$$

$$= \lambda(x, \alpha) + \frac{1}{t} \ln e_i \le \frac{1}{t} \ln(T_t \mathbf{1})(i).$$

Letting t tend to ∞ in this chain of inequalities, we obtain (4.9). Theorem 4.2 is proved. □

We note that in the case considered in Theorem 4.2, an equation can be written for the function $u(t, z)$ introduced above, without determining the eigenvalue $\lambda(x, \alpha)$ of $Q^{x,\alpha}$ which has the largest real part. Indeed, $\lambda(x, \alpha) = \lambda$ is a root of the characteristic equation

$$\det(q_{ij} + \delta_{ij}[(\alpha, b(x, i)) - \lambda]) = 0. \tag{4.11}$$

Since by (4.7) and Theorem 4.2, $u(t, z)$ satisfies the Hamilton–Jacobi equation $(\partial u/\partial t)(t, z) = \lambda(z, \nabla_z u)$, the function $u(t, z)$ also has to satisfy the equation

$$\det\left(q_{ij} - \delta_{ij}\left[(\nabla_z u(t, z), b(z, i)) - \frac{\partial u}{\partial t}\right]\right) = 0,$$

where we have to choose that solution of the equation for which $\partial u/\partial t$ is a root of equation (4.11) for $\alpha = \nabla_z u(t, z)$, having the largest real part.

§5. Large Deviations Continued

In this section we prove Theorem 4.1.

We choose a small number $\Delta > 0$ such that $T/\Delta = n$ is an integer. Let $\psi_t: [0, T] \to R^r$ be a piecewise constant right continuous function having points of discontinuity only at points of the form $k\Delta$, $k = 1, 2, \ldots, n - 1$. Let us consider the family of random processes

$$\tilde{X}_t^{\varepsilon, \psi} = x + \int_0^t b(\psi_s, \xi_{s/\varepsilon})\, ds$$

and write $\tilde{S}_{0T}(\varphi) = \tilde{S}^{\psi}_{0T}(\varphi) = \int_0^T L(\psi_s, \dot{\varphi}_s)\, ds$ if φ_s is absolutely continuous and $\tilde{S}_{0T}(\varphi) = +\infty$ for the remaining $\varphi \in \mathbf{C}_{0T}(R^r)$. The functional $\tilde{S}_{0T}(\varphi)$ is lower semicontinuous; the set $\Phi_x(s) = \{\varphi \in \mathbf{C}_{0T}(R^r): \tilde{S}_{0T}(\varphi) \le s, \varphi_0 = x\}$ is compact in $\mathbf{C}_{0T}(R^r)$. This can be proved in exactly the same way as in Lemma 4.2.

Moreover, we note that the functional $\tilde{S}^{\psi}(\varphi)$ is lower semicontinuous in ψ in the topology of uniform convergence for every φ. This follows easily from Fatou's lemma and the joint lower semicontinuity of $L(x, \beta)$ in all variables: if $\psi^n \to \psi$, then

$$\varliminf_{n\to\infty} \tilde{S}^{\psi(n)}_{0T}(\varphi) = \varliminf_{n\to\infty} \int_0^T L(\psi^n_s, \dot{\varphi}_s)\, ds$$

$$\ge \int_0^T \varliminf_{n\to\infty} L(\psi^n_s, \dot{\varphi}_s)\, ds \ge \int_0^T L(\psi_s, \dot{\varphi}_s)\, ds = \tilde{S}^{\psi}_{0T}(\varphi),$$

Lemma 5.1. *Let condition F be satisfied and let the function $H(x, \alpha)$ be differentiable with respect to the variables α.*

The functional $\tilde{S}^{\psi}(\varphi)$ is the normalized action functional in $\mathbf{C}_{0T}(R^r)$ for the family of processes $\tilde{X}^{\varepsilon,\psi}_t$ as $\varepsilon \to 0$, with normalizing coefficient $f(\varepsilon) = \varepsilon^{-1}$.

Proof. Let $\alpha_1, \alpha_2, \ldots, \alpha_n \in R^r$. We denote by $\alpha(s)$ the piecewise constant function on $[0, T]$ which assumes the value $\sum_{k=i}^n \alpha_k$ for $s \in ((i-1)\Delta, i\Delta]$, $i = 1, 2, \ldots, n$. The function $h^x_\varepsilon(\alpha_1, \alpha_2, \ldots, \alpha_n)$, where x is the initial condition $\tilde{X}^\varepsilon_0 = x$, is defined by the equality

$$h^x_\varepsilon(\alpha_1, \ldots, \alpha_n) = \varepsilon \ln \mathsf{M} \exp\left\{\varepsilon^{-1} \sum_{k=1}^n (\alpha_k, \tilde{X}^{\varepsilon,\psi}_{k\Delta})\right\}.$$

Taking account of the definition of $\tilde{X}^{\varepsilon,\psi}_t$, we may write

$$h^x_\varepsilon(\alpha_1, \ldots, \alpha_n) = \varepsilon \ln \mathsf{M} \exp\left\{\varepsilon^{-1} \int_0^T (\alpha(s), b(\psi_s, \xi_{s/\varepsilon}))\, ds\right\} + \left(x, \sum_{k=1}^n \alpha_k\right).$$

This and condition F imply the existence of the limit $h^x(\alpha_1, \ldots, \alpha_n) = \lim_{\varepsilon\to 0} h^x_\varepsilon(\alpha_1, \ldots, \alpha_n)$ and the equality

$$h^x(\alpha_1, \ldots, \alpha_n) = \int_0^T H(\psi_s, \alpha(s))\, ds + \left(x, \sum_{k=1}^n \alpha_k\right).$$

It is easy to see that the function $h^x(\alpha_1, \ldots, \alpha_n)$ is convex in the variables $\alpha_1, \ldots, \alpha_n$. The differentiability of $H(x, \alpha)$ with respect to the second argument implies the differentiability of $h^x(\alpha_1, \ldots, \alpha_n)$.

We denote by $l^x(\beta_1, \ldots, \beta_n)$, $\beta_k \in R^r$, the Legendre transform of $h^x(\alpha_1, \ldots, \alpha_n)$. The function $l^x(\beta_1, \ldots, \beta_n)$ can be expressed in terms of the

Legendre transform $L(x, \beta)$ of $H(x, \alpha)$ in the following way:

$$l^x(\beta_1, \ldots, \beta_n) = \int_0^T L(\psi_s, \dot{\beta}(s))\, ds, \tag{5.1}$$

where $\beta(s)$ is the piecewise linear function on $[0, T]$, having corners at the multiples of Δ and assuming the value β_k at $k\Delta$, $\beta_0 = x$. Indeed, if $x = 0$, then by the definition of Legendre's transformation we have

$$\begin{aligned}
l^0(\beta_1, \ldots, \beta_n) &= \sup_{\alpha_1, \ldots, \alpha_n} \left[\sum_{k=1}^n (\alpha_k, \beta_k) - \Delta \sum_{k=1}^n H\left(\psi_{(k-1)\Delta}, \sum_{i=k}^n \alpha_i\right) \right] \\
&= \Delta \sup_{\alpha_2, \ldots, \alpha_n} \left\{ \sup_{\alpha_1} \left[(\alpha_1, \beta_1/\Delta) - H\left(\psi_0, \alpha_1 + \sum_{i=2}^n \alpha_i\right) \right] \right. \\
&\quad \left. + \left[\sum_{k=2}^n (\alpha_k, \beta_k/\Delta) - \sum_{k=2}^n H\left(\psi_{(k-1)\Delta}, \sum_{i=k}^n \alpha_i\right) \right] \right\} \\
&= \Delta \sup_{\alpha_2, \ldots, \alpha_n} \left\{ L(\psi_0, \beta_1/\Delta) \right. \\
&\quad \left. + \left[\sum_{k=2}^n \left(\alpha_k, \frac{\beta_k - \beta_1}{\Delta}\right) - \sum_{k=2}^n H\left(\psi_{(k-1)\Delta}, \sum_{i=k}^n \alpha_i\right) \right] \right\} \\
&= \Delta L(\psi_0, \beta_1/\Delta) + \Delta \sup_{\alpha_2, \ldots, \alpha_n} \left[\sum_{k=2}^n \left(\alpha_k, \frac{\beta_k - \beta_1}{\Delta}\right) \right. \\
&\quad \left. - \sum_{k=2}^n H\left(\psi_{(k-1)\Delta}, \sum_{i=k}^n \alpha_i\right) \right].
\end{aligned}$$

Taking supremum with respect to α_2 in the last term, as it was taken with respect to α_1 earlier, and then taking supremum with respect to α_3 and so on, we arrive at the relation

$$l^0(\beta_1, \ldots, \beta_n) = \sum_{k=1}^n \Delta \cdot L\left(\psi_{(k-1)\Delta}, \frac{\beta_k - \beta_{k-1}}{\Delta}\right),$$

which is equivalent to (5.1) for $x = 0$. Taking into account that $l^x(\beta_1, \ldots, \beta_n) = l^0(\beta_1 - x, \ldots, \beta_n - x)$, we obtain (5.1) for an arbitrary initial condition $x \in R^r$.

Let us put $\boldsymbol{\eta}^\varepsilon = (\tilde{X}_\Delta^{\varepsilon,\psi}, \tilde{X}_{2\Delta}^{\varepsilon,\psi}, \ldots, \tilde{X}_{n\Delta}^{\varepsilon,\psi})$. If $\tilde{X}_t^{\varepsilon,\psi}$ is a process in the r-dimensional space, then $\boldsymbol{\eta}^\varepsilon$ is a random variable with values in $(R^r)^n$, the product of n copies of R^r. It follows from the definition of $h^x(\alpha_1, \ldots, \alpha_n)$ that $h^x(\alpha_1, \ldots, \alpha_n) = \lim_{\varepsilon \to 0} \varepsilon \ln \mathsf{M} \exp\{\varepsilon^{-1}(\boldsymbol{\alpha}, \boldsymbol{\eta}^\varepsilon)\}$, where $\boldsymbol{\alpha} = (\alpha_1, \alpha_2, \ldots, \alpha_n)$. We write $\Phi^\Delta(s) = \{\mathbf{e} \in (R^r)^n : l^x(e_1, \ldots, e_n) \le s\}$ for $s < \infty$ and $\bar{\rho}(\mathbf{e}, \mathbf{g}) = \max|e_k - g_k|$, where $\mathbf{e} = (e_1, \ldots, e_n)$ and $\mathbf{g} = (g_1, \ldots, g_n)$ are points of

$(R^r)^n$. It follows from Theorem 1.2 of Ch. 5 that for any $s, \delta, \gamma > 0$ and $\mathbf{e} \in (R^r)^n$ we have

$$\mathsf{P}\{\bar{\rho}(\boldsymbol{\eta}^\varepsilon, \mathbf{e}) < \delta\} \geq \exp\{-\varepsilon^{-1}(l^x(e_1, \ldots, e_n) + \gamma)\},$$
$$\mathsf{P}\{\bar{\rho}(\boldsymbol{\eta}^\varepsilon, \Phi^\Delta(s)) > \delta\} \leq \exp\{-\varepsilon^{-1}(s - \gamma)\}. \tag{5.2}$$

for sufficiently small ε. Let $\varphi \in \mathbf{C}_{0T}(R^r)$, $\varphi_0 = x$, $\tilde{S}_{0T}(\varphi) < \infty$. $\delta > 0$. We write $\boldsymbol{\varphi}^\Delta = (\varphi_\Delta, \varphi_{2\Delta}, \ldots, \varphi_{n\Delta}) \in (R^r)^n$. For sufficiently small $\delta' = \delta'(\delta)$ and $\Delta = \Delta(\delta)$ we have

$$\mathsf{P}\{\rho_{0T}(\tilde{X}^{\varepsilon, \psi}, \varphi) < \delta\} \geq \mathsf{P}\{\bar{\rho}(\boldsymbol{\eta}^\varepsilon, \boldsymbol{\varphi}^\Delta) < \delta'\}.$$

This inequality follows from the fact that the trajectories of $\tilde{X}_t^{\varepsilon, \psi}$ and any function φ for which $\tilde{S}^\psi(\varphi) < \infty$ satisfy a Lipschitz condition. Estimating the right side of the last inequality by means of the first of the inequalities (5.2), we obtain

$$\mathsf{P}|\rho_{0T}(\tilde{X}^{\varepsilon, \psi}, \varphi) < \delta\} \geq \exp\{-\varepsilon^{-1}(l^x(\varphi_\Delta, \ldots, \varphi_{n\Delta}) + \gamma)\}$$
$$= \exp\left\{-\varepsilon^{-1}\left(\int_0^T L(\psi_s, \dot{\bar{\varphi}}_s)\, ds + \gamma\right)\right\} \tag{5.3}$$

for every $\gamma > 0$ and sufficiently small ε, where $\bar{\varphi}_s$ is the piecewise linear function having corners at the multiples of Δ and coinciding with φ_s at these points.

Taking account of the absolute continuity of φ_s and the convexity of $L(x, \beta)$ in β, we obtain

$$\int_0^T L(\psi_s, \dot{\bar{\varphi}}_s)\, ds = \sum_{k=1}^n L\left(\psi_{(k-1)\Delta}, \frac{1}{\Delta}\int_{(k-1)\Delta}^{k\Delta} \dot{\varphi}_s\, ds\right)\Delta$$
$$\leq \sum_{k=1}^n \int_{(k-1)\Delta}^{k\Delta} L(\psi_{(k-1)\Delta}, \dot{\varphi}_s)\, ds = \int_0^T L(\psi_s, \dot{\varphi}_s)\, ds = \tilde{S}_{0T}^\psi(\varphi). \tag{5.4}$$

Relations (5.3) and (5.4) imply the first of the two inequalities in the definition of the action functional:

$$\mathsf{P}\{\rho_{0T}(\tilde{X}^{\varepsilon, \psi}, \varphi) < \delta\} \geq \exp\{-\varepsilon^{-1}(\tilde{S}_{0T}^\psi(\varphi) + \gamma)\}. \tag{5.5}$$

To prove the second inequality, we note that we have the inclusion

$$\{\omega \in \Omega\colon \rho_{0T}(\tilde{X}^{\varepsilon, \psi}, \tilde{\Phi}_x(s)) > \delta\} \subseteq \{\omega \in \Omega\colon \bar{\rho}(\boldsymbol{\eta}^\varepsilon, \Phi^\Delta(s)) > \delta'\}$$

for sufficiently small $\Delta(\delta)$ and $\delta'(\delta)$, where

$$\tilde{\Phi}_x(s) = \{\varphi \in \mathbf{C}_{0T}(R^r): \varphi_0 = x, \tilde{S}^{\psi}_{0T}(\varphi) \leq s\}.$$

This inclusion and the second estimate in (5.2) imply the required inequality

$$\mathsf{P}\{\rho_{0T}(\tilde{X}^{\varepsilon,\psi}, \tilde{\Phi}_x(s)) > \delta\} \leq \exp\{-\varepsilon^{-1}(s - \gamma)\}. \tag{5.6}$$

As has been remarked, the compactness of $\tilde{\Phi}_x(s)$ can be proved in the same way as in Lemma 4.2.

Lemma 5.1 is proved. □

As usual, this lemma and the lower semicontinuity of $S^{\psi}_{0T}(\varphi)$ imply that

$$\begin{aligned} -\inf_{\varphi \in (A)} \tilde{S}(\varphi) &\leq \varliminf_{\varepsilon \to 0} \varepsilon \ln \mathsf{P}\{\tilde{X}^{\varepsilon,\psi} \in A\} \\ &\leq \varlimsup_{\varepsilon \to 0} \varepsilon \ln \mathsf{P}\{\tilde{X}^{\varepsilon,\psi} \in A\} \leq -\inf_{\varphi \in [A]} \tilde{S}(\varphi) \end{aligned} \tag{5.7}$$

for every $A \subset \mathbf{C}^x_{0T}(R^r)$. Here (A) is the set of points of A, interior with respect to the space $\mathbf{C}^x_{0T}(R^r)$.

Lemma 5.2. *Suppose that the function $H(x, \alpha)$ is differentiable with respect to the variables α. Let $\psi^{(n)}: [0, T] \to R^r$ be a sequence of step functions uniformly converging to some function $\varphi \in \mathbf{C}_{0T}(R^r)$ as $n \to \infty$. Then there exists a sequence $\varphi^{(n)} \in \mathbf{C}_{0T}(R^r)$, uniformly converging to φ, such that*

$$\varlimsup_{n \to \infty} \int_0^T L(\psi^{(n)}_s, \dot{\varphi}^{(n)}_s)\, ds \leq S_{0T}(\varphi).$$

Proof. For an arbitrary partition $0 = t_0 < t_1 < t_2 < \cdots < t_{n-1} < t_n = T$ we have

$$\begin{aligned} \infty > S_{0T}(\varphi) &= \int_0^T L(\varphi_s, \dot{\varphi}_s)\, ds \\ &= \int_0^T \sup_\alpha [(\dot{\varphi}_s, \alpha) - H(\varphi_s, \alpha)]\, ds \\ &\geq \sum_{k=1}^n \sup_\alpha \int_{t_{k-1}}^{t_k} [(\dot{\varphi}_s, \alpha) - H(\varphi_s, \alpha)]\, ds \\ &= \sum_{k=1}^n \sup_\alpha \left[(\varphi_{t_k} - \varphi_{t_{k-1}}, \alpha) - \int_{t_{k-1}}^{t_k} H(\varphi_s, \alpha)\, ds\right]. \end{aligned} \tag{5.8}$$

We put

$$\gamma_k(\alpha) = \int_{t_{k-1}}^{t_k} H(\varphi_s, \alpha)\, ds; \qquad l_k(\beta) = \sup_\alpha [(\alpha, \beta) - \gamma_k(\alpha)].$$

It follows from the hypotheses of the lemma that the function $\gamma_k(\alpha)$ is convex and differentiable. The function $l_k(\beta)$ is convex, nonnegative, and lower semicontinuous. Relation (5.8) can be rewritten in the form

$$\infty > S_{0T}(\varphi) \geq \sum_{k-1}^{n} l_k(\varphi_{t_k} - \varphi_{t_{k-1}}).$$

It is known (cf. Rockafellar [1]) that if $l_k(\beta)$ is a lower semicontinuous convex function and $\beta^* \in \{\beta: l_k(\beta) < \infty\} = A_k$, then $l_k(\beta^*) = \underline{\lim}_{\beta \to \beta^*} l_k(\beta)$, where the points β belong to $\tilde{A}_k$, the interior of A_k with respect to its affine hull[1]). Therefore, for every $\delta > 0$ there exists a function $\tilde{\varphi}_t: [0, T] \to R^r$ such that

$$\sup_{0 \leq t \leq T} |\tilde{\varphi}_t - \varphi_t| < \delta, \quad \tilde{\varphi}_0 = \varphi_0, \qquad \tilde{\varphi}_{t_k} - \tilde{\varphi}_{t_{k-1}} \in \tilde{A}_k$$

and

$$\sum_{k=1}^{n} l_k(\varphi_{t_k} - \varphi_{t_{k-1}}) > \sum_{k=1}^{n} l_k(\tilde{\varphi}_{t_k} - \tilde{\varphi}_{t_{k-1}}) - \delta.$$

For such a function we have

$$S(\varphi) \geq \sum_{k=1}^{n} l_k(\tilde{\varphi}_{t_k} - \tilde{\varphi}_{t_{k-1}}) - \delta. \tag{5.9}$$

For points $\beta \in \tilde{A}_k$, the supremum in the definition of $l_k(\beta)$ is attained (cf. Rockafellar [1]). By virtue of this, there exist $\alpha_k \in R^r$ such that $l_k(\tilde{\varphi}_{t_k} - \tilde{\varphi}_{t_{k-1}}) = (\tilde{\varphi}_{t_k} - \tilde{\varphi}_{t_{k-1}}, \alpha_k) - \gamma_k(\alpha_k)$; the α_k satisfy the relation $\tilde{\varphi}_{t_k} - \tilde{\varphi}_{t_{k-1}} = \nabla\gamma_k(\alpha_k)$.

Let $\psi^{(m)}$ be any sequence of step functions uniformly converging to φ as $m \to \infty$. We choose $\varphi^{(m)}$ according to the conditions: $\varphi_0^{(m)} = \varphi_0$, $\dot{\varphi}_s^{(m)} = \nabla_\alpha H(\psi_s^{(m)}, \alpha_k)$ for $s \in (t_{k-1}, t_k)$ and $\varphi_{t_k+0}^{(m)} = \varphi_{t_k-0}^{(m)}$. Then $\varphi^{(m)}$ converges to $\tilde{\varphi}_s$ at the points $0, t_1, t_2, \ldots, t_n = T$ as $m \to \infty$. Indeed,

$$\begin{aligned} \varphi_{t_k}^{(m)} - \varphi_{t_{k-1}}^{(m)} &= \int_{t_{k-1}}^{t_k} \nabla_\alpha H(\psi_s^{(m)}, \alpha_k)\, ds \\ &\to \int_{t_{k-1}}^{t_k} \nabla_\alpha H(\varphi_s, \alpha_k)\, ds = \tilde{\varphi}_{t_k} - \tilde{\varphi}_{t_{k-1}}. \end{aligned} \tag{5.10}$$

[1] The affine hull (aff A) of a set $A \subseteq R^r$ is defined by the equality

$$\text{aff } A = \left\{\gamma^1 x_1 + \cdots + \gamma^m x_m : x_1, \ldots, x_m \in A, \sum_{k=1}^{m} \gamma^k = 1\right\}.$$

Here we have used the fact that the convergence of the convex functions $\gamma_k^m(\alpha) = \int_{t_{k-1}}^{t_k} H(\psi_s^{(m)}, \alpha)\, ds$ to the differentiable function $\gamma_k(\alpha)$ as $m \to \infty$ implies the convergence $\nabla\gamma_k^m(\alpha) \to \nabla\gamma_k(\alpha)$ (Rockafellar [1]).

Since $\dot\varphi_s^{(m)} = \nabla_\alpha H(\psi_s^{(m)}, \alpha_k)$ for $s \in (t_{k-1}, t_k)$, we have

$$\begin{aligned} L(\psi_s^{(m)}, \dot\varphi_s^{(m)}) &= \sup_\alpha[(\dot\varphi_s^{(m)}, \alpha) - H(\psi_s^{(m)}, \alpha)] \\ &= (\dot\varphi_s^{(m)}, \alpha_k) - H(\psi_s^{(m)}, \alpha_k). \end{aligned} \tag{5.11}$$

It follows from (5.10) and (5.11) that

$$\begin{aligned} \int_0^T L(\psi_s^{(m)}, \dot\varphi_s^{(m)})\, ds &= \sum_{k=1}^n \int_{t_{k-1}}^{t_k} \sup_\alpha[(\dot\varphi_s^{(m)}, \alpha) - H(\psi_s^{(m)}, \alpha)]\, ds \\ &= \sum_{k=1}^n \left[\int_{t_{k-1}}^{t_k} (\dot\varphi_s^{(m)}, \alpha_k)\, ds - \int_{t_{k-1}}^{t_k} H(\psi_s^{(m)}, \alpha_k)\, ds\right] \\ &\to \sum_{k=1}^n l_k(\tilde\varphi_{t_k} - \tilde\varphi_{t_{k-1}}) \le S_{0T}(\varphi) + \delta. \end{aligned} \tag{5.12}$$

as $m \to \infty$. Choosing δ and the intervals between the points t_k sufficiently small, we can bring $\varphi_t^{(m)}$ and φ_t arbitrarily close to each other. This and (5.12) imply the assertion of Lemma 5.2. □

Now we pass directly to the proof of Theorem 4.1. Let $\varphi \in \mathbf{C}_{0T}(R^r)$ and $S_{0T}(\varphi) < \infty$. We choose a step function ψ^λ and a function φ^λ such that $\rho_{0T}(\varphi^\lambda, \varphi) < \lambda$, $\sup_{0 \le t \le T} |\psi_t^\lambda - \varphi_t| < \lambda$ and $\int_0^T L(\psi_s^\lambda, \dot\varphi_s^\lambda)\, ds < S_{0T}(\varphi) + \gamma$. This can be done according to Lemma 5.2. It is easy to derive from the boundedness of the functions $b^k(x, y)$ and their derivatives that

$$\{\omega: \rho_{0T}(X^\varepsilon, \varphi) < \delta\} \supseteq \{\omega: \rho_{0T}(\tilde X^{\varepsilon, \psi^\lambda}, \varphi^\lambda) < \delta'\}$$

for any $\delta > 0$ provided that $\lambda = \lambda(\delta)$ and $\delta' = \delta'(\delta)$ are sufficiently small.

This inclusion and estimate (5.5) imply the inequality

$$\begin{aligned} \mathsf{P}\{\rho_{0T}(X^\varepsilon, \varphi) < \delta\} &\ge \mathsf{P}\{\rho_{0T}(\tilde X^{\varepsilon, \psi^\lambda}, \varphi^\lambda) < \delta'\} \\ &\ge \exp\{-\varepsilon^{-1}(\tilde S^{\psi^\lambda}(\varphi^\lambda) + \gamma)\} \\ &\ge \exp\{-\varepsilon^{-1}(S_{0T}(\varphi) + 2\gamma)\} \end{aligned} \tag{5.13}$$

for sufficiently small ε.

Now we prove the second inequality in the definition of the action functional. First of all we note that since $|b(x, y)|$ is bounded, the trajectory of X_t^ε, just as well as that of $\tilde{X}_t^{\varepsilon,\psi}$, issued from a point $x \in R^r$ belongs to some compactum $F \subset \mathbf{C}_{0T}(R^r)$ for $t \in [0, T]$. It follows from the semicontinuity of the functional $\tilde{S}_{0T}^{\psi}(\varphi)$ in ψ that for any $\gamma > 0$ there exists $\delta = \delta_\gamma(\varphi)$ such that $\tilde{S}^{\psi}(\varphi) > s - \gamma/2$ if $\rho_{0T}(\varphi, \psi) < \delta$ and $S_{0T}(\varphi) > s$. Relying on the semicontinuity of $\tilde{S}_{0T}^{\psi}(\varphi)$ in φ, we conclude that the functional $\delta_\gamma(\varphi)$ is lower semicontinuous in $\varphi \in \mathbf{C}_{0T}(R^r)$. Consequently, $\delta_\gamma(\varphi)$ attains its infimum on every compactum.

Let F_1 be the compactum obtained from F by omitting the $\delta/2$-neighborhood of the set $\Phi_x(s) = \{\varphi \in \mathbf{C}_{0T}(R^r): \varphi_0 = x, S_{0T}(\varphi) \le s\}$. Let us write $\bar{\delta}_\gamma = \inf_{\varphi \in F_1} \delta_\gamma(\varphi)$, $\delta' = \bar{\delta}_\gamma/(4KT + 2)$, where K is a Lipschitz constant of $b(x, y)$.

Let us choose a finite δ'-net in F and let $\varphi^{(1)}, \ldots, \varphi^{(N)}$ be the elements of this net, not belonging to $\Phi_x(s)$. It is obvious that

$$\mathsf{P}\{\rho_{0T}(X^\varepsilon, \Phi_x(s)) > \delta\} \le \sum_{i=1}^{N} \mathsf{P}\{\rho_{0T}(X^\varepsilon, \varphi^{(i)}) < \delta'\} \tag{5.14}$$

if $\delta' < \delta$. From the Lipschitz continuity of $b(x, y)$ for $\rho_{0T}(\varphi, \psi)$ we obtain the inclusion

$$\{\omega: \rho_{0T}(X^\varepsilon, \varphi) < \delta') \subseteq \{\omega: \rho_{0T}(\tilde{X}^{\varepsilon,\psi}, \varphi) < (2KT + 1)\delta'\}. \tag{5.15}$$

We choose step functions $\psi^{(1)}, \psi^{(2)}, \ldots, \psi^{(N)}$ such that $\rho_{0T}(\psi^{(i)}, \varphi^{(i)}) < \delta'/2$ for $i = 1, 2, \ldots, N$. Relations (5.14) and (5.15) imply the estimate

$$\mathsf{P}\{\rho_{0T}(X^\varepsilon, \Phi_x(s)) > \delta\} \le \sum_{i=1}^{n} \mathsf{P}\{\rho_{0T}(\tilde{X}^{\varepsilon,\psi^{(i)}}, \varphi^{(i)}) < (2KT + 1)\delta'\}. \tag{5.16}$$

Every term on the right side of the last inequality can be estimated by means of consequence (5.7) of Lemma 5.1. For every $i = 1, \ldots, N$ we have the estimate

$$\begin{aligned}&\mathsf{P}\{\rho_{0T}(\tilde{X}^{\varepsilon,\psi^{(i)}}, \varphi^{(i)}) < \bar{\delta}_\gamma/2\}\\ &\qquad \le \exp\{-\varepsilon^{-1}[\inf\{\tilde{S}^{\psi^{(i)}}(\varphi): \rho_{0T}(\varphi, \varphi^{(i)}) < \bar{\delta}_\gamma/2\} - \gamma/4]\}\end{aligned}$$

for sufficiently small ε. It follows from the definition of $\bar{\delta}_\gamma$ that the infimum on the right side of the last inequality is not less than $s - \gamma/2$, and consequently,

$$\mathsf{P}\{\rho_{0T}(\tilde{X}^{\varepsilon,\psi^{(i)}}, \varphi^{(i)}) < \bar{\delta}_\gamma/2\} \le \exp\{-\varepsilon^{-1}(s - \gamma)\}. \tag{5.17}$$

From (5.16) and (5.17) we obtain the lower estimate

$$\mathsf{P}\{\rho_{0T}(X^\varepsilon, \Phi_x(s)) > \delta\} \le \exp\{-\varepsilon^{-1}(s - \gamma)\}$$

for sufficiently small ε. Theorem 4.1 is completely proved. □

§6. The Behavior of the System on Large Time Intervals

As we have seen in Chapters 4 and 6, the behavior, on large time intervals, of a random process obtained as a result of small perturbations of a dynamical system is determined to a great extent by the character of large deviations. In this paragraph we discuss a series of results concerning the behavior of the process X_t^ε, the solution of system (1.2). These results are completely analogous to those expounded in Chapters 4 and 6. Because of this, we only give brief outlines of proofs, concentrating on the differences. We shall consider the case where the fast motion, i.e., ξ_t, is a Markov process with a finite number of states.

Let $O \in R^r$ be an asymptotically stable equilibrium position of the averaged system (2.2) and let D be a bounded domain containing O, whose boundary ∂D has a continuously rotating normal. We assume that the trajectories $\bar{x}_t$ beginning at points $\bar{x}_0 = x \in D \cup \partial D$ converge to O without leaving D as $t \to \infty$. We put $\tau^\varepsilon = \inf\{t: X_t^\varepsilon \notin D\}$. In contrast with the case considered in Ch. 4, τ^ε may be infinite with positive probability in general.

We introduce the function

$$V(x, y) = \inf_{\varphi \in H_{x,y}} S(\varphi),$$

where $H_{x,y}$ is the set of functions φ with values in R^r, defined on all possible intervals $[0, T]$, $T > 0$, for which $\varphi_0 = x$, $\varphi_T = y$. The function $V(x, y)$ can be expressed in terms of the function $u_x(t, y)$ appearing in §4: $V(x, y) = \inf_{t>0} u_x(t, y)$. It follows from the lower semicontinuity of $S(\varphi)$ that $V(x, y)$ is lower semicontinuous; it can be equal to $+\infty$. We write $V(x, \partial D) = \inf_{y \in \partial D} V(x, y)$.

Theorem 6.1 (Freidlin [11]). *Let the hypotheses of Theorem 4.2 be satisfied. Suppose that for every $y \in \partial D$ we have*

$$(b(y, i), n(y)) > 0 \tag{6.1}$$

for some $i = i(y)$, where $n(y)$ is the exterior normal vector to ∂D at y.

If $V(O, \partial D) < \infty$, then

$$\lim_{\varepsilon \downarrow 0} \varepsilon \ln \mathsf{M}_x \tau^\varepsilon = V(O, \partial D)$$

for any $x \in D$. If $V(O, \partial D) = +\infty$, then

$$\mathsf{P}_O\{\tau^\varepsilon = \infty\} = 1 \tag{6.2}$$

for any $\varepsilon > 0$.

For the proof of this theorem we need the following lemma.

Lemma 6.1. *Suppose that the hypotheses of Theorem* 6.1 *are satisfied. If* $V(x_0, \partial D) < \infty$ *for some* $x_0 \in D$, *then* $V(x, \partial D)$ *is continuous at* x_0. *If* $V(O, \partial D) < \infty$, *then* $V(x, \partial D) \leq V(O, \partial D) < \infty$ *for all* $x \in D$ *and* $V(x, \partial D)$ *is continuous.*

Proof. Let X_t^ε and Z_t^ε be solutions of equation (1.2) issued from x_0 and z, respectively, at time zero. For the difference $X_t^\varepsilon - Z_t^\varepsilon$ we have the inequality

$$|X_t^\varepsilon - Z_t^\varepsilon| \leq K \int_0^t |X_s^\varepsilon - Z_s^\varepsilon|\, ds + |x_0 - z|, \qquad t \geq 0.$$

From this by means of Lemma 1.1 of Ch. 2 we obtain

$$|X_t^\varepsilon - Z_t^\varepsilon| \leq e^{KT}|x_0 - z|. \tag{6.3}$$

If $V(x_0, \partial D) < \infty$, then taking account of condition (6.1), for any $\gamma > 0$ and sufficiently small $\delta > 0$ we can construct a function φ_t, $t \in [0, T]$, $\varphi_0 = x_0$, $\rho(\varphi_T, D) > \delta$, for which $S_{0T}(\varphi) < V(x_0, \partial D) + \gamma/4$. Relying on Theorem 4.1, from this we conclude that

$$\mathsf{P}\{\tau_{\delta/2}^\varepsilon < T\} \geq \exp\{-\varepsilon^{-1}(V(x_0, \partial D) + \gamma/2)\} \tag{6.4}$$

for sufficiently small $\varepsilon > 0$, where $\tau_{\delta/2}^\varepsilon = \inf\{t: \rho(X_t^\varepsilon, D) > \delta/2\}$. Let $|z - x_0| < e^{-KT}\delta/4$ and let A be the event that the trajectory of Z_t^ε leaves D until time T. It follows from (6.3) and (6.4) that

$$\mathsf{P}(A) \geq \exp\{-\varepsilon^{-1}(V(x_0, \partial D) + \gamma/2)\}.$$

On the other hand, $\mathsf{P}(A)$ can be estimated from above according to Theorem 4.1: we have

$$\mathsf{P}(A) \leq \exp\left\{-\varepsilon^{-1}\left(\inf_{\varphi \in \tilde{A}} S_{0T}(\varphi) - \gamma/4\right)\right\}$$

for sufficiently small $\varepsilon > 0$, where $\tilde{A}$ is the set of functions belonging to $\mathbf{C}_{0T}(R^r)$, such that $\varphi_0 = z$ and φ_t leaves D before time T. We obtain from the last two estimates that

$$V(z, \partial D) \leq \inf_{\varphi \in \tilde{A}} S_{0T}(\varphi) \leq V(x_0, \partial D) + \gamma$$

for $|x_0 - z| < e^{-KT}\delta/4$. On the other hand, the semicontinuity of $V(x, \partial D)$ implies the inequality

$$V(z, \partial D) > V(x_0, \partial D) - \gamma,$$

provided that δ is sufficiently small. Therefore, $|V(z, \partial D) - V(x_0, \partial D)| < \gamma$ for sufficiently small δ. The continuity at x_0 is proved.

If $V(O, \partial D) < \infty$, then $V(x, \partial D)$ is continuous at O. Let δ^1 be such that $|V(z, \partial D) - V(O, \partial D)| < \gamma$ for $|z - O| < \delta^1$. Since the trajectory of the averaged system, issued from a point $x \in D$, necessarily hits the δ^1-neighborhood of O and the action functional vanishes at trajectories of the averaged system, we have $V(x, \partial D) < V(O, \partial D) + \gamma$. By virtue of the arbitrariness of γ, this implies that $V(x, \partial D) < \infty$, and consequently, $V(x, \partial D)$ is continuous everywhere in D. Lemma 6.1 is proved. □

Now we outline the proof of Theorem 6.1. Due to the continuity of $V(x, \partial D)$, the proof of the first assertion of the theorem is completely analogous to that of Theorem 2.1 in Ch. 4. We only have to take into account that the process X_t^ε which we are now considering is not a Markov process in general, and therefore, we need to consider the pair $(X_t^\varepsilon, \xi_{t/\varepsilon})$. This pair forms a Markov process.

We prove the last assertion of the theorem. We assume the contrary. Without loss of generality, we may assume that (6.2) is not satisfied for $\varepsilon = 1$: the trajectories of X_t^1 issued from the point $X_0^1 = O$ leave D with positive probability. Then by condition (6.1) the trajectories of X_t^1 leave some δ-neighborhood $D_{+\delta}$ of D, $\delta > 0$, with positive probability: for some T we have

$$\mathsf{P}_O\{\tau_\delta^1 < T| > \alpha > 0, \tag{6.5}$$

where $\tau_\delta^1 = \inf\{t: X_t^1 \notin D_{+\delta}\}$.

Let $0 < t_1 < \cdots < t_{n-1} < T$, $\delta > 0$. $i_0, i_1, \ldots, i_{n-1}$ be a sequence of integers. We consider the following set of step functions on $[0, T]$:

$$\begin{aligned} A_{i_0, \ldots, i_{n-1}}^{t_1, \ldots, t_{n-1}}(\delta) = \{\psi\colon \psi(s) &= i_k \text{ for } s \in [s_k, s_{k+1}), \\ k &= 0, \ldots, n-1;\, s_0 = 0, |s_k - t_k| < \delta \\ &\text{for } 1 \le k \le n-1,\, s_n = T;\, \psi(T) = i_{n-1}\}. \end{aligned}$$

Relation (6.5) implies that there exist an integer $n > 0$, moments of time $t_1, \ldots, t_{n-1}$, integers $i_0, i_1, \ldots, i_{n-1}$ and $\delta' > 0$ such that the solution of the equation

$$\dot{y}_t = b(y_t, \psi(t)), \qquad y_0 = O,$$

goes out of $D_{+\delta/2}$ before time T for any function $\psi \in A_{i_0, \ldots, i_{n-1}}^{t_1, \ldots, t_{n-1}}(\delta')$. It is easy to derive from the Markov property of ξ_t that for some $c < \infty$ we have

$$\mathsf{P}\{\xi_{t/\varepsilon} \in A_{i_0, \ldots, i_{n-1}}^{t_1, \ldots, t_{n-1}}(\delta')\} > e^{-c\varepsilon^{-1}}.$$

Taking account of this estimate, we arrive at the conclusion:

$$\mathsf{P}\{\tau^\varepsilon < T\} > e^{-c\varepsilon^{-1}} \tag{6.6}$$

for any $\varepsilon > 0$.

On the other hand, by Theorem 4.1 we have

$$\overline{\lim_{\varepsilon \downarrow 0}}\, \varepsilon \ln \mathsf{P}\{\tau^\varepsilon < T\} \leq -\inf_{\varphi \in H_T^D} S_{0T}(\varphi), \tag{6.7}$$

where $H_T^D = \{\varphi \in \mathbf{C}_{0T}(R^r)\colon \varphi_0 = O, \varphi_s \in \partial D \text{ for some } s \in [0, T]\}$. Since $V(O, \partial D) \leq \inf_{\varphi \in H_T^D} S_{0T}(\varphi)$, we obtain from (6.6) and (6.7) that

$$V(O, \partial D) \leq \inf_{\varphi \in H_T^D} S(\varphi) \leq -\underline{\lim}_{\varepsilon \downarrow 0}\, \varepsilon \ln \mathsf{P}\{\tau^\varepsilon < T\} < c < \infty,$$

which contradicts the condition $V(O, \partial D) = +\infty$. The contradiction thus obtained proves the last assertion of Theorem 6.1. □

We note that for $V(O, \partial D)$ to be finite it is sufficient, for example, that for some $i = 1, 2, \ldots, N$ the solution of the equation

$$\dot{x}_t = b(x_t, i), \qquad x_0 = O,$$

leave D.

Analogously to §6, Ch. 6, for small ε we can solve the problem of exit of X^ε from a domain containing several equilibrium positions or limit sets of a more general form. We only have to take into account that in the situation considered in this section $V(x, y)$ may be equal to $+\infty$ and the corresponding passages may not be possible in general. In particular, as we have seen in Theorem 6.1, there may exist absorbing limit sets, i.e., limit sets with the property that if a trajectory hits the domain of attraction of the set, then it never leaves it. Analogously, there may exist absorbing cycles if, as was done in Ch. 6, we consider the decomposition of all limit sets into a hierarchy of cycles.

If the condition of connectedness: for any indices i, j of the limit sets $K_1, \ldots, K_l$ of system (2.2), $V(K_i, K_j) = V(x, y)|_{x \in K_i, y \in K_j} < \infty$, is satisfied, then all results of §6, Ch. 6 on fibering into cycles remain true.

In conclusion, we briefly treat invariant measures of X_t^ε and their behavior as $\varepsilon \downarrow 0$. If the projection of the averaged vector field $\bar{b}(x)$ onto the position vector connecting the origin of coordinates with the point x is negative and separated from zero uniformly in $x \notin F$, where F is a compactum in R^r, then it can be proved that every process X_t^ε has an invariant measure, at least for sufficiently small ε. This invariant measure is not unique in general. If we do not make any additional assumptions, then in general, the family μ^ε of invariant measures of X_t^ε has many cluster points (in the topology of weak convergence of measures) as $\varepsilon \downarrow 0$. If we assume that the condition of connectedness is satisfied, then there is only one cluster point in the case of general position. It can be described by means of the construction of $\{i\}$-graphs, considered in the preceding chapter (cf. Freidlin [11]).

§7. Not Very Large Deviations

We have already considered deviations of order $\varepsilon^{1/2}$ and of order 1 of X_t^ε from $\bar{x}_t$. Here we discuss briefly deviations of order $\varepsilon^\varkappa$, where $\varkappa \in (0, \frac{1}{2})$ (cf. Baier and Freidlin [1]). These deviations have common features with both deviations of order 1 (their probabilities converge to zero) and deviations of order $\varepsilon^{1/2}$ (they are determined by the behavior of the functions $b(x, y)$ near the averaged trajectory). In connection with this, deviations of order $\varepsilon^\varkappa$, $\varkappa \in (0, \frac{1}{2})$, are governed, as in the case of normal approximation, by the system obtained from (1.2) by linearization near $\bar{x}_t$.

For the sake of brevity, in this section we restrict ourselves to deviations from an equilibrium position of the averaged system, i.e., we shall assume that the initial condition in equation (1.2) is an equilibrium position of the averaged system and it coincides with the origin of coordinates: $X_0^\varepsilon = 0$, $\bar{b}(0) = 0$. We assume that this equilibrium position is asymptotically stable in the first approximation.

We introduce the notation

$$b(0, y) = f(y), \frac{\partial b^i}{\partial x^j}(0, y) = B_j^i(y), \qquad B(y) = (B_j^i(y)),$$

$$\lim_{T \to \infty} \frac{1}{T} \int_0^T B_j^i(\xi_s)\, ds = \bar{B}_j^i, \qquad \bar{B} = (\bar{B}_j^i).$$

We assume that the limits exist in probability and $\bar{B}_j^i = (\partial \bar{b}^i / \partial x^j)(0)$.

We shall say that condition $F^\varkappa$ is satisfied if for any step function α_s: $[0, T] \to R^r$ we have

$$\lim_{\varepsilon \downarrow 0} \varepsilon^{1-2\varkappa} \ln \mathsf{M} \exp\left\{\varepsilon^{\varkappa - 1} \int_0^T (\alpha_s, f(\xi_{s/\varepsilon}))\, ds\right\} = \frac{1}{2} \int_0^T (C\alpha_s, \alpha_s)\, ds, \quad (7.1)$$

where C is a constant symmetric matrix of order r. It is easy to see that (7.1) for $\alpha_s \equiv \alpha$ implies the equality

$$\lim_{T \to \infty} T^{2\varkappa - 1} \ln \mathsf{M} \exp\left\{T^{-\varkappa} \int_0^T (\alpha, f(\xi_s))\, ds\right\} = \tfrac{1}{2}(C\alpha, \alpha). \qquad (7.2)$$

Under certain assumptions, which we shall not specify, C is the matrix of second derivatives of the function $H(x, \alpha)$ introduced in §4 with respect to the variables α evaluated at $x = 0, \alpha = 0$.

In what follows, for simplicity, we shall assume that $\det C \neq 0$. If C is singular but has a nonzero eigenvalue, then all results can essentially be preserved; only their formulation becomes more complicated.

For functions belonging to $\mathbf{C}_{0T}(R^r)$ we define the functional

$$S(\varphi) = S_{0T}(\varphi) = \frac{1}{2}\int_0^T (C^{-1}(\dot\varphi_s - \bar B\varphi_s), \dot\varphi_s - \bar B\varphi_s)\,ds,$$

if φ_s is absolutely continuous; for the remaining $\varphi \in \mathbf{C}_{0T}(R^r)$ we put $S(\varphi) = +\infty$. We have already encountered this functional in Ch. 4; $S(\varphi)$ is the normalized action functional for the family of Gaussian processes ζ_t^λ defined by the differential equation

$$\dot\zeta_t^\lambda = \bar B\zeta_t^\lambda + \lambda C^{1/2}\dot w_t, \tag{7.3}$$

where $\dot w_t$ is a white noise process. Below we clarify the relationship between our problem and equation (7.3)

Theorem 7.1. *Suppose that condition $F^\varkappa$ is satisfied and* $\det C \neq 0$. *For some $\gamma > 1 - 2\varkappa$ let*

$$\lim_{\varepsilon\to 0} \varepsilon^\gamma \ln \mathbf{P}\left\{\sup_{0\le t\le T}\left|\varepsilon^{-\varkappa}\int_0^t (B(\xi_{s/\varepsilon}) - \bar B)\int_0^s e^{(s-u)\bar B} f(\xi_{u/\varepsilon})\,du\,ds\right| > \delta\right\} = -\infty \tag{7.4}$$

for any $\delta > 0$. Moreover, suppose that there exist $t_0 \in (0, T]$ and a function $\theta(t)$, $\theta(t) > 0$, $\lim_{t\downarrow 0}\theta(t) = 0$, such that

$$\varlimsup_{\varepsilon\downarrow 0}\ \sup_{\substack{\varepsilon\le t<t_0\\ 0\le h\le T-t}} \varepsilon^{1-2\varkappa}\left|\ln \mathbf{M}\exp\left\{\pm\frac{\varepsilon^{\varkappa-1}}{\theta(t)}\int_h^{h+t} f(\xi_{s/\varepsilon})\,ds\right\}\right| < \infty.$$

Then $\varepsilon^{2\varkappa-1}S_{0T}(\varphi)$ is the action functional for the family of processes $Z_t^\varepsilon = \varepsilon^{-\varkappa}X_t^\varepsilon$, $\varkappa \in (0, \frac{1}{2})$ in the space $\mathbf{C}_{0T}(R^r)$ as $\varepsilon \downarrow 0$, where X_t^ε is the solution of equation (1.2) *with the initial condition $X_0^\varepsilon = 0$* (we recall that $\bar b(0) = 0$).

The proof of this theorem is a combination of some arguments applied in the proofs of Theorems 3.1 and 4.1. Therefore, we only outline it. Firstly, it can be proved that $\varepsilon^{2\varkappa-1}S(\varphi)$ is the action functional for the family of processes Z_t^ε if $\varepsilon^{2\varkappa-1}S(\varphi)$ is the action functional for the processes $\tilde Z_t^\varepsilon = \varepsilon^{-\varkappa}\tilde X_t^\varepsilon$, where $\tilde X_t^\varepsilon$ is the solution of the linearized system

$$\dot{\tilde X}_t^\varepsilon = f(\xi_{t/\varepsilon}) + B(\xi_{t/\varepsilon})\tilde X_t^\varepsilon, \qquad \tilde X_0^\varepsilon = 0.$$

Then, using (7.4), it can be proved that the estimates of $\tilde{Z}_t^\varepsilon$, appearing in the definition of the action functional are satisfied if the same estimates hold for $\hat{Z}_t^\varepsilon = \varepsilon^{-\varkappa}\hat{X}_t^\varepsilon$, where $\hat{X}_t^\varepsilon$ is the solution of the equation

$$\dot{\hat{X}}_t^\varepsilon = f(\xi_{t/\varepsilon}) + \bar{B}\hat{X}_t^\varepsilon. \tag{7.5}$$

To obtain estimates for $\hat{X}_t^\varepsilon$, first we calculate the action functional for the family of processes $\eta_t^\varepsilon = \varepsilon^{-\varkappa}\int_0^t f(\xi_{s/\varepsilon})\,ds$. By means of Theorem 3.1 of Ch. 3, from this we calculate the action functional of $\hat{X}_t^\varepsilon$. Finally, the calculation of the action functional for the processes η_t^ε can be carried out in the same way as it was done in Lemma 4.3 (cf. also Gärtner [1]). In the course of this, we use the last condition of the theorem.

Consequently, the functional $S(\varphi)$ and the normalizing coefficient $f(\varepsilon) = \varepsilon^{2\varkappa-1}$ characterize deviations of order $\varepsilon^\varkappa$, $\varkappa \in (0, \frac{1}{2})$. In accordance with Theorem 3.1, the process $\varepsilon^{-1/2}X_t^\varepsilon$ converges to a Gaussian process ζ_t^0 as $\varepsilon \to 0$. As is easy to see, in our case this Gaussian process ζ_t^0 satisfies equation (7.3) for $\lambda = 1$. The ratio $X_t^\varepsilon/\varepsilon^\varkappa$ can be written in the form $\varepsilon^{1/2-\varkappa}\cdot\varepsilon^{-1/2}X_t^\varepsilon$ and we can interpret Theorem 3.1 in the following way: the probabilities of deviations of order $\varepsilon^\varkappa$, $\varkappa \in (0, \frac{1}{2})$, have the same asymptotics as deviations of order 1 caused by the Gaussian noise $\lambda C\dot{w}$, $\lambda = \varepsilon^{1/2-\varkappa}$. Of course, this circumstance is in perfect accordance with the fact that large but not very large deviations for sums of independent terms have the same asymptotics (in the principal terms) as the corresponding deviations of the normal approximation (cf. Ibragimov and Linnik [1]).

Now we consider some random processes for which the hypotheses of Theorem 7.1 can be verified and the matrix C can be calculated.

Theorem 7.2. *Let ξ_t be a stochastically continuous Markov process with a finite number of states, let $p_{ij}(t)$ be its transition probabilities, and let $q_{ij} = dp_{ij}(t)/dt|_{t=0}$. Suppose that all q_{ij} are different from zero. Let us denote by $\lambda(\alpha)$ the eigenvalue of the matrix $Q^\alpha = (q_{ij}^\alpha)$, $q_{ij}^\alpha = q_{ij} + \delta_{ij}(\alpha, b(0, i))$ with the largest real part. Then the hypotheses of Theorem 7.1 are satisfied for $\varkappa \in (0, \frac{1}{2})$ and $C = (C^{ij})$, where $C^{ij} = \partial^2\lambda(\alpha)/\partial\alpha_i\partial\alpha_j|_{\alpha=0}$.*

The proof of this theorem uses the construction applied in the proof of Theorem 4.2. First, in the same way as in Lemma 4.3, it can be proved that in the case of the process ξ_t, condition $F^\varkappa$ follows from relation (7.2). To prove (7.2), we note that

$$\mathsf{M}_i \exp\left\{t^{-\varkappa}\int_0^t (\alpha, f(\xi_s))\,ds\right\} = (T_t^{t^{-\varkappa}\alpha}\mathbf{1})(i), \tag{7.6}$$

where T_t^α is the semigroup of operators acting in the space of functions $g(i)$ defined on the phase space of ξ_t according to the formula

$$(T_t^\alpha g)(i) = \mathsf{M}_i g(\xi_t)\exp\left\{\int_0^t (\alpha, f(\xi_s))\,ds\right\}.$$

Using the notation introduced in the proof of Theorem 4.2, we obtain

$$\begin{aligned} t^{2\varkappa-1}\ln c + t^{2\varkappa-1}\ln(T_t^{t^{-\varkappa}\alpha}\mathbf{1})(i) &\le t^{2\varkappa-1}\ln(T_t^{t^{-\varkappa}\alpha}\mathbf{e})(i) \\ &= t^{2\varkappa}\lambda(t^{-\varkappa}\alpha) + t^{2\varkappa-1}\ln e_i \\ &\le t^{2\varkappa-1}\ln(T_t^{t^{-\varkappa}\alpha}\mathbf{1})(i). \end{aligned}$$

We note that 1 is the largest eigenvalue in absolute value of T_t^α for $\alpha = 0$. Therefore, $\lambda(0) = 0$. Taking into account that $\bar{b}(0) = 0$, it is easy to see that all first derivatives of $\lambda(\alpha)$ vanish at $\alpha = 0$. Therefore,

$$\lambda(\alpha) = \frac{1}{2}\sum_{ij}\frac{\partial^2\lambda}{\partial\alpha_i\partial\alpha_j}(0)\alpha_i\alpha_j + o(|\alpha|^2)$$

as $|\alpha| \to 0$. On account of the last equality, it follows from (7.7) that

$$\lim_{t\to\infty} t^{2\varkappa-1}\ln(T_t^{t^{-\varkappa}\alpha}\mathbf{1})(i) = \frac{1}{2}\sum_{i,j=1}^{r}\frac{\partial^2\lambda}{\partial\alpha_i\partial\alpha_j}(0)\alpha_i\alpha_j = \tfrac{1}{2}(C\alpha, \alpha).$$

This and (7.6) imply (7.2) and thus condition $F^\varkappa$. The verification of the remaining conditions of Theorem 7.1 is left to the reader.

An analogous result holds, of course, for some other Markov processes, for example, if ξ_t is a nondegenerate diffusion process on a compact space. Similar arguments enable us to verify conditions F and $F^\varkappa$ for some non-Markov processes with good mixing properties, as well (cf. Sinai [1]).

We write out the Hamilton–Jacobi equation for the function $u(t, x) = \inf\{S_{0t}(\varphi): \varphi_0 = 0, \varphi_t = x\}$. As has been said, the principal terms of the asymptotics of many quantities of interest can be expressed in terms of this function. By calculating the Legendre transform of the function $L(x, \beta) = \frac{1}{2}(C^{-1}(\beta - \bar{B}x), \beta - \bar{B}x)$, we arrive at the following equation for $u(t, x)$:

$$\frac{\partial u}{\partial t} = \tfrac{1}{2}(C\nabla_x u, \nabla_x u) + (\bar{B}x, \nabla_x u).$$

It is also possible to write out Euler's equations for the extremals; in the case under consideration, these equations are linear. Theorem 7.7 implies in particular that

$$\lim_{\varepsilon\to 0}\varepsilon^{1-2\varkappa}\ln \mathsf{P}\left\{\sup_{0\le t\le T}|X_t^\varepsilon| > \varepsilon^\varkappa\cdot d\right\} = -\min\{u(t, x): |x| = d, t\in[0, T]\}$$

for $\varkappa\in(0, \frac{1}{2})$.

The deviations of order $\varepsilon^\varkappa$ determine the average time needed by the trajectories of the process X_t^ε, $X_0^\varepsilon = 0$, to exit from a domain D^ε containing 0 if D^ε is obtained from a given domain D by a stretching with coefficient $\varepsilon^\varkappa$: $D^\varepsilon = \varepsilon^\varkappa \cdot D$. Let

$$\tau^\varepsilon = \inf\{t: X_t^\varepsilon \in D^\varepsilon\}, \qquad V(x) = \inf_{t \geq 0} u(t, x).$$

Theorem 7.3. *Suppose that the hypotheses of Theorem 7.2 are satisfied, the matrix C is nonsingular and $D^\varepsilon = \varepsilon^\varkappa \cdot D$, $\varkappa \in (0, \frac{1}{2})$, where D is a bounded domain with smooth boundary. Let us put $V_0 = \min_{x \in \partial D} V(x)$. Then*

$$\lim_{\varepsilon \downarrow 0} \varepsilon^{1-2\varkappa} \ln \mathsf{M}_0 \tau^\varepsilon = V_0,$$

$$\lim_{\varepsilon \downarrow 0} \mathsf{P}_0\{e^{\varepsilon^{2\kappa-1}(V_0-\gamma)} < \tau^\varepsilon < e^{\varepsilon^{2\kappa-1}(V_0+\gamma)}\} = 1$$

for any $\gamma > 0$.

The proof of this theorem is analogous to those of the corresponding results of Chapter 4 and we omit it.

We say a few words concerning the calculation of $V(x)$. Analogously to §3, Ch. 4, it can be proved that $V(x)$ is a solution of problem $\mathbf{R}_0$ for the equation

$$\tfrac{1}{2}(C\nabla V(x), \nabla V(x)) + (\bar{B}x, \nabla V(x)) = 0.$$

An immediate verification shows that if $C^{-1}\bar{B}$ is symmetric, then $V(x) = -(C^{-1}\bar{B}x, x)$.

Concluding this section, we note that analogous estimates can be obtained for not only deviations of order $\varepsilon^\varkappa$, $\varkappa \in (0, \frac{1}{2})$ from an equilibrium position but also deviations of order $\varepsilon^\varkappa$ from any averaged trajectory.

§8. Examples

Example 8.1. First we consider the case where the right sides of equations (1.2) do not depend on x and the process ξ_t is stationary. Then X_t^ε can be expressed in the form

$$X_t^\varepsilon = x + \int_0^t b(\xi_{s/\varepsilon})\, ds = x + t(t/\varepsilon)^{-1} \int_0^{t/\varepsilon} b(\xi_s)\, ds.$$

We write $m = \mathsf{M}b(\xi_s)$, $B^{ij}(\tau) = \mathsf{M}(b^i(\xi_{s+\tau}) - m^i)(b^j(\xi_s) - m^j)$ and assume that $\sum_{i=1}^r B^{ii}(\tau) \to 0$ as $\tau \to \infty$. By means of the Chebyshev inequality, we obtain from this for any $\delta > 0$ that

$$\mathsf{P}\left\{\left|\frac{1}{T}\int_t^{t+T} b(\xi_s)\,ds - m\right| > \delta\right\}$$
$$\leq \frac{1}{T^2\delta^2}\int_t^{t+T}\int_t^{t+T}\sum_{i=1}^r \mathsf{M}(b^i(\xi_s) - m^i)(b^i(\xi_u) - m^i)\,ds\,du$$
$$= \frac{1}{\delta^2 T^2}\int_t^{t+T}\int_t^{t+T}\sum_{i=1}^r B^{ii}(u - s)\,du\,ds \to 0$$

uniformly in $t \geq 0$ as T increases. Consequently, the hypotheses of Theorem 2.1 are satisfied in this case and $\sup_{0\leq t\leq T}|X_t^\varepsilon - x - mt| \to 0$ in probability as $\varepsilon \to 0$. If the process ξ_t has a mixing coefficient $\alpha(\tau)$ decreasing sufficiently fast and the functions $b(y)$ increase not too fast (for example if they are bounded) then Theorem 3.1 is applicable. In the case under consideration, this theorem is the assertion that the family of processes

$$\zeta_t^\varepsilon = \frac{1}{\sqrt{\varepsilon}}(X_t^\varepsilon - mt - x) = \sqrt{\varepsilon}\int_0^{t/\varepsilon}[b(\xi_s) - m]\,ds$$

converges weakly, as $\varepsilon \to 0$, to a Gaussian process ζ_t having mean zero and correlation matrix $(\mathsf{M}\zeta_t^i\zeta_s^j) = (t \wedge s)\cdot K = (t \wedge s)(K^{ij})$, where

$$K^{ij} = \int_{-\infty}^{\infty} B^{ij}(\tau)\,d\tau.$$

It is obvious that ζ_t has independent increments. The assertion that the distribution of ζ_t^ε converges to a Gaussian distribution (for a given t) constitutes the content of the central limit theorem for random processes (cf., for example, Rozanov [1], Ibragimov and Linnik [1]).

If condition F of §4 is satisfied for the process X_t^ε, then we can apply Theorem 4.1, which enables us to estimate large (or order 1) deviations of of X_t^ε from the linear function $mt + x, t \in [0, T]$. Since the right side of equation (1.2) does not depend on x, neither does the function $H(x, \alpha)$ defined by condition F, and the normalized action functional $S_{0T}(\varphi)$ has the form $S_{0T}(\varphi) = \int_0^T L(\dot{\varphi}_s)\,ds$, where $L(\beta)$ is the Legendre transform of $H(\alpha)$. As has been explained earlier, it is important to be able to determine the function

$$u(t, x) = \inf\{S_{0t}(\varphi)\colon \varphi_0 = x_0, \varphi_i = x\}.$$

for the calculation of the asymptotics of the probabilities of many interesting events. In the case considered here, the infimum and the extremal for which the infimum is attained can be calculated easily. We shall assume that $L(\beta)$ is strictly convex. Then Euler's equations

$$\frac{d}{dt}\nabla L(\dot{\varphi}) = 0$$

for the functional $S(\varphi)$ show that only the straight lines $\dot{\varphi} = c$, $c \in R^r$ are extremals. Using conditions at the endpoints of the interval $[0, t]$, we obtain that the infimum is attained for the function $\hat{\varphi}_s = x_0 + [(x - x_0)/t]s$, and $u(t, x) = tL((x - x_0)/t)$.

Suppose we would like to determine the asymptotics of $\ln \mathsf{P}\{X_t^\varepsilon \in D\}$ as $\varepsilon \downarrow 0$, where $X_t^\varepsilon = \int_0^t b(\xi_{s/\varepsilon})\, ds$, and D is a bounded set in R^r with boundary ∂D. It follows from Theorem 4.1 that

$$\lim_{\varepsilon\downarrow 0} \varepsilon \ln \mathsf{P}\{X_t^\varepsilon \in D\} = -t \cdot \inf_{x \in D \cup \partial D} L\left(\frac{x}{t}\right),$$

provided that the infimum coincides with the infimum taken over all interior points of D. If, moreover, the infimum is attained only at one point $\hat{x} \in D \cup \partial D$, then it is easy to show that

$$\lim_{\varepsilon\downarrow 0} \mathsf{P}\{|X_t^\varepsilon - \hat{x}| < \delta \,|\, X_t^\varepsilon \in D \cup \partial D\} = 1$$

for any $\delta > 0$.

For example, let ξ_t be defined by the equality

$$\xi_t = \eta_i \qquad \text{for } t \in [i, i+1),\ i \text{ is an integer}, \tag{8.1}$$

where $\eta_0, \eta_1, \ldots, \eta_n, \ldots$ is a sequence of independent variables with a common distribution function $F(x)$. Then condition F is satisfied and

$$H_\xi(\alpha) = \ln \int_{-\infty}^{\infty} e^{(\alpha, b(y))}\, dF(y),$$

provided that the integral under the logarithm is convergent. In this case Theorem 4.1 is close to theorems on large deviations for sums of independent terms. Theorem 4.1 is concerned with the rough, logarithmic, asymptotics of probabilities of large deviations, while theorems on large deviations for sums usually contain sharp asymptotics. On the other hand, Theorem 4.1 can be used to estimate probabilities of events concerning the course of a process X_t^ε on a whole interval $t \in [0, T]$ and not only for events related to a given moment of time.

EXAMPLE 8.2. Now let equation (1.2) have the form

$$\dot{X}_t^\varepsilon = b(X_t^\varepsilon) + \sigma(X_t^\varepsilon)\xi_{t/\varepsilon}, \qquad X_0^\varepsilon = x \in R^r, \tag{8.2}$$

where $b(x) = (b^1(x), \ldots, b^r(x))$, $\sigma(x) = (\sigma_j^i(x))$, and ξ_t is an r-dimensional random process with $\mathsf{M}\xi_t = 0$. The functions $b^i(x)$, $\sigma_j^i(x)$ are assumed to be bounded and sufficiently smooth. If the diagonal entries of the correlation matrix $B(s, t)$ of ξ_t converge to zero as $|t - s| \to \infty$, then by virtue of Theorem 2.1 we can conclude that X_t^ε converges in probability to the solution of the differential equation

$$\dot{\bar{x}}_t = b(\bar{x}_t), \qquad \bar{x}_0 = x, \tag{8.3}$$

uniformly on the interval $0 \le t \le T$ as $\varepsilon \downarrow 0$. If ξ_t has good mixing properties, then by means of Theorem 3.1 we can estimate the normal deviations from $\bar{x}_t$: we can calculate the characteristics of the Gaussian process ζ_t^0, the limit of $\zeta_t^\varepsilon = \varepsilon^{-1/2}(X_t^\varepsilon - \bar{x}_t)$.

We now assume that condition F is satisfied for the process ξ_t: there exists a function $H_\xi(\alpha)\colon R^r \to R^1$ such that for any step function $\alpha_s\colon [0, T] \to R^r$ we have

$$\lim_{\varepsilon \downarrow 0} \varepsilon \ln \mathsf{M} \exp\left\{\varepsilon^{-1} \int_0^T (\alpha_s, \xi_{s/\varepsilon})\, ds\right\} = \int_0^T H_\xi(\alpha_s)\, ds;$$

let $H_\xi(\alpha)$ be differentiable with respect to α. As is easy to see, condition F is satisfied for equation (8.2) and

$$H(x, \alpha) = (b(x), \alpha) + H_\xi(\sigma^*(x)\alpha). \tag{8.4}$$

The Legendre transform $L(x, \beta)$ of $H(x, \alpha)$ can be expressed simply in terms of the Legendre transform $L_\xi(\beta)$ of $H_\xi(\alpha)$:

$$L(x, \beta) = L_\xi(\sigma^{-1}(x)(\beta - b(x))),$$

provided that the matrix $\sigma(x)$ is nonsingular.

For example, let ξ_t be a Markov process taking two values $e_1, e_2 \in R^r$, let $(p_{ij}(t))$ be the matrix of transition probabilities and let $q_{ij} = (dp_{ij}/dt)(0)$. As is proved in §4, condition F is satisfied for ξ_t and $H_\xi(\alpha)$ is equal to the largest eigenvalue of

$$\begin{pmatrix} q_{11} + (\alpha, e_1) & q_{12} \\ q_{21} & q_{22} + (\alpha, e_2) \end{pmatrix}.$$

We consider the case where $q_{11} = q_{22} = -q$, $e_1 = -e_2 = e \in R^r$. Solving the characteristic equation, we find that

$$H_\xi(\alpha) = -q + \sqrt{q^2 + (\alpha, e)^2},$$

and by means of relation (8.4) we obtain the function $H(x, \alpha)$ for the family of processes X_t^ε.

We assume that 0 is an asymptotically stable equilibrium position for system (8.3). For the determination of the asymptotics of the mean exit time of a domain containing 0, of the point through which the exit takes place, of the asymptotics of the invariant measure of X_t^ε and of other interesting characteristics, we have to calculate the function

$$V(x) = \inf\left\{\int_0^T L(\varphi_s, \dot{\varphi}_s)\, ds \colon \varphi_0 = 0, \varphi_T = x, T > 0\right\},$$

as follows from §6. This function can be calculated as the solution of problem $\mathbf{R}_0$ (cf. §4, Ch. 5) for the equation

$$(b(x), \nabla V(x)) - q + \sqrt{q^2 + (\sigma^*(x)\nabla V(x), e)^2} = 0.$$

If system (8.3) has an asymptotically stable limit cycle Γ, then deviations from this cycle can be described by the quasipotential $V_\Gamma(x)$, which can be determined as the solution of problem $\mathbf{R}_\Gamma$ (cf. §4, Ch. 5) for the same equation.

In this example we now consider deviations of order $\varepsilon^\varkappa$, $\varkappa \in (0, \frac{1}{2})$, from the equilibrium position 0. If condition $F^\varkappa$ is satisfied for ξ_t, i.e., if the limit

$$\lim_{\varepsilon \downarrow 0} \varepsilon^{1-2\varkappa} \ln \mathsf{M} \exp\left\{\varepsilon^{\varkappa-1} \int_0^T (\alpha_s, \xi_{s/\varepsilon})\, ds\right\} = \frac{1}{2}\int_0^T (C_\xi \alpha_s, \alpha_s)\, ds$$

exists, where C_ξ is a symmetric matrix and α_s is any step function on $[0, T]$, then, as is easy to see, condition $F^\varkappa$ is also satisfied for the process X_t^ε defined by equation (8.2) and we have $C = \sigma(0)C_\xi\sigma^*(0)$. Let ξ_t be the Markov process with two states considered above. As follows from Theorem 7.2, condition $F^\varkappa$ is satisfied and $C_\xi = (\partial^2 H_\xi(\alpha)/\partial\alpha_i\partial\alpha_j|_{\alpha=0}) = (1/q)(e^i e^j)$, where $e^1, e^2, \ldots, e^r$ are the components of the vector e. Consequently, in this case for the family of processes X_t^ε we obtain $C = (1/q)\sigma(0)(e^i e^j)\sigma^*(0)$.

It is easy to prove that conditions F and $F^\varkappa$ are satisfied for equation (8.2) and in the case of the processes ξ_t defined by equality (8.1)

Example 8.3. Let us consider the van der Pol equation with random perturbations:

$$\ddot{x} + \omega^2 x = \varepsilon[f(x, \dot{x}, \nu t) + \varphi(x, \dot{x})\xi_t]. \tag{8.5}$$

Here $f(x, \dot{x}, vt)$ is a sufficiently smooth function, periodic in t with frequency v, ξ_t is a stationary process in R^1 with vanishing mathematical expectation and correlation function $K(\tau)$, and $\varphi(x, \dot{x})$ is a smooth bounded function.

As in the deterministic case, by introducing the van der Pol variables (r, θ), which are defined by the relations

$$x = r\cos(\omega t + \theta), \qquad \dot{x} = -r\omega\sin(\omega t + \theta),$$

and the "slow" time $s = \varepsilon t$, equation (8.5) can be rewritten as a system of two equations

$$\begin{aligned} \frac{dr^\varepsilon}{ds} &= F_1(\omega s/\varepsilon + \theta^\varepsilon, vs/\varepsilon, r^\varepsilon, \xi_{s/\varepsilon}), \\ \frac{d\theta^\varepsilon}{ds} &= F_2(\omega s/\varepsilon + \theta^\varepsilon, vs/\varepsilon, r^\varepsilon, \xi_{s/\varepsilon}), \end{aligned} \tag{8.6}$$

where

$$\begin{aligned} F_1(\tau, vt, r, \xi) &= -\frac{1}{\omega}[f(r\cos\tau, -r\omega\sin\tau, vt) \\ &\quad + \varphi(r\cos\tau, -r\omega\sin\tau)\xi]\sin\tau, \\ F_2(\tau, vt, r, \xi) &= -\frac{1}{r\omega}[f(r\cos\tau, -r\omega\sin\tau, vt) \\ &\quad + \varphi(r\cos\tau, -r\omega\sin\tau)\xi]\cos\tau. \end{aligned}$$

If $K(\tau)$ decreases with the increase of τ, then the right sides of equations (8.6) satisfy the hypotheses of Theorem 7.1. Moreover, as is easy to verify,

$$\begin{aligned} &\lim_{T\to\infty}\frac{1}{T}\int_0^T \varphi(r\cos(\omega s + \theta), -r\omega\sin(\omega s + \theta))\xi_s \sin(\omega s + \theta)\,ds \\ &\quad = \lim_{T\to\infty}\frac{1}{T}\int_0^T \varphi(r\cos(\omega s + \theta), -r\omega\sin(\omega s + \theta))\xi_s\cos(\omega s + \theta)\,ds \\ &\quad = 0, \end{aligned}$$

so that the terms containing the factors ξ_s in the expressions for the functions F_1, F_2 vanish upon averaging and the averaged equations have the same form as in the absence of stochastic terms. If the ratio ω/v is irrational (this case is said to be nonresonant), then the averaged system has the form

$$\frac{d\bar{r}}{ds} = \bar{F}_1(\bar{r}), \qquad \frac{d\bar{\theta}}{ds} = \bar{F}_2(\bar{r}),$$

where the functions $\bar{F}_1(r), \bar{F}_2(r)$ are given by formulas (1.9).

Now we assume, in addition, that ξ_t has finite moments up to order seven inclusive and satisfies the strong mixing condition with a sufficiently rapidly decreasing coefficient $\alpha(\tau)$. By Theorem 3.1 the family of random processes $\varepsilon^{-1/2}(r_t^\varepsilon - \bar{r}_t, \theta_t^\varepsilon - \bar{\theta}_t)$, $t \in [0, T]$ converges weakly to the Gaussian Markov process (ρ_t, ψ_t) which corresponds to the differential operator

$$Lu(\rho, \psi) = \frac{1}{2}\left(A^{11}(\bar{r}_t)\frac{\partial^2 u}{\partial \rho^2} + 2A^{12}(\bar{r}_t)\frac{\partial^2 u}{\partial \rho \partial \psi} + A^{22}(\bar{r}_t)\frac{\partial^2 u}{\partial \psi^2}\right) + \rho\frac{d\bar{F}_1}{dr}(\bar{r}_t)\frac{\partial u}{\partial \rho} + \rho\frac{d\bar{F}_2}{dr}(\bar{r}_t)\frac{\partial u}{\partial \psi},$$

where the functions $A^{ij}(r)$ are defined by the equalities

$$A^{11}(r) = \frac{1}{2\pi\omega^3}\int_0^{2\pi} dt \int_{-\infty}^{\infty} ds K\left(\frac{t-s}{\omega}\right)\varphi(r\cos t, -r\omega\sin t) \times \varphi(r\cos s, -r\omega\sin s)\sin s \cdot \sin t,$$

$$A^{22}(r) = \frac{1}{2\pi r^2\omega^3}\int_0^{2\pi} dt \int_{-\infty}^{\infty} ds K\left(\frac{t-s}{\omega}\right)\varphi(r\cos t, -r\omega\sin t) \times \varphi(r\cos s, -r\omega\sin s)\cos t\cos s,$$

$$A^{12}(r) = \frac{1}{2\pi r\omega^3}\int_0^{2\pi} dt \int_{-\infty}^{\infty} ds K\left(\frac{t-s}{\omega}\right)\varphi(r\cos t, -r\omega\sin t) \times \varphi(r\cos s, -r\omega\sin s)\cos t\sin s.$$

As was indicated in §1, if r_0 is the only root of the equation $\bar{F}_1(r) = 0$ and the function $\bar{F}_1(r)$ changes from positive to negative in passing through r_0, then independently of the initial conditions, periodic oscillations with amplitude close to r_0 and frequency close to ω are established in the system described by equation (8.5) without random perturbations ($\varphi = 0$) for sufficiently small ε. If random perturbations are present, then, as follows from Theorem 7.1, with probability converging to 1 as $\varepsilon \to 0$, the phase trajectories $(X_t^\varepsilon, \dot{X}_t^\varepsilon)$ approach a limit cycle of the form

$$\Gamma_{r_0} = \{(x, \dot{x}): x = r_0\cos(\omega t + \theta), \dot{x} = -r_0\omega\sin(\omega t + \theta)\},$$

over time of order ε^{-1} and they perform a motion close to the oscillations of the unperturbed system along Γ_{r_0}, deviating from these oscillations from time to time and returning to them again. We assume that at the initial moment the system was moving along on the cycle Γ_{r_0}. Then on time intervals of order ε^{-1}, deviations from periodic oscillations have order $\sqrt{\varepsilon}$ with overwhelming probability, according to Theorem 3.1. By means of this theorem we can calculate various probabilistic characteristics of these deviations.

For example, to calculate the probability that the amplitude r_t^ε deviates from r_0 by more than $h\sqrt{\varepsilon}$ at least once over the time $[0, T/\varepsilon]$, we need to solve the boundary value problem

$$\frac{\partial u}{\partial s}(s, r) = \tfrac{1}{2}A^{11}(r_0)\frac{\partial^2 u}{\partial r^2} + \frac{d\bar{F}_1}{dr}(r_0)r\frac{\partial u}{\partial r}, \qquad -h < r < h;$$

$$u(0, r) = 0, \qquad u(s, -h) = u(s, h) = 1.$$

The desired probability will be close to $u(T, 0)$ for $\varepsilon \ll 1$.

Deviations of order 1 take place over times of order greater than ε^{-1}. To study these deviations, we need to use the results of §4. For example, let $\xi_t = \eta_k$ for $t \in [\alpha - \pi k/\omega, \alpha + \pi k/\omega)$, where $\eta_0, \eta_1, \ldots, \eta_k, \ldots$ are identically distributed independent random variables and α is a random variable independent of $\{\eta_j\}$ and uniformly distributed in $[0, 2\pi/\omega]$. It is easy to see that condition F is satisfied for system (8.6) and

$$\begin{aligned} H(r, \theta, \alpha_1, \alpha_2) &= H(r, \alpha_1, \alpha_2) \\ &= \bar{F}_1(r)\alpha_1 + \bar{F}_2(r)\alpha_2 + H_\eta(\alpha_1\bar{\varphi}_1(r) + \alpha_2\bar{\varphi}_2(r)), \end{aligned}$$

where

$$\bar{\varphi}_1(r) = -\frac{1}{2\pi\omega}\int_0^{2\pi} \varphi(r\cos s, -r\omega\sin s)\sin s\, ds,$$

$$\bar{\varphi}_2(r) = -\frac{1}{2\pi r\omega}\int_0^{2\pi} \varphi(r\cos s, -r\omega\sin s)\cos s\, ds,$$

$$H_\eta(\tau) = \ln \mathsf{M}e^{\tau\eta_1}.$$

The functions $\bar{F}_1(r)$ and $\bar{F}_2(r)$ are defined by equalities (1.9). As is shown in §§4 and 6, the asymptotics of various probability theoretical characteristics of large deviations from the unperturbed motion can be calculated by means of the function $H(r, \alpha_1, \alpha_2)$. Let $(a, b) \ni r_0$ and let $\tau^\varepsilon = \min\{t : r_t^\varepsilon \notin (a, b)\}$. We calculate $\lim_{\varepsilon \downarrow 0} \varepsilon \ln \mathsf{M}_{r,\theta}\tau^\varepsilon$ for $r \in (a, b)$. It follows from the results of §6 that this limit is equal to $\min(u(a), u(b))$, where the function $u(r)$ is the quasi-potential of random perturbations on the half-line $r \geq 0$. It can be determined as the solution of problem $\mathbf{R}_{r_0}$ for the equation

$$H\left(r, \frac{du}{dr}, 0\right) = \bar{F}_1(r)\frac{du}{dr} + H_\eta\left(\bar{\varphi}_1(r)\frac{du}{dr}\right) = 0. \tag{8.7}$$

The solution of this problem obviously is reduced to the determination of the nonzero root of the equation $\bar{F}_1(r)z + H_\eta(\bar{\varphi}_1(r)z) = 0$ and a subsequent integration. For example, let the variables η_k have a Gaussian distribution,

$\mathsf{M}\eta_k = 0$ and $D\eta_k = \sigma^2$. In this case the results of §4 are applicable (cf. Grin' [3]). We have

$$H_\eta(\tau) = \frac{\sigma^2\tau^2}{2}, \qquad z(r) = -\frac{2\bar{F}_1(r)}{\sigma^2\bar{\varphi}_1^2(r)}, \quad \text{and} \quad u(r) = -\int_{r_0}^{r} \frac{2\bar{F}_1(\rho)}{\sigma^2\bar{\varphi}_1^2(\rho)}\,d\rho.$$

Now we assume that the equation $\bar{F}_1(r) = 0$ has several roots $r_0 < r_1 < \cdots < r_{2n}$ and the function $\bar{F}_1(r)$ changes sign from plus to minus at roots with even indices and from minus to plus at roots with odd indices. If $r_0^\varepsilon \in (r_{2k-1}, r_{2k+1})$, then oscillations with amplitude close to r_{2k} are established in the system without random perturbations for small ε. In general, random perturbations lead to passages between stable limit cycles. Let $u_{2k}(r)$ be the solution of problem $\mathbf{R}_{r_{2k}}$ for equation (8.7) on the interval $[r_{2k-1}, r_{2k+1}]$, $u_{2k}(r) < \infty$ and $u_{2k}(r_{2k+1}) < u_{2k}(r_{2k-1})$. Then with probability close to 1 for ε small, a passage takes place from the cycle $\Gamma_{r_{2k}} = \{(r, \theta): r = r_{2k}\}$ to the cycle $\Gamma_{r_{2(k+1)}}$ and the average time needed for the passage is logarithmically equivalent to $\exp\{\varepsilon^{-1}u_{2k}(r_{2k+1})\}$.

We define a function $V(r)$ on $(0, \infty)$ by the equalities

$$V(r) = u_{r_0}(r) \quad \text{for } r \in (0, r_1];$$
$$V(r) = V(r_{2k-1}) + u_{r_{2k}}(r) - u_{r_{2k}}(r_{2k-1}) \quad \text{for } r \in [r_{2k-1}, r_{2k+1}].$$

The function $V(r)$ has local minima at the points $r_0, r_2, \ldots, r_{2n}$. We assume that $V(r)$ attains its absolute minimum at a unique point r_{2k*}. Then, as follows from results of Ch. 6, the limit cycle $\Gamma_{r_{2k*}}$ is the "most stable": for small ε the trajectories $(r_t^\varepsilon, \theta_t^\varepsilon)$ spend most of the time in the neighborhood of $\Gamma_{r_{2k*}}$.

EXAMPLE 8.4. We consider the linear system

$$\dot{X}_t^\varepsilon = A(\xi_{t/\varepsilon})X_t^\varepsilon + b(\xi_{t/\varepsilon}), \qquad X_0^\varepsilon = x. \tag{8.8}$$

The entries of the matrix $A(y) = (A_j^i(y))$ and the components of the vector $b(y) = (b^1(y), \ldots, b^r(y))$ are assumed to be bounded. Concerning the process ξ_t we assume that it possesses a sufficiently rapidly decreasing mixing coefficient and $\mathsf{M}A(\xi_t) = \bar{A} = (\bar{A}_j^i)$, $\mathsf{M}b(\xi_t) = \bar{b}$. Relying on Theorem 2.1 we conclude that X_t^ε converges in probability to the solution of the differential equation

$$\dot{\bar{x}}_t = \bar{A}\bar{x}_t + \bar{b}, \qquad \bar{x}_0 = x,$$

uniformly on the interval $0 \le t \le T$ as $\varepsilon \downarrow 0$. The solution of this equation can be written as

$$\bar{x}_t = \int_0^t \exp\{\bar{A}(t-s)\}\bar{b}\,ds + e^{\bar{A}t}x.$$

To estimate normal deviations of X_t^ε from $\bar{x}_t$, we need to use Theorem 3.1. For the sake of simplicity we assume that ξ_t is stationary and $b(y) \equiv 0$. We denote by $K_{jm}^{in}(\tau)$ the joint correlation function of the processes $A_j^i(\xi_t)$ and $A_m^n(\xi_t)$ and write

$$\bar{K}_{jm}^{in} = \int_{-\infty}^{\infty} K_{jm}^{in}(\tau)\, d\tau, \qquad G^{in}(x) = \sum_{j,m} \bar{K}_{jm}^{in} x^j x^m.$$

Then by Theorem 3.1 the normalized difference $\zeta_t^\varepsilon = \varepsilon^{-1/2}(X_t^\varepsilon - \bar{x}_t)$ converges weakly to the Gaussian process $\zeta_t = \int_0^t e^{(t-s)\bar{A}} d\eta_s$ as $\varepsilon \to 0$, where η_s is a Gaussian process with independent increments and mean zero such that $\mathsf{M}\eta_\tau^i \eta_\tau^n = \int_0^\tau G^{in}(\bar{x}_s)\, ds$ (cf. Khas'minskii [4]).

It is easy to give examples showing that the system obtained from (8.8) upon averaging may have asymptotically stable stationary points even in cases where the vector fields $A(y)x + b(y)$ do not have equilibrium positions for any value of the parameter y or have unstable equilibrium positions. In the neighborhood of such points the process X_t^ε, $\varepsilon \ll 1$, spends a long time or is even attracted to them with great probability. For the sake of definiteness, let ξ_t be the Markov process with a finite number of states, considered in Theorem 4.2, let $\bar{b} = 0$ and let all eigenvalues of the matrix $\bar{A}$ have negative real parts. Then the origin of coordinates 0 is an asymptotically stable equilibrium position of the averaged system. Let D be a bounded domain containing the origin and having boundary ∂D and let $V(x, \partial D)$ be the function introduced in §6. If $V(0, \partial D) < \infty$, then in accordance with Theorem 6.1, the trajectories X_t^ε, $X_0^\varepsilon = x \in D$, leave D over time τ^ε going to infinity in probability as $\varepsilon \to 0$ and $\ln \mathsf{M}_x \tau^\varepsilon \sim \varepsilon^{-1} V(0, \partial D)$. The case $V(0, \partial D) = +\infty$ is illustrated by the following example. Let ξ_t be a process with two states and let $A(y)$ be equal to

$$A_1 = \begin{pmatrix} -\bar{a} & 0 \\ 0 & \underline{a} \end{pmatrix}, \qquad A_2 = \begin{pmatrix} \underline{a} & 0 \\ 0 & -\bar{a} \end{pmatrix}.$$

in these states, respectively. We consider the system $\dot{X}_t^\varepsilon = A(\xi_{t/\varepsilon}) X_t^\varepsilon$. If ξ_t did not pass from one state to another, then for any initial state ξ_0, the origin of coordinates would be an unstable stationary point—a saddle. Let the matrix Q for the process ξ_t have the form

$$Q = \begin{pmatrix} -1 & 1 \\ 1 & -1 \end{pmatrix}.$$

The stationary distribution of this process is the uniform distribution, that is $(\frac{1}{2}, \frac{1}{2})$. The averaged system

$$\dot{\bar{x}}^1 = \tfrac{1}{2}(\underline{a} - \bar{a})\bar{x}^1, \qquad \dot{\bar{x}}^2 = \tfrac{1}{2}(\underline{a} - \bar{a})\bar{x}^2$$

has an asymptotically stable equilibrium position at 0 if $\bar{a} > \underline{a}$. It is easy to prove that in this case the trajectories of X_t^ε converge to 0 with probability 1 as $t \to 0$ for any $\varepsilon > 0$, i.e., due to random passages of ξ_t, the system acquires stability. By means of results of the present chapter we can calculate the logarithmic asymptotics of $\mathsf{P}_x\{\tau^\varepsilon < \infty\}$ as $\varepsilon \to 0$, where $\tau^\varepsilon = \min\{t: X_t^\varepsilon \notin D\}$ (D is a neighborhood of the equilibrium position) as well as the asymptotics of this probability for $x \to 0$, $\varepsilon = \text{const}$.

EXAMPLE 8.5. Consider again equation (8.2). Let $r = 2$, $b(x) = \overline{\nabla} H(x) = (\partial H(x)/\partial x^2, -\partial H(x)/\partial x^1)$, and ξ_t, $-\infty < t < \infty$, be a two-dimensional mean zero stationary process with the correlation matrix $B(\tau)$. The averaged system is now a Hamiltonian one:

$$\dot{\bar{X}}_t = \overline{\nabla} H(\bar{X}_t), \qquad \bar{X}_0 = x \in R^2, \tag{8.9}$$

and the Hamiltonian function $H(x)$ is a first integral for (8.9). Assume that the function $H(x)$, the matrix $\sigma(x)$, and the process ξ_t are such that conditions 1–5 of Theorem 3.2 are satisfied. Here

$$\begin{aligned} g(x,z) &= \sigma(x)z, \\ F(x,z) &= (\nabla H(x), \sigma(x)z), \\ D(x,s) &= \mathsf{M}(\nabla H(x), \sigma(x)\xi_s)(\nabla H(x), \sigma(x)\xi_0), \\ Q(x,s) &= \mathsf{M}(\nabla(\nabla H(x), \sigma(x)\xi_s), \sigma(x)\xi_0); \qquad x, z \in R^2, s > 0. \end{aligned} \tag{8.10}$$

Let $B(\tau) = (B^{ij}(\tau))$, $B^{ij}(\tau) = \mathsf{M}\xi_\tau^i \xi_0^j$,

$$\bar{B} = \int_0^\infty B(\tau)\, d\tau.$$

The convergence of this integral follows from our assumptions. One can derive from (8.10) that

$$D(x) = 2\int_0^\infty D(x,s)\, ds = (\sigma(x)\bar{B}\sigma^*(x)\nabla H(x), \nabla H(x)).$$

Since $B(\tau)$ is a positive definite function, $\bar{B}$ and $\sigma(x)\bar{B}\sigma^*(x)$ are also positive definite. Thus one can introduce $\sigma(y)$ such that

$$\sigma^2(y) = \left(\int_{C(y)} \frac{dl}{|\nabla H(x)|}\right)^{-1} \int_{C(y)} \frac{(\sigma(x)\bar{B}\sigma^*(x)\nabla H(x), \nabla H(x))\, dl}{|\nabla H(x)|}.$$

To write down the drift coefficient for the limiting process, we need some notations. For any smooth vector field $e(x)$ in R^2 denote by $\nabla e(x)$ the matrix $(e_{ij}(x))$, $e_{ij}(x) = \partial e_j(x)/\partial x^i$. Simple calculations show that

$$Q(x) = \int_0^\infty Q(x, s)\, ds = \operatorname{tr}(\sigma^*(x) \cdot \nabla(\sigma^*(x)\nabla H(x)) \cdot \bar{B}),$$

and we have the following expression for the drift,

$$B(y) = \left(\int_{C(y)} \frac{d\ell}{|\nabla H(x)|}\right)^{-1} \int_{C(y)} \frac{\operatorname{tr}(\sigma^*(x) \cdot \nabla(\sigma^*(x)\nabla H(x)) \cdot \bar{B})\, d\ell}{|\nabla H(x)|}.$$

Let, for example, $\sigma(x)$ be the unit matrix. Then

$$D(x) = (\bar{B}\nabla H(x), \nabla H(x)), \qquad Q(x) = \operatorname{tr}(\hat{H}(x)\bar{B}),$$

where $\hat{H}(x)$ is the Hessian matrix for $H(x) : \hat{H}_{ij}(x) = \partial^2 H(x)/\partial x^i \partial x^j$.

§9. The Averaging Principle for Stochastic Differential Equations

We consider the system of differential equations

$$\begin{aligned} \dot{X}^\varepsilon &= b(X^\varepsilon, Y^\varepsilon) + \sigma(X^\varepsilon, Y^\varepsilon)\dot{w}, & X_0^\varepsilon &= x, \\ \dot{Y}^\varepsilon &= \varepsilon^{-1} B(X^\varepsilon, Y^\varepsilon) + \varepsilon^{-1/2} C(X^\varepsilon, Y^\varepsilon)\dot{w}, & Y_0^\varepsilon &= y, \end{aligned} \tag{9.1}$$

where

$$x \in R^r,\ y \in R^l,\ b(x, y) = (b^1(x, y), \ldots, b^r(x, y)),$$
$$B(x, y) = (B^1(x, y), \ldots, B^l(x, y)),$$

w_t is an n-dimensional Wiener process and $\sigma(x, y) = (\sigma_j^i(x, y))$, $C(x, y) = (C_j^i(x, y))$ are matrices transforming R^n into R^r and R^l, respectively. The functions $b^i(x, y)$, $B^i(x, y)$, $\sigma_j^i(x, y)$, $C_j^i(x, y)$ are assumed to be bounded and satisfy a Lipschitz condition. By this example we illustrate equations of the type (1.5) where the velocity of fast motion depends on the slow variables. We also note that in contrast to the preceding sections, the slow variables in (9.1) form a random process even for given Y_t^ε, $t \in [0, t]$.

We introduce a random process Y_t^{xy}, $x \in R^r$, $y \in R^l$, which is defined by the stochastic differential quation

$$\dot{Y}_t^{xy} = B(x, Y_t^{xy}) + C(x, Y_t^{xy})\dot{w}_t, \qquad Y_0^{xy} = y. \tag{9.2}$$

The solutions of this equation form a Markov process in R^l, depending on $x \in R^r$ as a parameter.

First we formulate and prove the averaging principle in the case where the entries of the matrix $\sigma(x, y)$ do not depend on y and then we indicate the changes necessary for the consideration of the general case.

We assume that there exists a function $\bar{b}(x) = (\bar{b}^1(x), \bar{b}^2(x), \dots, \bar{b}^r(x))$, $x \in R^r$, such that for any $t \geq 0$, $x \in R^r$, $y \in R^l$ we have

$$\mathsf{M}\left|\frac{1}{T}\int_t^{t+T} b(x, Y_s^{xy})\, ds - \bar{b}(x)\right| < \varkappa(T), \tag{9.3}$$

where $\varkappa(T) \to 0$ as $T \to \infty$.

Theorem 9.1. *Let the entries of $\sigma(x, y) = \sigma(x)$ be independent of y and let condition (9.3) be satisfied. Let us denote by $\bar{X}_t$ the random process determined in R^r by the differential equation*[1]

$$\dot{\bar{X}}_t = \bar{b}(\bar{X}_t) + \sigma(\bar{X}_t)\dot{w}_t, \qquad \bar{X}_0 = x.$$

Then for any $T > 0$, $\delta > 0$, $x \in R^r$ and $y \in R^r$ we have

$$\lim_{\varepsilon \to 0} \mathsf{P}\left\{\sup_{0 \leq t \leq T} |X_t^\varepsilon - \bar{X}_t| > \delta\right\} = 0.$$

Proof. We consider a partition of $[0, T]$ into intervals of the same length Δ. We construct auxiliary processes $\hat{Y}_t^\varepsilon$ and $\hat{X}_t^\varepsilon$ by means of the relations

$$\hat{Y}_t^\varepsilon = Y_{k\Delta}^\varepsilon + \frac{1}{\varepsilon}\int_{k\Delta}^t B(X_{k\Delta}^\varepsilon, \hat{Y}_s^\varepsilon)\, ds + \frac{1}{\sqrt{\varepsilon}}\int_{k\Delta}^t C(X_{k\Delta}^\varepsilon, \hat{Y}_s^\varepsilon)\, dw_s,$$

$$t \in [k\Delta, (k+1)\Delta],$$

$$\hat{X}_t^\varepsilon = x + \int_0^t b(X_{[s/\Delta]\Delta}^\varepsilon, \hat{Y}_s^\varepsilon)\, ds + \int_0^t \sigma(X_s^\varepsilon)\, dw_s.$$

We show that the intervals $\Delta = \Delta(\varepsilon)$ can be chosen such that $\varepsilon^{-1}\Delta(\varepsilon) \to \infty$, $\Delta(\varepsilon) \to 0$ as $\varepsilon \to 0$ and

$$\mathsf{M}|Y_t^\varepsilon - \hat{Y}_t^\varepsilon|^2 \to 0 \tag{9.4}$$

uniformly in $x \in R^r$, $y \in R^l$ and $t \in [0, T]$. It follows from the definition of Y_t^ε and $\hat{Y}_t^\varepsilon$ that for t belonging to $[k\Delta, (k+1)\Delta]$ we have

$$\mathsf{M}|Y_t^\varepsilon - \hat{Y}_t^\varepsilon|^2 = \mathsf{M}\left|\frac{1}{\varepsilon}\int_{k\Delta}^t [B(X_s^\varepsilon, Y_s^\varepsilon) - B(X_{k\Delta}^\varepsilon, \hat{Y}_s^\varepsilon)]\, ds \right.$$
$$\left. + \frac{1}{\sqrt{\varepsilon}}\int_{k\Delta}^t [C(X_s^\varepsilon, Y_s^\varepsilon) - C(X_{k\Delta}^\varepsilon, \hat{Y}_s^\varepsilon)]\, dw_s\right|^2$$

[1] In this equation, w_t is the same Wiener process as in (9.1). Since $b(x, y)$ satisfies a Lipschitz condition, $\bar{b}(x)$ also satisfies a Lipschitz condition, so that the solution of the equation exists and is unique.

$$
\begin{aligned}
&\le C_1 \frac{\Delta}{\varepsilon^2}\left(\int_{k\Delta}^t \mathsf{M}|X_s^\varepsilon - X_{k\Delta}^\varepsilon|^2\,ds + \int_{k\Delta}^t \mathsf{M}|Y_s^\varepsilon - \hat{Y}_s^\varepsilon|^2\,ds\right)\\
&\quad + \frac{C_2}{\varepsilon}\left(\int_{k\Delta}^t \mathsf{M}|X_s^\varepsilon - X_{k\Delta}^\varepsilon|^2\,ds + \int_{k\Delta}^t \mathsf{M}|Y_s^\varepsilon - \hat{Y}_s^\varepsilon|^2\,ds\right)\\
&\le C_3\left(\frac{\Delta}{\varepsilon^2} + \frac{1}{\varepsilon}\right)\int_{k\Delta}^t \mathsf{M}|X_s^\varepsilon - X_{k\Delta}^\varepsilon|^2\,ds\\
&\quad + C_4\left(\frac{\Delta}{\varepsilon^2} + \frac{1}{\varepsilon}\right)\int_{k\Delta}^t \mathsf{M}|Y_s^\varepsilon - \hat{Y}_s^\varepsilon|^2\,ds. \qquad (9.5)
\end{aligned}
$$

Here and in what follows, we denote by C_i constants depending only on thc Lipschitz coefficients of $(b^i(x, y), B^l(x, y), \sigma_j^i(x, y), C_j^i(x, y)$, the maximum of the absolute values of these coefficients and the dimension of the space.

It follows from the boundedness of the coefficients of the stochastic equation for X_s^ε that for $\Delta < 1$ we have the estimate

$$
\mathsf{M}|X_s^\varepsilon - X_{k\Delta}^\varepsilon|^2 \le C_5|s - k\Delta|^2 \qquad (9.6)
$$

for $s \in [k\Delta, (k+1)\Delta]$. We obtain from this inequality and (9.5) that

$$
\mathsf{M}|Y_t^\varepsilon - \hat{Y}_t^\varepsilon|^2 \le C_6\left(\frac{1}{\varepsilon} + \frac{\Delta}{\varepsilon^2}\right)\Delta^2 + C_6\left(\frac{\Delta}{\varepsilon^2} + \frac{1}{\varepsilon}\right)\int_{k\Delta}^t \mathsf{M}|Y_s^\varepsilon - \hat{Y}_s^\varepsilon|^2\,ds
$$

for $t \in [k\Delta, (k+1)\Delta]$, from which we arrive at the relation

$$
\mathsf{M}|Y_t^\varepsilon - \hat{Y}_t^\varepsilon|^2 \le C_6\left(\frac{\Delta}{\varepsilon} + \frac{\Delta^2}{\varepsilon^2}\right)\Delta \exp\left\{C_6\left(\frac{\Delta^2}{\varepsilon^2} + \frac{\Delta}{\varepsilon}\right)\right\}.
$$

From this we conclude that (9.4) is satisfied if we put

$$
\Delta = \Delta(\varepsilon) = \varepsilon\sqrt[4]{\ln \varepsilon^{-1}}.
$$

Now we show that for any $\delta > 0$ we have

$$
\mathsf{P}\left\{\sup_{0 \le t \le T} |X_t^\varepsilon - \hat{X}_t^\varepsilon| > \delta\right\} \to 0 \qquad (9.7)
$$

as $\varepsilon \to 0$ and $\Delta = \Delta(\varepsilon) = \varepsilon\sqrt[4]{\ln \varepsilon^{-1}}$ uniformly in $x \in R^r$, $y \in R^l$. Indeed, it follows from the definition of X_t^ε and $\hat{X}_t^\varepsilon$ that

$$
\mathsf{P}\left\{\sup_{0 \le t \le T} |X_t^\varepsilon - \hat{X}_t^\varepsilon| > \delta\right\} \le \mathsf{P}\left\{\int_0^T |b(X_s^\varepsilon, Y_s^\varepsilon) - b(X_{[s/\Delta]\Delta}^\varepsilon, \hat{Y}_s^\varepsilon)|\,ds > \delta\right\}.
$$

Estimating the probability on the right side by means of Chebyshev's inequality and taking account of (9.4) and (9.6), we obtain (9.7). It is also easy to obtain from (9.4) and (9.6) that

$$\sup_{0 \le t \le T} \mathsf{M}|X_t^\varepsilon - \hat{X}_t^\varepsilon|^2 \to 0 \tag{9.8}$$

as $\varepsilon \to 0$.

Now we show that $\sup_{0 \le s \le T}|\hat{X}_t^\varepsilon - \bar{X}_s|$ converges to zero in probability as $\varepsilon \to 0$. The assertion of the theorem will obviously follow from this and (9.7).

First we note that it follows from the definition of $\hat{Y}_t^\varepsilon$ that for $s \in [0, \Delta]$ the process $Z_s = \hat{Y}_{k\Delta+\varepsilon}^\varepsilon$ coincides in distribution with the process $X_{s/\varepsilon}^{X_{k\Delta}^\varepsilon Y_{k\Delta}^\varepsilon}$ defined by equation (9.2). We only have to choose the Wiener process w_t in (9.2) independent of $X_{k\Delta}^\varepsilon$, $Y_{k\Delta}^\varepsilon$.

Taking into account that $\varepsilon^{-1}\Delta(\varepsilon) \to \infty$, we obtain, relying on (9.3), that

$$\mathsf{M}\left|\int_{k\Delta}^{(k+1)\Delta} b(X_{k\Delta}^\varepsilon, \hat{Y}_t^\varepsilon)\,dt - \Delta\bar{b}(X_{k\Delta}^\varepsilon)\right|$$
$$= \Delta\mathsf{M}\left|\frac{\varepsilon}{\Delta}\int_0^{\Delta\varepsilon}[b(X_{k\Delta}^\varepsilon, Z_s) - \bar{b}(X_{k\Delta}^\varepsilon)]\,ds\right| \le \Delta\cdot\varkappa(\Delta/\varepsilon).$$

Using this estimate, we arrive at the relation

$$\mathsf{M}\left\{\sup_{0\le t\le T}\left|\int_0^t b(X_{[s/\Delta]\Delta}^\varepsilon, \hat{Y}_s^\varepsilon)\,ds - \int_0^t \bar{b}(X_s^\varepsilon)\,ds\right|\right\}$$
$$\le \mathsf{M}\left\{\max_{0\le l\le[T/\Delta]}\left|\sum_{k=0}^{l}\int_{k\Delta}^{(k+1)\Delta}[b(X_{k\Delta}^\varepsilon, \hat{Y}_s^\varepsilon) - \bar{b}(X_{k\Delta}^\varepsilon)]\,ds\right|\right\} + C_7\Delta$$
$$\le \sum_{k=0}^{[T/\Delta]}\mathsf{M}\left|\int_{k\Delta}^{(k+1)\Delta}[b(X_{k\Delta}^\varepsilon, \hat{Y}_s^\varepsilon) - \bar{b}(X_{k\Delta}^\varepsilon)]\,ds\right| + C_7\Delta$$
$$\le C_7\Delta + T\varkappa(\Delta/\varepsilon) \to 0 \tag{9.9}$$

as $\varepsilon \to 0$, since $\Delta(\varepsilon) \to 0$, $\varkappa(\Delta(\varepsilon)/\varepsilon) \to 0$ as $\varepsilon \to 0$.

We estimate $m^\varepsilon(t) = \mathsf{M}|\hat{X}_t^\varepsilon - \bar{X}_t|^2$. It follows from the definition of $\bar{X}_t$ and $\hat{X}_t^\varepsilon$ that

$$\hat{X}_t^\varepsilon - \bar{X}_t = \int_0^t [b(X_{[s/\Delta]\Delta}^\varepsilon, \hat{Y}_s^\varepsilon) - \bar{b}(X_s^\varepsilon)]\,ds$$
$$+ \int_0^t [\bar{b}(X_s^\varepsilon) - \bar{b}(\bar{X}_s)]\,ds + \int_0^t [\sigma(X_s^\varepsilon) - \sigma(\bar{X}_s)]\,dw_s.$$

Upon squaring both sides and using some elementary inequalities and the

Lipschitz condition, we arrive at the relation

$$m^\varepsilon(t) \le C_8 t \int_0^t m^\varepsilon(s)\,ds + C_9 \int_0^t m^\varepsilon(s)\,ds + 3\mathsf{M}\left|\int_0^t b(X^\varepsilon_{[s/\Delta]\Delta}, \hat{Y}^\varepsilon_s)\,ds - \int_0^t \bar{b}(X^\varepsilon_s)\,ds\right|^2.$$

We obtain from this relation that

$$m^\varepsilon(t) \le 3\mathsf{M}\left|\int_0^t b(X^\varepsilon_{[s/\Delta]\Delta}, \hat{Y}_s)\,ds - \int_0^t \bar{b}(X^\varepsilon_s)]\,ds\right|^2 \cdot e^{C_{10}(T+T^2)}$$

for $t \in [0, T]$. This implies by (9.9) that $m^\varepsilon(t) \to 0$ uniformly on $[0, T]$ as $\varepsilon \to 0$.

For $\delta > 0$ we have the inequality

$$\begin{aligned}
&\mathsf{P}\left\{\sup_{0\le t\le T} |\hat{X}^\varepsilon_t - \bar{X}_t| > \delta\right\} \\
&\quad\le \mathsf{P}\left\{\sup_{0\le t\le T}\left|\int_0^t [b(X^\varepsilon_{[s/\Delta]\Delta}, \hat{Y}^\varepsilon_s) - \bar{b}(\bar{X}^\varepsilon_s)]\,ds\right| > \delta/6\right\} \\
&\quad\quad + \mathsf{P}\left\{\int_0^T |\bar{b}(\bar{X}_s) - \bar{b}(\hat{X}^\varepsilon_s)|\,ds > \delta/6\right\} \\
&\quad\quad + \mathsf{P}\left\{\int_0^T |\bar{b}(\hat{X}^\varepsilon_s) - \bar{b}(X^\varepsilon_s)|\,ds > \delta/6\right\} \\
&\quad\quad + \mathsf{P}\left\{\sup_{0\le t\le T}\left|\int_0^t [\sigma(X^\varepsilon_s) - \sigma(\hat{X}^\varepsilon_s)]\,dw_s\right| > \delta/6\right\} \\
&\quad\quad + \mathsf{P}\left\{\sup_{0\le t\le T}\left|\int_0^t [\sigma(\hat{X}^\varepsilon_s) - \sigma(\bar{X}_s)]\,dw_s\right| > \delta/6\right\}
\end{aligned}$$

The first term on the right side converges to zero by virtue of (9.9). To prove that the second and third terms also converge to zero, we need to use Chebyshev's inequality, relation (9.8) and the fact that $m^\varepsilon(t) \to 0$ as $\varepsilon \downarrow 0$. The fourth and fifth terms can be estimated by means of Kolmogorov's inequality and also converge to zero. Consequently, we obtain that $\sup_{0\le t\le T} |\hat{X}^\varepsilon_t - \bar{X}_t| \to 0$ in probability as $\varepsilon \downarrow 0$. The assertion of Theorem 9.1 follows from this and (9.7). □

Now we briefly discuss the case where the entries of σ depend on x and y. We assume that condition (9.3) is satisfied, and moreover, there exists a matrix $\bar{a}(x) = (\bar{a}^{ij}(x))$ such that for any $t \ge 0$, $x \in R^r$ and $y \in R^l$ we have

$$\max_{i,j} \mathsf{M}\left|\frac{1}{T}\int_t^{t+T} \sum_k \sigma^i_k(x, Y^{xy}_s)\sigma^j_k(x, Y^{xy}_s)\,ds - \bar{a}^{ij}(x)\right| < \varkappa(T), \quad (9.10)$$

where $\varkappa(T) \to 0$ as $T \to \infty$ and Y_s^{xy} is the solution of equation (9.2). Let $\tilde{X}_t$ be the solution of the stochastic differential equation

$$\dot{\tilde{X}}_t = \bar{b}(\tilde{X}_t) + \bar{\sigma}(\tilde{X}_t)\dot{w}_t, \qquad \tilde{X}_0 = x_1$$

where $\bar{\sigma}(x) = (\bar{a}(x))^{1/2}$ and w_t is an r-dimensional Wiener process. It turns out that in this case X_t^ε converges to $\tilde{X}_t$. However, the convergence has to be understood in a somewhat weaker sense. Namely, the measure corresponding to X_t^ε in the space of trajectories converges weakly to the measure corresponding to $\tilde{X}_t$ as $\varepsilon \downarrow 0$. We shall not include the proof of this assertion here but rather refer the reader to Khas'minskii [6] for a proof. We note that in that work the averaging principle is proved under assumptions allowing some growth of the coefficients. Conditions (9.3) and (9.10) are also replaced by less stringent ones.

We consider an example. Let $(r^\varepsilon, \varphi^\varepsilon)$ be the two-dimensional Markov process governed by the differential operator

$$L^\varepsilon = \frac{1}{2}\frac{\partial^2}{\partial r^2} + b(r, \varphi)\frac{\partial}{\partial r} + \frac{1}{\varepsilon}\left[B(r, \varphi)\frac{\partial}{\partial \varphi} + \tfrac{1}{2}C^2(r, \varphi)\frac{\partial}{\partial \varphi^2}\right].$$

This process can also be described by the stochastic equations

$$\dot{r}^\varepsilon = b(r^\varepsilon, \varphi^\varepsilon) + \dot{w}^1,$$
$$\dot{\varphi}^\varepsilon = \varepsilon^{-1}B(r^\varepsilon, \varphi^\varepsilon) + \varepsilon^{-1/2}C(r^\varepsilon, \varphi^\varepsilon)\dot{w}^2.$$

We assume that the functions $b(r, \varphi)$, $B(r, \varphi)$ and $C(r, \varphi)$ are 2π-periodic in φ and $C(r, \varphi) \geq c_0 > 0$. In this case the process $\tilde{\varphi}_t^{(r_0)}$ obtained from the process

$$\varphi_t^{(r_0)} = \varphi_0^{(r_0)} + \int_0^t B(r_0, \varphi_s^{(r_0)})\, ds + \int_0^t C(r_0, \varphi_s^{(r_0)})\, dw_s$$

by identifying the values differing by an integral multiple of 2π is a nondegenerate Markov process on the circle. This process has a unique invariant measure on the circle with density $m(r_0, \varphi)$ and there exist $C, \lambda > 0$ such that for any bounded measurable function $f(\varphi)$ on the circle

$$\left| \mathsf{M}_{\varphi_0} f(\tilde{\varphi}_t^{(r_0)}) - \int_0^{2\pi} f(\varphi)m(r_0, \varphi)\, d\varphi \right| \leq Ce^{-\lambda t} \sup_{0 \leq \varphi \leq 2\pi} |f(\varphi)|$$

(cf., for example, Freidlin [3]). This implies relation (9.3). Indeed, putting $\bar{b}(r) = \int_0^{2\pi} b(r, \varphi)m(r, \varphi)d\varphi$, we obtain

$$\left(\mathsf{M}_{\varphi_0}\left|\frac{1}{T}\int_0^T b(r, \varphi_s^{(r)})\, ds - \bar{b}(r)\right|\right)^2$$
$$\leq \mathsf{M}_{\varphi_0}\left|\frac{1}{T}\int_0^T b(r, \varphi_s^{(r)})\, ds - \bar{b}(r)\right|^2$$

$$= \frac{1}{T^2} \int_0^T \int_0^T \mathsf{M}_{\varphi_0}(b(r, \varphi_s^{(r)}) - \bar{b}(r))(b(r, \varphi_t^{(r)}) - \bar{b}(r))\, ds\, dt$$

$$= \frac{2}{T^2} \int_0^T ds \int_s^T \mathsf{M}_{\varphi_0}(b(r, \varphi_s^{(r)}) - \bar{b}(r))\mathsf{M}_{\varphi_s}(b(r, \varphi_{t-s}^{(r)}) - \bar{b}(r))\, dt$$

$$\leq \max_{r, \varphi} |b(r, \varphi)| \frac{C}{T^2} \int_0^T ds \int_s^T e^{-\lambda(t-s)}\, dt \to 0$$

as $T \to \infty$.

Hence by Theorem 9.1, r_t^ε converges in probability to the diffusion process $\bar{r}_t$ satisfying the differential equation

$$\dot{\bar{r}}_t = \bar{b}(\bar{r}_t) + \dot{w}_t^1, \qquad \bar{r}_0 = r_0^\varepsilon = r,$$

uniformly on the interval $0 \leq t \leq T$ as $\varepsilon \downarrow 0$.

If a functional $F[r_t^\varepsilon, 0 \leq t \leq T]$ of the process r_t^ε is continuous in $\mathbf{C}_{0T}(R^1)$, then what has been said implies that $F[r_t^\varepsilon, 0 \leq t \leq T] \to F[\bar{r}_t, 0 \leq t \leq T]$ in probability as $\varepsilon \downarrow 0$. The convergence is preserved if F has discontinuities such that the set of functions at which F is discontinuous has measure zero with respect to the measure induced by the limit process $\bar{r}_t$ in the function space.

The results discussed in this section can be used for the study of the behavior, as $\varepsilon \downarrow 0$, of solutions of some elliptic or parabolic equations with a small parameter.

Consider, for example, the Dirichlet problem

$$\begin{aligned} L^\varepsilon u^\varepsilon(r, \varphi) &= 0, \qquad r \in (r_1, r_2); \\ u^\varepsilon(r_1, \varphi) &= C_1, \qquad u^\varepsilon(r_2, \varphi) = C_2 \end{aligned} \tag{9.11}$$

in the domain $D = \{(r, \varphi): 0 < r_1 < r < r_2\}$. If (r, φ) are interpreted as polar coordinates in the plane, then the above domain is the ring bounded by the concentric circles of radii r_1 and r_2, respectively, and center at the origin of coordinates. As is known, the solution of this problem can be written in the form

$$u^\varepsilon(r, \varphi) = C_1 \mathsf{P}_{r, \varphi}\{r_{\tau^\varepsilon}^\varepsilon = r_1\} + C_2 \mathsf{P}_{r, \varphi}\{r_{\tau^\varepsilon}^\varepsilon = r_2\},$$

where $\tau^\varepsilon = \inf\{t: r_t^\varepsilon \notin (r_1, r_2)\}$. We write $\tau = \inf\{t > 0: \bar{r}_t \notin (r_1, r_2)\}$. It is easy to verify that $\max_{r_1 \leq r \leq r_2} \mathsf{P}_r\{\tau > T\} \to 0$ as $T \to \infty$ and that the boundary points of $[r_1, r_2]$ are regular for $\bar{r}_t$ in $[r_1, r_2]$, i.e., that $\mathsf{P}_{r_i}\{\tau = 0\} = 1, i = 1, 2$ (cf. Wentzell [1]). This and the uniform convergence in probability of r_t^ε to $\bar{r}_t$ on every finite interval $[0, T]$ imply that

$$\lim_{\varepsilon \downarrow 0} u^\varepsilon(r, \varphi) = \bar{u}(r) = C_1 \mathsf{P}_r(\bar{r}_\tau = r_1\} + C_2 \mathsf{P}_r\{\bar{r}_\tau = r_2\}.$$

The function $\bar{u}(r)$ can be determined as the solution of the problem

$$\tfrac{1}{2}\bar{u}''(r) + \bar{b}(r)\bar{u}'(r) = 0, \qquad r \in (r_1, r_2).$$

$$\bar{u}(r_1) = C_1, \qquad \bar{u}(r_2) = C_2.$$

Solving this problem, we obtain that

$$\lim_{\varepsilon \downarrow 0} u^\varepsilon(r, \varphi) = \bar{u}(r) = C_1 + (C_2 - C_1) \int_{r_1}^{r} \exp\left\{-2 \int_0^y \bar{b}(x)\, dx\right\} dy$$
$$\times \left(\int_{r_1}^{r_2} \exp\left\{-2 \int_0^y \bar{b}(x)\, dx\right\} dy\right)^{-1}.$$

Some examples of a more general character can be found in Khas'minskii [6].

Now we consider large deviations in systems of the type (9.1). We restrict ourselves to the case where there is no diffusion in the slow motion and the fast motion takes place on a compact manifold and the diffusion coefficients with respect to the fast variables do not depend on the slow variables.

Let M, E be two Riemannian manifolds of class $\mathbf{C}^\infty$. Suppose that E is compact and $\dim M = r$, $\dim E = l$. We denote by TM_x and TE_y the tangent spaces of M and E at $x \in M$ and $y \in E$, respectively. We consider a family of vector fields $b(x, y)$ on M, depending on $y \in E$ as a parameter and a family $B(x, y)$ of fields on E, depending on $x \in M$. On E we consider an elliptic differential operator L of the second order, mapping constants to zero. The functions $b(x, y)$, $B(x, y)$ as well as the coefficients of L are assumed to be infinitely differentiable with respect to their variables.

On the direct product $M \times E$ we consider the family of Markov processes $Z_t^\varepsilon = (X_t^\varepsilon, Y_t^\varepsilon)$ governed by the operators

$$\mathscr{L}^\varepsilon f(x, y) = (b(x, y), \nabla_x f(x, y)) + \varepsilon^{-1}[L_y f(x, y) + (B(x, y), \nabla_y f(x, y))],$$

where ∇_x, ∇_y are the gradient operators on M and E, respectively. In coordinate form the trajectory of the process, $Z_t^\varepsilon = (X_t^\varepsilon, Y_t^\varepsilon)$ can be given by the system of stochastic equations

$$\begin{aligned} \dot{X}_t^\varepsilon &= b(X_t^\varepsilon, Y_t^\varepsilon), \\ \dot{Y}_t^\varepsilon &= \varepsilon^{-1}[B(X_t^\varepsilon, Y_t^\varepsilon) + g(Y_t^\varepsilon)] + \varepsilon^{-1/2} C(Y_t^\varepsilon)\dot{w}_t, \end{aligned} \tag{9.12}$$

where $g(y) = (g^1(y), \dots, g^1(y))$ are the coefficients of first order derivatives in L, the matrix $C(y)$ is connected with the coefficients $a^{ij}(y)$ of higher order differentiations in L by the relation $C(y)C^*(y) = (a^{ij}(y))$ and w_t is an l-dimensional Wiener process.

If M is not compact, then we have to impose an assumption on $b(x, y)$ that it does not grow too fast: for any $T > 0$ and $x \in M$ there exists a compactum $F \subset M$, $x \in F$, such that $\mathsf{P}_{xy}\{X_t^\varepsilon \in F$ for $t \in [0, T]\} = 1$ for every $y \in E$ and $\varepsilon > 0$.

Let α be an element of the dual T^*M_x of TM_x. We introduce the differential operator $R = R(x, z, \alpha)$ acting on functions $f(y)$, $y \in E$, according to the formula

$$R(x, z, \alpha)f(y) = Lf(y) + (B(z, y), \nabla_y f(y)) + (\alpha, b(x, y))f(y);$$

$x, z \in M$ and $\alpha \in T^*M_x$ are parameters. For all values of the parameters, $R(x, z, \alpha)$ is an elliptic differential operator in the space of functions defined on E. It is the restriction, to smooth functions, of the infinitesimal generator of a positive semigroup. Analogously to §4, we can deduce from this that $R(x, z, \alpha)$ has a simple eigenvalue $\mu(x, z, \alpha)$ with largest real part. This eigenvalue is real and by virtue of its simplicity, it is differentiable with respect to the parameters x, z, α.

We introduce the diffusion process Y_t^z, $z \in M$, on E, governed by the operator

$$N^z = L + (B(z, y), \nabla_y).$$

Lemma 9.1. *Let $\alpha \in T^*M_x$ and let F be a compactum in M. The limit*

$$\lim_{T\to\infty} \frac{1}{T} \ln \mathsf{M}_y \exp\left\{\int_0^T (\alpha, b(x, Y_\varepsilon^z))\, ds\right\} = \mu(x, z, \alpha)$$

exists uniformly in $x, z \in F$ and $y \in E$. The function $\mu(x, z, \alpha)$ is convex downward in the variables α.

Proof. Let us write $V(x, y, \alpha) = (\alpha, b(x, y))$. The family of operators T_t^V acting in the space of bounded measurable functions on E according to the formula

$$T_t^V f(y) = \mathsf{M}_y f(Y_t^z) \exp\left\{\int_0^t V(x, Y_s^z, \alpha)\, ds\right\}$$

forms a positive semigroup. The assertion of Lemma 9.1 can be derived from this analogously to the proof of Theorem 4.2. □

Let us denote by $L(x, z, \beta)(x, z \in M, \beta \in TM_x)$ the Legendre transform of the function $\mu(x, z, \alpha)$ with respect to the last argument:

$$L(x, z, \beta) = \sup_\alpha [(\alpha, \beta) - \mu(x, z, \alpha)].$$

We shall sometimes consider $L(x, z, \beta)$ with coinciding first two arguments. We write $L(x, x, \beta) = L(x, \beta)$. This function is obviously the Legendre transform of $\mu(x, x, \alpha)$. We note that $L(x, z, \alpha)$ is lower semicontinuous.

Theorem 9.2 (Freidlin [11]). *Let X_t^ε be the first component of the Markov process Z_t^ε governed by the operator $\mathcal{L}^\varepsilon$ on $M \times E$. Let us put*

$$S_{0T}(\varphi) = \int_0^T L(\varphi_s, \dot{\varphi}_s)\, ds$$

for absolutely continuous functions $\varphi \in \mathbf{C}_{0T}(M)$; for the remaining $\varphi \in \mathbf{C}_{0T}(M)$ we set $S_{0T}(\varphi) = +\infty$.

The functional $\varepsilon^{-1} S_{0T}(\varphi)$ is the action functional for the family of processes X_t^ε, $t \in [0, T]$ in $\mathbf{C}_{0T}(M)$ as $\varepsilon \downarrow 0$.

Proof. Together with the process $Z_t^\varepsilon = (X_t^\varepsilon, Y_t^\varepsilon)$, we consider the process $\tilde{Z}_t^\varepsilon = (\tilde{X}_t^\varepsilon, \tilde{Y}_t^\varepsilon)$, where

$$\dot{\tilde{X}}_t^\varepsilon = b(\tilde{X}_t^\varepsilon, \tilde{Y}_t^\varepsilon),$$
$$\dot{\tilde{Y}}_t^\varepsilon = \varepsilon^{-1} g(\tilde{Y}_t^\varepsilon) + \varepsilon^{-1/2} C(\tilde{Y}_t^\varepsilon)\dot{w}_t.$$

We write $e(x, y) = C^{-1}(y)B(x, y)$. The process $\tilde{Z}_t^\varepsilon$ differs from Z_t^ε by a change of the drift vector in the variables in which there is a nondegenerate diffusion, so that the measures corresponding to these processes in the space of trajectories are absolutely continuous with respect to each other. Taking account of this observation, we obtain for any function $\varphi\colon [0, T] \to M$ and $\delta > 0$ that

$$\begin{aligned} \mathsf{P}_{x,y}\{\rho_{0T}(X^\varepsilon, \varphi) < \delta\} = \mathsf{M}_{x,y}\Big\{\rho_{0T}(\tilde{X}^\varepsilon, \varphi) < \delta; \\ \exp\Big\{\varepsilon^{-1/2}\int_0^T (e(\tilde{X}_s^\varepsilon, \tilde{Y}_s^\varepsilon), dw_s) - (2\varepsilon)^{-1}\int_0^T |e(\tilde{X}_s^\varepsilon, \tilde{Y}_s^\varepsilon)|^2\, ds\Big\}\Big\}. \end{aligned} \tag{9.13}$$

Let $\psi^{(n)}\colon [0, T] \to M$ be a step function such that $\rho_{0T}(\varphi, \psi^{(n)}) < 1/n$. For any $\gamma, C > 0$ we have

$$\mathsf{P}\Big\{\varepsilon^{-1/2}\Big|\int_0^T (e(\tilde{X}_s^\varepsilon, \tilde{Y}_s^\varepsilon), dw_s) - \int_0^T (e(\psi_s^{(n)} \tilde{Y}_s^\varepsilon), dw_s)\Big| > \frac{\gamma}{4\varepsilon}; \rho_{0T}(\tilde{X}^\varepsilon, \varphi) < \delta\Big\}$$
$$< \exp\{-C\varepsilon^{-1}\}$$

for sufficiently small δ and $1/n$. This estimate can be verified by means of the exponential Chebyshev inequality. We obtain from this and (9.13) that for any $\varphi \in \mathbf{C}_{0T}(M)$ and $\gamma > 0$ we have

$$\mathsf{M}_{x,y}\left\{\rho_{0T}(\tilde{X}^\varepsilon, \varphi) < \delta; \exp\left\{\varepsilon^{-1/2}\int_0^T (e(\psi_s^{(n)}, \tilde{Y}_s^\varepsilon), dw_s) - (2\varepsilon)^{-1}\int_0^T |e(\psi_s^{(n)}, \tilde{Y}_s^\varepsilon)|^2\, ds - \frac{\gamma}{3\varepsilon}\right\}\right\}$$
$$\leq \mathsf{P}_{x,y}\{\rho_{0T}(X^\varepsilon, \varphi) < \delta\}$$
$$\leq \mathsf{M}_{x,y}\left\{\rho_{0T}(\tilde{X}^\varepsilon, \varphi) < \delta; \exp\left\{\varepsilon^{-1/2}\int_0^T (e(\psi_s^{(n)}, \tilde{Y}_s^\varepsilon), dw_s) - (2\varepsilon)^{-1}\int_0^T |e(\psi_s^{(n)}, \tilde{Y}_s^\varepsilon)|^2\, ds + \frac{\gamma}{3\varepsilon}\right\}\right\} \tag{9.14}$$

for sufficiently small δ and $1/n$.

We introduce still another process, $\hat{Z}_t^\varepsilon = (\hat{X}_t^\varepsilon, \hat{Y}_t^\varepsilon)$, which is defined by the stochastic equations

$$\dot{\hat{X}}_t^\varepsilon = b(\hat{X}_t^\varepsilon, \hat{Y}_t^\varepsilon),$$
$$\dot{\hat{Y}}_t^\varepsilon = \varepsilon^{-1}g(\hat{Y}_t^\varepsilon) + \varepsilon^{-1}B(\psi_t^{(n)}, \hat{Y}_t^\varepsilon) + \varepsilon^{-1/2}C(\hat{Y}_t^\varepsilon)\dot{w}_t$$

in coordinate form.

Taking account of the absolute continuity of the measures corresponding to Z_t^ε and $\hat{Z}_t^\varepsilon$, it follows from inequality (9.14) that

$$e^{-\gamma/2\varepsilon} < \frac{\mathsf{P}_{x,y}\{\rho_{0T}(X^\varepsilon, \varphi) < \delta\}}{\mathsf{P}_{x,y}\{\rho_{0T}(\hat{X}^\varepsilon, \varphi) < \delta\}} < e^{\gamma/2\varepsilon} \tag{9.15}$$

for sufficiently small δ and $1/n$.

Let $t_1 < t_2 < \cdots < t_{m-1}$ be the points where $\psi_t^{(n)}$ has jumps, $t_0 = 0$, $t_m = T$ and $\psi_t^{(n)} = \psi(k)$ for $t \in [t_k, t_{k+1})$, $k = 0, 1, \ldots, m-1$. The process $\hat{X}_t^\varepsilon$ satisfies the hypotheses of Theorem 4.1 on every interval $[t_k, t_{k+1})$. The rọle of $\xi_{t/\varepsilon}$ is played by $\hat{Y}_t^\varepsilon$, which can be represented in the form $\hat{Y}_t^\varepsilon = \hat{Y}_{t/\varepsilon}^1$, $\dot{\hat{Y}}_t^1 = g(\hat{Y}_t^1) + B(\psi(k), \hat{Y}_t^1) + C(\hat{Y}_t^1)\dot{w}_t$. The fulfilment of condition F follows from Lemmas 4.2 and 9.1. The corresponding functional has the form $\int_{t_k}^{t_{k+1}} L(\psi(k), \varphi_s, \dot{\varphi}_s)\, ds$ for absolutely continuous functions. It follows from this and (9.15) that for any $\gamma > 0$ and sufficiently small δ and $1/n$ there exists $\varepsilon_0 > 0$ such that for $\varepsilon < \varepsilon_0$ we have the estimates

$$\exp\left\{-\varepsilon^{-1}\left(\int_0^T L(\psi_s^{(n)}, \varphi_s, \dot{\varphi}_s)\, ds + \frac{\gamma}{2}\right)\right\}$$
$$\leq \mathsf{P}_{x,y}\{\rho_{0T}(X^\varepsilon, \varphi) < \delta\}$$
$$\leq \exp\left\{-\varepsilon^{-1}\left(\int_0^T L(\psi_s^{(n)}, \varphi_s, \dot{\varphi}_s)\, ds - \frac{\gamma}{2}\right)\right\}. \tag{9.16}$$

The concluding part of the proof of this theorem can be carried out in the same way as the end of the proof of Theorem 4.1. We leave it to the reader. □

Remark 1. The assertion of Theorem 9.2 remains true if the manifold E is replaced by a compact manifold $\tilde{E}$ with boundary and as the process $Z_t^\varepsilon = (X_t^\varepsilon, Y_t^\varepsilon)$, we choose a process in $M \times \tilde{E}$, which is governed by the operator $\mathscr{L}^\varepsilon$ at interior points and for y belonging to the boundary $\partial\tilde{E}$ of $\tilde{E}$ it satisfies some boundary conditions, for example, the condition of reflection along a field $n(y)$, where $n(y)$ is a vector field on $\partial\tilde{E}$, tangent to $\tilde{E}$ but not tangent to $\partial\tilde{E}$. The study of such a process can be reduced to the study of a process on the manifold $M \times E'$, where E' is the manifold without boundary, obtained by pasting two copies of $\tilde{E}$ along the boundary $\partial\tilde{E}$ (cf. Freidlin [3]).

Remark 2. It is easy to write out the action functional characterizing deviations of order $\varepsilon^\varkappa$, $\varkappa \in (0, \frac{1}{2})$ of the process X_t^ε defined by equations (9.12) from the averaged system. For example, let the origin of coordinates 0 be an equilibrium position of the averaged system. Then the action functional for the process $Z_t^\varepsilon = \varepsilon^{-\varkappa} X_t^\varepsilon$ has the same form as in Theorem 7.1; as the matrix C we have to take $(\partial^2 \mu(0, 0, \alpha)/\partial\alpha_i\, \partial\alpha_j)$ for $\alpha = 0$.

As a majority of results related to diffusion processes, the theorem given here on large deviations is closely connected with some problems for differential equations of the second order with a nonnegative characteristic form. We consider an example. Let $Y_t^\varepsilon = \xi_{t/\varepsilon}$, where ξ_t is a Wiener process on the interval $[-1, 1]$ with reflection at the endpoints. We define a process X_t^ε in R^r by the differential equation $\dot{X}_t^\varepsilon = b(X_t^\varepsilon, Y_t^\varepsilon)$. To write out the action functional for the family of processes X_t^ε, in accordance with Theorem 9.2 and Remark 1 (or according to Theorem 4.1; the fast and slow motions are separated here), we need to consider the eigenvalue problem

$$Nu(y) = \frac{1}{2}\frac{d^2u}{dy^2} + (\alpha, b(x, y))u = \lambda u(y), \qquad \left.\frac{du}{\partial y}\right|_{y=\pm 1} = 0.$$

Let $\lambda = \lambda(x, \alpha)$ be the eigenvalue of the operator N with largest real part and let $L(x, \beta)$ be the Legendre transform of $\lambda(x, \alpha)$ with respect to the second argument. Then the normalized action functional for the processes X_t^ε as $\varepsilon \downarrow 0$ has the form $S_{0T}(\varphi) = \int_0^T L(\varphi_s, \dot{\varphi}_s)\, ds$ for absolutely continuous functions. As was explained earlier, to determine the asymptotics of $\mathsf{P}_{x,y}\{X_t^\varepsilon \in D\}$, $D \subset R^r$, as $\varepsilon \downarrow 0$ and the asymptotics of the probabilities of other events connected with X_t^ε, we need to calculate $u_x(t, z) = \inf\{S_{0t}(\varphi): \varphi \in H_t(x, z)\}$, where $H_t(x, z)$ is the set of functions φ such that $\varphi_0 = x$, $\varphi_t = z$. The initial point x is assumed to be given and we sometimes omit it in our notation. As follows from Theorem 4.1, the action functional vanishes for trajectories of the averaged system and only for them. It can be proved that in our case the averaged system has the form

$$\dot{\bar{x}}_t = \bar{b}(\bar{x}_t), \qquad \bar{b}(x) = \frac{1}{2}\int_{-1}^{1} b(x, y)\,dy, \tag{9.17}$$

so that if the point z lies on the trajectory $\bar{x}_t$ of system (9.17) for which $\bar{x}_0 = x$ and $\bar{x}_{t_0} = z$, then $u(t_0, z) = 0$. For the determination of $u(t, z)$ we may use the Hamilton–Jacobi equation. In the case being considered it has the form

$$\frac{\partial u}{\partial t} = \lambda(z, \nabla_z u). \tag{9.18}$$

Since N is a self-adjoint semibounded operator, for its largest eigenvalue $\lambda(x, \alpha)$ we have the representation

$$\lambda(x, \alpha) = \min_{\Sigma}\left\{\frac{1}{2}\int_{-1}^{1} [f'(y)|^2\,dy - \int_{-1}^{1} (\alpha, b(x, y))f^2(y)\,dy\right\},$$

where $\sum = \{f: \int_{-1}^{1} |f(y)|^2\,dy = 1, f'(1) = f'(-1) = 0\}$. It follows from this that the solutions of equation (9.18) satisfy the following nonlinear differential equation:

$$\frac{\partial u}{\partial t}(t, z) = \min_{\Sigma}\left\{\frac{1}{2}\int_{-1}^{1} [f'(y)]^2\,dy - \int_{-1}^{1} (\nabla_z u, b(z, y))f^2(y)\,dy\right\}.$$

Consequently, an equation can be given for $u(t, z)$ without determining the first eigenvalue of N.

Suppose we would like to determine the asymptotics, as $\varepsilon \downarrow 0$, of $v^\varepsilon(t, x, y) = \mathsf{M}_{xy} f(X_t^\varepsilon)$, where $f(x): R^r \to R^1$ is a smooth nonnegative function, different from zero on a set $G \subset R^r$. It is easy to see that $v^\varepsilon(t, x, y) \to f(\bar{x}_t(x))$, where $\bar{x}_t(x)$ is the solution of the averaged system (9.17) with initial condition $\bar{x}_0(x) = x$. In particular, if $\bar{x}_t(x) \notin G$, then $v^\varepsilon(t, x, y) \to 0$ as $\varepsilon \downarrow 0$. The rate of convergence of $v^\varepsilon(t, x, y)$ to zero can be estimated by means of the theorem on large deviations for the family of processes X_t^ε:

$$\lim_{\varepsilon \downarrow 0} \varepsilon \ln v^\varepsilon(t, x, y) = - \inf_{z \in G \cup \partial G} u_x(t, z). \tag{9.19}$$

As is known, $v^\varepsilon(t, x, y)$ is a solution of the problem

$$\frac{\partial v^\varepsilon}{\partial t} = \frac{1}{2\varepsilon}\frac{\partial^2 v^\varepsilon}{\partial y^2} + \sum_{i=1}^{r} b^i(x, y)\frac{\partial v^\varepsilon}{\partial x^i}, \qquad x \in R^r, \quad y \in (-1, 1), \quad t > 0,$$

$$v^\varepsilon(0, x, y) = f(x), \qquad \left.\frac{\partial v^\varepsilon}{\partial y}(t, x, y)\right|_{y = \pm 1} = 0. \tag{9.20}$$

The function $\bar{v}(t, x) = f(\bar{x}_t(x))$ obviously satisfies the equation

$$\frac{\partial \bar{v}}{\partial t}(t, x) = \sum_{i=1}^{r} \bar{b}^i(x) \frac{\partial \bar{v}}{\partial x^i}, \qquad t > 0, \quad x \in R^r;$$
$$\bar{v}(0, x) = f(x), \tag{9.21}$$

where $\bar{b}^i(x) = \frac{1}{2}\int_{-1}^{1} b^i(x, y)\,dy$. Therefore, the convergence of X_t^ε to the trajectories of system (9.17) implies the convergence of the solution of problem (9.20) to the solution of problem (9.21) as $\varepsilon \downarrow 0$. Relation (9.18) enables us to estimate the rate of this convergence.

As we have seen in Chapters 4 and 6, in stationary problems with a small parameter, large deviations may determine the principal term of asymptotics, not only the terms converging to zero with ε. We consider the stationary problem corresponding to equation (9.20). For the sake of brevity, we shall assume that $r = 1$:

$$L^\varepsilon w^\varepsilon(x, y) = \frac{1}{2\varepsilon}\frac{\partial^2 w^2}{\partial y^2} + b(x, y)\frac{\partial w^\varepsilon}{\partial x} = 0, \qquad x \in (-1, 1), \quad y \in (-1, 1),$$
$$\left.\frac{\partial w^\varepsilon}{\partial y}(x, y)\right|_{y = \pm 1} = 0, \qquad w^\varepsilon(1, y)|_{y \in \Gamma_+} = \psi(1, y), \tag{9.22}$$
$$w^\varepsilon(-1, y)|_{y \in \Gamma_-} = \psi(-1, y),$$

where $\Gamma_+ = \{y \in [-1, 1]: b(1, y) > 0\}$, $\Gamma_- = \{y \in [-1, 1]: b(-1, y) < 0\}$, $\psi(1, y)$ and $\psi(-1, y)$ are continuous functions defined for $y \subset [\ 1, 1]$. A solution, at least a generalized solution, of problem (9.22) always exists but without additional assumptions it is not unique (cf. Freidlin [1], [3]). We require that for every $x_0 \in [-1, 1]$ there exists $y_0 = y_0(x_0)$ such that either $b(x, y_0) > 0$ for $x \geq x_0$ or $b(x, y_0) < 0$ for $x \leq x_0$. This condition ensures the uniqueness of the solution of problem (9.22). If $Z_t^\varepsilon = (X_t^\varepsilon, Y_t^\varepsilon)$ is the process in the strip $|y| \leq 1$ with reflection along the normal to the boundary, governed by the operator L^ε for $|y| < 1, x \in (-\infty, \infty)$ and $\tau^\varepsilon = \min\{t: |X_t^\varepsilon| = 1\}$, then the unique solution of problem (9.22) can be written in the form $w^\varepsilon(x, y) = \mathsf{M}_{x, y}\psi(X_{\tau^\varepsilon}^\varepsilon, Y_{\tau^\varepsilon}^\varepsilon)$.

Now we consider the averaged dynamical system (9.17) on the real line. We assume that the trajectories of the averaged equation, beginning at any point $x \in [-1, 1]$, leave $[-1, 1]$. It is then obvious that $\bar{b}(x)$ does not change sign. For the sake of definiteness, let $\bar{b}(x) > 0$ for $x \in [-1, 1]$. If we assume in addition that $b(1, y) \geq 0$ for $y \in [-1, 1]$, then it is easy to prove that

$$\lim_{\varepsilon \downarrow 0} w^{\varepsilon}(x, y) = \int_{-1}^{1} \psi(1, y) b(1, y)\, dy \left(\int_{-1}^{1} b(1, y)\, dy \right)^{-1}.$$

If we do not assume that $b(1, y) \geq 0$ for $y \in [-1, 1]$ in (1.9), then the situation becomes much more complicated. Concerning this, cf. Sarafyan, Safaryan, and Freidlin [1]. This work also discusses the case where the trajectories of the averaged motion do not leave $(-1, 1)$.

Chapter 8

Random Perturbations of Hamiltonian Systems

§1. Introduction

Consider the dynamical system in R^r defined by a smooth vector field $b(x)$:

$$\dot{x}_t = b(x_t), \qquad x_0 = x \in R^r. \tag{1.1}$$

In Chapters 2–6 we have considered small random perturbations of system (1.1) described by the equation

$$\dot{\tilde{X}}_t^\varepsilon = b(\tilde{X}_t^\varepsilon) + \varepsilon \dot{\tilde{w}}_t, \qquad \tilde{X}_0^\varepsilon = x, \tag{1.2}$$

where $\tilde{w}_t$ is a Wiener process, and ε a small parameter (we reserve the notation without ˜ for another Wiener process and another stochastic equation). The long-time behavior of the process $\tilde{X}_t^\varepsilon$ was described in Chapter 6 in terms that presupposed identification of points x, y of the space that are equivalent under the equivalence relation that was introduced in Section 1 of that chapter. But for some dynamical systems *all* points of the phase space are equivalent—e.g., if the trajectories of the dynamical system are of the form shown in Fig. 10. An important class of such systems is provided by Hamiltonian systems.

A dynamical system (1.1) in R^{2n} is called a Hamiltonian system if there exists a smooth function $H(x)$, $x \in R^{2n}$, such that

$$b(x) = \bar{\nabla} H(x) = \left(\frac{\partial H(p,q)}{\partial q_1}, \ldots, \frac{\partial H(p,q)}{\partial q_n}; -\frac{\partial H(p,q)}{\partial p_1}, \ldots, -\frac{\partial H(p,q)}{\partial p_n} \right),$$

$x = (p,q) = (p_1, \ldots, p_n; q_1, \ldots, q_n) \in R^{2n}$.

The function $H(p,q)$ is called the Hamiltonian, and n is the number of degrees of freedom. It is well known that $H(x)$ is an integral of motion: $H(x_t) = H(p_t, q_t) = H(p_0, q_0)$ is a constant; and the flow (p_t, q_t) preserves the Euclidean volume in R^{2n}. The invariant measure concentrated on a trajectory of the dynamical system is proportional to $dl/|b(x)|$, where dl is the length along the trajectory.

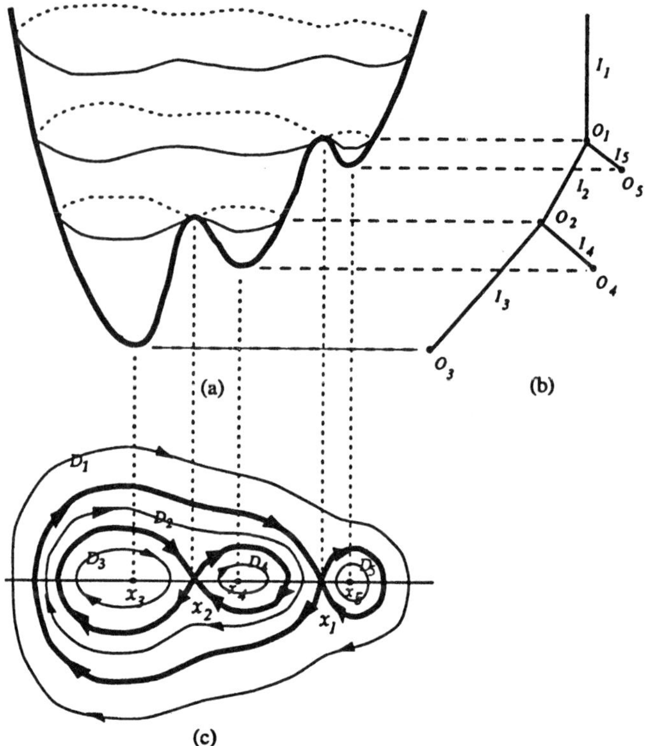

Figure 18.

We consider Hamiltonian systems with one degree of freedom:

$$\dot{x}_t(x) = \overline{\nabla} H(x_t(x)), \qquad x_0(x) = x \in R^2. \tag{1.3}$$

We assume that $H(x)$ is smooth enough and $\lim_{|x|\to\infty} H(x) = +\infty$. A typical example is shown in Fig. 18(a). The trajectories of the corresponding system are shown in Fig. 18(b). These trajectories consist of five families of periodic orbits and separatrices dividing these families: three families of closed trajectories encircling exactly one of the stable equilibrium points x_3, x_4, or x_5 (the regions covered by the trajectories are denoted by D_3, D_4, and D_5, respectively); the family of closed trajectories encircling x_3 and x_4 (they cover the region D_2); and the family of closed trajectories encircling all equilibrium points (the orbits of this family cover the region D_1). These families are separated by two ∞-shaped curves with crossing points x_1 and x_2. Each ∞-shaped curve consists of an equilibrium point (x_1 or x_2) and two trajectories approaching the equilibrium point as $t \to \pm\infty$.

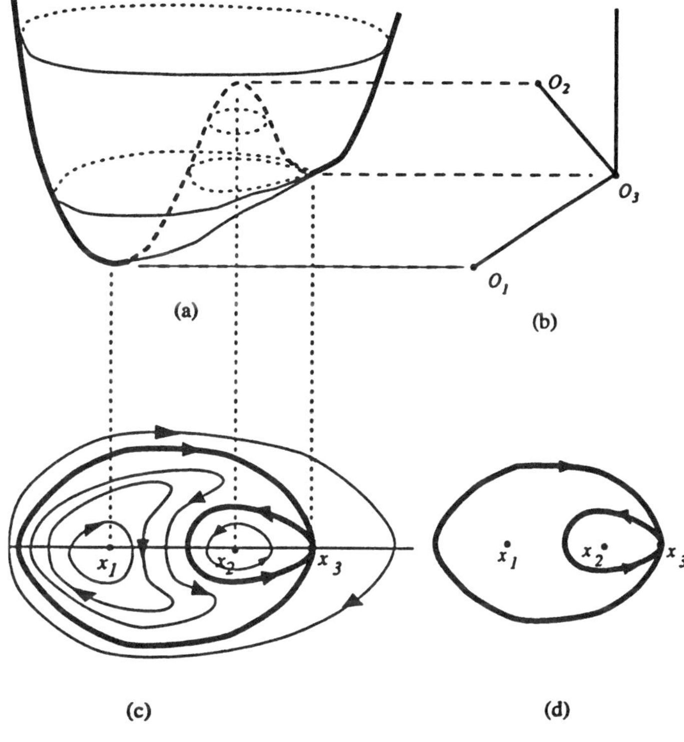

Figure 19.

Note that if the Hamiltonian has the shape shown in Fig. 19(a), the self-intersecting separatrix may look like that in Fig. 19 (d). Still in this case we call it loosely an ∞-shaped curve.

Consider the perturbed system corresponding to (1.3):

$$\dot{\tilde{X}}_t^\varepsilon = \overline{\nabla} H(\tilde{X}_t^\varepsilon) + \varepsilon \dot{\tilde{w}}_t, \qquad \tilde{X}_0^\varepsilon = x \in R^2. \tag{1.4}$$

Here $\tilde{w}_t$ is a two-dimensional Wiener process, and ε, a small parameter. The motion $\tilde{X}_t^\varepsilon$, roughly speaking, consists of fast rotation along the non-perturbed trajectories and slow motion across them; so this is a situation where the averaging principle is to be expected to hold (see Chapter 7).

To study slow motion across the deterministic orbits it is convenient to rescale the time. Consider, along with $\tilde{X}_t^\varepsilon$, the process X_t^ε described by

$$\dot{X}_t^\varepsilon = \frac{1}{\varepsilon^2} \overline{\nabla} H(X_t^\varepsilon) + \dot{w}_t, \qquad X_0^\varepsilon = x \in R^2, \tag{1.5}$$

where w_t is a two-dimensional Wiener process. It is easy to see that the process X_t^ε coincides, in the sense of probability distributions, with $\tilde{X}_{t/\varepsilon^2}^\varepsilon$. We denote by P_x^ε the probabilities evaluated under the assumption that $\tilde{X}_0^\varepsilon = x$ or $X_0^\varepsilon = x$. The diffusions $\tilde{X}_t^\varepsilon$, X_t^ε defined by (1.4) and (1.5) have generating differential operators

$$\tilde{L}^\varepsilon = b(x) \cdot \nabla + \frac{\varepsilon^2}{2} \Delta, \qquad L^\varepsilon = \frac{1}{\varepsilon^2} b(x) \cdot \nabla + \tfrac{1}{2}\Delta,$$

where $b(x) = \bar{\nabla} H(x)$, and the Lebesgue measure is invariant for these processes, as it is for system (1.3). (The same is true for other forms of perturbations of the system (1.3) that lead to diffusions with generators

$$\tilde{L}^\varepsilon = b(x) \cdot \nabla + \varepsilon^2 L_0, \qquad L^\varepsilon = \frac{1}{\varepsilon^2} b(x) \cdot \nabla + L_0,$$

where L_0 is a self-adjoint elliptic operator; but for simplicity we stick to the case of $L_0 = \frac{1}{2}\Delta$.)

If the diffusions $\tilde{X}^\varepsilon$, X^ε are moving in a region covered by closed trajectories $x_t(x)$ (in one of the regions D_k in Fig. 18(b)), the results of Khas'minskii [6] can be applied, and the averaging principle holds: for small ε the motion in the direction of these trajectories is approximately the same as the nonrandom motion along the same closed trajectory, and the process makes very many rotations before it moves to another trajectory at a significant distance from the initial one; the motion from one trajectory to another ("across" the trajectories) is approximately a diffusion whose characteristics at some point (trajectory) are obtained by averaging with respect to the invariant measure concentrated on this closed trajectory. This diffusion is "slow" in the case of the process $\tilde{X}_t^\varepsilon$, but it has a "natural" time scale in the case of X_t^ε.

As for the characteristics of this diffusion on the trajectories, let us use the function H as the coordinate of a trajectory (H can take the same value on different trajectories, so it is only a *local* coordinate). Let us apply Itô's formula to $H(X_t^\varepsilon)$:

$$dH(X_t^\varepsilon) = \nabla H(X_t^\varepsilon) \cdot \frac{1}{\varepsilon^2} b(X_t^\varepsilon)\, dt + \nabla H(X_t^\varepsilon) \cdot dw_t + \tfrac{1}{2}\Delta H(X_t^\varepsilon)\, dt.$$

The dot product $\nabla H \cdot b = 0$ since $b(x) = \bar{\nabla} H(x)$, so

$$H(X_t^\varepsilon) = H(X_0^\varepsilon) + \int_0^t \nabla H(X_s^\varepsilon) \cdot dw_s + \int_0^t \tfrac{1}{2}\Delta H(X_s^\varepsilon)\, ds. \tag{1.6}$$

Since X_t^ε rotates many times along the trajectories of the Hamiltonian system before $H(X_t^\varepsilon)$ changes considerably, the integral of $\frac{1}{2}\nabla H(X_s^\varepsilon)$ is approximately equal to the integral $\int_0^t B(H(X_s^\varepsilon))\,ds$, where

$$B(H) = \frac{\oint(\frac{1}{2}\Delta H(x)/|b(x)|)\,dl}{\oint(1/|b(x)|)\,dl}, \tag{1.7}$$

the integrals being taken over the connected component of the level curve $\{x : H(x) = H\}$ lying in the region considered. As for the stochastic integral in (1.6), it is well known that it can be represented as

$$\int_0^t \nabla H(W_s^\varepsilon) \cdot dw_s = W\left(\int_0^t |\nabla H(X_s^\varepsilon)|^2\,ds\right), \tag{1.8}$$

where $W(\cdot)$ is a one-dimensional Wiener process, $W(0) = 0$ (see, e.g., Freidlin [15], Chapter 1). The argument in this Wiener process is approximately $\int_0^t A(H(X_s^\varepsilon))\,ds$, where

$$A(H) = \frac{\oint(|\nabla H(x)|^2/|b(x)|)\,dl}{\oint(1/|b(x)|)\,dl}, \tag{1.9}$$

the integrals being taken over the same curve as in (1.7) (we may mention that $|b(x)| = |\nabla H(x)|$, so the integrand in the upper integral is equal to $|\nabla H(x)|$). This suggests that the "slow" process on the trajectories is, for small ε, approximately the same as the diffusion corresponding to the differential operator

$$Lf(H) = \tfrac{1}{2}A(H)f''(H) + B(H)f'(H).$$

This describes the "slow" process in the space obtained by identifying all points on the same trajectory of (1.3); but *only while* the process X_t^ε is moving in a region covered by closed trajectories. But this process can move from one region covered by closed trajectories to another, and such regions are separated by the components of level curves $\{x : H(x) = H\}$ that are not closed trajectories (see Fig. 18(b)).

If we identify all points belonging to the same component of a level curve $\{x : H(x) = H\}$, we obtain a graph consisting of several segments, corresponding: I_1, to the trajectories in the domain D_1 outside the outer ∞-curve; I_2, to the trajectories in D_2 between the outer and the inner ∞-curve; I_3 and I_4, to the trajectories inside the two loops of the inner ∞-curve (domains D_3 and D_4), and I_5, to those inside the right loop of the outer ∞-curve (domain D_5) (see Fig. 18).

The ends of the segments are vertices O_1 and O_2 corresponding to the ∞-curves, O_3, O_4, O_5 corresponding to the extrema x_3, x_4, x_5 (it happens that the curves corresponding to O_1, O_2 each contain one critical point of H: a saddle point). Let us complement our graph by a vertex O_∞ being the end of the segment I_1 that corresponds to the point at infinity. Let us denote the graph thus obtained by Γ.

Let $Y(x)$ be the identification mapping ascribing to each point $x \in R^2$ the corresponding point of the graph. We denote the function H carried over to the graph Γ under this mapping also by H (we take $H(O_\infty) = +\infty$). The function H can be taken as the local coordinate on this graph. Couples (i, H), where i is the number of the segment I_i, define global coordinates on the graph. Several such couples may correspond to a vertex.

The graph has the structure of a tree in the case of a system in R^2; but for a system on a different manifold it may have loops.

Note that the function $i(x)$, $x \in R^2$, the number of the segment of the graph Γ containing the point $Y(x)$, is preserved along each trajectory of the unperturbed dynamical system: $i(X_t^x) \equiv i(x)$ for every t. This means that $i(x)$ is a first integral of the system (1.1)—a discrete one. If the Hamiltonian has more than one critical point, then there are points $x, y \in R^2$ such that $H(x) = H(y)$ and $i(x) \neq i(y)$. In this case the system (1.1) has two independent first integrals, $H(x)$ and $i(x)$. This is, actually, the reason why $H(X_t^\varepsilon)$ does not converge to a Markov process as $\varepsilon \to 0$ in the case of several critical points. If there is more than one first integral, we have to take a larger phase space, including all first integrals as the coordinates in it.

In the present case, it is two coordinates (i, H).

The couple $(i(X_t^\varepsilon), H(X_t^\varepsilon)) = Y(X_t^\varepsilon)$ is a stochastic process on the graph Γ, and it is reasonable to expect that as $\varepsilon \to 0$, this process converges in a sense to some diffusion process Y_t on the graph Γ.

The problem about diffusions on graphs arising as limits of fast motion of dynamical systems was first proposed in Burdzeiko, Ignatov, Khas'minskii, Shakhgil'dyan [1]. Some results concerning random perturbations of Hamiltonian systems with one degree of freedom leading to random processes on graphs were considered by G. Wolansky [1], [2]. We present here the results of Freidlin and Wentzell [2]. Some other asymptotic problems for diffusion processess in which the limiting process is a diffusion process on a graph were considered in Freidlin and Wentzell [3].

What can we say about the limiting process Y_t?

First of all, with each segment I_i of the graph we can associate a differential operator acting on functions on this segment:

$$L_i f(H) = \tfrac{1}{2} A_i(H) f''(H) + B_i(H) f'(H); \tag{1.10}$$

$A_i(H)$, $B_i(H)$ are defined by formulas (1.9) and (1.7) with the integrals taken over the connected component of the level curve $\{x : H(x) = H\}$ lying in the region D_i corresponding to I_i. These differential operators determine the behavior of the limiting process Y_t on Γ, but only as long as it moves inside one segment of the graph. What can we say about the process after it leaves the segment of the graph where it started?

The problem naturally arises of describing the class of diffusions on a graph that are governed by the given differential operators while inside its segments. This problem was considered by W. Feller [1], for *one* segment. It turned out that in order to determine the behavior of the process after it goes out of the interior of the segment, some *boundary conditions* must be given, but only for those ends of the segment that are accessible from the inside. Criteria of accessibility of an end from the inside, and also of reaching the insider from an end were given. One of them: if the integral

$$\int \exp\left\{-\int \frac{2B(H)}{A(H)}\,dH\right\} dH \tag{1.11}$$

diverges at the end H_k, then H_k is not accessible from the inside. These criteria, and also the boundary conditions, are formulated in a simpler way if we represent the differential operator Lf as a generalized second derivative $(d/dv)((d/du)f)$ with respect to two increasing functions $u(H)$, $v(H)$ (see Feller [2] and Mandl [1]). For example, the condition of the integral (1.11) diverging at H_k is replaced by unboundedness of $u(H)$ near H_k (in fact, (1.11) is an expression for $u(H)$); the end H_k is accessible from inside and the inside is accessible from H_k if and only if $u(H)$, $v(H)$ are bounded at H_k. Another condition for inaccessibility that we need: if $\int v(H)\,du(H)$ diverges at H_k, then the end H_k is not accessible from the inside.

Representing the differential operator in the form of a generalized second derivative is especially convenient in our case, because the operators (1.10) degenerate at the ends of the segment I_i: its coefficients $A_i(H)$, $B_i(H)$ given by formulas (1.9) and (1.7) have finite limits, but $A_i(H) \to 0$ at these ends (in the case of nondegenerate critical points, at an inverse logarithmic rate at an end corresponding to a level curve that contains a saddle point, and linearly at an end corresponding to an extremum).

Feller's result can be carried over to diffusions on graphs; if some segments I_i meet at a vertex O_k (which we write as $I_i \sim O_k$) and O_k is accessible from the inside of at least one segment, then some "interior boundary" conditions (or "gluing" conditions) have to be prescribed at O_k. (If the vertex O_k is inaccessible from any of the segments, no condition has to be given.) If for all segments $I_i \sim O_k$ the end of I_i corresponding to O_k is accessible from the inside of I_i, and the inside of I_i can be reached from this end, then the general interior boundary condition can be written in the form

$$\alpha_k Lf(O_k) = \sum_{i:I_i \sim O_k} (\pm\beta_{ki}) \frac{df}{du_i}(O_k), \tag{1.12}$$

where $Lf(O_k)$ is the common limit at O_k of the functions $L_i f$ defined by (1.10) for all segments $I_i \sim O_k$; u_i is the function on I_i used in the representation $L_i = (d/dv_i)(d/du_i)$; $\alpha_i \geq 0$, $\beta_{ki} \geq 0$, and the β_{ki} is taken with + if the function u_i has its minimum at O_k, and with − if it has its maximum there; and $\alpha_k + \sum_{i:I_i \sim O_k} \beta_{ki} > 0$ (otherwise the condition (1.12) is reduced to

$0 = 0$). The coefficient α_k is not zero if and only if the process spends a positive time at the point O_k. (Theorem 2.1 of the next section is formulated only in the case $\alpha_k = 0$, and only as a theorem about the existence and uniqueness of a diffusion corresponding to given gluing conditions; in Freidlin and Wentzell [3] the formulation is more extensive.)

Before considering our problem in a more concrete way, let us introduce some notations:

The indicator function of a set A is denoted by $\chi_A(\cdot)$; the supremum norm in the function space, by $\| \|$; the closure of a set A, by $\bar{A}$;

D_i denotes the set of all points $x \in R^2$ such that $Y(x)$ belongs to the interior of the segment I_i;

$C_k = \{x : Y_x = O_k\}$;

$C_{ki} = C_k \cap \partial D_i$;

for H being one of the values of the function $H(x)$,

$C(H) = \{x : H(x) = H\}$;

for H being one of the values of the function $H(x)$ on $\bar{D}_i$,

$C_i(H) = \{x \in \bar{D}_i : H(x) = H\}$;

for two such numbers $H_1 < H_2$,

$D_i(H_1, H_2) = D_i(H_2, H_1) = \{x \in D_i : H_1 < H(x) < H_2\}$;

for a vertex O_k and a small number $\delta > 0$,

$D_k(\pm\delta)$ is the connected component of the set $\{H(O_k) - \delta < x < H(O_k) + \delta\}$ containing C_k;

$D(\pm\delta) = \bigcup_k D_k(\pm\delta)$;

for a vertex O_k, a segment $I_i \sim O_k$, and a small $\delta > 0$,

$C_{ki}(\delta) = \{x \in D_i : H(x) = H(O_k) \pm \delta\}$

(the sets $C_{ki}(\delta)$ are the connected components of the boundary of $D_k(\pm\delta)$);

if D with some subscripts, and the like denotes a region in R^2, τ^ε with the same subscripts, and the like denotes the first time at which the process X_t^ε leaves the region; for example, $\tau_k^\varepsilon(\pm\delta) = \min\{t : X_t^\varepsilon \notin D_k(\pm\delta)\}$; and $\tilde{\tau}^\varepsilon$ with the same subscripts and the like denotes the corresponding time for the process $\tilde{X}_t^\varepsilon$.

The pictures of the domains $D_k(\pm\delta)$ and $D_i(H(O_k) \pm \delta, H')$ are different for a vertex corresponding to an extremum point x_k and for one corresponding to a level curve containing a saddle point x_k; they are shown in Fig. 20 and 21.

In the notations we introduced, the coefficients $A_i(H)$, $B_i(H)$ can be written as

$$
\begin{aligned}
A_i(H) &= \frac{\oint_{C_i(H)} (|\nabla H(x)|^2 / |b(x)|)\, dl}{\oint_{C_i(H)} (1/|b(x)|)\, dl}, \\
B_i(H) &= \frac{\oint_{C_i(H)} (\frac{1}{2} \Delta H(x) / |b(x)|)\, dl}{\oint_{C_i(H)} (1/|b(x)|)\, dl},
\end{aligned}
\tag{1.13}
$$

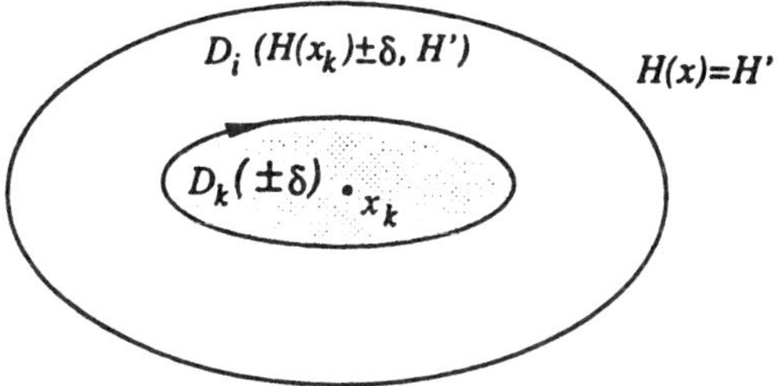

Figure 20. Case of x_k being an extremum.

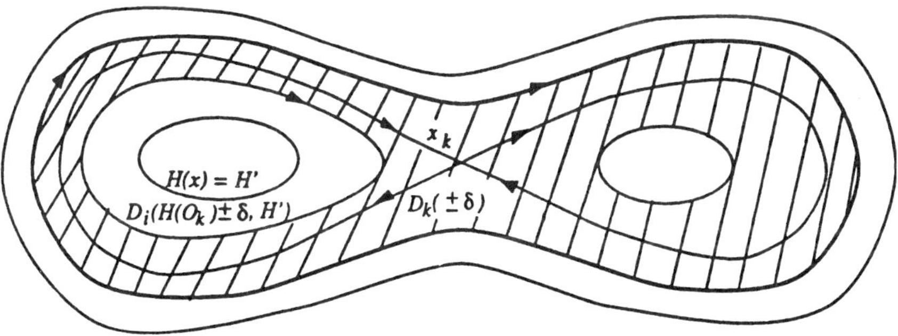

Figure 21. Case of x_k being a saddle point.

Integrals of the kind used in (1.13) can be expressed in another form, and differentiated with respect to the variable H using the following lemma.

Lemma 1.1. *Left f be a function that is continuously differentiable in the closed region $\overline{D}_i(H_1, H_2)$. Then for II, $H_0 \subset [H_1, H_2]$,*

$$\oint_{C_i(H)} f(x)|\nabla H(x)|\,dl = \oint_{C_i(H_0)} f(x)|\nabla H(x)|\,dl$$

$$\pm \iint_{D_i(H_0,H)} [\nabla f(x)\cdot\nabla H(x) + f(x)\Delta H(x)]\,dx,$$

where the sign + or − is taken according to whether the gradient $\nabla H(x)$ at $C_i(H)$ is pointing outside the region $D_i(H_0, H)$ or inside it; and

$$\frac{d}{dH}\oint_{C_i(H)} f(x)|\nabla H(x)|\,dl = \oint_{C_i(H)} \left[\frac{\nabla f(x)\cdot\nabla H(x)}{|\nabla H(x)|} + f(x)\frac{\Delta H(x)}{|\nabla H(x)|}\right] dl.$$

The integral in the denominator in (1.13) is handled by means of Lemma 1.1 with $f(x) = 1/|\nabla H(x)|^2$; those in the numerators, with $f(x) \equiv 1$ and with $f(x) = \Delta H(x)/|\nabla H(x)|^2$.

By Lemma 1.1, the coefficients $A_i(H)$, $B_i(H)$ are continuously differentiable at the interior points of the interval $H(I_i)$ at least $k-1$ times if $H(x)$ is continuously differentiable k times.

Let us consider the behavior of these coefficients as H approaches the ends of the interval $H(I_i)$, restricting ourselves to the case of the function H having only nondegenerate critical points (i.e., with nondegenerate matrix $(\partial^2 H/\partial x^i \partial x^j)$ of second derivatives).

As H approaches an end of the interval $H(I_i)$ corresponding to a nondegenerate extremum x_k of the Hamiltonian, the integral $\int_{C_i(H)} (1/|\nabla H|)\, dl$, which is equal to the period of the orbit $C_i(H)$, converges to

$$T_k = 2\pi \Bigg/ \sqrt{\frac{\partial^2 H}{\partial p^2}(x_k)\frac{\partial^2 H}{\partial q^2}(x_k) - \left(\frac{\partial^2 H}{\partial p \partial q}(x_k)\right)^2} > 0;$$

$$\oint_{C_i(H)} |\nabla H|\, dl \sim \text{const} \cdot (H - H(O_k)) \to 0;$$

and

$$\oint_{C_i(H)} (\Delta H/|\nabla H|)\, dl \to T_k \cdot \Delta H(x_k).$$

So $A_i(H) \to 0$ as $H \to H(O_k) = H(x_k)$, $B_i(H) \to \frac{1}{2}\Delta H(x_k)$ (positive if x_k is a minimum, and negative in the case of a maximum).

If H approaches $H(O_k)$, where O_k corresponds to a level curve containing a nondegenerate saddle point, we have $\oint_{C_i(H)} (1/|\nabla H|)\, dl \sim \text{const} \cdot |\ln|H - H(O_k)|| \to \infty$; $\oint_{C_i(H)} |\nabla H|\, dl \to \oint_{C_{ki}} |\nabla H|\, dl > 0$. The coefficient $A_i(H) \to 0$ (at an inverse logarithmic rate); and it can be shown that in this case too $B_i(H) \to \frac{1}{2}\Delta H(x_k)$ (which can be positive, negative, or zero).

One can obtain some accurate estimates of the derivatives $A_i'(H)$, $B_i'(H)$ as H approaches the ends of $H(I_i)$ corresponding to critical points of the Hamiltonian, but we do not need such. What we need, and what is easily obtained from Lemma 1.1, is that

$$|A_i'(H)|, |B_i'(H)| \leq |H - H(O_k)|^{-A_0} \tag{1.14}$$

for sufficiently small $|H - H(O_k)|$, where A_0 is some positive constant.

A second corollary of Lemma 1.1 is that the derivatives of the functions $v_i(H)$, $u_i(H)$ in the representation $L_i = (d/dv_i)(d/du_i)$ can be chosen as follows:

$$v_i'(H) = \oint_{C_i(H)} \frac{1}{|\nabla H(x)|}\, dl, \tag{1.15}$$

$$u_i'(H) = 2\left(\oint_{C_i(H)} |\nabla H(x)|\, dl\right)^{-1}. \tag{1.16}$$

Indeed,

$$\frac{d}{dv_i}\left(\frac{df}{du_i}\right) = \frac{1}{v_i'(H)u_i'(H)} \cdot f''(H) + \frac{1}{v_i'(H)}\left(\frac{1}{u_i'(H)}\right)' \cdot f'(H),$$

the first coefficient is $A_i(H)$, and the derivative of $\oint_{C_i(H)} |\nabla H|\, dl$ is equal to $\oint_{C_i(H)}(\nabla H/|\nabla H|)\, dl$. The function $v_i(H)$ can be taken equal to $\pm$ the area enclosed by $C_i(H)$.

Let us consider the vertices O_k of the graph from the point of view of their accessibility. The functions $v_i(H)$ are bounded at all vertices O_k except at O_∞. Now,

$$\lim_{H\to H(O_k)} u_i'(H) = 2\left(\oint_{C_{ki}} |\nabla H(x)|\, dl\right)^{-1} \tag{1.17}$$

(if $u_i(H(O_k))$ is finite, this limit is the one-sided derivative of u_i at $H(O_k)$). The limit (1.17) is finite for a vertex O_k corresponding to a separatrix containing a saddle point, and the function u_i is bounded at the end corresponding to O_k; the point O_k is accessible. Since $u_i'(H)$ has a finite positive limit at the end $H(O_k)$ corresponding to O_k, we can rewrite the interior boundary condition (1.12) in the form

$$\alpha_k Lf(O_k) = \sum_{i:I_i \sim O_k} (\pm\beta_{ki}) f_i'(H(O_k)), \tag{1.18}$$

where f_i' denotes the derivative with respect to the local coordinate H on the ith segment, the coefficients β_{ki} (that are different from those in the formulas (1.12)) are taken with + if $H \geq H(O_k)$ on I_i, and with – if $H \leq H(O_k)$ on I_i.

As for a vertex O_k corresponding to an extremum x_k, we have $\oint_{C_i(H)} |\nabla H|\, dl \sim \text{const} \cdot (H - H(x_k))^2$, $u_i(H) \sim -\text{const} \cdot (H - H(x_k))^{-1}$ as $H \to H(x_k)$, so the end O_k corresponding to an extremum x_k is inaccessible.

For the vertex O_∞, we have: $v_i(H) = 2(u_i'(H))^{-1}$, the integral $\int_{H_0}^{\infty} v_i(H)\, du_i(H) = \int_{H_0}^{\infty} v_i(H) u_i'(H)\, dH = \int_{H_0}^{\infty} 2\, dH$ diverges, and this vertex is inaccessible.

We prove (Theorem 2.2) that the process $Y(X_t^\varepsilon)$ converges to a diffusion process Y_t on the graph Γ in the sense of weak convergence of distributions in the space of continuous functions with values in the graph Γ; and we find the interior boundary conditions at the vertices O_k that correspond to curves with saddle points on them. The coefficients α_k in the conditions (1.18) turn out to be 0.

Let us outline a plan of the proof.

The proof consists of several parts ensuring that:

(1) a continuous limiting process Y_t exists;
(2) it is the diffusion process corresponding to the operator L_i before its leaving the interior of the segment I_i;
(3) it spends zero time at the vertices O_k;
(4) the behavior of the process after it reaches a vertex O_k does not depend on where it came from (so that it has a strong Markov character also with respect to the times of reaching a vertex);
(5) the coefficients β_{ki} in the interior boundary conditions (1.18) are such and such.

A more precise formulation is as follows.

(1) The family of distributions of $Y(X_\bullet^\varepsilon)$ in the space of continuous functions is tight (weakly precompact). This is the statement of Lemma 3.2.

We can reformulate the points (2) and (3) of our plan in terms of averaging in every finite time interval and spending little time in neighborhoods of C_k in every finite time interval; but instead we use the Laplace transforms:

(2) for every $\lambda > 0$ for every smooth function f on an interval $[H_1, H_2]$ consisting of interior points of the interval $H(I_i)$ we have

$$\mathbf{M}_x^\varepsilon\left[e^{-\lambda\tau_i^\varepsilon(H_1,H_2)}f(H(X^\varepsilon_{\tau_i^\varepsilon(H_1,H_2)}))+\int_0^{\tau_i^\varepsilon(H_1,H_2)}e^{-\lambda t}[\lambda f(H(X_t^\varepsilon))-L_i f(H(X_t^\varepsilon))]\,dt\right]$$
$$-f(H(x)) = O(k(\varepsilon)), \tag{1.19}$$

uniformly with respect to $x \in D_i(H_1, H_2)$, where $k(\varepsilon) \to 0$ as $\varepsilon \to 0$. This is the statement of Lemma 3.3.

An intermediate stage in proving (1.19) is estimating

$$\mathbf{M}_x^\varepsilon \int_0^{\tau_{ki}^\varepsilon(\delta,H)} e^{-\lambda t} g(X_t^\varepsilon)\,dt \tag{1.20}$$

for functions g in the region $D_i(H_1, H_2)$ such that the integral $\oint g(x)/|b(x)|\,dl$ over each closed trajectory in the region in equal to 0 (Lemma 4.4).

(3) We prove that there exists a function $h(\delta)$, $\lim_{\delta\to 0} h(\delta) = 0$, such that for every fixed $\delta > 0$,

$$\mathbf{M}_x^\varepsilon \int_0^\infty e^{-\lambda t}\chi_{D_k(\pm\delta)}(X_t^\varepsilon)\,dt \le h(\delta) \tag{1.21}$$

for sufficiently small ε and for all $x \in R^2$.

The form (1.21) of this property is easier to understand (little time is spent by the process $Y(X_t^\varepsilon)$ in a neighborhood of any vertex O_k), but it is easier to prove and to use in the proof of our main result in the form of an estimate for $\mathsf{M}_x^\varepsilon \int_0^{\tau_k^\varepsilon(\pm\delta)} e^{-\lambda t}\,dt$ for all $x \in D_k(\pm\delta)$. Such estimates turn out to be different for vertices corresponding to extremum points and to those that correspond to curves containing saddle points; they are given in Lemmas 3.4 and 3.5.

(4) We prove that for every small $\delta > 0$ there exists δ', $0 < \delta' < \delta$, such that the probabilities $\mathsf{P}_x^\varepsilon\{X_{\tau_k^\varepsilon(\pm\delta)}^\varepsilon \in C_{kj}(\delta)\}$ almost do not depend on x if it changes on $\bigcup_i C_{ki}(\delta')$ (the beginning of the proof of Lemma 3.6).

(5) To find the coefficients β_{ki} we use the fact that the invariant measure μ of the process X_t^ε is the Lebesgue measure, so the limiting process for $Y(X_t^\varepsilon)$ must have the invariant measure $\mu \circ Y^{-1}$.

One can find the discussion of some generalizations of the problem described in this section in Section 7. In particular, there we consider biefly perturbations of systems with conservation laws. Perturbations of Hamiltonian systems on tori may lead to processes on graphs that spend a positive time at some vertices. Degenerate perturbations of Hamiltonian systems, in particular, perturbations of a nonlinear oscillator with one degree of freedom, are also considered in Section 7. In addition, we briefly consider there the multidimensional case and the case of several conservation laws.

§2. Main Results

Theorem 2.1. *Let Γ be a graph consisting of closed segments $I_1, \dots, I_N$ and vertices $O_1, \dots, O_M$. Let a coordinate be defined in the interior of each segment I_i; let $u_i(y)$, $v_i(y)$, for every segment I_i, be two functions on its interior that increase (strictly) as the coordinate increases; and let u_i be continuous. Suppose that the vertices are divided into two classes: interior vertices, for which $\lim_{y\to O_k} u_i(y)$, $\lim_{y\to O_k} v_i(y)$ are finite for all segments I_i meeting at O_k (notation: $I_i \sim O_k$); and exterior vertices, such that only one segment I_l enters O_k, and $\int (c + v_l(y))\,du_l(y)$ diverges at the end O_k for some constant c.*

For each interior vertex O_k, let β_{ki} be nonnegative constants defined for i such that $I_i \sim O_k$; $\sum_{i:I_i\sim O_k} \beta_{ki} > 0$. Consider the set $D(A) \subset \mathbf{C}(\Gamma)$ consisting of all functions f such that

f has a continuous generalized derivative $(d/dv_i)(d/du_i)f$ in the interior of each segment I_i;

finite limits $\lim_{y\to O_k} (d/dv_i)(d/du_i)f(y)$ exist at every vertex O_k, and they do not depend on the segment $I_i \sim O_k$;

for each interior vertex O_k,

$$\sum_{i:I_i\sim O_k}' \beta_{ki} \lim_{y\to O_k} \frac{df}{du_i}(y) - \sum_{i:I_i\sim O_k}'' \beta_{ki} \lim_{y\to O_k} \frac{df}{du_i}(y) = 0, \tag{2.1}$$

where the sum $\sum'$ *contains all i such that the coordinate on the ith segment has a minimum at* O_k, *and* $\sum''$, *those for which it has a maximum.*

Define the operator A *with domain of definition* $D(A)$ *by* $Af(y) = (d/dv_i)(d/du_i)f(y)$ *in the interior of every segment* I_i, *and at the vertices, as the limit of this expression.*

Then there exists a strong Markov process (y_t, P_y) *on* Γ *with continuous trajectories whose infinitesimal operator is* A.

If we take the sapce $\mathbf{C}[0, \infty)$ *of all continuous functions on* $[0, \infty)$ *with values in* Γ *as the sample space for this process, with* y_t *being the value of a function of this space at the point* t, *such a process is unique.*

If O_k *is an exterior vertex, and* $y \neq O_k$, *then with* P_y*-probability* 1 *the process never reaches* O_k.

The proof can be carried out similarly to that of the corresponding result for diffusions on an interval. First we verify the fulfillment of the conditions of the Hille–Yosida theorem; then existence of a continuous-path version of the Markov process is proved; and so on: see Mandl [1] or the original papers by Feller [1], [2]. Diffusion processes on graphs have been considered in some papers, in particular, in Baxter and Chacon [1]. The corresponding Theorem 3.1 in Freidlin and Wentzell [3] allows considering only interior vertices.

In the situation of a graph associated with a Hamiltonian system, the vertices corresponding to level curves of the Hamiltonian containing saddle points are interior vertices, and those corresponding to extrema and to the point at infinity, exterior.

Theorem 2.2. *Let the Hamiltonian* $H(x)$, $x \in R^2$, *be four times continuously differentiable with bounded second derivatives;* $H(x) \geq A_1|x|^2$, $|\nabla H(x)| \geq A_2|x|$, $\Delta H(x) \geq A_3$ *for sufficiently large* $|x|$, *where* A_1, A_2, A_3 *are positive constants. Let* $H(x)$ *have a finite number of critical points* $x_1, \ldots, x_N$, *at which the matrix of second derivatives is nondegenerate. Let every level curve* C_k (*see notations in Section* 1) *contain only one critical point* x_k.

Let $(X_t^\varepsilon, \mathsf{P}_x^\varepsilon)$ *be the diffusion process on* R^2 *corresponding to the differential operator* $L^\varepsilon f(x) = \frac{1}{2}\Delta f(x) + \varepsilon^{-2}\bar{\nabla} H(x) \cdot \nabla f(x)$. *Then the distribution of the process* $Y(X_t^\varepsilon)$ *in the space of continuous functions on* $[0, \infty)$ *with values in* $Y(R^2)(\subset \Gamma)$ *with respect to* P_x^ε *converges weakly to the probability measure* $\mathsf{P}_{Y(x)}$, *where* (y_t, P_y) *is the process on the graph whose existence is stated in Theorem* 2.1, *corresponding to the functions* u_i, v_i *defined by formulas* (1.15) *and* (1.16), *and to the coefficients* β_{ki} *given by*

$$\beta_{ki} = \oint_{C_{ki}} |\nabla H(x)|\, dl. \tag{2.2}$$

The proof is given in Sections 3–6, and now we give an application to partial differential equations.

Let G be a bounded region in R^2 with smooth boundary ∂G. Consider the Dirichlet problem

$$
\begin{aligned}
L^\varepsilon f^\varepsilon(x) = \tfrac{1}{2}\Delta f^\varepsilon(x) + \tfrac{1}{\varepsilon^2}\,\overline{\nabla} H(x)\cdot \nabla f^\varepsilon(x) = -g(x), \qquad & x \in G,\\
f^\varepsilon(x) = \psi(x), \qquad x \in \partial G. &
\end{aligned}
\tag{2.3}
$$

Here $H(x)$ is the same as in Theorem 2.2, $\psi(x)$ and $g(x)$ are continuous functions on ∂G and on $G \cup \partial G$, respectively, and ε is a small parameter.

It is well known that the behavior of $f^\varepsilon(x)$ as $\varepsilon \to 0$ depends on the behavior of the trajectories of the dynamical system $\dot{x}_t = \overline{\nabla} H(x_t)$: If the trajectory $x_t(x)$, $t \geq 0$, starting at $x_0(x) = x$ hits the boundary ∂G at a point $z(x) \in \partial G$, and $\overline{\nabla} H(z(x)) \cdot n(z(x)) \neq 0$ ($n(z(x))$ is the outward normal vector to ∂G at the point $z(x)$), then $f^\varepsilon(x) \to \psi(z(x))$ as $\varepsilon \to 0$ (see, e.g., Freidlin [15]; no integral of the function g is involved in this formula, because, in the case we are considering, the trajectory X_t^ε leaves the region G in a time of order ε^2). If $x_t(x)$ does not leave the region G, the situation is more complicated.

Let $\hat{G}$ be the subset of G covered by the orbits $C_i(H)$ that belong entirely to $G : \hat{G} = \bigcup_{i,H:C_i(H) \subset G} C_i(H)$; and let Γ_G be the subset of the graph Γ that corresponds to $\hat{G} \subset R^2 : \Gamma_G = Y(\hat{G}) = \{(i,H) : C_i(H) \subset G\}$.

(i) Assume that the set Γ_G is connected (otherwise one should consider its connected components separately). Suppose that the boundary of Γ_G consists of the points (i_k, H_k), $k = 1, \ldots, l$. Each of the curves $C_{i_k}(H_k)$, $(i_k, H_k) \in \partial\Gamma_G$, has a nonempty intersection with ∂G.

(ii) Assume that for each $(i_k, H_k) \in \partial\Gamma_G$ the intersection $C_{i_k}(H_k) \cap \partial G$ consists of exactly one point z_k. This should be considered as the main case. Later we discuss the case of $C_{i_k}(H_k) \cap \partial G$ consisting of more than one point.

(iii) Assume that $\partial\Gamma_G$ contains no vertices. Let Λ be the set of all vertices that belong to Γ_G. Let $\Lambda = \Lambda_1 \cup \Lambda_2$, where Λ_1 consists interior vertices, and Λ_2, of exterior vertices.

Theorem 2.3. *If for a point $x \in G$ the trajectory $x_t(x)$, $t \geq 0$, hits the boundary ∂G at a point $z(x) \in \partial G$, and $\overline{\nabla} H(z(x)) \cdot n(z(x)) \neq 0$, then*

$$\lim_{\varepsilon \to 0} f^\varepsilon(x) = \psi(z(x)).$$

If $x \in \hat{G}$, and if the Hamiltonian $H(x)$ satisfies the conditions of Theorem 2.2, *and conditions* (i), (ii), *and* (iii) *are fulfilled, then*

$$\lim_{\varepsilon \to 0} f^\varepsilon(x) = f(i(x), H(x)),$$

where $f(i,H)$ is the solution of the following Dirichlet problem on Γ_G,

$$\tfrac{1}{2}A_i(H)f''(i,H) + B_i(H)f'(i,H) = -\hat{g}(i,H), \qquad (i,H) \in \Gamma_G,$$
$$f(i_k, H_k) = \psi(z_k) \quad \textit{for } (i_k, H_k) \in \partial\Gamma_G,$$
$$f(i,H) \textit{ is continuous on } \Gamma_G, \tag{2.4}$$
$$\sum_{i:I_i \in O_k} (\pm\beta_{ki}) f'(i, H(O_k)) = 0 \quad \textit{for } O_k \in \Lambda_1.$$

Here

$$A_i(H) = \frac{\oint_{C_i(H)} |\nabla H(x)|\, dl}{\oint_{C_i(H)} |\nabla H(x)|^{-1}\, dl}, \qquad B_i(H) = \frac{\oint_{C_i(H)} \Delta H(x)/2|\nabla H(x)|\, dl}{\oint_{C_i(H)} |\nabla H(x)|^{-1}\, dl},$$

β_{ki} *are defined by* (2.2) *and taken with* + *if* $H \geq H(O_k)$ *for* $(i,H) \in I_i$, *and with* − *if* $H \leq H(O_k)$ *for* $(i,H) \in I_i$; *and*

$$\hat{g}(i,H) = \frac{\oint_{C_i(H)} g(x)|\nabla H(x)|^{-1}\, dl}{\oint_{C_i(H)} |\nabla H(x)|^{-1}\, dl}.$$

The solution of problem (2.4) *exists and is unique. If the functions* $u_i(H)$, $v_i(H)$ *are defined by* (1.15), (1.16), *then*

$$\tilde{f}(i,H) = \tilde{f}(i,H) + a_i^{(1)} u_i(H) + a_i^{(2)}, \tag{2.5}$$

where

$$\tilde{f}(i,H) = \int_{H(O_k)}^{H} \left[\int_{H(O_k)}^{z} \hat{g}(i,y) v_i'(y)\, dy \right] u_i'(z)\, dz,$$

O_k *being an end of the segment* I_i. *The constants* $a_i^{(1)}$, $a_i^{(2)}$ *are determined uniquely by the boundary conditions on* ∂G, *the continuity, and the gluing conditions* (2.1) *at the vertices belonging to* Λ_1.

The proof of this theorem is based on Theorem 2.2 and includes the following statements.

1. The solutions of problems (2.3) and (2.4) can be represented as follows.

$$f^\varepsilon(x) = \mathsf{M}_x^\varepsilon \psi(X_{\tau^\varepsilon}^\varepsilon) + \mathsf{M}_x^\varepsilon \int_0^{\tau^\varepsilon} g(X_s^\varepsilon)\, dx,$$

$$f(i,H) = \mathsf{M}_{(i,H)} \psi(Y_{\tau^0}) + \mathsf{M}_{(i,H)} \int_0^{\tau^0} \hat{g}(Y_s)\, ds,$$

where τ^ε is the first exit time from G for $X_t^\varepsilon : \tau^\varepsilon = \min\{t : X_t^\varepsilon \notin G\}$, τ^0, the first exit time from Γ_G for the process y_t of Theorem 2.2: $\tau^0 = \min\{t : y_t \notin \Gamma_G\}$, and $\mathsf{M}_{(i,H)} = \mathsf{M}_y$ is the expectation corresponding to the probability P_y associated with the process y_t on the graph.

2. $\mathsf{M}_x^\varepsilon \tau^\varepsilon \leq A < \infty$ for every small $\varepsilon \neq 0$ and for every $x \in G$. The proof of this statement consists of two main parts. First we get a bound for the mean exit time from a region U such that the closure of $Y(U)$ contains no vertices, and this is done in the standard way. Another bound is obtained for the expected exit time from a region D such that $Y(D)$ belongs to a neighborhood of a vertex of Γ; for this Lemmas 3.4 and 3.5 are used.

3. $A_{i_k}(H_k) > 0$ for all k, $(i_k, H_k) \in \partial\Gamma_G$, and thus every (i_k, H_k) is a regular point of the boundary $\partial\Gamma_G$ for the process (y_t, P_y).

4. If we denote $\alpha(k, H) = \inf\{|\nabla H(x)| : x \in C_k(H)\}$, then for every positive δ,

$$\lim_{h\to 0} \lim_{\varepsilon\to 0} \mathsf{P}_x^\varepsilon \left\{ \left| \frac{1}{h} \int_0^h g(X_t^\varepsilon)\, ds - \hat{g}(i(x), H(x)) \right| > \delta \right\} = 0$$

uniformly in x such that $\alpha(Y(x)) \geq \alpha > 0$.

Example. Let us consider the boundary-value problem (2.3) with $g(x) \equiv 0$, $H(x)$ as shown in Fig. 18, and the region G as shown in Fig. 22(a).

The region G has two holes. Its boundary touches the orbits at the points z_1, z_2, z_3. In the part of G outside the region bounded by the dotted line the trajectories of the dynamical system leave the region G. For $x \in G \setminus (\hat{G} \cup \partial\hat{G})$ the limit of $f^\varepsilon(x)$ is equal to the value of the boundary function at the point at which the trajectory $x_t(x)$ first leaves G. To evaluate $\lim_{\varepsilon\to 0} f^\varepsilon(x)$ for $x \in \hat{G}$ ($\hat{G}$ is shown separately in Fig. 22(c)) one must consider the graph Γ corresponding to our Hamiltonian and its part $\Gamma_G \subset \Gamma$ corresponding to the region G (see Fig. 22(b)). In our example $\partial\Gamma_G$ consists of three points $(4, H_1)$, $(5, H_2)$, and $(1, H_3)$. The boundary conditions at $\partial\Gamma_G$ are:

$$f(4, H_1) = \psi(z_1), \qquad f(5, H_2) = \psi(z_2), \qquad f(1, H_3) = \psi(z_3).$$

Since the right-hand side of the equation in problem (2.4) is equal to 0, on every segment I_i the limiting solution $f(i, H)$ is a linear function of the function $u_i(H)$. Since the functions $u_i(H)$ are unbounded at the vertices of the second class (in particular, at O_3), and the solution $f(i, H)$ is bounded, this solution is constant on the segment I_3. On the remaining segments the formulas (2.5) give the limiting solution up to two constants on each of the segments I_1, I_2, I_4, I_5. So we have nine unknown constants. To determine them we have three boundary conditions, four equations arising from continuity conditions at O_1 and O_2, and two gluing conditions at these vertices. So we have a system of nine linear equations with nine unknowns. This system has a unique solution.

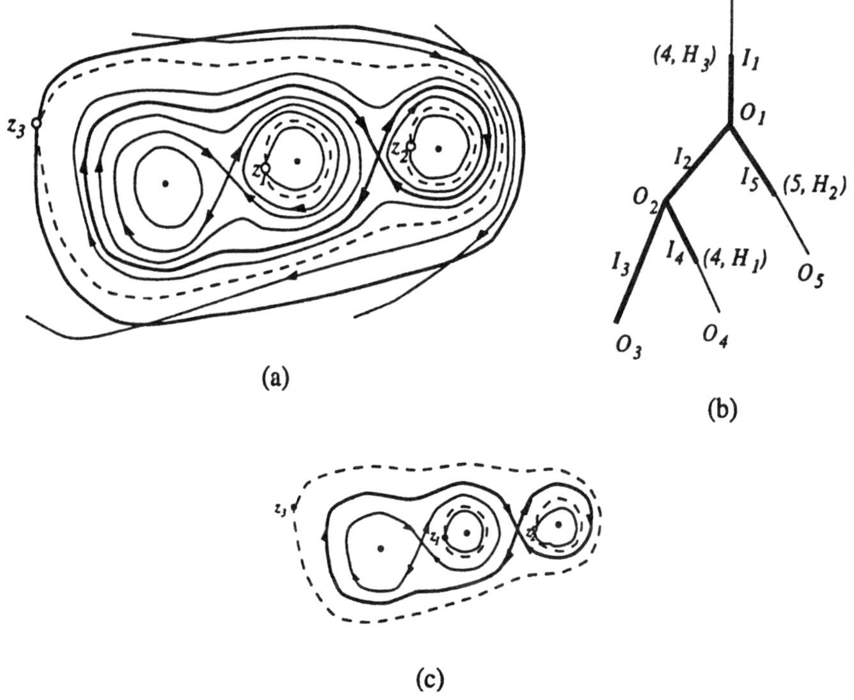

Figure 22.

Another class of examples is that with $g(x) \equiv 1$ and $\psi(x) \equiv 0$. The solution of the corresponding problem (2.3) is the expectation $\mathsf{M}_x^\varepsilon \tau^\varepsilon$. So Theorem 2.3 provides the method of finding the limit of the expected exit time of the process X_t^ε from a region G (or the main term of the asymptotic of the expectation $\mathsf{M}_x^\varepsilon \tilde{\tau}^\varepsilon$ of the exit time for the process $\tilde{X}_t^\varepsilon$ as $\varepsilon \to 0$).

Remark. We assumed in Theorem 2.3 that for each $(i_k, H_k) \in \partial\Gamma_G$ the set $C_{i_k}(H_k) \cap \partial G$ consists of exactly one point z_k. The opposite case is that this intersection is equal to $C_{i_k}(H_k)$. In this case one should replace the boundary condition $f(i_k, H_k) = \psi(z_k)$ by

$$f(i_k, H_k) = \bar{\psi}_k,$$

where

$$\bar{\psi}_k = \frac{\oint_{C_{i_k}(H_k)} \psi(x) \cdot |\nabla H(x)|\, dl}{\oint_{C_{i_k}(H_k)} |\nabla H(x)|\, dl}.$$

If $\partial G \cap C_{i_k}(H_k)$ consists of more than one point but does not coincide with $C_{i_k}(H_k)$, the situation is more complicated.

§3. Proof of Theorem 2.2

Before proving Theorem 2.2 we formulate the necessary lemmas and introduce some notations. The proof of the lemmas is given in Sections 4 and 5.

Lemma 3.1. *Let M be a metric space; Y, a continuous mapping $M \mapsto Y(M)$, $Y(M)$ being a complete separable metric space. Let $(X_t^\varepsilon, \mathsf{P}_x^\varepsilon)$ be a family of Markov processes in M; suppose that the process $Y(X_t^\varepsilon)$ has continuous trajectories* (in Theorem 2.2, the process X_t^ε itself has continuous paths; but we want to formulate our lemma so that it can be applied in a wider class of situations). *Let (y_t, P_y) be a Markov process with continuous paths in $Y(M)$ whose infinitesimal operator is A with domain of definition $D(A)$. Let us suppose that the space $\mathbf{C}[0, \infty)$ of continuous functions on $[0, \infty)$ with values in Γ is taken as the sample space, so that the distribution of the process in the space of continuous functions is simply P_y. Let Ψ be a subset of the space $\mathbf{C}(Y(M))$ such that for measures μ_1, μ_2 on $Y(M)$ the equality $\int f\,d\mu_1 = \int f\,d\mu_2$ for all $f \in \Psi$ implies $\mu_1 = \mu_2$. Let D be a subset of $D(A)$ such that for every $f \in \Psi$ and $\lambda > 0$ the equation $\lambda F - AF = f$ has a solution $F \in D$.*

Suppose that for every $x \in M$ the family of distributions Q_x^ε of $Y(X_\bullet^\varepsilon)$ in the space $\mathbf{C}[0, \infty)$ corresponding to the probabilities P_x^ε for all ε is tight; and that for every compact $K \subseteq Y(M)$, for every $f \in D$ and every $\lambda > 0$,

$$\mathsf{M}_x^\varepsilon \int_0^\infty e^{-\lambda t}[\lambda f(Y(X_t^\varepsilon)) - Af(Y(X_t^\varepsilon))]\,dt \to f(Y(x)) \tag{3.1}$$

as $\varepsilon \to 0$, uniformly in $x \in Y^{-1}(K)$.

Then Q_x^ε converges weakly as $\varepsilon \to 0$ to the probability measure $\mathsf{P}_{Y(x)}$.

The lemmas that follow are formulated under the assumption that the conditions of Theorem 2.2 are satisfied.

Lemma 3.2. *The family of distributions Q_x^ε (those of $Y(X_\bullet^\varepsilon)$ with respect to the probability measures P_x^ε in the space $\mathbf{C}[0, \infty)$) with small nonzero ε is tight.*

It can also be proved that the family $\{\mathsf{Q}_x^\varepsilon\}$ with small ε and x changing in an arbitrary compact subset of R^2 is tight.

Lemma 3.3. *For every fixed $H_1 < H_2$ belonging to the interior of the interval $H(I_i)$, for every function f on $[H_1, H_2]$ that is three times continuously differentiable on this closed interval, and for every positive number λ,*

$$\mathsf{M}_x^\varepsilon\left[e^{-\lambda\tau_i^\varepsilon(H_1,H_2)} f(H(X^\varepsilon_{\tau_i^\varepsilon(H_1,H_2)})) + \int_0^{\tau_i^\varepsilon(H_1,H_2)} e^{-\lambda t}[\lambda f(H(X_t^\varepsilon)) - L_i f(H(X_t^\varepsilon))]\,dt\right]$$
$$\to f(H(x)) \tag{3.2}$$

as $\varepsilon \to 0$, uniformly with respect to $x \in \bar{D}_i(H_1, H_2)$.

Lemma 3.4. *Let O_k be an exterior vertex (one corresponding to an extremum of the Hamiltonian). Then for every positive λ and κ there exists $\delta_{3.4} > 0$ such that for sufficiently small ε for all $x \in D_k(\pm\delta_{3.4})$,*

$$\mathsf{M}_x^\varepsilon \int_0^{\tau_k^\varepsilon(\pm\delta_{3.4})} e^{-\lambda t}\, dt < \kappa. \tag{3.3}$$

Lemma 3.5. *Let O_k be an interior vertex (corresponding to a curve containing a saddle point). Then for every positive λ and κ there exists $\delta_{3.5} > 0$ such that for $0 < \delta \le \delta_{3.5}$ for sufficiently small ε for all $x \in D_k(\pm\delta)$,*

$$\mathsf{M}_x^\varepsilon \int_0^{\tau_k^\varepsilon(\pm\delta)} e^{-\lambda t}\, dt < \kappa \cdot \delta. \tag{3.3$'$}$$

Remark. Formulas (3.3) and (3.3') are almost the same as $\mathsf{M}_x^\varepsilon \tau_k^\varepsilon(\pm\delta) < \kappa$ or $\kappa\delta$, and such inequalities can also be proved; but it is these formulas rather than the more natural estimates for $\mathsf{M}_x^\varepsilon \tau_k^\varepsilon(\pm\delta)$ that are used in the proof of Theorem 2.2.

Lemma 3.6. *Define, for $I_i \sim O_k$, $p_{ki} = \beta_{ki} / \left(\sum_{i: I_i \sim O_k} \beta_{ki}\right)$, where β_{ki} are defined by formula (2.2). For any positive κ there exists a positive $\delta_{3.6}$ such that for $0 < \delta \le \delta_{3.6}$ there exists $\delta'_{3.6} = \delta'_{3.6}(\delta) > 0$ such that for sufficiently small ε,*

$$|\mathsf{P}_x^\varepsilon\{X^\varepsilon_{\tau_k^\varepsilon(\pm\delta)} \in C_{ki}(\delta)\} - p_{ki}| < \kappa \tag{3.4}$$

for all $x \in \bar{D}_k(\pm\delta'_{3.6})$.

Proof of Theorem 2.2. Make use of Lemma 3.1. As the set Ψ we take the subset of $\mathbf{C}(\Gamma)$ consisting of all functions that are twice continuously differentiable with respect to the local coordinate H inside each segment I_i; D is the subset of $D(A)$ consisting of functions with four continuous derivatives inside each segment.

The tightness required in Lemma 3.1 is the statement of Lemma 3.2; so what remains is to prove that the difference of both sides in (3.1) is smaller than an arbitrary positive number η for sufficiently small ε for every $x \in R^2$.

Let a function $f \in D$, a number $\lambda > 0$, and a point $x \in R^2$ be fixed.

Choose $H_0 > \max_k H(O_k)$ so that

$$\mathsf{M}_x^\varepsilon e^{-\lambda T^\varepsilon} < \frac{\eta}{2(\|f\| + \lambda^{-1}\|\lambda f - Af\|)} \tag{3.5}$$

for all sufficiently small ε, where $T^\varepsilon = T^\varepsilon(H_0) = \min\{t : H(X_t^\varepsilon) \ge H_0\}$ (this is possible by Lemma 3.2).

To prove that

$$\left|\mathsf{M}_x^\varepsilon \int_0^\infty e^{-\lambda t}[\lambda f(Y(X_t^\varepsilon)) - Af(Y(X_t^\varepsilon))]\,dt - f(Y(x))\right| < \eta$$

it is sufficient to prove that

$$\left|\mathsf{M}_x^\varepsilon \left[e^{-\lambda T^\varepsilon} f(Y(X_{T^\varepsilon}^\varepsilon)) + \int_0^{T^\varepsilon} e^{-\lambda t}[\lambda f(Y(X_t^\varepsilon)) - Af(Y(X_t^\varepsilon))]\,dt\right] - f(Y(x))\right| < \eta/2. \tag{3.6}$$

Now we take small positive δ and $\delta' < \delta$ and consider cycles between the succesive times of leaving the set $\bigcup_k D_k(\pm\delta)$ and reaching the set $\bigcup_{k,i} C_{ki}(\delta')$, before the process reaches the level curve $C(H_0)$: define Markov times $\tau_0 \leq \sigma_1 \leq \tau_1 \leq \sigma_2 \leq \cdots$ by $\tau_0 = 0$,

$$\sigma_n = \min\left\{t \geq \tau_{n-1} : X_t^\varepsilon \notin \bigcup_k D_k(\pm\delta)\right\},$$

$$\tau_n = \min\left\{t \geq \sigma_n : X_t^\varepsilon \in \bigcup_{k,i} C_{ki}(\delta') \cup C(H_0)\right\}$$

(after the process reaches $C(H_0)$, all τ_n and σ_n are equal to T^ε). The difference in (3.6) can be represented as the sum over time intervals from τ_n to σ_{n+1} and from σ_n to τ_n: it is equal to

$$\begin{aligned}
&\mathsf{M}_x^\varepsilon \sum_{n=0}^\infty \left[e^{-\lambda\sigma_{n+1}} f(Y(X_{\sigma_{n+1}}^\varepsilon)) - e^{-\lambda\tau_n} f(Y(X_{\tau_n}^\varepsilon))\right.\\
&\qquad \left.+ \int_{\tau_n}^{\sigma_{n+1}} e^{-\lambda t}[\lambda f(Y(X_t^\varepsilon)) - Af(Y(X_t^\varepsilon))]\,dt\right]\\
&+ \mathsf{M}_x^\varepsilon \sum_{n=1}^\infty \left[e^{-\lambda\tau_n} f(Y(X_{\tau_n}^\varepsilon)) - e^{-\lambda\sigma_n} f(Y(X_{\sigma_n}^\varepsilon))\right.\\
&\qquad \left.+ \int_{\sigma_n}^{\tau_n} e^{-\lambda t}[\lambda f(Y(X_t^\varepsilon)) - Af(Y(X_t^\varepsilon))]\,dt\right]
\end{aligned} \tag{3.7}$$

(the formally infinite sums are finite for every trajectory for which $T^\varepsilon < \infty$).

We can write the expectations of the sums as infinite sums of expectations:

$$\sum_{n=0}^{\infty} \mathsf{M}_x^\varepsilon \Bigg[e^{-\lambda\sigma_{n+1}} f(Y(X_{\sigma_{n+1}}^\varepsilon)) - e^{-\lambda\tau_n} f(Y(X_{\tau_n}^\varepsilon)) + \int_{\tau_n}^{\sigma_{n+1}} e^{-\lambda t}[\lambda f(Y(X_t^\varepsilon)) - Af(Y(X_t^\varepsilon))]\,dt \Bigg] + \sum_{n=1}^{\infty} \mathsf{M}_x^\varepsilon \Bigg[e^{-\lambda\tau_n} f(Y(X_{\tau_n}^\varepsilon)) - e^{-\lambda\sigma_n} f(Y(X_{\sigma_n}^\varepsilon)) + \int_{\sigma_n}^{\tau_n} e^{-\lambda t}[\lambda f(Y(X_t^\varepsilon)) - Af(Y(X_t^\varepsilon))]\,dt \Bigg] \qquad (3.8)$$

if the sums $\sum_{n=0}^{\infty} \mathsf{M}_x^\varepsilon\{\tau_n < T^\varepsilon; e^{-\lambda\tau_n}\}$, $\sum_{n=1}^{\infty} \mathsf{M}_x^\varepsilon\{\sigma_n < T^\varepsilon; e^{-\lambda\sigma_n}\} < \infty$. So the first thing is to estimate these sums (which are approximately the same as the expected number of cycles—to be precise, of cycles that contribute a significant amount to the sum).

Let us denote by $D(H_0, \delta')$ the region

$$\{x : H(x) < H_0\} \Big\backslash \bigcup_k \bar{D}_k(\pm\delta')(D(H_0,\delta)$$

($D(H_0,\delta)$ is the same set with δ in lieu of δ'). The random time τ_n, $n \geq 1$, is the first exit time from $D(H_0,\delta')$ after the time σ_n. Use the strong Markov property with respect to σ_n:

$$\mathsf{M}_x^\varepsilon\{\tau_n < T^\varepsilon; e^{-\lambda\tau_n}\} = \mathsf{M}_x^\varepsilon\{\sigma_n < T^\varepsilon; e^{-\lambda\sigma_n}\phi_1^\varepsilon(X_{\sigma_n}^\varepsilon)\}, \qquad (3.9)$$

where

$$\phi_1^\varepsilon(z) = \mathsf{M}_z^\varepsilon\left\{ X_{\tau^\varepsilon(H_0,\delta')}^\varepsilon \in \bigcup_{k,i} C_{ki}(\delta'); e^{-\lambda\tau^\varepsilon(H_0,\delta')} \right\} \qquad (3.10)$$

(according to our system, we use the notation $\tau^\varepsilon(H_0,\delta')$ for the first exit time from the region $D(H_0,\delta')$). It is clear that $X_{\sigma_n}^\varepsilon \in \bar{D}(H_0,\delta)$ for every n, so

$$\begin{aligned} \mathsf{M}_x^\varepsilon\{\tau_n < T^\varepsilon; e^{-\lambda\tau_n}\} &\leq \mathsf{M}_x^\varepsilon\{\sigma_n < T^\varepsilon; e^{-\lambda\sigma_n}\} \cdot \max\{\phi_1^\varepsilon(z) : z \in \bar{D}(H_0,\delta)\} \\ &\leq \mathsf{M}_x^\varepsilon\{\tau_{n-1} < T^\varepsilon; e^{-\lambda\tau_{n-1}}\} \cdot \max\{\phi_1^\varepsilon(z) : z \in \bar{D}(H_0,\delta)\}, \end{aligned}$$

and by induction

$$\mathsf{M}_x^\varepsilon\{\tau_n < T^\varepsilon; e^{-\lambda\tau_n}\} \leq [\max\{\phi_1^\varepsilon(z) : z \in \bar{D}(H_0,\delta)\}]^n. \qquad (3.11)$$

Since in defining the set $D(H_0,\delta)$ we delete neighborhoods of all separatrices and of extremum points, the set $\bar{D}(H_0,\delta)$ splits into pieces that lie each within its own region D_i. Suppose the ends of the segment I_i are the vertices O_k, $O_{k'}$, $H_{i1} = H(O_k) < H_{i2} = H(O_{k'})$. Define, for every small $\delta \geq 0$,

$$H_{i1}^{\delta} = H_{i1} + \delta, \quad H_{i2}^{\delta} = \begin{cases} H_{i2} - \delta & \text{if } H_{i2} \neq +\infty, \text{ i.e., if } O_{k'} \neq O_{\infty}, \\ H_0 & \text{if one of the ends of } I_i \text{ is the vertex } O_{\infty}. \end{cases}$$

Then $\bar{D}(H_0, \delta) = \bigcup_i \bar{D}_i(H_{i1}^{\delta}, H_{i2}^{\delta})$.

Suppose the process X_t^{ε} starts at a point $z \in \bar{D}_i(H_{i1}^{\delta}, H_{i2}^{\delta})$. In the region $D_i(H_{i1}^{\delta'}, H_{i2}^{\delta'})$ averaging takes place: we can apply Lemma 3.3 to estimate the expectation

$$\mathsf{M}_z^{\varepsilon}\left\{ X_{\tau^{\varepsilon}(H_0,\delta')}^{\varepsilon} \in \bigcup_{k,i} C_{ki}(\delta'); e^{-\lambda\tau^{\varepsilon}(H_0,\delta')} \right\}.$$

The idea is to evaluate the corresponding expectation for the limiting (the averaged) process, and to use it as the approximation for ϕ_1^{ε}.

Let $U_i^{\delta'}(H)$ be the solution of the boundary-value problem

$$L_i U_i^{\delta'}(H) - \lambda U_i^{\delta'}(H) = 0, \qquad H_{i1}^{\delta'} < H < H_{i2}^{\delta'}, \; U_i^{\delta'}(H_{i1}^{\delta'}) = 1,$$

$$U_i^{\delta'}(H_{i2}^{\delta'}) = \begin{cases} 1 & \text{if } O_{k'} \neq O_{\infty}, \\ 0 & \text{if } O_{k'} = O_{\infty} \text{ (and } H_{i2}^{\delta'} = H_0). \end{cases}$$

Applying Lemma 3.3 to this function, we obtain that

$$\phi_1^{\varepsilon}(z) \to U_i^{\delta'}(H(z)), \tag{3.12}$$

uniformly in $z \in \bar{D}_i(H_{i1}^{\delta'}, H_{i2}^{\delta'})$.

It is clear that the solution $U_i^{\delta'}(H)$ is strictly less than 1 for $H_{i1}^{\delta'} < H < H_{i2}^{\delta'}$; and it is equal to 0 at the boundary point H_0 if one of the ends of I_i is O_{∞}. Therefore

$$\max\{U_i^{\delta'}(H) : H_{i1}^{\delta} \leq H \leq H_{i2}^{\delta}\}$$

is strictly less than 1. For sufficiently small ε the maximum

$$\max\{\phi_1^{\varepsilon}(z) : z \in D_i(H_{i1}^{\delta}, H_{i2}^{\delta})\}$$

is also strictly smaller than 1; so

$$\max\{\phi_1^{\varepsilon}(z) \colon z \in \bar{D}(H_0, \delta)\} > 1,$$

$$\sum_{n=0}^{\infty} \mathsf{M}_x^{\varepsilon}\{\tau_n > T^{\varepsilon}; e^{-\lambda\tau_n}\} \leq \sum_{n=0}^{\infty} [\max\{\phi_1^{\varepsilon}(z) : z \in \bar{D}(H_0, \delta)\}]^n$$
$$= (1 - \max\{\phi_1^{\varepsilon}(z) : z \in \bar{D}(H_0, \delta)\})^{-1} < \infty,$$

and

$$\sum_{n=1}^{\infty} \mathsf{M}_x^\varepsilon\{\sigma_n < T^\varepsilon; e^{-\lambda\sigma_n}\} \leq \sum_{n=0}^{\infty} \mathsf{M}_x^\varepsilon\{\tau_n < T^\varepsilon; e^{-\lambda\tau_n}\}$$

is estimated by the same quantity. So the transition from (3.7) to (3.8) is possible.

We need to know how these sums depend on δ. By (3.12) we have for sufficiently small ε:

$$\sum_{n=0}^{\infty} \mathsf{M}_x^\varepsilon\{\tau_n < T^\varepsilon; e^{-\lambda\tau_n}\} \leq 2(1 - \max\{U_i^{\delta'}(H) : H_{i1}^\delta \leq H \leq H_{i2}^\delta\})^{-1}. \quad (3.13)$$

If we make $\delta' \to 0$, the solution $U_i^{\delta'}(H)$ converges to $U_i^0(H)$, also a solution of the same equation on the interval (H_{i1}^0, H_{i2}^0), with boundary values 1 at the ends that are interior vertices, and with limits < 1 at any other end (0 at the right-hand end if this is H_0). The function $U_i^0(H)$ is downward convex with respect to the function $u_i(H)$; so

$$\max\{U_i^0(H) : H_{i1}^\delta \leq H \leq H_{i2}^\delta\} \leq \max(U_i^0(H_{i1}^\delta), U_i^0(H_{i2}^\delta)).$$

If $H_{ij}^0 = H_{ij}$ is an end corresponding to an interior vertex, we have $1 - U^0(H_{ij}^\delta) \sim A_{ij}|u(H_{ij}^\delta) - u(H_{ij}^0)|$ as $\delta \to 0$, where A_{ij} is a positive constant. Since the one-sided derivative of $u_i(H)$ with respect to H exists and is positive at such ends (see formula (1.17)), we have $U^0(H_{ij}^\delta) < 1 - \tilde{A}_{ij}\delta$ for sufficiently small δ, where $\tilde{A}_{ij} > 0$. At all other ends we have $U^0(H_{ij}^\delta) < 1 - B_{ij}$ for small δ, $B_{ij} > 0$. So we obtain $\max\{U_i^0(H) : H_{i1}^\delta \leq H \leq H_{i2}^\delta\} \leq 1 - A_i\delta$, $A_i > 0$.

Now, for every small positive δ there exists δ_0', $0 < \delta_0' < \delta$, such that for all δ', $0 < \delta' \leq \delta_0'$,

$$\max\{U_i^{\delta'}(H) : H_{i1}^\delta \leq H \leq H_{i2}^\delta\} \leq 1 - \frac{A_i}{2}\delta,$$

and by (3.13) we have

$$\sum_{n=0}^{\infty} \mathsf{M}_x^\varepsilon\{\tau_n < T^\varepsilon; e^{-\lambda\tau_n}\} < C\delta^{-1}, \quad (3.14)$$

where C is a constant (there exists a constant $\delta_0 > 0$ such that, for every positive $\delta \leq \delta_0$, this holds for all δ', $0 < \delta' \leq \delta_0' = \delta_0'(\delta)$, for sufficiently small ε and for all x).

Now we return to the sums (3.8) (being the expression for the difference in (3.6)). Using the strong Markov property with respect to the Markov times

τ_n, σ_n, we can write:

$$\mathsf{M}_x^\varepsilon\left[e^{-\lambda T^\varepsilon} f(Y(X_{T^\varepsilon}^\varepsilon)) + \int_0^{T^\varepsilon} e^{-\lambda t}[\lambda f(Y(X_t^\varepsilon)) - Af(Y(X_t^\varepsilon))]\,dt\right] - f(Y(x))$$
$$= \sum_{n=0}^{\infty} \mathsf{M}_x^\varepsilon\{\tau_n < T^\varepsilon; e^{-\lambda\tau_n}\phi_2^\varepsilon(X_{\tau_n}^\varepsilon)\} + \sum_{n=1}^{\infty} \mathsf{M}_x^\varepsilon\{\sigma_n < T^\varepsilon; e^{-\lambda\sigma_n}\phi_3^\varepsilon(X_{\sigma_n}^\varepsilon)\}, \tag{3.15}$$

where

$$\phi_2^\varepsilon(z) = \mathsf{M}_z^\varepsilon\left[e^{-\lambda\tau^\varepsilon(\pm\delta)} f(Y(X_{\tau^\varepsilon(\pm\delta)}^\varepsilon)) + \int_0^{\tau^\varepsilon(\pm\delta)} e^{-\lambda t}[\lambda f(Y(X_t^\varepsilon)) - Af(Y(X_t^\varepsilon))]\,dt\right]$$
$$- f(Y(z)) \tag{3.16}$$

($\tau^\varepsilon(\pm\delta)$ being the time at which the process X_t^ε leaves $D(\pm\delta) = \bigcup_k D_k(\pm\delta)$),

$$\phi_3^\varepsilon(z) = \mathsf{M}_z^\varepsilon\left[e^{-\lambda\tau^\varepsilon(H_0,\delta')} f(Y(X_{\tau^\varepsilon(H_0,\delta')}^\varepsilon))\right.$$
$$\left. + \int_0^{\tau^\varepsilon(H_0,\delta')} e^{-\lambda t}[\lambda f(Y(X_t^\varepsilon)) - Af(Y(X_t^\varepsilon))]\,dt\right] - f(Y(z)). \tag{3.17}$$

The argument $X_{\sigma_n}^\varepsilon$ in ϕ_3^ε, as that in ϕ_1^ε in (3.9), belongs to $\bar{D}(\delta)$. As for $X_{\tau_n}^\varepsilon$, it is the initial point x for $n = 0$, but for $n \ge 1$ it belongs to $\bigcup_{k,i} C_{ki}(\delta') \cup C(H_0)$. Clearly the function $\phi_2^\varepsilon(z) = 0$ for $z \notin \bigcup_k D_k(\pm\delta)$, in particular, on $C(H_0)$, so the absolute value of the expression (3.15) does not exceed

$$|\phi_2^\varepsilon(x)| + \sum_{n=1}^{\infty} \mathsf{M}_x^\varepsilon\{\tau_n < T^\varepsilon; e^{-\lambda\tau_n}\} \cdot \max\left\{|\phi_2^\varepsilon(z)| : z \in \bigcup_{k,i} C_{ki}(\delta')\right\}$$
$$+ \sum_{n=1}^{\infty} \mathsf{M}_x^\varepsilon\{\sigma_n < T^\varepsilon; e^{-\lambda\sigma_n}\} \cdot \max\{|\phi_3^\varepsilon(z)| : z \in \bar{D}(H_0,\delta)\}. \tag{3.18}$$

By Lemma 3.3, $|\phi_3^\varepsilon(z)|$ is arbitrarily small for sufficiently small ε, uniformly in $z \in \bar{D}(H_0,\delta') \supset \bar{D}(H_0,\delta)$, so

$$\sum_{n=1}^{\infty} \mathsf{M}_x^\varepsilon\{\sigma_n < T^\varepsilon; e^{-\lambda\sigma_n}\} \cdot \max\{|\phi_3^\varepsilon(z)| : z \in \bar{D}(H_0,\delta)\} < \eta/20 \tag{3.19}$$

for sufficiently small ε, and this is true for every function f on Γ that is smooth on the segments I_i, even if it does not satisfy the gluing conditions.

As for ϕ_2^ε, we cannot see at once that it converges to 0 as $\varepsilon \to 0$, but we can make it small by choosing δ and δ'. This is not enough, because $\max\{|\phi_2^\varepsilon(z)| : z \in \bigcup k, i\, C_{ki}(\delta')\}$ is multiplied by the sum that increases at the rate of δ^{-1} (see (3.14)) as $\delta \to 0$; so we have to obtain an estimate for $\phi_2^\varepsilon(z)$ for an arbitrary $z \in D(\pm\delta)$, and a sharper one for $z \in \bigcup_{k,i} C_{ki}(\delta')$.

We estimate $\phi_2^\varepsilon(z)$ in different ways for an arbitrary z belonging to $D(\pm\delta)$ (which is used to estimate the first summand in (3.18)), in the case of $z \in C_{ki}(\delta')$, where O_k is an interior vertex, and in that of an exterior O_k.

Clearly, for $z \in D_k(\pm\delta)$,

$$|\phi_2^\varepsilon(z)| \leq |\mathsf{M}_z^\varepsilon f(Y(X_{\tau_k^\varepsilon(\pm\delta)}^\varepsilon)) - f(Y(z))| + [\lambda\|f\| + \|\lambda f - Af\|] \cdot \mathsf{M}_z^\varepsilon \int_0^{\tau_k^\varepsilon(\pm\delta)} e^{-\lambda t}\, dt. \tag{3.20}$$

Choose a positive δ_1 so that for every vertex O_k

$$|f(y) - f(O_k)| < \eta/20 \quad \text{for all } y \in \bar{D}_k(\pm\delta_1)$$

(we are using continuity of the function f) and that

$$\mathsf{M}_z^\varepsilon \int_0^{\tau_k^\varepsilon(\pm\delta_1)} e^{-\lambda t}\, dt < \frac{\eta}{20[\lambda\|f\| + \|\lambda f - Af\|]}$$

for sufficiently small ε and all $z \in \bar{D}_k(\pm\delta_1)$ (we are using Lemmas 3.4 and 3.5). Then for $\delta \leq \delta_1$ and for sufficiently small ε

$$|\phi_2^\varepsilon(z)| < \eta/10 \quad \text{for all } z \in \bar{D}(\pm\delta). \tag{3.21}$$

Next case: $z \in C_{ki}(\delta')$; the vertex O_k is an interior one.

Using the fact that the one-sided derivatives $f_i'(O_k)$ exist, choose a positive δ_2 so that

$$\left|\frac{f(y) - f(O_k)}{H(y) - H(O_k)} - f_i'(O_k)\right| < \frac{\eta}{20\, C}$$

for $y \in I_i \cap \bar{D}(\pm\delta_2)$ for all segments I_i meeting at O_k, where C is the constant of the estimate (3.14).

Choose $\delta_3 > 0$, by Lemma 3.5, so that for $0 < \delta \leq \delta_3$ for sufficiently small ε and for all $z \in D_k(\pm\delta)$,

$$\mathsf{M}_x^\varepsilon \int_0^{\tau_k(\pm\delta)} e^{-\lambda t}\, dt < \delta \cdot \frac{\eta}{20\, C[\lambda\|f\| + \|\lambda f - Af\|]}. \tag{3.22}$$

Choose a positive δ_4, by Lemma 3.6, so that for $0 < \delta \le \delta_4$ there exists a $\delta_4' = \delta_4'(\delta)$, $0 < \delta_4' < \delta$, such that

$$|\mathsf{P}_z^\varepsilon\{X_{\tau^\varepsilon(\pm\delta)}^\varepsilon \in C_{ki}(\delta)\} - p_{ki}| < \frac{\eta}{20\, C \sum_{i:I_i \sim O_k} |f_i'(O_k)|}$$

for sufficiently small ε and for all $z \in \bar{D}_k(\pm\delta_k')$.

Using continuity of f again, choose, for every $\delta > 0$, a positive $\delta_5' = \delta_5'(\delta) < \delta$ so that

$$|f(y) - f(O_k)| < \delta \cdot \eta/20 \tag{3.23}$$

for all $y \in \bar{D}_k(\pm\delta_5')$.

Now for $0 < \delta \le \min(\delta_2, \delta_3, \delta_4)$, $0 < \delta' \le \min(\delta_4'(\delta), \delta_5'(\delta))$, for sufficiently small ε and for all $z \in C_{ki}(\delta') \subset \bar{D}_k(\pm\delta_4') \cap \bar{D}_k(\pm\delta_5')$ we have

$$\begin{aligned}|\mathsf{M}_z^\varepsilon f(Y(X_{\tau_k^\varepsilon(\pm\delta)}^\varepsilon)) - f(Y(z))| \le{}& \left|\sum_{i:I_i \sim O_k} p_{ki} \cdot [f(i, H(O_k) \pm \delta) - f(O_k)]\right| \\ &+ \sum_{i:I_i \sim O_k} |\mathsf{P}_z^\varepsilon\{X_{\tau^\varepsilon(\pm\delta)}^\varepsilon \in C_{ki}(\delta)\} - p_{ki}| \cdot |f(i, H(O_k) \pm \delta) - f(O_k)| \\ &+ |f(O_k) - f(Y(z))|. \end{aligned} \tag{3.24}$$

In the first summand on the right-hand side the quantity in the brackets differs from $\delta \cdot f_i'(O_k)$ by not more than $\delta.\ \eta/20\, C$, so the first summand does not exceed

$$\left|\sum_{i:I_i \sim O_k} p_{ki} \cdot \delta \cdot f_i'(O_k)\right| + \sum_{i:I_i \sim O_k} p_{ki} \cdot \delta \cdot \frac{\eta}{20\, C}.$$

The first summand here is equal to 0 because the gluing condition is satisfied at O_k, and the second one is equal to $\delta \cdot \eta/20\, C$.

Using the inequality

$$|f(i, H(O_k) \pm \delta) - f(O_k)| \le \delta \cdot |f_i'(O_k)| + \delta \cdot \eta/20\, C,$$

we see that the second summand on the right-hand side of (3.24) is not greater than

$$\sum_{i:I_i \sim O_k} \frac{\eta}{20\, C \sum_{i:I_i \sim O_k} |f_i'(O_k)|} \cdot \delta \cdot |f_i'(O_k)| + 3\delta \cdot \eta/20\, C = \delta \cdot 0.2\eta/C$$

($3\delta \cdot \eta/20\, C$ because the number of segments meeting at O_k is equal to 3 if the level curves do not contain more than one critical point, and $|\mathsf{P}_z^\varepsilon\{X_{\tau^\varepsilon(\pm\delta)}^\varepsilon \in C_{ki}(\delta)\} - p_{ki}| \le 1$).

The third summand on the right-hand side of (3.24) is not greater than $\delta \cdot \eta/20\,C$ by (3.23).

By (3.22), the second summand on the right-hand side of (3.20) is not greater than $\delta \cdot \eta/20\,C$, and

$$|\phi_2^\varepsilon(z)| \le \delta \cdot 0.3\eta/C \quad \text{for } z \in C_{ki}(\delta') \tag{3.25}$$

(for interior vertices O_k).

The last case is that of $z \in C_{ki}(\delta')$, O_k being an exterior vertex (remember, only one segment I_i enters a vertex corresponding to an extremum point).

Let us use again the technique by which we obtained formulas (3.15)–(3.18). Choose a positive $\delta'' < \delta'$, and consider smaller cycles between reaching two sets, taking δ' instead of δ, δ'' instead of δ', and $C_{ki}(\delta)$ instead of $C(H_0)$ (everything in a neighborhood of an extremum point x_k). Namely, let us introduce Markov times $\sigma_1' \le \tau_1' \le \sigma_2' \le \tau_2' \le \cdots$ by $\sigma_1' = 0$, $\tau_n' = \min\{t \ge \sigma_n' : X_t^\varepsilon \in C_{ki}(\delta'') \cup C_{ki}(\delta)\}$, $\sigma_n' = \min\{t \ge \tau_{n-1}' : X_t^\varepsilon \notin D_k(\pm\delta')\}$ (after the process reaches $C_{ki}(\delta)$, all τ_n' and σ_n' are equal to $\tau_k^\varepsilon(\pm\delta)$; the difference is that, since we are starting from a point $z \in C_{ki}(\delta')$, we have $\sigma_1' = 0$, so we do not need to consider τ_0'). We have

$$\phi_2^\varepsilon(z) = \sum_{n=1}^\infty \mathsf{M}_z^\varepsilon\{\tau_n' < \tau_k^\varepsilon(\pm\delta); e^{-\lambda\tau_n'}\phi_4(X_{\tau_n'}^\varepsilon)\}$$

$$+ \sum_{n=1}^\infty \mathsf{M}_z^\varepsilon\{\sigma_n' < \tau_k^\varepsilon(\pm\delta); E^{-\lambda\sigma_n'}\phi_5^\varepsilon(X_{\tau_n'}^\varepsilon)\}, \tag{3.26}$$

$$\phi_4^\varepsilon(z') = \mathsf{M}_{z'}^\varepsilon\Bigg[e^{-\lambda\tau_k^\varepsilon(\pm\delta')}f(Y(X_{\tau_k^\varepsilon(\pm\delta')}^\varepsilon)) + \int_0^{\tau_k^\varepsilon(\pm\delta')} e^{-\lambda t}[\lambda f(Y(X_t^\varepsilon)) - Af(Y(X_t^\varepsilon))]\,dt\Bigg] - f(Y(z')), \tag{3.27}$$

$$\phi_5^\varepsilon(z') = \mathsf{M}_{z'}^\varepsilon\Bigg[e^{-\lambda\tau_i^\varepsilon(H(O_k)\pm\delta, H(O_k)\pm\delta'')}f(Y(X_{\tau_i^\varepsilon(H(O_k)\pm\delta, H(O_k)\pm\delta'')}^\varepsilon)) + \int_0^{\tau_i^\varepsilon(H(O_k)\pm\delta, H(O_k)\pm\delta'')} e^{-\lambda t}[\lambda f(Y(X_t^\varepsilon)) - Af(Y(X_t^\varepsilon))]\,dt\Bigg] - f(Y(z')), \tag{3.28}$$

where $\tau_i^\varepsilon(H(O_k) \pm \delta, H(O_k) \pm \delta'')$, according to our system of notations, is the time at which the process leaves the region $D_i(H(O_k) \pm \delta, H(O_k) \pm \delta'')$ (this region is the same as the difference $D_k(\pm\delta)\setminus\bar{D}_k(\pm\delta'')$); and for $z \in C_{ki}(\delta')$,

$$|\phi_2^\varepsilon(z)| \le \sum_{n=1}^{\infty} \mathsf{M}_z^\varepsilon\{\tau_n' < \tau_k^\varepsilon(\pm\delta); e^{-\lambda\tau_n'}\} \cdot \max\left\{|\phi_4^\varepsilon(z')| : z \in \bigcup_{i:I_i\sim O_k} C_{ki}(\delta'')\right\}$$
$$+ \sum_{n=1}^{\infty} \mathsf{M}_z^\varepsilon\{\sigma_n' < \tau_k^\varepsilon(\pm\delta); e^{-\lambda\sigma_n'}\} \cdot \max\left\{|\phi_5^\varepsilon(z')| : z \in \bigcup_{i:I_i\sim O_k} C_{ki}(\delta')\right\}. \tag{3.29}$$

Again $\phi_5^\varepsilon(z')$ can be made arbitrarily small by Lemma 3.3; similarly to (3.19), for sufficiently small ε,

$$\sum_{n=1}^{\infty} \mathsf{M}_z^\varepsilon\{\sigma_n' < \tau_k^\varepsilon(\pm\delta); e^{-\lambda\sigma_n'}\} \cdot \max\left\{|\phi_5^\varepsilon(z')| : z' \in \bigcup_{i:I_i\sim O_k} C_{ki}(\delta')\right\} < \delta\cdot\eta/20\,C \tag{3.30}$$

for all $z \in C_{ki}(\delta')$.

Similarly to the way we obtained (3.21), we can, for every $\delta > 0$, choose a positive $\delta_6' = \delta_6'(\delta) < \delta$ so small that for sufficiently small ε for all $z' \in \bar{D}_k(\pm\delta')$,

$$|\phi_4^\varepsilon(z')| < \delta\cdot\eta/10\,C; \tag{3.31}$$

and similarly to (3.9)–(3.11), we obtain:

$$\mathsf{M}_z^\varepsilon\{\tau_n' < \tau_k^\varepsilon(\pm\delta); e^{-\lambda\tau_n'}\} \le [\max\{\phi_6^\varepsilon(z') : z' \in C_{ki}(\delta')\}]^n, \tag{3.32}$$

$$\phi_6^\varepsilon(z') = \mathsf{M}_{z'}^\varepsilon\{X^\varepsilon_{\tau_i^\varepsilon(H(O_k)\pm\delta, H(O_k)\pm\delta'')} \in C_{ki}(\delta'); e^{-\lambda\tau_i^\varepsilon(H(O_k)\pm\delta, H(O_k)\pm\delta'')}\}. \tag{3.33}$$

Suppose the process X_t^ε starts from a point $z \in C_{ki}(\delta')$. Let us consider a solution $V_i^\delta(H)$ of the differential equation $L_i V_i^\delta(H) - \lambda V_i^\delta(H) = 0$ with $V_i^\delta(H(O_k)\pm\delta) = 0$. This solution is determined uniquely up to a multiplicative constant. It does not change the sign between $H(O_k)$ and $H(O_k)\pm\delta$, and it converges to $+\infty$ or to $-\infty$ as $H \to H(O_k)$ (since the function $u_i(H)$ has an infinite limit at $H(O_k)$). Applying Lemma 3.3 to the function $V_i^\delta(H)$, we obtain that

$$\phi_6^\varepsilon(z') \to \frac{V_i^\delta(H(z'))}{V_i^\delta(H(O_k)\pm\delta)} \tag{3.34}$$

as $\varepsilon \to 0$, uniformly in $z' \in D_i(H(O_k)\pm\delta'', H(O_k)\pm\delta)$. Since $|V_i^\delta(H(O_k)\pm\delta'')| \to \infty$ as $\delta'' \to 0$, for every $\delta' > 0$ we can choose $\delta_i'' = \delta''(\delta,\delta')$, $0 < \delta_i'' < \delta'$, so that $V_i^\delta(H(O_k)\pm\delta')/V_i^\delta(H(O_k)\pm\delta'') < \frac{1}{3}$ for $0 <$

$\delta'' \le \delta_i''$. Then for $\delta'' \le \min_i \delta_i''$ and for sufficiently small ε,

$$\phi_6^\varepsilon(z') < \tfrac{2}{3} \quad \text{for } z' \in C_{ki}(\delta').$$

Using this together with (3.29)–(3.31), we obtain that for sufficiently small δ, for $\delta' \le \min_i(\delta_i'(\delta)) < \delta$, for $0 < \delta'' \le \min_i(\delta_i''(\delta, \delta')) < \delta'$, and for sufficiently small ε we have:

$$|\phi_2^\varepsilon(z)| \le \delta \cdot 0.25\eta/C \quad \text{for } z \in \bigcup_{i: I_i \sim O_k} C_{ki}(\delta'),$$

and together with (3.25) this yields $\max\{|\phi_2^\varepsilon(z)| : z \in \bigcup_{k,i} C_{ki}(\delta')\} \le \delta \cdot 0.4\eta/C$. Using this and estimates (3.18), (3.19), (3.14), and (3.21), we obtain that the difference in (3.6) is smaller than $0.1\eta + (C/\delta) \cdot \delta \cdot 0.3\eta/C + 0.05\eta < \eta/2$.

This proves the theorem.

§4. Proof of Lemmas 3.1 to 3.4

Proof of Lemma 3.1. The tool we use to establish weak convergence is martingale problems. See Stroock and Varadhan [1, 2], where martingale problems are formulated in terms of the space $\mathbf{C}^\infty$ of infinitely differentiable functions; we use some other sets of functions.

Let us denote the difference of the left-hand side of (3.1) and its limit in the right-hand side by $\Delta^\varepsilon(x)$. Let $G(y_1, \ldots, y_n)$, $y_i \in Y(M)$, be a bounded measurable function. Then, by the Markov property of the process $(X_t^\varepsilon, \mathsf{P}_x^\varepsilon)$, for any $0 \le t_1 < \cdots < t_n \le t_0$ we have

$$\mathsf{M}_x^\varepsilon G(Y(X_{t_1}^\varepsilon), \ldots, Y(X_{t_n}^\varepsilon)) \cdot \left[\int_{t_0}^\infty e^{-\lambda t}[\lambda f(Y(X_t^\varepsilon)) - Af(Y(X_t^\varepsilon))]\,dt \right.$$
$$\left. - e^{-\lambda t_0} f(Y(X_{t_0}^\varepsilon))\right] = \mathsf{M}_x^\varepsilon G(Y(X_{t_1}^\varepsilon), \ldots, Y(X_{t_n}^\varepsilon)) \cdot e^{-\lambda t_0} \Delta^\varepsilon(X_{t_0}^\varepsilon).$$

We can represent this expectation as the sum of expectations taken over the event $\{Y(X_{t_0}^\varepsilon) \in K\}$ and over its complement, obtaining

$$\mathsf{M}_x^\varepsilon\{Y(X_{t_0}^\varepsilon) \in K; G(Y(x_{t_1}^\varepsilon), \ldots, Y(X_{t_n}^\varepsilon)) \cdot e^{-\lambda t_0} \Delta^\varepsilon(X_{t_0}^\varepsilon)\}$$
$$+ \mathsf{M}_x^\varepsilon\{Y(X_{t_0}^\varepsilon) \notin K; G(Y(X_{t_1}^\varepsilon), \ldots, Y(X_{t_n}^\varepsilon)) \cdot e^{-\lambda t_0} \Delta^\varepsilon(X_{t_0}^\varepsilon)\}.$$

The second expectation does not exceed $\|G\| \cdot e^{-\lambda t_0} \cdot \sup_{z,\varepsilon} \Delta^\varepsilon(z) \cdot \mathsf{P}_x^\varepsilon\{Y(X_{t_0}^\varepsilon) \notin K\} \le \|G\| \cdot e^{-\lambda t_0} \cdot (2\|f\| + \lambda^{-1}\|Af\|) \cdot \mathsf{P}_x^\varepsilon\{Y(X_{t_0}^\varepsilon) \notin K\}$, and can be made arbitrarily small for all ε by choosing a compact $K \subseteq Y(M)$.

The first expectation is not greater than $\|G\| \cdot e^{-\lambda t_0} \cdot \sup_{z \in Y^{-1}(K)} \Delta^\varepsilon(z)$, and can be made small by choosing ε sufficiently small. So we obtain

$$\mathsf{M}_x^\varepsilon G(Y(X_{t_1}^\varepsilon), \ldots, Y(X_{t_n}^\varepsilon)) \cdot \left[\int_{t_0}^\infty e^{-\lambda t}[\lambda f(Y(X_t^\varepsilon)) - Af(Y(X_t^\varepsilon))]\, dt \right.$$
$$\left. - e^{-\lambda t_0} f(Y(X_{t_0}^\varepsilon))\right] \to 0 \qquad (\varepsilon \to 0).$$

We can rewrite this formula using the expectations with respect to the measure Q_x^ε in the space $\mathbf{C}[0, \infty)$, which we denote by $\tilde{\mathsf{M}}_x^\varepsilon$:

$$\tilde{\mathsf{M}}_x^\varepsilon G(y_{t_1}, \ldots, y_{t_n}) \cdot \left[\int_{t_0}^\infty e^{-\lambda t}[\lambda f(y_t) - Af(y_t)]\, dt - e^{-\lambda t_0} f(y_{t_0})\right] \to 0 \quad (\varepsilon \to 0). \tag{4.1}$$

Since the family of distributions $\{\mathsf{Q}_x^\varepsilon\}$ is tight, there exists a sequence $\varepsilon_n \to 0$ such that $\mathsf{Q}_x^{\varepsilon_n}$ converges weakly to a probability measure $\tilde{\mathsf{P}}$. Let us denote the expectation corresponding to this probability by $\tilde{\mathsf{M}}$.

If the function G is continuous, the functional of $y_\bullet \in \mathbf{C}[0, \infty)$ under the expectation sign in (4.1) is also continuous, and we can write:

$$\tilde{\mathsf{M}} G(y_{t_1}, \ldots, y_{t_n}) \cdot \left[\int_{t_0}^\infty e^{-\lambda t}[\lambda f(y_t) - Af(y_t)]\, dt - e^{-\lambda t_0} f(y_{t_0})\right] = 0.$$

Changing the order of integration and integrating by parts, we obtain

$$\int_{t_0}^\infty \lambda e^{-\lambda t} \tilde{\mathsf{M}} G(y_{t_1}, \ldots, y_{t_n}) \cdot \left[f(y_t) - f(y_{t_0}) - \int_{t_0}^t Af(y_s)\, ds\right] = 0.$$

Since a continuous function is determined uniquely by its Laplace transform, this means that $\tilde{\mathsf{M}} G(y_{t_1}, \ldots, y_{t_n}) \cdot [f(y_t) - f(y_{t_0}) - \int_{t_0}^t Af(y_s)\, ds] = 0$ for all n and $0 \le t_1 < \cdots < t_n \le t_0$; so the random function $f(y_t) - \int_0^t Af(y_s)\, ds$ is a martingale with respect to the family of σ-algebras $\mathscr{F}_{[0,t]}$ generated by the functionals $y_\bullet \mapsto y_s$, $s \in [0, t]$, and the probability measure $\tilde{\mathsf{P}}$.

In other words, the probability measure is the solution of a martingale problem corresponding to the contraction of the operator A defined on the set $D \subseteq \mathbf{C}(Y(M))$, with the starting point $Y(x)$ (which is clear because $Y(X_0^\varepsilon) = Y(x)$ almost surely with respect to each of the probabilities P_x^ε).

Now, the condition of existence of a solution $F \in D$ of the equation $\lambda F - AF = f$ for $f \in \Psi$ ensures uniqueness of the solution of the martingale problem. This statement is a variant of Theorem 6.3.2 (with condition (ii)) of Stroock and Varadhan [2], only freed from reference to the space $\mathbf{C}^\infty$ of infinitely differentiable functions and adapted to time-homogeneous processes.

So $\tilde{\mathsf{P}} = \mathsf{P}_{Y(x)}$. The fact that not only the sequence $\mathsf{Q}_x^{\varepsilon_n}$ converges weakly to $\mathsf{P}_{Y(x)}$, but also Q_x^ε as $\varepsilon \to 0$, is proved in the usual way.

The remaining lemmas involve many estimates with various positive constants, which we denote by the letter A with some subscript.

We very often make use of Itô's formula applied to $e^{-\lambda t} f(H(X_t^\varepsilon))$, where f is a smooth function, $\lambda \geq 0$:

$$\begin{aligned} e^{-\lambda t} f(H(X_t^\varepsilon)) &= f(H(X_0^\varepsilon)) + \int_0^t e^{-\lambda s} f'(H(X_s^\varepsilon)) \nabla H(X_s^\varepsilon) \cdot dw_s \\ &\quad + \int_0^t e^{-\lambda s} [-\lambda f(H(X_s^\varepsilon)) + \frac{1}{2} f''(H(X_s^\varepsilon)) |\nabla H(X_s^\varepsilon)|^2 \\ &\quad + \frac{1}{2} f'(H(X_s^\varepsilon)) \Delta H(X^\varepsilon)] \, ds, \end{aligned} \tag{4.2}$$

$$\begin{aligned} \mathsf{M}_x^\varepsilon e^{-\lambda \tau} f(H(X_\tau^\varepsilon)) = f(H(x)) + \mathsf{M}_x^\varepsilon \int_0^\tau e^{-\lambda s} [-\lambda f(H(X_s^\varepsilon)) \\ + \frac{1}{2} f''(H(X_s^\varepsilon)) |\nabla H(X_s^\varepsilon)|^2 \\ + \frac{1}{2} f'(H(X_s^\varepsilon)) \Delta H(X^\varepsilon)] \, ds \end{aligned} \tag{4.3}$$

for the time τ of going out of an arbitrary bounded region. In particular, for $\lambda = 0, f(H) = H$,

$$\mathsf{M}_x^\varepsilon H(X_t^\varepsilon) = H(x) + \mathsf{M}_x^\varepsilon \int_0^t \frac{1}{2} \Delta H(X_s^\varepsilon) \, ds, \tag{4.4}$$

$$\mathsf{M}_x^\varepsilon H(X_\tau^\varepsilon) = H(x) + \mathsf{M}_x^\varepsilon \int_0^\tau \frac{1}{2} \Delta H(X_s^\varepsilon) \, ds. \tag{4.5}$$

Proof of Lemma 3.2. Introducing the graph, we did not describe the metric on it. Now we do this: for any two points $y, y' \in Y(R^2)$, consider all paths on the graph that connect these points following the segments $I_i : y = (i_0, H_0) \leftrightarrow (i_0, H_1) = O_{k_1} = (i_1, H_1) \leftrightarrow (i_1, H_2) = O_{k_2} = (i_2, H_2) \leftrightarrow \cdots \leftrightarrow (i_{l-1}, H_{l-1}) = O_{k_{l-1}} = (i_l, H_{l-1}) \leftrightarrow (i_l, H_l) = y'$, and take

$$\rho(y, y') = \min \sum_{i=0}^{l-1} |H_{i+1} - H_i|,$$

where the minimum is taken over all such paths.

In order to prove the tightness it is sufficient to prove that:

(1) for every $T > 0$ and $\delta > 0$ there exists a number H_0 such that

$$\mathsf{P}_x^\varepsilon \left\{ \max_{0 \leq t \leq T} H(X_t^\varepsilon) \geq H_0 \right\} < \delta \tag{4.6}$$

for all ε;

(2) for every compact subset $K \subset R^2$ and for every sufficiently small $\rho > 0$ there exists a constant A_ρ such that for every $a \in K$ there exists a function $f_\rho^a(y)$ on $Y(R^2)$ such that $f_\rho^a(a) = 1$, $f_\rho^a(y) = 0$ for $\rho(y, a) \geq \rho$, $0 \leq f_\rho^a(y) \leq 1$ everywhere, and $f_\rho^a(Y(X_t^\varepsilon)) + A_\rho t$ is a submartingale for all ε (see Šroock and Varadhan [2]).

As for (1), we can write, using formula (4.4),

$$\mathsf{M}_x^\varepsilon H(X_t^\varepsilon) \leq H(x) + \frac{A_4}{2}\, t, \tag{4.7}$$

where $A_4 \geq \sup \Delta H(x)$ (the constant denoted A_0 was introduced in (1.14), and A_1, A_2, A_3 in the formulation of Theorem 2.2). Using Kolmogorov's inequality, we obtain for $H_0 > H(x) + (A_4/2)T$,

$$\begin{aligned}\mathsf{P}_x^\varepsilon\left\{\max_{0 \leq t \leq T} H(X_t^\varepsilon) \geq H_0\right\} &\leq \frac{\mathsf{M}_x^\varepsilon \int_0^T |\nabla H(X_t^\varepsilon)|^2\, dt}{H_0 - H(x) - (A_4/2)T} \\ &\leq \frac{\int_0^T [A_5 + A_6 \mathsf{M}_x^\varepsilon H(X_t^\varepsilon)]\, dt}{H_0 - H(x) - (A_4/2)T},\end{aligned}$$

and using (4.7), we obtain (4.6).

Let us prove (2).

Let $h(x)$ be a fixed smooth function such that $0 \leq h(x) \leq 1$, $h(x) = 1$ for $x \leq 0$, and $h(x) = 0$ for $x \geq 1$.

Let us take a positive ρ smaller than half the length of the shortest segment of the graph. Let a be a point of $Y(R^2)$. If the distance from a to the nearest vertex of the graph is greater than $2\rho/5$, we put $f_\rho^a(y) = h(5\rho(y, a)/\rho)$; if $\rho(a, O_k) \leq 2\rho/5$, we put $f_\rho^a(y) = h(5\rho(y, O_k)/\rho - 2)$. In both cases the function $f_\rho^a(y) = 0$ outside the ρ-neighborhood of the point a.

Now let us consider the functions $f_\rho^a(Y(x))$, $x \in R^2$. These functions are expressed by the formulas $f_\rho^a(Y(x)) = H(5|H(x) - H(a)|/\rho)$, or $f_\rho^a(Y(x)) = H(5|H(x) \quad H(O_k)|/\rho - 2)$, or the identical 0 in different regions, and the regions in which different formulas are valid overlap. The functions $|H(x) - H(a)|$, $|H(x) - H(O_k)|$ are smooth except on the lines where $H(x) = H(a)$ or $H(O_k)$; so the functions $f_\rho^a(Y(x))$ are smooth outside these lines. But also they are smooth in some neighborhoods of these lines, because $h'(0) = h''(0) = h'(-2) = h''(-2) = 0$, and the functions $h(|x|)$, $h(|x| - 2)$ are smooth. So the functions $f_\rho^a(Y(x))$ are twice continuously differentiable everywhere, and their gradients are orthogonal to $\bar{\nabla} H(x)$.

Apply Itô's formula:

$$f_\rho^a(Y(X_t^\varepsilon)) = f_\rho^a(Y(X_0^\varepsilon)) + \int_0^t \nabla f_\rho^a(Y(X_s^\varepsilon)) \cdot dw_s + \int_0^t \frac{1}{2} \Delta f_\rho^a(Y(X_s^\varepsilon))\, ds,$$

and take

$$A_\rho = \sup_{a \in K, x \in R^2} \frac{1}{2} \Delta f_\rho^a(Y(x))$$
$$\leq \frac{1}{2}\left[\sup_{x \in [0,1]} |h'(x)| \cdot \frac{5}{\rho} \cdot \sup_{x \in K} |\Delta H(x)| + \sup_{x \in [0,1]} |h''(x)| \cdot \frac{25}{\rho^2} \cdot \sup_{x \in K} |\nabla H(x)|^2\right].$$

The lemma is proved. □

Proof of Lemma 3.4. Let us use formula (4.5) with $\tau = \tau_k^\varepsilon(\pm\delta)$. The boundary $\partial D_k(\pm\delta)$ consists of one component, and either $H(X^\varepsilon_{\tau_k^\varepsilon(\pm\delta)})$ is always equal to $H(x_k) + \delta$, or always to $H(x_k) - \delta$ (depending on whether x_k is a minimum or a maximum of the Hamiltonian). Since the point x_k is a nondegenerate extremum point, we have $\Delta H(x_k) \neq 0$. Of course, $|\Delta H(x)| > \frac{1}{2}|\Delta H(x_k)|$ in a neighborhood of this point, so for sufficiently small $\delta > 0$,

$$\mathsf{M}_x^\varepsilon \tau_k^\varepsilon(\pm\delta) \leq \frac{|H(x_k) \pm \delta - H(x)|}{\inf\{|\frac{1}{2}\Delta H(x)| : x \in D_k(\pm\delta)\}} \leq \frac{4\delta}{|\Delta H(x_k)|}$$

for all $x \in D_k(\pm\delta)$ and for all ε (and the expectation on the left-hand side of (3.3) is smaller still).

Lemmas 3.3 and 3.5 are formulated for domains $D_i(H_1, H_2)$, $D_k(\pm\delta)$ that do not depend on ε; but we obtain versions of these lemmas with domains depending on ε : $D_k(\pm\delta)$ with $\delta = \delta(\varepsilon) > 0$ decreasing with ε (but not too fast), and for $D_i(H_1, H_2)$ with one of the ends H_1, H_2—or both—being possibly at a distance $\delta(\varepsilon)$ from an end of the interval $H(I_i)$. Lemma 4.2 is the ε-dependent version of Lemma 3.3, including this lemma as a particular case; and Lemma 5.1 is the time-dependent version of Lemma 3.5, which *does not* include it as a particular case. In the proof of Lemma 3.5 both Lemmas 4.2 and 5.1 are used. The proofs rely on a rather complicated system of smaller lemmas.

Let us formulate two ε-dependent versions of Lemma 3.3.

Let $\mathfrak{G}_i(H_1, H_2)$ be the class of all functions g defined in the closed region $\bar{D}_i(H_1, H_2)$, satisfying a Lipschitz condition $|g(x') - g(x)| \leq \mathrm{Lip}(g) \cdot |x' - x|$, and such that

$$\oint_{C_i(H)} g(x) \frac{dl}{|b(x)|} = 0$$

for all $H \in (H_1, H_2)$.

Lemma 4.1. *There exist (small) positive constants A_7, A_8 such that for every positive A_9 for every H_1, $H_2 \leq A_9$ in the interval $H(I_i)$ at a distance not less than $\delta = \delta(\varepsilon) = \varepsilon^{A_7}$ from the ends of $H(I_i)$, for every positive λ, for sufficiently* *small ε, for every $g \in \mathfrak{G}_i(H_1, H_2)$, and for every $x \in D_i(H_1, H_2)$,*

$$\left|\mathsf{M}_x^\varepsilon \int_0^{\tau_i^\varepsilon(H_1,H_2)} e^{-\lambda t} g(X_t^\varepsilon)\, dt\right| \le (\|g\| + \mathrm{Lip}(g)) \cdot \varepsilon^{A_8}. \tag{4.8}$$

Lemma 4.2. *Under the conditions of the previous lemma, there exist such positive constants A_{10}, A_{11} that for sufficiently small ε, for every $H_1, H_2 \le A_9$ in the interval $H(I_i)$ at a distance not less than $\delta = \delta(\varepsilon) = \varepsilon^{A_{10}}$ from its ends, for every $x \in D_i(H_1, H_2)$, and for every smooth function f on the interval $[H_1, H_2]$,*

$$\Bigg|\mathsf{M}_x^\varepsilon\Bigg[e^{-\lambda\tau_i^\varepsilon(H_1,H_2)} f(H(X^\varepsilon_{\tau_i^\varepsilon(H_1,H_2)})) + \int_0^{\tau_i^\varepsilon(H_1,H_2)} e^{-\lambda t}[\lambda f(H(X_t^\varepsilon)) - L_i f(H(X_t^\varepsilon))]\, dt\Bigg] - f(H(x))\Bigg| \le (\|f'\| + \|f''\| + \|f'''\|)\varepsilon^{A_{11}}. \tag{4.9}$$

Lemma 3.3 is a particular case of this lemma.

Lemma 4.2 is proved simply enough if Lemma 4.1 is proved. To prove Lemma 4.1 we use the fact that our process X_t^ε moves fast along the trajectories of the Hamiltonian system (and more slowly across them). It is convenient for us to consider the "slow" process $\tilde{X}_t^\varepsilon$. The following lemma makes precise the statement about our process moving along the trajectories.

Lemma 4.3. *For every positive η,*

$$\mathsf{P}_x^\varepsilon\Big\{\max_{0\le t\le T} |\tilde{X}_t^\varepsilon - x_t(x)| \ge \eta\Big\} \le 3\left(\frac{e^{2LT}-1}{2L}\right)^2 \frac{\varepsilon^4}{\eta^4}, \tag{4.10}$$

where L is the Lipschitz constant of the function $\bar{\nabla} H$ (the same as that of ∇H); and

$$\mathsf{M}_x^\varepsilon|\tilde{X}_t^\varepsilon - x_t(x)| \le \left(\frac{e^{2Lt} - 1}{2L}\right)^{1/2} \varepsilon. \tag{4.11}$$

Proof. Applying Itô's formula to the random function $|\tilde{X}_t^\varepsilon - x_t(x)|^2$, we obtain:

$$\begin{aligned} &|\tilde{X}_t^\varepsilon - x_t(x)|^2 \\ &= \int_0^t [2(\tilde{X}_s^\varepsilon - x_s(x)) \cdot (b(\tilde{X}_s^\varepsilon) - b(x_s(x))) + \varepsilon^2]\, ds + \int_0^t 2\varepsilon(\tilde{X}_s^\varepsilon - x_s(x)) \cdot dw_s \\ &\le \int_0^t 2L|\tilde{X}_s^\varepsilon - x_s(x)|^2\, ds + \varepsilon^2 t + \int_0^t 2\varepsilon(\tilde{X}_s^\varepsilon - x_s(x)) \cdot dw_s, \end{aligned}$$

$$\mathsf{M}_x^\varepsilon|\tilde{X}_t^\varepsilon - x_t(x)|^2 \le 2L\int_0^t \mathsf{M}_x^\varepsilon|\tilde{X}_s^\varepsilon - x_s(x)|^2\, ds + \varepsilon^2 t.$$

Using the Gronwall–Bellman inequality, we have:

$$\mathsf{M}_x^\varepsilon|\tilde{X}_t^\varepsilon - x_t(x)|^2 \le \frac{e^{2Lt}-1}{2L}\,\varepsilon^2.$$

The estimate (4.11) follows from this at once.

Applying Itô's formula to $|\tilde{X}_t^\varepsilon - x_t(x)|^4$, we obtain in the same way:

$$\begin{aligned} |\tilde{X}_t^\varepsilon - x_t(x)|^4 \le \int_0^t [4L|\tilde{X}_s^\varepsilon - x_s(x)|^4 + 6\varepsilon^2|\tilde{X}_s^\varepsilon - x_s(x)|^2]\,ds \\ + \int_0^t 4\varepsilon|\tilde{X}_s^\varepsilon - x_s(x)|^2(\tilde{X}_s^\varepsilon - x_s(x))\cdot dw_s, \qquad (4.12) \end{aligned}$$

$$\mathsf{M}_x^\varepsilon|\tilde{X}_t^\varepsilon - x_t(x)|^4 \le 3\varepsilon^4\left(\frac{e^{2Lt}-1}{2L}\right)^2.$$

Putting the Markov time $\tau = \min\{t : |\tilde{X}_t^\varepsilon - x_t(x)| \ge \eta\} \wedge T$ instead of t in (4.12) and taking the expectation (the idea of the Kolmogorov inequality), we obtain

$$\begin{aligned} \eta^4 \cdot \mathsf{P}_x^\varepsilon\Big\{\max_{0\le t\le T}|\tilde{X}_t^\varepsilon - x_t(x)| \ge \eta\Big\} &\le \mathsf{M}_x^\varepsilon|\tilde{X}_\tau^\varepsilon - x_\tau(x)|^4 \\ &\le \mathsf{M}_x^\varepsilon\int_0^\tau [4L|\tilde{X}_s^\varepsilon - x_s(x)|^4 + 6\varepsilon^2|\tilde{X}_s^\varepsilon - x_s(x)|^2]\,ds \\ &\le \int_0^T \mathsf{M}_x^\varepsilon[4L|\tilde{X}_s^\varepsilon - x_s(x)|^4 + 6\varepsilon^2|\tilde{X}_s^\varepsilon - x_s(x)|^2]\,ds \le 3\varepsilon^4\left(\frac{e^{2LT}-1}{2L}\right)^2. \end{aligned}$$

Note that Lemma 4.3 implies not only convergence in probability $\tilde{X}_t^\varepsilon \to_{\mathsf{P}} x_t(x)$ that is uniform in every finite time interval $[0, T]$, but also in the interval $[0, c|\ln\varepsilon|]$ that grows indefinitely as ε decreases, where the constant $c < L^{-1}$.

The most likely behavior of the process $\tilde{X}_t^\varepsilon$ (or X_t^ε) in the region $D_i(H_1, H_2)$ is rotating many times along the closed trajectories filling this region. We have to devise some way to count these rotations, adopting some definition of *what a rotation is*. No problem arises in the case of rotations of a solution $x_t(x)$ of the unperturbed system $\dot{x}_t = \bar{\nabla} H(x_t)$: we can consider the first time at which $x_t(x)$ returns to the same point $x_0(x) = x$. But the diffusion process $\tilde{X}_t^\varepsilon$ never returns to $\tilde{X}_0^\varepsilon$. Let us take a curve ∂ in $\bar{D}_i(H_1, H_2)$ running across all trajectories $x_t(x)$ in this region so that every trajectory (level curve of the Hamiltonian) intersects ∂ at one point. Unfortunately, the time at which $\tilde{X}_t^\varepsilon$ returns to this curve cannot serve as the definition of a rotation time, because a trajectory $\tilde{X}_t^\varepsilon$ starting at some point of ∂ returns to this curve infinitely many times at arbitrarily small positive times. So to define a "rotation," we take two curves ∂ and ∂' running across the trajectories at two different places, and define $\tilde{\tau}_0 < \tilde{\sigma}_1 < \tilde{\tau}_1 < \tilde{\sigma}_2 < \tilde{\tau}_2 < \cdots$ by

$$\tilde{\tau}_0 = 0, \qquad \tilde{\sigma}_n = \min\{t \geq \tilde{\tau}_{n-1} : \tilde{X}_t^\varepsilon \in \partial'\}, \qquad \tilde{\tau}_n = \min\{t \geq \tilde{\sigma}_n : \tilde{X}_t^\varepsilon \in \partial\}. \tag{4.13}$$

For the process $\tilde{X}_t^\varepsilon$ starting at the curve ∂, the times $\tilde{\tau}_1, \tilde{\tau}_2, \ldots$ (up to the time $\tilde{\tau}_i^\varepsilon(H_1, H_2)$ at which the process leaves $D_i(H_1, H_2)$) can be considered as the times at which the first rotation ends, the second one ends, and so on.

Let $T(x)$ be the period of the trajectory starting at the point x; $t(x) = \min\{t > 0 : x_t(x) \in \partial\}$, $t'(x) = \min\{t > 0 : x_t(x) \in \partial'\}$. For $x \in \partial$ we have $t'(x) < t(x) = T(x)$.

Lemma 4.4. *Let $H_1 < H_1' < H_2' < H_2$ be some numbers in the interval $H(I_i)$. Let the closed region $\bar{D}_i(H_i', H_2')$ lie at a distance greater than some positive d from the complement of the region $D_i(H_1, H_2)$; and let the distance between the curves ∂, ∂' be greater than $2d$. Let there exist positive constants $\underline{B}$ and h such that for all $x \in \partial \cap \bar{D}_i(H_1', H_2')$ the distance of $x_t(x)$ from the curve ∂,*

$$\rho(x_t(x), \partial) > \begin{cases} \underline{B}t & \text{for } 0 < t \leq h, \\ d & \text{for } h \leq t \leq T(x) - h, \\ \underline{B}(T(x) - t) & \text{for } T(x) - h \leq t < T(x) \end{cases}$$

(it is clear that $\underline{B} \leq \min\{|b(x)| : x \in \bar{D}_i(H_1', H_2')\}$).

Suppose that $\bar{B}h \leq d$, where $\bar{B} = \max\{|b(x)| : x \in \bar{D}_i(H_1', H_2')\}$.

Then for $x \in \partial \cap \bar{D}_i(H_1', H_2')$ and $0 < \Delta t \leq h$,

$$\mathsf{P}_x^\varepsilon\{|\tilde{\tau}_1 - T(x)| \geq \Delta t\} \leq 3\left(\frac{e^{2L(T(x)+h)} - 1}{2L}\right)^2 \cdot \frac{\varepsilon^4}{\underline{B}^4 \Delta t^4}, \tag{4.14}$$

and for every $\lambda > 0$,

$$\mathsf{M}_x^\varepsilon \left| \int_{\tilde{\tau}_1}^{T(x)} e^{-\lambda\varepsilon^2 t}\, dt \right|$$

$$\leq \frac{4}{3^{3/4}} \left(\frac{e^{2L(T(x)+h)} - 1}{2L}\right)^{1/2} \cdot \frac{\varepsilon}{\underline{B}} + 3\left(\frac{e^{2L(T(x)+h)} - 1}{2L}\right)^2 \cdot \frac{\varepsilon^2}{\underline{B}^4 h^4 \lambda}, \tag{4.15}$$

$$\mathsf{M}_x^\varepsilon e^{-\lambda\varepsilon^2\tilde{\tau}_1} \leq e^{-\lambda\varepsilon^2(T(x)-h)} + 3\left(\frac{e^{2L(T(x)+h)} - 1}{2L}\right)^2 \cdot \frac{\varepsilon^4}{\underline{B}^4 h^4}. \tag{4.16}$$

For all $x \in \bar{D}_i(H_1', H_2')$,

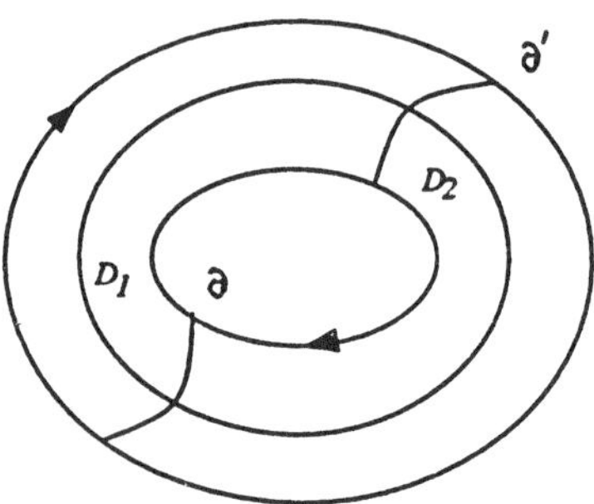

Figure 23.

$$\mathsf{P}_x^\varepsilon\{\tilde{\tau}_1 \geq 2T(x) + h\} \leq 3\left(\frac{e^{2L(T(x)+h)} - 1}{2L}\right)^2 \cdot \frac{\varepsilon^4}{\underline{B}^4 h^4}, \tag{4.17}$$

$$\mathsf{M}_x^\varepsilon \int_0^{\tilde{\tau}_1} e^{-\lambda\varepsilon^2 t}\, dt \leq 2T(x) + h + 3\left(\frac{e^{2L(2T(x)+h)} - 1}{2L}\right)^2 \cdot \frac{\varepsilon^2}{\underline{B}^4 h^4 \lambda}. \tag{4.18}$$

Proof. Let us introduce two regions: $D_1 = \{x \in D_i(H_1, H_2) \setminus (\partial \cup \partial') : t'(x) < t(x)\}$, $D_2 = \{x \in D_i(H_1, H_2) \setminus (\partial \cap \partial') : t(x) < t'(x)\}$ (see Fig. 23).

Let us consider a point $x \in \partial \cap \bar{D}_i(H_1', H_2')$. For $0 < t \leq \Delta t$ we have $x_t(x) \in D_1$; this point lies at a distance not greater than $\bar{B}\Delta t \leq d$ from ∂, and so is at a distance greater than d from ∂'. For $T(x) - \Delta t \leq t < T(x)$ the point $x_t(x)$ lies in D_2, and for $T(x) < t \leq T(x) + \Delta t$ again in D_1. For $\Delta t \leq t \leq T(x) - \Delta t$ the point $x_t(x)$ lies at a distance at least $\underline{B}\Delta t$ from the curve ∂.

Now suppose a trajectory $\tilde{X}_t^\varepsilon$ starting at $\tilde{X}_0^\varepsilon = x$ (remember, $x \in \partial \cap \bar{D}_i(H_1', H_2')$) is such that $\max_{0 \leq t \leq T(x)+h} |\tilde{X}_t^\varepsilon - x_t(x)| < \underline{B}\Delta t$. First of all, this trajectory does not leave $D_i(H_1, H_2)$ before the time $T(x) + h$. It does not intersect ∂' for $0 \leq t \leq \Delta t$, and so $\tilde{\sigma}_1 > \Delta t$. Since $x_{\Delta t}(x) \in D_1$ lies at a distance at least $\underline{B}\Delta t$ from ∂, we have also $\tilde{X}_{\Delta t}^\varepsilon \in D_1$. For t between Δt and $T(x) - \Delta t$, the trajectory $\tilde{X}_t^\varepsilon$ does not intersect ∂. Since the points $x_{T(x)-\Delta t}(x) \in D_2$, $x_{T(x)+\Delta t}(x) \in D_1$ lie at a distance at least $\underline{B}\Delta t$ from the curve ∂, we have also $\tilde{X}_{T(x)-\Delta t}^\varepsilon \in D_2$, $\tilde{X}_{T(x)+\Delta t}^\varepsilon \in D_1$. So between the times Δt and $T(x) - \Delta t$ the trajectory $\tilde{X}_t^\varepsilon$ passes from D_1 to D_2, intersecting ∂' (and therefore $\tilde{\sigma}_1$ is between Δt and $T(x) - \Delta t$), and between $T(x) - \Delta t$ and $T(x) + \Delta t$ (but not between $\tilde{\sigma}_1$ and $T(x) - \Delta t$) the trajectory passes from D_2 to D_1, intersecting ∂. So, finally, $T(x) - \Delta t < \tilde{\tau}_1 < T(x) + \Delta t$.

So we have proved the inclusion of the events:

$$\left\{\max_{0 \leq t \leq T(x)+h} |\tilde{X}_t^\varepsilon - x_t(x)| < \underline{B}\Delta t\right\} \subseteq \left\{|\tilde{\tau}_1 - T(x)| < \Delta t\right\};$$

and the inequality

$$\mathsf{P}_x^\varepsilon\{|\tilde{\tau}_1 - T(x)| \geq \Delta t\} \leq \mathsf{P}_x^\varepsilon\left\{\max_{0 \leq t \leq T(x)+h} |\tilde{X}_t^\varepsilon - x_t(x)| \geq \underline{B}\Delta t\right\}$$

together with Lemma 4.3 yields (4.14).

Now,

$$\mathsf{M}_x^\varepsilon\left|\int_{\tilde{\tau}_1}^{T(x)} e^{-\lambda\varepsilon^2 t}\,dt\right| \leq \mathsf{M}_x^\varepsilon\{|\tilde{\tau}_1 - T(x)| < h; |\tilde{\tau}_1 - T(x)|\}$$
$$+ \mathsf{M}_x^\varepsilon\left\{|\tilde{\tau}_1 - T(x)| \geq h; \int_0^\infty e^{-\lambda\varepsilon^2 t}\,dt\right\}.$$

The second term is equal to $(\lambda\varepsilon^2)^{-1} \cdot \mathsf{P}_x^\varepsilon\{|\tilde{\tau}_1 - T(x)| \geq h\}$ and is estimated using (4.14); the first one is also estimated by (4.14) and the following elementary lemma applied to the random variable $|\tilde{\tau}_1 - T(x)| \wedge h$.

Lemma 4.5. *If a nonnegative random variable ξ is such that $\mathsf{P}\{\xi \geq x\} \leq C/x^4$ for every $x > 0$, then $\mathsf{M}\xi \leq \frac{4}{3}C^{1/4}$.*

So we have (4.15). The estimate (4.16) follows from (4.14) and the fact that $\mathsf{M}_x^\varepsilon e^{-\lambda\varepsilon^2\tilde{\tau}_1} = \mathsf{M}_x^\varepsilon\{\tilde{\tau}_1 \leq T(x) - h; e^{-\lambda\varepsilon^2\tilde{\tau}_1}\} + \mathsf{M}_x^\varepsilon\{\tilde{\tau}_1 > T(x) - h; e^{-\lambda\varepsilon^2\tilde{\tau}_1}\}$.

As for the initial point $\tilde{X}_0^\varepsilon = x \notin \partial$, the trajectory $\tilde{X}_t^\varepsilon$ that is close to $x_t(x)$ need not cross the curve ∂' and ∂ after this within the time $T(x) + h$, but is bound to do so within *two* rotations plus h, and the estimate (4.17) is proved the same way as the more precise (4.14); (4.18) follows immediately.

Proof of Lemma 4.1. The expectation in (4.8) can be written as

$$\varepsilon^2\mathsf{M}_x^\varepsilon \int_0^{\tilde{\tau}_i^\varepsilon(H_1,H_2)} e^{-\lambda\varepsilon^2 t} g(\tilde{X}_t^\varepsilon)\,dt.$$

Take two curves ∂, ∂' in a slightly larger region $\bar{D}_i(H_1 - \delta/2, H_2 + \delta/2)$, and consider the sequence of Markov times $\tilde{\tau}_k$ defined by (4.13). Let $\nu = \max\{k : \tilde{\tau}_k < \tilde{\tau}_i^\varepsilon(H_1, H_2)\}$. We have

$$\int_0^{\tau_i^\varepsilon(H_1,H_2)} e^{-\lambda\varepsilon^2 t} g(X_t^\varepsilon)\,dt = \sum_{k=0}^{\nu-1} \int_{\tilde{\tau}_k}^{\tilde{\tau}_{k+1}} e^{-\lambda\varepsilon^2 t} g(X_t^\varepsilon)\,dt + \int_{\tilde{\tau}_\nu}^{\tau_i^\varepsilon(H_1,H_2)} e^{-\lambda\varepsilon^2 t} g(X_t^\varepsilon)\,dt. \tag{4.19}$$

We are going to apply Lemma 4.3 with $D_i(H_1 - \delta/2, H_2 + \delta/2)$ instead of $D_i(H_1, H_2)$ and $\bar{D}_i(H_1, H_2)$ instead of $\bar{D}_i(H_1', H_2')$. We have to estimate $T(x)$, $\bar{B}$, choose the curves ∂, ∂', and estimate the constants $\underline{B}$, h, d.

We have $\bar{B} \leq \max\{|\nabla H(x)| : H(x) \leq A_9\}$; the distance from $\bar{D}_i(H_1, H_2)$ to the complement of $D_i(H_1 - \delta/2, H_2 + \delta/2)$ is not less than $\delta/2\bar{B}$. Let us take $d = \delta/2\bar{B}$ (so the first condition imposed on d is satisfied; the rest is taken care of later).

The period $T(x)$ of an orbit $x_t(x)$ going through a point $x \in D_i$ is positive, it has a finite limit as x approaches an extremum point x_k (such that the corresponding vertex O_k is one of the ends of the segment I_i), and it grows at the rate of const $\cdot\, |\ln|H(x) - H(O_k)||$ as x approaches the curve C_{ki} if it contains a saddle point of the Hamiltonian. In any case, we have $A_{12} \leq T(x) \leq A_{13}|\ln\delta|$ for $x \in \bar{D}_i(H_1, H_2)$.

Take two points x_0 and x_0' on a fixed level curve $C_i(H_0)$ in the region D_i, and consider solutions $y_t(x_0)$, $y_t(x_0')$, $-\infty < t < \infty$, of $\dot{y}_t = \nabla H(y_t)$ going through the points x_0, x_0'; the curves described by these solutions are orthogonal to the level curves of the Hamiltonian. Let us take ∂, ∂' being the parts of the curves described by $y_t(x_0)$, $y_t(x_0')$ in the region $\bar{D}_i(H_1 - \delta/2, H_2 + \delta/2)$.

The curves described by $y_t(x_0)$, $y_t(x_0')$ may enter the same critical point x_k as $t \to \infty$ or $t \to -\infty$ (they certainly enter the same critical point if one of the ends of the segment I_i corresponds to an extremum of the Hamiltonian), but from different directions. The distance from $D_i(H_1 - \delta/2, H_2 + \delta/2)$ to a critical point is clearly not less than $A_{14}\sqrt{\delta}$, and it is easy to see that the distance between the curves ∂ and ∂' is at least $A_{15}\sqrt{\delta}$. This distance is greater than $d = \delta/2\bar{B}$ for sufficiently small δ (i.e., for sufficiently small ε).

Also we can take $\underline{B} = A_{16}\sqrt{\delta}$. We could have taken $h =$ const as far as it concerns the requirements $\rho(x_t(x), \partial) > \underline{B}t$ for $x \in \partial \cap \bar{D}_i(H_1, H_2)$, $0 < t \leq h$, $\rho(x_t(x), \partial) > \underline{B}(T(x) - t)$ for the same x and $T(x) - h \leq t < T(x)$; but we also have to ensure that $\bar{B}h < d$, so we take $h = A_{17}\sqrt{\delta}$.

Now, if we choose the constant A_7 so that $A_7 \cdot (2LA_{13} + 1) < \frac{1}{2}$, then for sufficiently small ε we obtain from (4.15), (4.16), and (4.18),

$$\mathsf{M}_x^\varepsilon \left| \int_{\tilde{\tau}_1}^{T(x)} e^{-\lambda\varepsilon^2 t}\, dt \right| \leq \varepsilon^{A_{18}} \tag{4.20}$$

for $x \in \partial \cap \bar{D}_i(H_1, H_2)$, where $A_{18} < \min(1 - A_7 \cdot (LA_{13} + \frac{1}{2}),\ 2 - 4A_7 \cdot (LA_{13} + 1))$,

$$\mathsf{M}_x^\varepsilon e^{-\lambda\varepsilon^2 \tilde{\tau}_1} \leq 1 - \frac{\lambda}{2} A_{12} \cdot \varepsilon^2 \tag{4.21}$$

for $x \in \partial \cap \bar{D}_i(H_1, H_2)$, and

$$\mathsf{M}_x^\varepsilon \int_0^{\tilde{\tau}_1} e^{-\lambda\varepsilon^2 t}\, dt \leq 2A_{13}A_7 \cdot |\ln\varepsilon| + 1 \tag{4.22}$$

for all $x \in \bar{D}_i(H_1, H_2)$.

Now let us extend the function g to the whole plane so that $\|g\|$ and the Lipschitz constant $\mathrm{Lip}(g)$ do not increase. Let us add to the right-hand side of (4.19) and subtract from it $\int_{\tilde{\tau}_i^\varepsilon(H_1,H_2)}^{\tilde{\tau}_{\nu+1}} e^{-\lambda\varepsilon^2 t} g(\tilde{X}_t^\varepsilon)\,dt$; then we can represent this right-hand side as

$$\sum_{k=0}^{\infty}\left[\chi_{\{\tilde{\tau}_k<\tilde{\tau}_i^\varepsilon(H_1,H_2)\}}\cdot\int_{\tilde{\tau}_k}^{\tilde{\tau}_{k+1}} e^{-\lambda\varepsilon^2 t} g(\tilde{X}_t^\varepsilon)\,dt\right]-\int_{\tilde{\tau}_i^\varepsilon(H_1,H_2)}^{\tilde{\tau}_{\nu+1}} e^{-\lambda\varepsilon^2 t} g(\tilde{X}_t^\varepsilon)\,dt.$$

Since $\mathsf{M}_x^\varepsilon \sum_{k=0}^{\infty} |I_{\{\tilde{\tau}_k<\tilde{\tau}_i^\varepsilon(H_1,H_2)\}}\cdot\int_{\tilde{\tau}_k}^{\tilde{\tau}_{k+1}} e^{-\lambda\varepsilon^2 t} g(\tilde{X}_t^\varepsilon)\,dt| \le \|g\|/\lambda\varepsilon^2 < \infty$, we can write the expectation of the infinite sum as the sum of expectations:

$$\begin{aligned}\mathsf{M}_x^\varepsilon \int_0^{\tilde{\tau}_i^\varepsilon(H_1,H_2)} e^{-\lambda\varepsilon^2 t} g(\tilde{X}_t^\varepsilon)\,dt &= \sum_{k=0}^{\infty} \mathsf{M}_x^\varepsilon\left\{\tilde{\tau}_k<\tilde{\tau}_i^\varepsilon(H_1,H_2);\int_{\tilde{\tau}_k}^{\tilde{\tau}_{k+1}} e^{-\lambda\varepsilon^2 t} g(\tilde{X}_t^\varepsilon)\,dt\right\}\\ &\quad - \mathsf{M}_x^\varepsilon \int_{\tilde{\tau}_i^\varepsilon(H_1,H_2)}^{\tilde{\tau}_{\nu+1}} e^{-\lambda\varepsilon^2 t} g(\tilde{X}_t^\varepsilon)\,dt.\end{aligned} \tag{4.23}$$

The last integral does not exceed $\|g\|$ times the integral of $e^{-\lambda\varepsilon^2 t}$ from the time $\tilde{\tau}_i^\varepsilon(H_1,H_2)$ until the end of the cycle starting at this time of reaching ∂' and ∂ after that. Applying the strong Markov property with respect to the Markov time $\tilde{\tau}_i^\varepsilon(H_1,H_2)$ and making use of (4.22), we obtain:

$$\begin{aligned}\left|\mathsf{M}_x^\varepsilon \int_{\tilde{\tau}_i^\varepsilon(H_1,H_2)}^{\tilde{\tau}_{\nu+1}} e^{-\lambda\varepsilon^2 t} g(\tilde{X}_t^\varepsilon)\,dt\right| &\le \|g\|\cdot \mathsf{M}_x^\varepsilon\left[\mathsf{M}_z^\varepsilon \int_0^{\tilde{\tau}_1} e^{-\lambda\varepsilon^2 t}\,dt\right]_{z=\tilde{X}^\varepsilon_{\tilde{\tau}_i^\varepsilon(H_1,H_2)}}\\ &\le (2A_{13}A_7\cdot|\ln\varepsilon|+1)\|g\|.\end{aligned} \tag{4.24}$$

Applying the strong Markov property with respect to $\tilde{\tau}_k$ to the kth summand in the infinite sum in (4.23), we obtain:

$$\begin{aligned}&\mathsf{M}_x^\varepsilon\left\{\tilde{\tau}_k<\tilde{\tau}_i^\varepsilon(H_1,H_2);\int_{\tilde{\tau}_k}^{\tilde{\tau}_{k+1}} e^{-\lambda\varepsilon^2 t} g(\tilde{X}_t^\varepsilon)\,dt\right\}\\ &\quad = \mathsf{M}_x^\varepsilon\left\{\tilde{\tau}_k<\tilde{\tau}_i^\varepsilon(H_1,H_2); e^{-\lambda\varepsilon^2\tilde{\tau}_k}\mathsf{M}_z^\varepsilon \int_0^{\tilde{\tau}_1} e^{-\lambda\varepsilon^2 t} g(\tilde{X}_t^\varepsilon)\,dt\bigg|_{z=\tilde{X}^\varepsilon_{\tilde{\tau}_k}}\right\}.\end{aligned}$$

So we can write an estimate for the infinite sum in (4.23) that is quite similar to the estimate (3.18):

$$\left|\sum_{k=0}^{\infty} \mathsf{M}_x^\varepsilon\left\{\tilde{\tau}_k < \tilde{\tau}_i^\varepsilon(H_1, H_2); \int_{\tilde{\tau}_k}^{\tilde{\tau}_{k+1}} e^{-\lambda\varepsilon^2 t} g(\tilde{X}_t^\varepsilon)\, dt\right\}\right|$$

$$\leq \left|\mathsf{M}_x^\varepsilon \int_0^{\tilde{\tau}_1} e^{-\lambda\varepsilon^2 t} g(\tilde{X}_t^\varepsilon)\, dt\right| + \max_{z \in \partial \cap \bar{D}_i(H_1,H_2)} \left|\mathsf{M}_z^\varepsilon \int_0^{\tilde{\tau}_1} e^{-\lambda\varepsilon^2 t} g(\tilde{X}_t^\varepsilon)\, dt\right|$$

$$\cdot \sum_{k=1}^{\infty} \mathsf{M}_x^\varepsilon\{\tilde{\tau}_k < \tilde{\tau}_i^\varepsilon(H_1, H_2); e^{-\lambda\varepsilon^2 \tilde{\tau}_k}\}. \tag{4.25}$$

Similarly to (3.11), the sum on the right-hand side is not greater than

$$\sum_{k=0}^{\infty}\left[\max_{z \in \partial \cap \bar{D}_i(H_1,H_2)} \mathsf{M}_z^\varepsilon e^{-\lambda^2 \tilde{\tau}_1}\right]^k = \left[1 - \max_{z \in \partial \cap \bar{D}_i(H_1,H_2)} \mathsf{M}_z^\varepsilon e^{-\lambda\varepsilon^2 \tilde{\tau}_1}\right]^{-1} \leq \frac{2}{\lambda A_{12}} \varepsilon^{-2} \tag{4.26}$$

for sufficiently small ε (see (4.21)).

As for the integral $\int_0^{\tilde{\tau}_1} e^{-\lambda\varepsilon^2 t} g(\tilde{X}_t^\varepsilon)\, dt$, it is approximately equal to

$$\int_0^{T(z)} g(x_t(z))\, dt = \oint_{C_i(H(z))} g(x) \frac{dl}{|b(x)|} = 0.$$

Let us estimate the expectation of the difference: for $z \in \partial \cap D_i(H_1, H_2)$,

$$\left|\mathsf{M}_z^\varepsilon \int_0^{\tilde{\tau}_1} e^{-\lambda\varepsilon^2 t} g(\tilde{X}_t^\varepsilon)\, dt - \int_0^{T(z)} g(x_t(z))\, dt\right|$$

$$\leq \|g\| \cdot \mathsf{M}_z^\varepsilon \left|\int_{\tilde{\tau}_1}^{T(z)} e^{-\lambda\varepsilon^2 t}\, dt\right|$$

$$+ \|g\| \cdot \int_0^{T(z)} (1 - e^{-\lambda\varepsilon^2 t})\, dt + \operatorname{Lip}(g) \cdot \int_0^{T(z)} \mathsf{M}_z^\varepsilon |\tilde{X}_t^\varepsilon - x_t(z)|\, dt.$$

The first expectation on the right-hand side is estimated by (4.20); the second integral is not greater than $\lambda\varepsilon^2 \cdot T(z)^2/2 \leq (\lambda/2) A_{13} A_7 \varepsilon^2 \ln^2 \varepsilon < \varepsilon^{A_{18}}$; and the last integral, by (4.9), does not exceed

$$\varepsilon \int_0^{T(z)} \sqrt{\frac{e^{2Lt} - 1}{2L}}\, dt < \varepsilon \cdot \frac{e^{LT(z)}}{L} < \text{const} \cdot \varepsilon^{1 - L A_{13} A_7} < \varepsilon^{A_{18}}.$$

So we have for sufficiently small ε for all $z \in \partial \cap D_i(H_1, H_2)$:

$$\left|\mathsf{M}_z^\varepsilon \int_0^{\tilde{\tau}_1} e^{-\lambda\varepsilon^2 t} g(\tilde{X}_t^\varepsilon)\, dt\right| \leq 2(\|g\| + \operatorname{Lip}(g)) \cdot \varepsilon^{A_{18}}. \tag{4.27}$$

The first term in (4.25) is estimated by means of (4.22). Putting this estimate together with (4.24)–(4.27), and remembering to multiply the expectation by ε^2, we obtain for small ε:

$$\left|\mathsf{M}_x^\varepsilon \int_0^{\tau_i^\varepsilon(H_1,H_2)} e^{-\lambda t} g(X_t^\varepsilon)\,dt\right| \leq (4A_{13}A_7 + 2)\varepsilon^2|\ln\varepsilon| + \frac{4}{\lambda A_{12}}\,\varepsilon^{A_{18}}.$$

So the estimate of the lemma is true if $A_7 < 1/(4LA_{13} + 2)$, and as A_8 can take any positive number that is smaller than $\min(1 - A_7 \cdot (LA_{13} + \frac{1}{2}),\, 2 - 4A_7 \cdot (LA_{13} + 1))$.

Proof of Lemma 4.2. Apply formula (4.3) to $\tau_i^\varepsilon(H_1, H_2)$:

$$\begin{aligned}
&\mathsf{M}_x^\varepsilon e^{-\lambda\tau_i^\varepsilon(H_1,H_2)} f(H(X^\varepsilon_{\tau_i^\varepsilon(H_1,H_2)})) - f(H(x)) \\
&\quad = \mathsf{M}_x^\varepsilon \int_0^{\tau_i^\varepsilon(H_1,H_2)} e^{-\lambda t}[-\lambda f(H(X_t^\varepsilon)) \\
&\qquad + \frac{1}{2} f''(H(X_t^\varepsilon))|\nabla H(X_t^\varepsilon)|^2 + \frac{1}{2} f'(H(X_t^\varepsilon))\Delta H(X_t^\varepsilon)]\,dt.
\end{aligned}$$

The right-hand side differs from the expectation of the integral in (4.9) only by

$$\mathsf{M}_x^\varepsilon \int_0^{\tau_i^\varepsilon(H_1,H_2)} e^{-\lambda t} g(X_t^\varepsilon)\,dt,$$

where

$$g(x) - \tfrac{1}{2} f''(H(x))[|\nabla H(x)|^2 - A_i(H(x))] + f'(H(x))[\tfrac{1}{2}\Delta H(x) - B_i(H(x))].$$

The definition (1.13) of the coefficients $A_i(H)$, $B_i(H)$ implies $g \in \mathfrak{G}_i(H_1, H_2)$. In order to apply Lemma 4.1, we have to estimate the supremum norm of g and its Lipshitz constant.

Since $|\nabla H(x)|^2$, $\Delta H(x)$ are bounded in every bounded region, and the coefficients $A_i(H)$, $B_i(H)$ in every finite subinterval of $H(I_i)$, we have $\|g\| \leq A_{19}(\|f'\| + \|f''\|)$. As for the Lipschitz condition, we have $\mathrm{Lip}(f^{(k)}(H)) \leq A_{20}\|f^{(k+1)}\|$, where $A_{20} = \max\{|\nabla H(x)| : H(x) \leq A_9\}$;

$$\begin{aligned}
&\mathrm{Lip}(\tfrac{1}{2} f''(H)|\nabla H|^2 + \tfrac{1}{2} f'(H)\Delta H) \\
&\quad \leq \tfrac{1}{2}[A_{20}^3\|f'''\| + A_{21}\|f''\| + A_{20} \cdot A_{22}\|f''\| + A_{23}\|f'\|],
\end{aligned}$$

where

$$A_{21} = \max\{|\nabla(|\nabla H(x)|^2)| : H(x) \le A_9\},$$
$$A_{22} = \max\{|\Delta H(x)| : H(x) \le A_9\},$$
$$A_{23} = \max\{|\nabla(\Delta H(x))| : H(x) \le A_9\};$$

$$\begin{aligned}\operatorname{Lip}(\tfrac{1}{2} f''(H)A_i(H) + f'(H)B_i(H)) \\ \le \tfrac{1}{2}[A_{20}\|f'''\| \cdot \max_{H_1 \le H \le H_2} A_i(H) + \|f''\| \cdot \max_{H_1 \le H \le H_2} |A_i'(H)|] \\ + [A_{20}\|f''\| \cdot \max_{H_1 \le H \le H_2} |B_i(H)| + \|f'\| \cdot \max_{H_1 \le H \le H_2} |B_i'(H)|].\end{aligned}$$

Using the estimate (1.14), we obtain

$$\operatorname{Lip}(g) \le A_{24}(\|f'\| + \|f''\| + \|f'''\|)\delta^{-A_0} = A_{24}(\|f'\| + \|f''\| + \|f'''\|)\varepsilon^{-A_0 A_{10}}.$$

Taking positive $A_{10} < \min(A_8/A_0, A_7)$, $A_{11} < A_8 - A_0 A_{10}$, and applying Lemma 4.1, we prove Lemma 4.2 and with it, also Lemma 3.3.

Now let us formulate a lemma that follows from Lemma 4.2 and that is used in the proof of Lemma 3.6.

Lemma 4.6. *Let $[H_1, H_2]$ be a subinterval of the interval $H(I_i)$. Let g be a continuous function on $[H_1, H_2]$; φ, a function defined only at the points H_1, H_2. Then, for the time $\tau_i^\varepsilon(H_1, H_2)$ of leaving the region $D_i(H_1, H_2)$,*

$$\lim_{\varepsilon \to 0} \mathsf{M}_x^\varepsilon \left[\varphi(H(X^\varepsilon_{\tau_i^\varepsilon(H_1,H_2)})) + \int_0^{\tau_i^\varepsilon(H_1,H_2)} g(H(X_t^\varepsilon))\, dt \right] = f(H(x)) \tag{4.28}$$

uniformly in $x \in \bar{D}_i(H_1, H_2)$, where

$$\begin{aligned} f(H) = {} & \frac{u_i(H_2) - u_i(H)}{u_i(H_2) - u_i(H_1)} \left[\varphi(H_1) + \int_{H_1}^{H} (u_i(H) - u_i(H_1)) g(H)\, dv_i(H) \right] \\ & + \frac{u_i(H) - u_i(H_1)}{u_i(H_2) - u_i(H_1)} \left[\varphi(H_2) + \int_{H}^{H_2} (u_i(H_2) - u_i(H)) g(H)\, dv_i(H) \right]. \end{aligned} \tag{4.29}$$

Proof. Lemma 4.2 has to do with functions of the value of the process multiplied by an exponentially decreasing function. To be able to tell anything about expectations without such multiplication, we need an estimate for the expectation $\mathsf{M}_x^\varepsilon \tau_i^\varepsilon(H_1, H_2)$.

Lemma 4.7. *For all* $x \in \bar{D}_i(H_1, H_2)$,

$$\mathsf{M}_x^\varepsilon \tau_i^\varepsilon(H_1, H_2) \le A_{25} = \frac{2b}{B^2} \exp\left\{\frac{B(H_2 - H_1)}{b}\right\}, \tag{4.30}$$

where

$$b = \min\{|\nabla H(x)| : x \in \bar{D}_i(H_1, H_2)\} > 0,$$

$$B = \max\{|\Delta H(x)| : x \in \bar{D}_i(H_1, H_2)\} < \infty.$$

Proof. Use formula (4.3) with $\lambda = 0$ and

$$f(H) = \cosh((2B/b)(H - (H_1 + H_2)/2)).$$

Now let us prove Lemma 4.6. The function f is the solution of the boundary-value problem

$$L_i f(x) = -g(H), \qquad H_1 < H < H_2,$$
$$f(H_j) = \varphi(H_j), \qquad j = 1, 2.$$

It is enough to prove (4.28) for smooth g and f (because a continuous g can be uniformly approximated by smooth functions, and $\mathsf{M}_x^\varepsilon \tau_i^\varepsilon(H_1, H_2)$ is uniformly bounded by Lemma 4.7). By Lemma 4.2,

$$\left|\mathsf{M}_x^\varepsilon\left[e^{-\lambda\tau_i^\varepsilon(H_1,H_2)}\varphi(H(X^\varepsilon_{\tau_i^\varepsilon(H_1,H_2)})) + \int_0^{\tau_i^\varepsilon(H_1,H_2)} e^{-\lambda t}[\lambda f(H(X_t^\varepsilon)) + g(H(X_t^\varepsilon))]\,dt\right]\right.$$
$$\left. - f(H(x))\right| \le A_{26}\varepsilon^{A_{11}}, \tag{4.31}$$

where the constant A_{26} may depend on $\lambda > 0$.

The difference between the expectation in (4.28) and that in (4.30) does not exceed

$$\|\varphi\| \cdot \mathsf{M}_x^\varepsilon(1 - e^{-\lambda\tau_i^\varepsilon(H_1,H_2)}) + \|f\| \cdot \mathsf{M}_x^\varepsilon \int_0^{\tau_i^\varepsilon(H_1,H_2)} \lambda e^{-\lambda t}\,dt$$
$$+ \|g\| \cdot \mathsf{M}_x^\varepsilon \int_0^{\tau_i^\varepsilon(H_1,H_2)} (1 - e^{-\lambda t})\,dt$$
$$\le 2\|f\| \cdot \lambda \mathsf{M}_x^\varepsilon \tau_i^\varepsilon(H_1, H_2) + \|g\| \cdot \lambda \mathsf{M}_x^\varepsilon(\tau_i^\varepsilon(H_1, H_2))^2$$
$$\le 2\lambda(\|f\| \cdot A_{25} + \|g\| \cdot A_{25}^2),$$

because $\mathsf{M}^\varepsilon(\tau_i^\varepsilon(H_1, H_2))^2/2 \le \mathsf{M}_x^\varepsilon \tau_i^\varepsilon(H_1, H_2) \cdot \sup_y \mathsf{M}_y^\varepsilon \tau_i^\varepsilon(H_1, H_2)$.

Taking λ small enough, and then ε small enough, we make the difference between the expectation in (4.29) and $f(H(x))$ small.

Lemma 4.6 is proved. □

§5. Proof of Lemma 3.5

Let us formulate the ε-dependent version of Lemma 3.5 (not implying this lemma immediately).

Lemma 5.1. *Let O_k be an interior vertex corresponding to a level curve C_k that contains a saddle point x_k. There exists a positive constant $A_{27} < \frac{1}{2}$ such that for $\delta = \delta(\varepsilon)$ such that $\delta(\varepsilon)|\ln \varepsilon| \to 0$ as $\varepsilon \to 0$, $\delta(\varepsilon) \geq \varepsilon^{A_{27}}$ we have*

$$\mathsf{M}_x^\varepsilon \tau_k^\varepsilon(\pm\delta) = O(\delta^2 |\ln \varepsilon|)$$

as $\varepsilon \to 0$, uniformly in $x \in D_k(\pm\delta)$.

Proof. Let us return to formulas (1.6), (1.8):

$$H(X_t^\varepsilon) = H(X_0^\varepsilon) + W\left(\int_0^t |\nabla H(X_s^\varepsilon)|^2 \, ds\right) + \int_0^t \frac{1}{2} \Delta H(X_s^\varepsilon) \, ds.$$

The time $\tau_k^\varepsilon(\pm\delta)$ comes at the latest when $|H(X_t^\varepsilon) - H(X_0^\varepsilon)|$ has reached the level 2δ. It occurs, in particular, if the integral $\int_0^t \frac{1}{2} \Delta H(X_s^\varepsilon) \, ds$ is small in absolute value, and the stochastic integral (the value of $W(\int_0^t |\nabla H(X_s^\varepsilon)|^2 \, ds)$) is large. A precise formulation is as follows:

$$\left\{\int_0^{t_0} |\frac{1}{2} \Delta H(X_s^\varepsilon)| \, ds < \delta, \max_{0 \leq t \leq t_0} \left| W\left(\int_0^t |\nabla H(X_s^\varepsilon)^2 \, ds\right)\right| > 3\delta \right\} \subseteq \{\tau_k^\varepsilon(\pm\delta) < t_0\}.$$

We are going to use the fact that if $C\delta^2 \leq \int_0^{t_0} |\nabla H(X_s^\varepsilon)|^2 \, ds$, we have

$$\max_{0 \leq t \leq t_0} \left| W\left(\int_0^t |\nabla H(X_s^\varepsilon)|^2 \, ds\right)\right| \geq |W(C\delta^2)|,$$

and that the random variable $W(C\delta^2)$ has a normal distribution with zero mean and variance $C\delta^2$. We have

$$\begin{aligned}
\{\tau_k^\varepsilon(\pm\delta) \geq t_0\} \subseteq & \left\{\tau_k^\varepsilon(\pm\delta) \geq t_0, \int_0^{t_0} |\nabla H(X_s^\varepsilon)|^2 \, ds < C\delta^2 \right\} \\
& \cup \left\{\tau_k^\varepsilon(\pm\delta) \geq t_0, \int_0^{t_0} |\nabla H(X_s^\varepsilon)|^2 \, ds \geq C\delta^2, |W(C\delta^2)| \geq 3\delta \right\} \\
& \cup \left\{\tau_k^\varepsilon(\pm\delta) \geq t_0, \int_0^{t_0} |\nabla H(X_s^\varepsilon)|^2 \, ds \geq C\delta^2, |W(C\delta^2)| < 3\delta \right\}.
\end{aligned}$$

Suppose $\int_0^{t_0} |\frac{1}{2}\Delta H(X_s^\varepsilon)|\,ds < \delta$ for all trajectories X_t^ε such that $\tau_k^\varepsilon(\pm\delta) \geq t_0$. Then the second event on the right-hand side cannot occur, and the third one is a part of the event $\{W(C\delta^2) < 3\delta\}$. Using the normal distribution, we have for all $x \in D_k(\pm\delta)$,

$$\mathrm{P}_x^\varepsilon\{\tau_k^\varepsilon(\pm\delta) \geq t_0\} \leq \mathrm{P}_x^\varepsilon\left\{\tau_k^\varepsilon(\pm\delta) \geq t_0, \int_0^{t_0} |\nabla H(X_s^\varepsilon)|^2\,ds < C\delta^2\right\} + \int_{-3/\sqrt{C}}^{3/\sqrt{C}} \frac{1}{\sqrt{2\pi}} e^{-y^2/2}\,dy. \tag{5.1}$$

It does not matter much for our proof what C we take in this estimate. Let us take $C = 9$; then the last integral is equal to 0.6826.

Of course, $|\frac{1}{2}\Delta H| \leq A_{28}$ in some neighborhood of the curve C_k; so if, for small δ, we take $t_0 < \delta/A_{28}$, the inequality $\int_0^{t_0} |\frac{1}{2}\Delta H(X_s^\varepsilon)|\,ds < \delta$ is guaranteed. So we have to estimate, for some $t_0 = t_0(\delta) < \delta/A_{28}$, the first summand on the right-hand side of (5.1).

To do this, we consider once again cycles between reaching two lines. Just as in the proof of Lemma 4.1, introduce curves ∂, ∂' in a slightly larger region $\bar{D}_k(\pm 2\delta)$:

$$\partial = \{x \in \bar{D}_k(\pm 2\delta) : |x - x_k| = A_{29}\sqrt{\delta}\},$$
$$\partial' = \{x \in \bar{D}_k(\pm 2\delta) : |x - x_k| = A_{30}\sqrt{\delta}\}.$$

Just as in the proof of Lemma 4.1, let us consider the "slow" process $\tilde{X}_t^\varepsilon$. Define Markov times $\tilde{\tau}_0 < \tilde{\sigma}_1 < \tilde{\tau}_1 < \tilde{\sigma}_2 < \cdots$ by

$$\tilde{\tau}_0 = 0, \quad \tilde{\sigma}_i = \min\{t \geq \tilde{\tau}_{i-1} : \tilde{X}_t^\varepsilon \in \partial'\}, \quad \tilde{\tau}_i = \min\{t \geq \tilde{\sigma}_i : \tilde{X}_t^\varepsilon \in \partial\}.$$

For sufficiently large positive constants $A_{29} < A_{30}$ the curves ∂, ∂' each consist of four arcs: two across the "sleeves" of $D_k(\pm 2\delta)$ in which the solutions of the Hamiltonian system go *out of* the neighborhood of the saddle point, and two across the "sleeves" through which the trajectories go *into* the central, cross-like part of $D_k(\pm 2\delta)$. Let us denote the corresponding parts by ∂_{out}, ∂'_{out}, ∂_{in}, ∂'_{in} (they each consist of two arcs; see Fig. 24).

Before the time $\tilde{\tau}_k^\varepsilon(\pm\delta)$, the times $\tilde{\sigma}_i$ are those at which the process $\tilde{X}_t^\varepsilon$ goes out of the central part of the region $D_k(\pm\delta)$; and $\tilde{\tau}_i$, $i > 0$, are those at which it, after going through one of the two "handles" of this region, reaches its boundary. Lemmas 5.2 and 5.3 are taking care of the spaces between these times.

Let D_c be one of the two outer "handles" of the region $D_k(\pm\delta)$; e.g., its left part between the lines ∂_{out} and ∂_{in} (or the right part); $\tilde{\tau}_c$ will denote the first time at which the process $\tilde{X}_t^\varepsilon$ leaves this region.

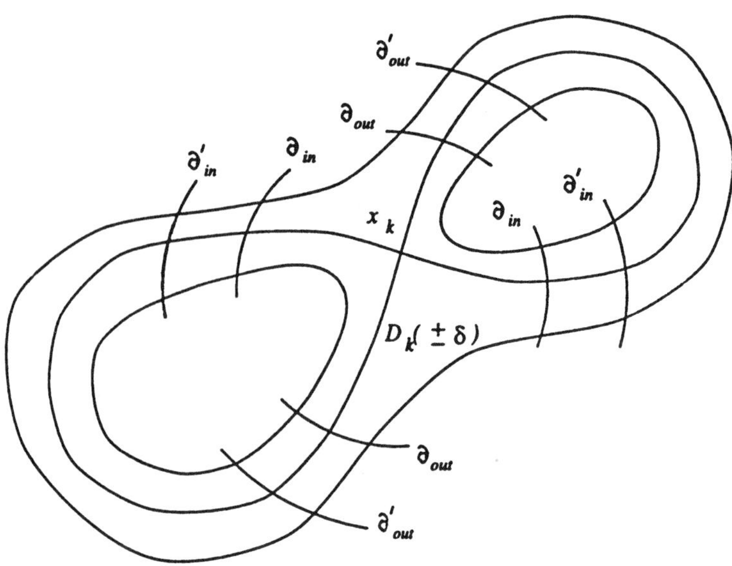

Figure 24.

Lemma 5.2. *There exist positive constants A_{31}, A_{32}, A_{33} such that for sufficiently small ε and small $\delta \geq \varepsilon^{A_{31}}$,*

$$\mathsf{M}_x^\varepsilon \tilde{\tau}_c^\varepsilon \leq A_{32} \cdot |\ln \delta|, \tag{5.2}$$

for every $x \in D_c$;

$$\mathsf{P}_x^\varepsilon \{ \tilde{X}_{\tilde{\tau}_c^\varepsilon}^\varepsilon \in \partial_{\text{in}} \} \to 1 \quad (\varepsilon \to 0), \tag{5.3}$$

uniformly in $x \in \partial'$; and

$$\mathsf{P}_x^\varepsilon \left\{ \int_0^{\tilde{\tau}_c^\varepsilon} |\nabla H(\tilde{X}_s^\varepsilon)|^2 \, ds > A_{33} \right\} \to 1 \qquad (\varepsilon \to 0), \tag{5.4}$$

uniformly in $x \in \partial'_{\text{out}}$.

Proof. The proof is similar to those of Lemmas 4.4 and 4.1, and relies on the same Lemma 4.3. As in the beginning of the proof of Lemma 4.1, we have $\bar{B} = \max\{|b(x)| : x \in D_k(\pm 2\delta)\} \leq A_{34}$; again we take $d = \delta/2B$. The time $T_c(x)$ in which the trajectory $x_t(x)$ leaves the region D_c (through the line ∂_{in}, of course) is bounded by $A_{35} \cdot |\ln \delta|$, and the time the trajectory starting on ∂'_{out} spends in the set $\{x : |x - x_k| \geq A_{36}\}$ is at least A_{37} if we choose the constant A_{36} small enough and if δ is small. We can take $\underline{B} = A_{38}\sqrt{\delta}$ and

$h = A_{39}\sqrt{\delta}$ so that for all $x \in D_c$,

$$\rho(x_t(x), \partial) > \begin{cases} \underline{B}t & \text{for } 0 < t \leq h, \\ d & \text{for } h \leq t \leq T_c(x) - h, \\ \underline{B}(T_c(x) - t) & \text{for } T_c(x) - h \leq t < T_c(x). \end{cases}$$

We obtain, by Lemma 4.3, the estimates (5.3), (5.4), and

$$\sup\{\mathsf{P}_z^\varepsilon\{\tau_c^\varepsilon > A_{35}|\ln\delta| + 1\} : z \in D_c\} \leq \sup\{\mathsf{P}_z^\varepsilon\{\tau_c^\varepsilon > T_c(z) + h\} : z \in D_c\} \to 0$$
$$(\varepsilon \to 0),$$
$$\mathsf{P}_x^\varepsilon\{\tau_c^\varepsilon > n \cdot (A_{35}|\ln\delta| + 1)\} \leq [\sup\{\mathsf{P}_z^\varepsilon\{\tau_c^\varepsilon > A_{35}|\ln\delta| + 1\} : z \in D_c\}]^n,$$
$$\mathsf{M}_x^\varepsilon\tau_c^\varepsilon \leq \frac{A_{35}|\ln\delta| + 1}{1 - \sup\{\mathsf{P}_z^\varepsilon\{\tau_c^\varepsilon > A_{35}|\ln\delta| + 1\} : z \in D_c\}}$$
$$\leq A_{32} \cdot |\ln\delta|$$

for sufficiently small ε.

Now let $D_\times$ be the cross-like part of $D_k(\pm\delta)$ between the lines $\partial' : D_\times = \{x \in D_k(\pm\delta) : |x - x_k| < A_{30}\sqrt{\delta}\}$ (see Fig. 25). Let $\tilde{\tau}_x^\varepsilon = \min\{t : \tilde{X}_t^\varepsilon \notin D_\times\}$.

Lemma 5.3. *For some choice of positive constants $A_{29} < A_{30}$, if $\delta = \delta(\varepsilon)$ is such that $\delta(\varepsilon) \to 0$, $\delta(\varepsilon)/\varepsilon^2 \to \infty$ as $\varepsilon \to 0$, we have*

$$\mathsf{M}_x^\varepsilon\tilde{\tau}_x^\varepsilon = O(\ln(\delta/\varepsilon^2)) \tag{5.5}$$

uniformly in $x \in D_\times$;

$$\mathsf{P}_x^\varepsilon\{\tilde{\tau}_x^\varepsilon < \tilde{\tau}_k^\varepsilon(\pm\delta) \text{ or } \tilde{X}_{\tilde{\tau}_x^\varepsilon}^\varepsilon \in \partial'_{\text{out}}\} \to 1 \tag{5.6}$$

as $\varepsilon \to 0$, uniformly in $x \in \partial$.

Proof. By the Morse lemma, one can introduce new coordinates ξ, η (in lieu of p, q) in a neighborhood of x_k so that the point x_k is the origin, and the two branches of the ∞-shaped curve are the axes ξ and η. Let the ξ-axis be the branch along which the saddle point x_k is unstable for the dynamical system. In the new coordinates the operator $\tilde{L}^\varepsilon$ has the form:

$$\tilde{L}^\varepsilon f(x) = \frac{\varepsilon^2}{2}\left[a^{11}(x)\frac{\partial^2 f(x)}{\partial\xi^2} + 2a^{12}(x)\frac{\partial^2 f(x)}{\partial\xi\partial\eta} + a^{22}(x)\frac{\partial^2 f(x)}{\partial\eta^2}\right]$$
$$+ B^1(x)\frac{\partial f(x)}{\partial\xi} + B^2(x)\frac{\partial f(x)}{\partial\eta} + \varepsilon^2\left[\tilde{b}^1(x)\frac{\partial f(x)}{\partial\xi} + \tilde{b}^2(x)\frac{\partial f(x)}{\partial\eta}\right], \tag{5.7}$$

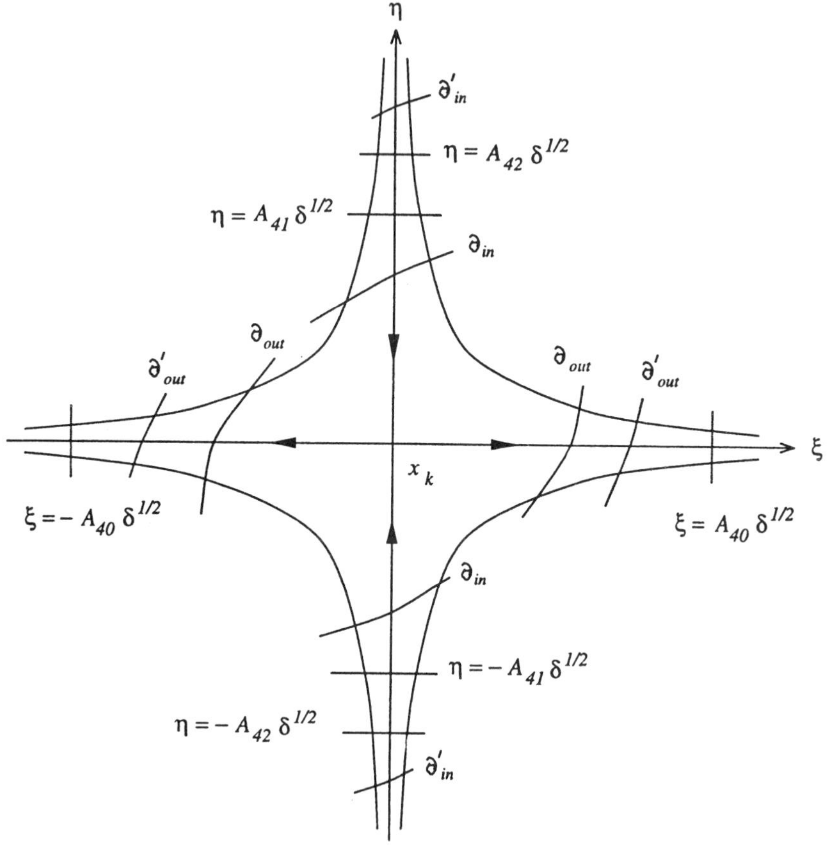

Figure 25.

where the coefficients $a^{ij}(x)$, $B^i(x)$, $\tilde{b}^i(x)$ are smooth, and the symmetric matrix $(a^{ij}(x))$ is positive definite. The second sum here is equal to $\bar{\nabla}H(x) \cdot \nabla f(x)$. Because of the fact that x_k is a nondegenerate saddle point of the Hamiltonian, and because of our choice of the axes, we have

$$B^1(0,\eta) = B^2(\xi,0) = 0; \qquad \frac{\partial B^1}{\partial \xi} > 0, \quad \frac{\partial B^2}{\partial \eta} < 0.$$

The functions $a^{ij}(\xi,\eta)$, $B^i(\xi,\eta)$, $\tilde{b}^i(\xi,\eta)$ are defined in a neighborhood of the point (0,0); let us extend these functions of the whole plane so that a^{ij}, $\tilde{b}^i$ are bounded,

$$a^{ii}(\xi,\eta) \geq a_0 > 0, \quad B^1(0,\eta) = B^2(\xi,0) = 0, \quad \frac{\partial B^1}{\partial \xi} \geq B_0 > 0, \quad \frac{\partial B^2}{\partial \eta} \leq -B_0$$

everywhere.

We use the notation $(\tilde{\xi}_t^\varepsilon, \tilde{\eta}_t^\varepsilon)$ for the diffusion in the plane governed by the operator (5.7) with coefficients extended as described above. Before the process $\tilde{X}_t^\varepsilon$ leaves a neighborhood of x_k, $\tilde{\xi}_t^\varepsilon$ and $\tilde{\eta}_t^\varepsilon$ are the ξ- and η-coordinates of this process.

The curves ∂, ∂' are no longer parts of circles in our new coordinates, but for small δ they are approximately parts of ellipses the ratio of whose sizes is $\sqrt{A_{30}/A_{29}}$. We can choose positive constants $A_{29} < A_{30}$, A_{40}, and $A_{41} < A_{42}$ so that, for small positive δ, $|\xi| < A_{40}\sqrt{\delta}$ for $x \in \partial'$, $|\eta| < A_{41}\sqrt{\delta}$ for $x \in \partial$, and $|\eta| > A_{42}\sqrt{\delta}$ for $x \subset \partial'_{\text{in}}$.

Let us consider the times $\tilde{\tau}_{\text{out}}^\varepsilon = \min\{t : |\tilde{\xi}_t^\varepsilon| = A_{40}\sqrt{\delta}\}$ and $\tilde{\tau}_{\text{in}}^\varepsilon = \min\{t : |\tilde{\eta}_t^\varepsilon| = A_{42}\sqrt{\delta}\}$. It is clear that $\tilde{\tau}_x^\varepsilon \le \tilde{\tau}_{\text{out}}^\varepsilon$; and if the process $\tilde{X}_t^\varepsilon$ starts from a point $\tilde{X}_0^\varepsilon = x \in \partial$, and $\tilde{\tau}_{\text{out}}^\varepsilon < \tilde{\tau}_{\text{in}}^\varepsilon$, then it leaves the region D_x either through one of its sides (that are trajectories of the system), or through ∂'_{out}.

It turns out that the time $\tilde{\tau}_{\text{out}}^\varepsilon$ is of order of $\ln(\delta/\varepsilon^2)$, whereas $\tilde{\tau}_{\text{in}}^\varepsilon$ is of order at least δ/ε^2, i.e., infinitely large compared with $\tilde{\tau}_{\text{out}}^\varepsilon$ (one can prove using large-deviation estimates similar to Theorem 4.4.2 that $\tilde{\tau}_{\text{in}}^\varepsilon$ is, in fact, of order $\exp\{\text{const}\,\delta/\varepsilon^2\}$, but we do not need that). The proof is based on comparing the processes $\tilde{\xi}_t^\varepsilon$, $\tilde{\eta}_t^\varepsilon$ (that need not be Markov ones if taken separately) with one-dimensional diffusions.

Let us introduce the function $F(x) = \int_0^x e^{-y^2}\,[\int_0^y e^{z^2}\,dz]\,dy$. Using l'Hôpital's rule, we prove that $F(x) \sim \frac{1}{2}\ln x$ as $x \to \infty$. Let us consider the function.

$$u^\varepsilon(\xi) = 2B_0^{-1}[F(\sqrt{B_0}A_{40}\sqrt{\delta}/\varepsilon) - F(\sqrt{B_0}\xi/\varepsilon)].$$

This function is the solution of the boundary-value problem

$$\frac{\varepsilon^2}{2}\frac{d^2u^\varepsilon}{d\xi^2} + B_0\xi\,\frac{du^\varepsilon}{d\xi} = -1, \qquad u^\varepsilon(\pm A_{40}\sqrt{\delta}) = 0;$$

so $u^\varepsilon(\xi)$ is the expectation of the time of leaving the interval $(-A_{40}\sqrt{\delta}, A_{40}\sqrt{\delta})$ for the one-dimensional diffusion process with generator $(\varepsilon^2/2)(d^2/d\xi^2) + B_0\xi(d/d\xi)$, starting from a point $\xi \in (-A_{40}\sqrt{\delta}, A_{40}\sqrt{\delta})$.

Let us apply the operator $\tilde{L}^\varepsilon$ to the function u^ε on the plane, depending on the ξ-coordinate only:

$$\begin{aligned}\tilde{L}^\varepsilon u^\varepsilon(x) &= a^{11}(x)\cdot(2(\sqrt{B_0}\xi/\varepsilon)F'(\sqrt{B_0}\xi/\varepsilon) - 1\\ &\quad - B^1(x)\frac{2}{B_0}(\sqrt{B_0}/\varepsilon)F'(\sqrt{B_0}\xi/\varepsilon) - \frac{2\varepsilon}{B_0}\tilde{b}^1(x)F'(\sqrt{B_0}\xi/\varepsilon).\end{aligned}$$

We have $B^1(\xi,\eta) = B^1(0,\eta) + (\partial B^1/\partial\xi)(\tilde{\xi},\eta)$, $\tilde{L}^\varepsilon v^\varepsilon(x) \le -a_0 + (2\varepsilon/C)\cdot \sup|\tilde{b}^1(x)|\cdot\max|F'(x)| \le -a_0/2$ for sufficiently small ε, and we obtain by formula (5.1) of Chapter 1:

$$\mathsf{M}_x^\varepsilon \tilde{\tau}_{\text{out}}^\varepsilon \leq \frac{u^\varepsilon(\xi)}{a_0/2} \leq A_{43} \cdot \ln(\delta/\varepsilon^2).$$

By Chebyshev's inequality,

$$\mathsf{P}_x^\varepsilon\{\tilde{\tau}_{\text{out}}^\varepsilon \geq (\delta/\varepsilon^2)^{1/2}\} \to 0$$

as $\varepsilon \to 0$, uniformly in x.

Now let us consider the function $v^\varepsilon(\eta) = e^{B_0(\eta/\varepsilon)^2}$. This function is a positive solution of the equation

$$\frac{\varepsilon^2}{2}\frac{d^2 v^\varepsilon}{d\eta^2} - B_0\eta\frac{dv^\varepsilon}{d\eta} - B_0 v^\varepsilon = 0,$$

so the $(-B_0)$-exponential moment of the exit time for the one-dimensional diffusion with generator $((\varepsilon^2/2)(d^2/d\eta^2)) - B_0\eta(d/d\eta)$ can be expressed through it. Applying the operator $\tilde{L}^\varepsilon$ to the function $v^\varepsilon(\eta)$, we obtain:

$$\tilde{L}^\varepsilon v^\varepsilon(x) = \left[a^{22}(x) \cdot \left(B_0 + \frac{2B_0^2\eta^2}{\varepsilon^2}\right) + B^2(x) \cdot 2\,\frac{B_0\eta}{\varepsilon^2} + \tilde{b}^2(x) \cdot 2B_0\eta\right] e^{B_0(\eta/\varepsilon)^2}.$$

For sufficiently small $|\eta|$ we have $\tilde{L}^\varepsilon v^\varepsilon - B_0 v^\varepsilon \leq 0$, and by formula (5.2) of Chapter 1,

$$\mathsf{M}_x^\varepsilon e^{-B_0\tilde{\tau}_{\text{in}}^\varepsilon} \leq \frac{v^\varepsilon(\eta)}{v^\varepsilon(A_{42}\sqrt{\delta})}.$$

For $x \in \partial$ we have $\mathsf{M}_x^\varepsilon e^{-B_0\tilde{\tau}_{\text{in}}^\varepsilon} \leq e^{B_0(A_{41}^2 - A_{42}^2)\delta/\varepsilon^2}$, and by Chebyshev's inequality

$$\begin{aligned}\mathsf{P}_x^\varepsilon\{\tilde{\tau}_{\text{in}}^\varepsilon \leq (\delta/\varepsilon^2)^{1/2}\} &= \mathsf{P}_x^\varepsilon\{e^{-B_0\tilde{\tau}_{\text{in}}^\varepsilon} \geq e^{-B_0(\delta/\varepsilon^2)^{1/2}}\} \\ &\leq \frac{\mathsf{M}_x^\varepsilon e^{-B_0\tilde{\tau}_{\text{in}}^\varepsilon}}{e^{-B_0(\delta/\varepsilon^2)^{1/2}}} \leq e^{-B_0(A_{42}^2 - A_{41}^2)\delta/\varepsilon^2 + B_0(\delta/\varepsilon^2)^{1/2}} \to 0\end{aligned}$$

as $\varepsilon \to 0$, uniformly in $x \in \partial$.

This proves Lemma 5.3. □

Let us finish the proof of Lemma 5.1. Return to Fig. 24. The first probability on the right-hand side of (5.1), in which we have taken $C = 9$, is equal to

$$\mathsf{P}_x^\varepsilon\left\{\tilde{\tau}_k^\varepsilon(\pm\delta) \geq t_0/\varepsilon^2, \int_0^{t_0/\varepsilon^2} |\nabla H(\tilde{X}_s^\varepsilon)|^2\,ds < 9\delta^2/\varepsilon^2\right\}. \tag{5.8}$$

For an arbitrary natural n this probability does not exceed

$$\mathsf{P}_x^\varepsilon\left\{\tilde{\tau}_n < t_0/\varepsilon^2 \le \tilde{\tau}^\varepsilon(\pm\delta), \int_0^{\tilde{\tau}_n} |\nabla H(\tilde{X}_s^\varepsilon)|^2\, ds < 9\delta^2/\varepsilon^2\right\}$$
$$+ \mathsf{P}_x^\varepsilon\{\tilde{\tau}_n \wedge \tilde{\tau}_k^\varepsilon(\pm\delta) \ge t_0/\varepsilon^2\}. \tag{5.9}$$

The second probability is estimated by Chebyshev's inequality:

$$\mathsf{P}_x^\varepsilon\{\tilde{\tau}_n \wedge \tilde{\tau}_k^\varepsilon(\pm\delta) \ge t_0/\varepsilon^2\} \le \frac{\mathsf{M}_x^\varepsilon(\tau_n \wedge \tau_k^\varepsilon(\pm\delta))}{t_0/\varepsilon^2}. \tag{5.10}$$

Using the strong Markov property with respect to the Markov times $\tilde{\tau}_i$, $\tilde{\sigma}_i$, we obtain, by Lemmas 5.2 and 5.3, for sufficiently small ε,

$$\mathsf{M}_x^\varepsilon(\tilde{\tau}_n \wedge \tilde{\tau}_k^\varepsilon(\pm\delta)) \le n\cdot[\sup\{\mathsf{M}_z^\varepsilon\tilde{\tau}_x^\varepsilon : z \in D_x\} + \sup\{\mathsf{M}_z^\varepsilon\tilde{\tau}_c^\varepsilon : z \in D_c\}]$$
$$\le n\cdot A_{44}|\ln\varepsilon|. \tag{5.11}$$

To estimate the first probability in (5.9), we use the exponential Chebyshev inequality: for $a > 0$,

$$\mathsf{P}_x^\varepsilon\left\{\tilde{\tau}_n < t_0/\varepsilon^2 \le \tilde{\tau}^\varepsilon(\pm\delta), \exp\left\{-a\int_0^{\tilde{\tau}_n} |\nabla H(\tilde{X}_s^\varepsilon)|^2\, ds\right\} \ge e^{-9a\delta^2/\varepsilon^2}\right\}$$
$$\le \frac{\mathsf{M}_x^\varepsilon\left\{\tilde{\tau}_n < \tilde{\tau}^\varepsilon(\pm\delta); \exp\left\{-a\int_0^{\tilde{\tau}_n} |\Delta H(\tilde{X}_s^\varepsilon)|^2\, ds\right\}\right\}}{e^{-9a\delta^2/\varepsilon^2}}. \tag{5.12}$$

Using the strong Markov property with respect to the times $\tilde{\tau}_i$, we can write

$$\mathsf{M}_x^\varepsilon\left\{\tilde{\tau}_n < \tilde{\tau}^\varepsilon(\pm\delta); \exp\left\{-a\int_0^{\tilde{\tau}_n} |\nabla H(\tilde{X}_s^\varepsilon)|^2\, ds\right\}\right\}$$
$$\le \left[\sup_{z\in\partial} \mathsf{M}_z^\varepsilon\left\{\tilde{\tau}_1 < \tilde{\tau}^\varepsilon(\pm\delta); \exp\left\{-a\int_0^{\tilde{\tau}_1} |\nabla H(\tilde{X}_s^\varepsilon)|^2\, ds\right\}\right\}\right]^{n-1}.$$

Let us use the strong Markov property with respect to the Markov time $\tilde{\sigma}_1$ (which, for the process starting at a point $\tilde{X}_0^\varepsilon = z \in \partial$, coincides with the time $\tilde{\tau}_x^\varepsilon$ if it is smaller than $\tilde{\tau}_k^\varepsilon(\pm\delta)$):

$$\mathsf{M}_z^\varepsilon\left\{\tilde{\tau}_1 < \tilde{\tau}^\varepsilon(\pm\delta); \exp\left\{-a\int_0^{\tilde{\tau}_1} |\nabla H(\tilde{X}_s^\varepsilon)|^2\, ds\right\}\right\}$$
$$= \mathsf{M}_z^\varepsilon\left\{\tilde{\tau}_x^\varepsilon < \tilde{\tau}_k^\varepsilon(\pm\delta); \exp\left\{-a\int_0^{\tilde{\tau}_x^\varepsilon} |\nabla H(\tilde{X}_s^\varepsilon)|^2\, ds\right\}\cdot\phi_7^\varepsilon(\tilde{X}_{\tilde{\tau}_x^\varepsilon}^\varepsilon)\right\}, \tag{5.13}$$

where

$$\phi_7^\varepsilon(z') = \mathsf{M}_z^\varepsilon\left\{\tilde\tau_c^\varepsilon < \tilde\tau_k^\varepsilon(\pm\delta); \exp\left\{-a\int_0^{\tilde\tau_c^\varepsilon} |\nabla H(\tilde X_s^\varepsilon)|^2\,ds\right\}\right\}.$$

The argument in ϕ_7^ε always belongs to ∂'. For $z' \in \partial'_{\text{in}}$, we have $\phi_7^\varepsilon(z') < 1$, and for $z' \in \partial'_{\text{out}}$ we use the inequality

$$\phi_7^\varepsilon(z') < \mathsf{P}_{z'}^\varepsilon\left\{\int_0^{\tilde\tau_c^\varepsilon} |\nabla H(\tilde X_s^\varepsilon)|^2\,ds \le A_{33}\right\} + \mathsf{P}_{z'}^\varepsilon\left\{\int_0^{\tilde\tau_c^\varepsilon} |\Delta H(\tilde X_s^\varepsilon)|^2\,ds > A_{33}\right\}\cdot e^{-aA_{33}}.$$

Again, it does not matter much what a we take; let us take $a = A_{33}^{-1}$. By Lemma 5.2, this expression, for sufficiently small ε and for every $z' \in \partial'$, is smaller than e^{-1} plus an arbitrarily small quantity.

Now, the expression (5.13) does not exceed

$$\mathsf{P}_z^\varepsilon\{\tilde\tau_k^\varepsilon < \tilde\tau_k^\varepsilon(\pm\delta), \tilde X_{\tilde\tau_x^\varepsilon}^\varepsilon \in \partial'_{\text{in}}\} + \mathsf{P}_z^\varepsilon\{\tilde\tau_x^\varepsilon < \tilde\tau_k^\varepsilon(\pm\delta), \tilde X_{\tilde\tau_x^\varepsilon}^\varepsilon \in \partial'_{\text{out}}\}\cdot \sup\{\phi_7^\varepsilon(z') : z' \in \partial'_{\text{out}}\},$$

and by Lemma 5.3, for sufficiently small ε and for every $z \in \partial$, is smaller than e^{-1} plus an arbitrarily small quantity.

Now let us take $n = n(\varepsilon) \to \infty$ that is equivalent to $10A_{33}^{-1}\delta(\varepsilon)^2\varepsilon^2$ as $\varepsilon \to 0$. Then, by the estimate (5.12) and the following, the first summand in (5.9) goes to 0 as $\varepsilon \to 0$, uniformly in $x \in D_k(\pm\delta)$. If we take $t_0 = 100A_{44}A_{33}^{-1}\delta^2|\ln\varepsilon|$, the second summand in (5.9) is estimated by a quantity that coverges to 0.1 as $\varepsilon \to 0$ (see (5.10) and (5.11)), and the probability (5.8) is less than 0.11 for sufficiently small ε and all $x \in D_k(\pm\delta)$.

Remember also that for (5.1) to be satisfied we postulated that t_0 should be smaller than δ/A_{28}: this is satisfied for sufficiently small ε, because we require that $\delta(\varepsilon)|\ln\varepsilon| \to 0$. So by (5.1) we have $\mathsf{P}_x^\varepsilon\{\tau_k^\varepsilon(\pm\delta) \ge t_0\} \le 0.8$ for sufficiently small ε and $x \in D_k(\pm\delta)$.

Application of the Markov property yields $\mathsf{P}_x^\varepsilon\{\tau_k^\varepsilon(\pm\delta) \ge nt_0\} \le 0.8^n$, and

$$\mathsf{M}_x^\varepsilon \tau_k^\varepsilon(\pm\delta) \le \frac{t_0}{1-0.8} = 500A_{44}A_{33}^{-1}\delta^2|\ln\varepsilon|.$$

Lemma 5.1 is proved. □

Proof of Lemma 3.5. We are going to prove that there exist positive constants A_{45}, A_{46}, A_{47} such that

$$\mathsf{M}_x^\varepsilon \int_0^{\tau_k^\varepsilon(\pm\delta)} e^{-\lambda t}\,dt \le A_{45}\delta^2|\ln\delta| \tag{5.14}$$

for all sufficiently small ε, $\varepsilon^{A_{46}} \le \delta \le A_{47}$, and $x \in D_k(\pm\delta)$.

Take $\delta_0 = \delta_0(\varepsilon) = \varepsilon^{A_{46}}/3$, and consider the cycles between the Markov times $\tau_0 \le \sigma_1 \le \tau_1 \le \sigma_2 \le \cdots$, where

$$\tau_0 = 0, \sigma_n = \min\{t \geq \tau_{n-1} : X_t^\varepsilon \notin D_k(\pm\delta_0)\},$$

and $\tau_n = \min\{t \geq \sigma_n : X_t^\varepsilon \in \bigcup_i C_{ki}(\delta_0) \cup \bigcup_i C_{ki}(\delta)\}$. The expectation in (5.14) is equal to

$$\sum_{n=0}^{\infty} \mathsf{M}_x^\varepsilon\left\{\tau_n < \tau_k^\varepsilon(\pm\delta); \int_{\tau_n}^{\sigma_{n+1}} e^{-\lambda t}\, dt\right\} + \sum_{n=1}^{\infty} \mathsf{M}_x^\varepsilon\left\{\sigma_n < \tau_k^\varepsilon(\pm\delta); \int_{\sigma_n}^{\tau_n} e^{-\lambda t}\, dt\right\}$$

$$\leq \sum_{n=0}^{\infty} \mathsf{M}_x^\varepsilon\{\tau_n < \tau_k^\varepsilon(\pm\delta); e^{-\lambda\tau_n}\} \cdot \sup\{\mathsf{M}_z^\varepsilon \tau_k^\varepsilon(\pm 2\delta_0) : z \in D_k(\pm 2\delta_0)\}$$

$$+ \sum_{n=1}^{\infty} \mathsf{M}_x^\varepsilon\{\sigma_n < \tau_k^\varepsilon(\pm\delta); e^{-\lambda\sigma_n}\} \cdot \sup\left\{\mathsf{M}_z^\varepsilon \int_0^{\tau_1} e^{-\lambda t}\, dt : z \in \bigcup_i C_{ki}(2\delta_0)\right\} \tag{5.15}$$

(the strong Markov property is used).

The first supremum is estimated by Lemma 5.1; it is $O(\delta_0^2|\ln\varepsilon|) \leq A_{48}\delta_0\delta|\ln\delta|$. For a path starting at a point $z \in C_{ki}(2\delta_0)$, the time τ_1 is nothing but $\tau_i^\varepsilon(H(O_k) \pm \delta_0,\ H(O_k) \pm \delta)$. To estimate its M_z^ε-expectation, we apply Lemma 4.2 to the solution of the boundary-value problem

$$\lambda f_i(H) - L_i f_i(H) = 1,$$
$$f_i(H(O_k) \pm \delta_0) = f_i(H(O_k) \pm \delta) = 0.$$

It is easy to see that $f_i(H) \leq A_{49}\delta^2|\ln\delta|$, $f_i(H(O_k) \pm 2\delta_0) \leq A_{50}\delta_0\delta|\ln\delta|$, and $\|f_i^{(m)}\| \leq A_m\delta^{2-m}|\ln\delta|$. By Lemma 4.2 we have

$$\mathsf{M}_z^\varepsilon \int_0^{\tau_1} e^{-\lambda t}\, dt = \mathsf{M}_z^\varepsilon \int_0^{\tau_i^\varepsilon(H(O_k)\pm\delta_0, H(O_k)\pm\delta)} e^{-\lambda t}[\lambda f_i(H(X_t^\varepsilon)) - L_i f_i(H(X_t^\varepsilon))]\, dt$$
$$\leq f_i(H(z)) + (\|f_i'\| + \|f_i''\| + \|f_i'''\|)\varepsilon^{A_{11}}.$$

The first term is not greater than $A_{50}\delta_0\delta|\ln\delta|$; the second is infinitely small compared to $\delta_0\delta|\ln\delta|$ if we choose A_{46} small enough. So both supremums in (5.15) are $O(\delta_0\delta|\ln\delta|)$.

Now, using the strong Markov property, we obtain the estimates:

$$\sum_{n=1}^{\infty} \mathsf{M}_x^\varepsilon\{\sigma_n < \tau_k^\varepsilon(\pm\delta); e^{-\lambda\sigma_n}\} \leq \sum_{n=0}^{\infty} \mathsf{M}_x^\varepsilon\{\tau_n < \tau_k^\varepsilon(\pm\delta); e^{-\lambda\tau_n}\}$$
$$\leq \sum_{n=0}^{\infty}\left[\sup\left\{\mathsf{M}_z^\varepsilon\left\{X_{\tau_1}^\varepsilon \in \bigcup_i C_{ki}(\delta_0); e^{-\lambda\tau_1}\right\}\right\}\right]^n. \tag{5.16}$$

Let F_i be the solution of the boundary-value problem

$$\lambda F_i(H) - L_i F_i(H) = 0,$$
$$F_i(H(O_k) \pm \delta_0) = 1, \qquad F_i(H(O_k) \pm \delta) = 0.$$

For this solution $F_i(H(O_k) \pm 2\delta_0) \le 1 - A_{ki}\delta_0/\delta$; so, applying Lemma 4.2, we get for sufficiently small ε and for $z \in C_{ki}(2\delta_0)$,

$$\mathsf{M}_z^\varepsilon\left\{X_{\tau_1}^\varepsilon \in \bigcup_i C_{ki}(\delta_0); e^{-\lambda\tau_1}\right\} \le 1 - A_{ki}\delta_0/(2\delta),$$

and the sum (5.16) does not exceed $2\delta/(\delta_0 \min_i A_{ki})$. This yields (5.14). □

Remark. It is easy to deduce from formula (5.14) that

$$\mathsf{M}_x^\varepsilon \tau_k^\varepsilon(\pm\delta) \le A_{51}\delta^2|\ln\delta| \tag{5.17}$$

for all sufficiently small ε, $\varepsilon^{A_{52}} \le \delta \le A_{53}$, and $x \in D_k(\pm\delta)$. To do this, we use the fact that $\tau_k(\pm\delta) \le (1 - e^{-1})^{-1} \int_0^{\tau_k(\pm\delta)} e^{-\lambda t}\, dt$ for $\tau_k(\pm\delta) < 1/\lambda$,

$$\mathsf{P}_x^\varepsilon\{\tau_k(\pm\delta) \ge 1/\lambda\} \le A_{45}\delta^2|\ln\delta|/(1 - e^{-1}),$$
$$\mathsf{P}_x^\varepsilon\{\tau_k(\pm\delta) \ge n/\lambda\} \le [A_{45}\delta^2|\ln\delta|/(1 - e^{-1})]^n.$$

§6. Proof of Lemma 3.6

We use a result in partial differential equations, which we give here without proof, reformulated to fit out particular problem.

Let $\mathscr{L}^\varepsilon$ be a two-dimensional second-order differential operator in the domain $D = (H_1, H_2) \times (-\infty, \infty)$ given by

$$\mathscr{L}^\varepsilon = \varepsilon^{-2}\frac{\partial}{\partial\theta} + a^{11}(H,\theta)\frac{\partial^2}{\partial H^2} + 2a^{12}(H,\theta)\frac{\partial^2}{\partial H\partial\theta} + a^{22}(H,\theta)\frac{\partial^2}{\partial\theta^2}$$
$$+ b^1(H,\theta)\frac{\partial}{\partial H} + b^2(H,\theta)\frac{\partial}{\partial\theta}, \tag{6.1}$$

where $\underline{A}(\xi^2 + \eta^2) \le a^{11}(H,\theta)\xi^2 + 2a^{12}(H,\theta)\xi\eta + a^{22}(H,\theta)\eta^2 \le \bar{A}(\xi^2 + \eta^2)$ for all $(H,\theta) \in D$ and real ξ, η, $0 < \underline{A} \le \bar{A} < \infty$, and $|b^i(H,\theta)| \le \bar{B}$.

Lemma 6.1 (see Krylov and Safonov [1]). *For every $\delta > 0$ there exist constants $\alpha = \alpha(\underline{A}, \bar{A}) \in (0,1)$ and $N = N(\underline{A}, \bar{A}, \bar{B}, \delta) < \infty$ such that*

$$|u(H',\theta') - u(H,\theta)| \le N(|H' - H| + \varepsilon^2|\theta' - \theta|)^\alpha \cdot \sup_D |u| \tag{6.2}$$

for every solution u of the equation $\mathscr{L}^\varepsilon u = 0$, *for all* H, $H' \in (H_1 + \delta, H_2 - \delta)$, *and all* θ, θ'. *Neither* α *nor* N *depend on* ε.

Using this result, we obtain a lemma that is important in the proof of Lemma 3.6 (in which we apply it to the numbers $H(O_k) \pm \delta'/2$, $H(O_k) \pm \delta'$, and $H(O_k) \pm 2\delta'$).

Lemma 6.2. *For a segment* I_i *of our graph, let* $H_1 < H < H_2$ *be numbers in the interior of the interval* $H(I_i)$. *Then*

$$\lim_{\varepsilon \to 0} \max_{x_1, x_2 \in C_i(H)} \max_{f: \|f\| \le 1} |\mathsf{M}^\varepsilon_{x_1} f(X^\varepsilon_{\tau^\varepsilon_i(H_1, H_2)}) - \mathsf{M}^\varepsilon_{x_2} f(X^\varepsilon_{\tau^\varepsilon_i(H_1, H_2)})| = 0 \tag{6.3}$$

(the functions f over which the maximum is taken are defined on $\partial D_i(H_1, H_2)$).

Proof. The function $u^\varepsilon(x) = \mathsf{P}^\varepsilon_x\{X^\varepsilon_{\tau^\varepsilon_i(H_1, H_2)} \in \gamma\}$ is a solution of the equation $L^\varepsilon u^\varepsilon(x) = 0$ in the domain $D_i(H_1, H_2)$. Introducing new coordinates in this annulus-shaped domain: the value of the Hamiltonian H, and the periodic "angular" coordinate θ, we have $L^\varepsilon = \varepsilon^{-2} B(H, \theta) \cdot (\partial/\partial\theta) + \frac{1}{2}\Delta$ with positive $B(H, \theta)$, and dividing the equation by $B(H, \theta)$ brings the operator to the form (6.1). The estimate (6.2) with $H' = H$ proves (6.3) (the rate of convergence is that of const $\cdot\, \varepsilon^{2\alpha}$, where the positive exponent depends on H_1, H_2, and the constant also on H).

Now let us prove Lemma 3.6.

We want to prove that for $I_j \sim O_k$ for every $\delta > 0$, for sufficiently small δ', $0 < \delta' < \delta$, the difference

$$|\mathsf{P}^\varepsilon_x\{X^\varepsilon_{\tau^\varepsilon_k(\pm\delta)} \in C_{kj}(\delta)\} - p_{kj}|$$

is small for sufficiently small ε and for all $x \in \bar{D}_k(\pm\delta')$, where p_{kj} are given by $p_{kj} = \beta_{kj} / \sum_{i: I_i \sim O_k} \beta_{ki}$.

The first step is proving that for every i such that $I_i \sim O_k$ and for every function f on $\partial D_k(\pm\delta)$ with $\|f\| \le 1$ (in particular, for $f(z) = \chi C_{kj}(\delta)(z)$),

$$\lim_{\varepsilon \to 0} \max_{x_1, x_2 \in C_{ki}(\delta')} |F^\varepsilon(x_1) - F^\varepsilon(x_2)| = 0, \tag{6.4}$$

where

$$F^\varepsilon(x) = \mathsf{M}^\varepsilon_x f(X^\varepsilon_{\tau^\varepsilon_k(\pm\delta)}). \tag{6.5}$$

Take δ'' and δ''' so that $0 < \delta'' < \delta' < \delta''' < \delta$. Let us apply the strong Markov property with respect to the time $\tau^\varepsilon_i(H(O_k) \pm \delta'', H(O_k) \pm \delta''') <$

$\tau_k^\varepsilon(\pm\delta)$: for $x_m \in C_{ki}(\delta')$, $m = 1, 2$, we have

$$F^\varepsilon(x_m) = \mathsf{M}_{x_m}^\varepsilon F^\varepsilon(X^\varepsilon_{\tau_i^\varepsilon(H(O_k)\pm\delta'', H(O_k)\pm\delta''')}).$$

Now (6.4) follows from Lemma 6.2.

The second step is proving that for every $\delta > 0$ and every $\kappa > 0$ there exists δ', $0 < \delta' < \delta$, such that for sufficiently small ε,

$$\max_{x_1, x_2 \in \bar{D}_k(\pm\delta')} |F^\varepsilon(x_1) - F^\varepsilon(x_2)| < \kappa. \tag{6.6}$$

There are exactly three regions, D_{i_0}, D_{i_1}, and D_{i_2}, that are separated by the separatrix C_k, one of them, D_{i_0}, adjoining the whole curve C_k, and D_{i_1}, D_{i_2} adjoining only parts of it (so that $C_{ki_0} = C_k = C_{ki_1} \cup C_{ki_2}$). We have either $H(x) > H(O_k)$ for $x \in D_{i_0}$ and $H(x) < H(O_k)$ for $x \in D_{i_1} \cup D_{i_2}$, or all signs are opposite.

Let τ be the first time at which the process X_t^ε reaches $C_{ki_0}(\delta')$ or $C_{ki_1}(\delta) \cup C_{ki_2}(\delta)$. Since $\tau \le \tau_k^\varepsilon(\pm\delta)$ for the process starting at a point $x \in \bar{D}_k(\pm\delta')$, we can apply the strong Markov property with respect to τ. We have for $x \in \bar{D}_k(\pm\delta')$,

$$F^\varepsilon(x) = \mathsf{M}_x^\varepsilon\{X_\tau^\varepsilon \in C_{ki_1}(\delta) \cup C_{ki_2}(\delta); f(X_\tau^\varepsilon)\} + \mathsf{M}_x^\varepsilon\{X_\tau^\varepsilon \in C_{ki_0}(\delta); F^\varepsilon(X_\tau^\varepsilon)\},$$

$$\mathsf{P}_x^\varepsilon\{X_\tau^\varepsilon \in C_{ki_1}(\delta) \cup C_{ki_2}(\delta)\} \cdot \inf_{z \in \partial D_k(\pm\delta)} f(z) + \mathsf{P}_x^\varepsilon\{X_\tau^\varepsilon \in C_{ki_0}(\delta)\} \cdot \inf_{z \in C_{ki_0}(\delta')} F^\varepsilon(z)$$

$$\le F^\varepsilon(x)$$

$$\le \mathsf{P}_x^\varepsilon\{X_\tau^\varepsilon \in C_{ki_1}(\delta) \cup C_{ki_2}(\delta)\} \cdot \sup_{z \in \partial D_k(\pm\delta)} f(z) + \mathsf{P}_x^\varepsilon\{X_\tau^\varepsilon \in C_{ki_0}(\delta)\}$$

$$\cdot \sup_{z \in C_{ki_0}(\delta')} F^\varepsilon(z),$$

therefore for x_1, $x_2 \in \bar{D}_k(\pm\delta')$,

$$|F^\varepsilon(x_1) - F^\varepsilon(x_2)| \le \left[\sup_{z \in \partial D_k(\pm\delta)} f(z) - \inf_{z \in \partial D_k(\pm\delta)} f(z)\right]$$

$$\times \max(\mathsf{P}_{x_1}^\varepsilon\{X_\tau^\varepsilon \in C_{ki_1}(\delta) \cup C_{ki_2}(\delta)\}, \mathsf{P}_{x_2}^\varepsilon\{X_\tau^\varepsilon \in C_{ki_1}(\delta) \cup C_{ki_2}(\delta)\})$$

$$+ \left[\sup_{z \in C_{ki_0}(\delta')} F^\varepsilon(z) - \inf_{z \in C_{ki_0}(\delta')} F^\varepsilon(z)\right].$$

The second difference, by (6.4), is arbitrarily small if ε is small enough. Let us estimate $\mathsf{P}_x^\varepsilon\{X_\tau^\varepsilon \in C_{ki_1}(\delta) \cup C_{ki_2}(\delta)\}$ for $x \in \bar{D}_k(\pm\delta')$.

Apply formula (4.5):

$$\mathsf{M}_x^\varepsilon H(X_\tau^\varepsilon) - H(x) = \mathsf{M}_x^\varepsilon \int_0^\tau \frac{1}{2} \Delta H(X_t^\varepsilon)\, dt. \tag{6.7}$$

Here $X_\tau^\varepsilon \in C_{ki_0}(\delta') \cup C_{ki_1}(\delta) \cup C_{ki_2}(\delta)$. The function H takes the value $H(O_k) \pm \delta'$ on $C_{ki_0}(\delta')$, and the value $H(O_k) \mp \delta$ on $C_{ki_1}(\delta) \cup C_{ki_2}(\delta)$. The left-hand side in (6.7) is equal to

$$\begin{aligned}&[H(O_k) \pm \delta'] \cdot [1 - \mathsf{P}_x^\varepsilon\{X_\tau^\varepsilon \in C_{ki_1}(\delta) \cup C_{ki_2}(\delta)\}]\\ &\quad + [H(O_k) \mp \delta] \cdot \mathsf{P}_x^\varepsilon\{X_\tau^\varepsilon \in C_{ki_1}(\delta) \cup C_{ki_2}(\delta)\},\end{aligned}$$

so (6.7) can be rewritten as

$$\begin{aligned}&(\delta + \delta') \cdot \mathsf{P}_x^\varepsilon\{X_\tau^\varepsilon \in C_{ki_1}(\delta) \cup C_{ki_2}(\delta)\\ &\quad = \delta' \pm [H(O_k) - H(x)] \mp \mathsf{M}_x^\varepsilon \int_0^\tau \frac{1}{2} \Delta H(X_t^\varepsilon)\, dt\\ &\quad \le 2\delta' + \left|\mathsf{M}_x^\varepsilon \int_0^\tau \frac{1}{2} \Delta H(X_t^\varepsilon)\, dt\right|,\end{aligned}$$

and the expectation of the integral does not exceed

$$\tfrac{1}{2} \sup|\Delta H| \cdot \mathsf{M}_x^\varepsilon \tau \le \tfrac{1}{2} \sup|\Delta H| \cdot \mathsf{M}_x^\varepsilon \tau_k^\varepsilon(\pm\delta) \le A_{54} \cdot \delta^2 |\ln \delta|$$

for sufficiently small ε by (5.17). So for sufficiently small ε,

$$\mathsf{P}_x^\varepsilon\{X_\tau^\varepsilon \in C_{ki_1}(\delta) \cup C_{ki_1}(\delta)\} \le \frac{2\delta'}{\delta} + A_{54}\delta|\ln \delta|,$$

which can be made arbitrarily small by choosing small positive δ, and then small positive $\delta' < \delta$. So the second step of our proof is complete.

It is true that we formulated our statement for an arbitrary positive δ, and finished with establishing it only for sufficiently small δ. But, first of all, small δ are enough for our purpose of proving our main theorem; and secondly, this happened because we used the very rough inequality $\mathsf{M}_x^\varepsilon \tau \le \mathsf{M}_x^\varepsilon \tau_k^\varepsilon(\pm\delta)$, and it is very easy to verify (6.6) for arbitrary positive δ, making use of the strong Markov property.

So $\mathsf{P}_x^\varepsilon\{X_{\tau_k^\varepsilon(\pm\delta)}^\varepsilon \in C_{kl}(\delta)\}$ has approximately the same value for all $x \in \bar{D}_k(\pm\delta')$ (for δ' small enough and ε small enough). What remains is to prove that this value is approximately $p_{kj} - \beta_{kj}/\sum_{i:I_i \sim O_k} \beta_{ki}$, where β_{ki} are given by formula (2.2). To do this, we use the fact that the invariant measure μ for the process $(X_t^\varepsilon, \mathsf{P}_x^\varepsilon)$, which is the Lebesgue measure, can be written as an integral with respect to the invariant measure of the imbedded Markov chain. Let us adduce this result in our specific case.

Let us choose a number H_0 that is greater than all $H(x_k) + 2\delta$. Let $C_\delta = \bigcup_{k,i} C_{ki}(\delta) \cup C(H_0 - \delta)$, $C_{\delta'} = \bigcup_{k,i} C_{ki}(\delta') \cup C(H_0 - \delta')$. Introduce the random times $\tau_0 \le \sigma_0 < \tau_1 < \sigma_1 < \cdots < \tau_n < \sigma_n < \cdots$ by $\tau_0 = 0$, $\sigma_k = \min\{t \ge \tau_k : X_t^\varepsilon \in C_\delta\}$, $\tau_k = \min\{t \ge \sigma_{k-1} : X_t^\varepsilon \in C'_\delta\}$. The sequence $X_{\tau_k}^\varepsilon$, $k = 0, 1, 2, \ldots, n, \ldots$, is a Markov chain (for $k \ge 1$, all $X_{\tau_k}^\varepsilon \in C_{\delta'}$); and if $X_0^\varepsilon \in C_\delta$, the sequence $X_{\sigma_k}^\varepsilon$, $k = 0, 1, 2, \ldots, n, \ldots$, is also a Markov chain on C_δ.

Every invariant measure μ of the process $(X_t^\varepsilon, \mathbf{P}_x^\varepsilon)$ can be represented in the form

$$\mu(A) = \int_{C_\delta} \nu^\varepsilon(dx) \mathbf{M}_x^\varepsilon \int_0^{\sigma_1} \chi_A(X_t^\varepsilon)\, dt = \int_{C_{\delta'}} \nu'^\varepsilon(dx) \mathbf{M}_x^\varepsilon \int_0^{\tau_1} \chi_A(X_t^\varepsilon)\, dt, \tag{6.8}$$

where ν^ε, ν'^ε are measures on C_δ, $C_{\delta'}$ satisfying the system of integral equations

$$\nu^\varepsilon(B) = \int_{C_{\delta'}} \nu'^\varepsilon(dx) \mathbf{P}_x^\varepsilon\{X_{\sigma_0}^\varepsilon \in B\}, \quad \nu'^\varepsilon(C) = \int_{C_\delta} \nu^\varepsilon(dx) \mathbf{P}_x^\varepsilon\{X_{\tau_1}^\varepsilon \in C\} \tag{6.9}$$

(see Khas'minskii [7], also Wentzell [5]). These measures are invariant measures of the Markov chains $X_{\sigma_k}^\varepsilon$, $X_{\tau_k}^\varepsilon$.

The measures ν^ε, ν'^ε are nonzero (since $\mu \ne 0$), and for every $I_i \sim O_k$,

$$\nu^\varepsilon(C_{ki}(\delta)) > 0, \quad \nu'^\varepsilon(C_{ki}(\delta')) > 0,$$

because starting from each of these sets, or from $C(H_0 - \delta)$ (or $C_{ki}(\delta')$) it is possible, with positive probability, to reach every other $C_{ki}(\delta)$, $C_{ki}(\delta')$ in a finite number of steps (cycles). So we can introduce the averages:

$$p_{kj}^\varepsilon = \frac{\int_{\bigcup_{i:I_i \sim O_k} C_{ki}(\delta')} \nu'^\varepsilon(dx) \mathbf{P}_x^\varepsilon\{X_{\tau_k^\varepsilon(\pm\delta)}^\varepsilon \in C_{kj}(\delta)\}}{\nu'^\varepsilon(\bigcup_{i:I_i \sim O_k} C_{ki}(\delta'))}. \tag{6.10}$$

According to (6.6),

$$|\mathbf{P}_x^\varepsilon\{X_{\tau_k^\varepsilon(\pm\delta)}^\varepsilon \in C_{kj}(\delta)\} - p_{kj}^\varepsilon| < \kappa \tag{6.11}$$

for all $x \in \bar{D}_k(\pm\delta')$.

If the process X_t^ε starts in the set $C_{\delta'}$, it cannot be in $C_{kj}(\delta)$ at time τ_1 unless it started in $\bigcup_{i:I_i \sim O_k} C_{ki}(\delta')$. Therefore, by the first equation in (6.9), $\nu^\varepsilon\left(\bigcup_{i:I_i \sim O_k} C_{ki}(\delta)\right) = \nu'^\varepsilon(\bigcup_{i:I_i \sim O_k} C_{ki}(\delta'))$, and the numerator in (6.10) is nothing but $\nu^\varepsilon(C_{kj}(\delta))$. So (6.11) can be rewritten as

$$\left|\mathbf{P}_x^\varepsilon\{X_{\tau_k^\varepsilon(\pm\delta)}^\varepsilon \in C_{kj}(\delta)\} - \frac{\nu^\varepsilon(C_{kj}(\delta))}{\sum_{i:I_i \sim O_k} \nu^\varepsilon(C_{ki}(\delta))}\right| < \kappa \tag{6.12}$$

for sufficiently small positive $\delta' < \delta$, sufficiently small ε, and $x \in \bar{D}_k(\pm\delta')$.

So the next step is to estimate $\nu^{\varepsilon}(C_{kj}(\delta))$ for small $\delta > 0$, small δ', $0 < \delta' < \delta$, and small ε.

We have not used formula (6.8) yet. This formula implies the corresponding equality for the integrals with respect to the measure μ (the Lebesgue measure): if $G(x)$, $x \in R^2$, is integrable, we have

$$\iint_{R^2} G(x)\mu(dx) = \iint_{R^2} G(x)\,dx = \int_{C_\delta} \nu^{\varepsilon}(dx)\mathsf{M}_x^{\varepsilon} \int_0^{\sigma_1} G(X_t^{\varepsilon})\,dt.$$

If the function $G(x)$ is equal to 0 in the region $\{x : H(x) > H_0 - \delta\}$ and in all regions $D_k(\pm\delta)$, then $\int_{\tau_1}^{\sigma_1} G(X_t^{\varepsilon})\,dt = 0$, and

$$\iint_{R^2} G(x)\,dx = \int_{C_\delta} \nu^{\varepsilon}(dx)\mathsf{M}_x^{\varepsilon} \int_0^{\tau_1} G(X_t^{\varepsilon})\,dt. \tag{6.13}$$

Let us consider an arbitrary segment I_j of the graph with ends O_{k_1} and O_{k_2}. Let us denote $H_{k_1} = H(O_{k_1})$, $H_{k_2} = H(O_{k_2})$, except if the vertex O_{k_2} corresponds to the point at infinity, in which case we take $H_{k_2} = H_0$. Let $g(H)$ be an arbitrary continuous function on the interval $H(I_j)$ that is equal to 0 outside its subinterval $(H_{k_1} + \delta, H_{k_2} - \delta)$. Consider the function $G(x)$ that is equal to $g(H(x))$ in the domain D_j, and to 0 outside D_j. The left-hand side of (6.13) can be rewritten as

$$\iint_{D_j} g(H(x))\,dx = \int_{H_{k_1}+\delta}^{H_{k_2}-\delta} g(H)\,dv_j(H),$$

where $u_j(H)$ is the function whose derivative is given by formula (1.15).

For the expectation in the right-hand side of (6.13), formulas (4.28) and (4.29) hold, and we can write

$$\begin{aligned}
\int_{H_{k_1}+\delta}^{H_{k_2}-\delta} g(H)\,dv_j(H) &= \nu^{\varepsilon}(C_{k_1j}(\delta))\left[\frac{u_j(H_{k_1}+\delta) - u_j(H_{k_1}+\delta')}{u_j(H_{k_2}-\delta') - u_j(H_{k_1}+\delta')}\right.\\
&\quad\times \left.\int_{H_{k_1}+\delta}^{H_{k_2}-\delta} (u_j(H_{k_2}-\delta') - u_j(H))g(H)\,dv_j'(H) + o(1)\right]\\
&\quad+ \nu^{\varepsilon}(C_{k_2j}(\delta))\left[\frac{u_j(H_{k_2}-\delta') - u_j(H_{k_2}-\delta)}{u_j(H_{k_2}-\delta') - u_j(H_{k_1}+\delta')}\right.\\
&\quad\times \left.\int_{H_{k_1}+\delta}^{H_{k_2}-\delta} (u_j(H) - u_j(H_{k_1}+\delta'))g(H)\,dv_j'(H) + o(1)\right],
\end{aligned}$$

where the $o(1)$ go to 0 as $\varepsilon \to 0$.

In order for this to be true for an arbitrary continuous function g, it is necessary that the limits of $\nu^{\varepsilon}(C_{k_1j}(\delta))$, $v^{\varepsilon}(C_{k_2j}(\delta))$ exist as $\varepsilon \to 0$, namely,

$$\lim_{\varepsilon\to 0} \nu^\varepsilon(C_{k_1 j}(\delta)) = \frac{1}{u_j(H_{k_1}+\delta) - u_j(H_{k_1}+\delta')},$$

$$\lim_{\varepsilon\to 0} \nu^\varepsilon(C_{k_2 j}(\delta)) = \frac{1}{u_j(H_{k_2}-\delta') - u_j(H_{k_2}-\delta)}.$$

But by formula (1.17), $u_j(H') - u_j(H) \sim 2(\oint_{C_{k_i j}} |\nabla H(x)|\,dl)^{-1} \cdot (H' - H)$ as $H, H' \to H_{k_i}$ (we disregard the case of $H_{k_2} = H_0$). This means that, for sufficiently small δ and ε,

$$\left| \nu^\varepsilon(C_{kj}(\delta)) - \frac{\beta_{kj}}{2(\delta-\delta')} \right| < \frac{\kappa}{\delta-\delta'}.$$

Together with formula (6.12), this yields

$$|\mathsf{P}_x^\varepsilon\{X^\varepsilon_{\tau_k^\varepsilon(\pm\delta)} \in C_{kj}(\delta)\} - p_{kj}| < 2\kappa,$$

and Lemma 3.6 is proved. □

§7. Remarks and Generalizations

1. Let a system

$$\dot{X}_t = b(X_t), \quad X_0 = x \in R^2, \tag{7.1}$$

in the plane have a smooth first integral $H(x) : H(X_t) = H(x)$ for $t \geq 0$. Assume that $H(x)$ has a finite number of critical points and $\lim_{|x|\to\infty} H(x) = \infty$. Since $H(x)$ is a first integral, $\nabla H(x) \cdot b(x) = 0$, $x \in R^2$, and thus

$$b(x) = \beta(x)\overline{\nabla} H(x),$$

where $\beta(x)$ is a scalar. Consider small white noise perturbations of such a system:

$$\dot{\tilde{X}}_t^\varepsilon = \beta(\tilde{X}_t^\varepsilon)\overline{\nabla} H(\tilde{X}_t^\varepsilon) + \varepsilon \dot{\tilde{W}}_t, \quad \tilde{X}_0^\varepsilon = x. \tag{7.2}$$

Here $\tilde{W}_t$ is the Wiener process in R^2, $0 < \varepsilon \ll 1$. If $\beta(x) \equiv$ constant, system (7.1) is Hamiltonian. If $\beta(x) \not\equiv$ constant but does not change its sign, the situation is similar to the Hamiltonian case. But if $\beta(x)$ changes its sign, the problem becomes more complicated: even if $H(x)$ has just one critical point, the slow component of the perturbed process may not converge to a Markov process as $\varepsilon \to 0$. To have a Markov limit, it may be impossible to avoid consideration of processes on graphs. We restrict ourselves to just an example (see Freidlin [19]).

Let $H(x)$, $x \in R^2$, have just one minimum at the origin and $H(0) = 0$, and

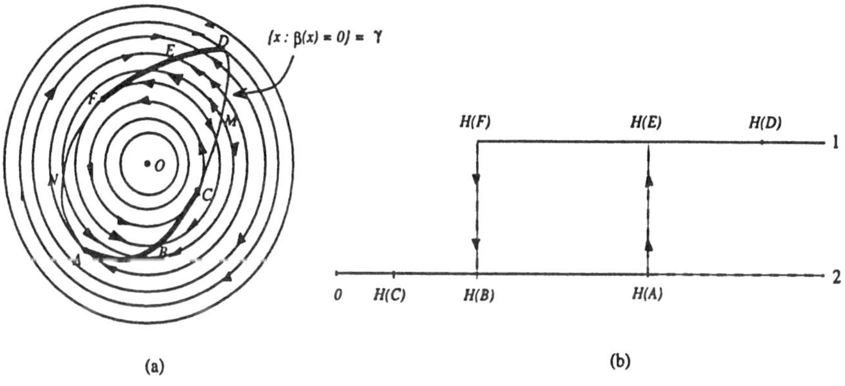

Figure 26.

let the level sets $C(y) = \{x \in R^2 : H(x) = y\}$, $y > 0$, be smooth curves homeomorphic to the circle (see Fig. 26). Suppose $\beta(x)$ is negative inside the loop $ABCDEFA$ and positive outside this loop. Here, A, C, D, F are the points where the loop $\{x \in R^2 : \beta(x) = 0\}$ is tangent to the level curve of $H(x)$; B is the point where the trajectory starting at F crosses AC for the first time, and E is the point at which the trajectory starting at A crosses FD for the first time.

The dynamical system on a level set $C(y)$ has a unique invariant measure if $y \notin (H(C), H(D))$. The density of this measure with respect to the length element on $C(y)$ is const $\times$ $(|\beta(x)|\,|\nabla H(x)|)^{-1}$. But, for $y \in (H(C), H(D))$, the dynamical system has more than one equilibrium point on $C(y)$. An invariant measure is concentrated at each of these equilibrium points. The dynamical system has four rest points when $y \in (H(B),\ H(A))$: two stable and two unstable. This results in the formation of a new "stable" first integral independent of $H(x)$.

To define this new first integral $k(x)$, denote by $x_0^{(1)}(y)$ the point of intersection of $C(y)$ and the arc FED of the curve $\gamma = \{x \in R^2 : \beta(x) = 0\}$. Denote the intersection point of $C(y)$ and of the arc ABC by $x_0^{(2)}(y)$. We assume that there exists at most one such point $x_0^{(1)}(y)$ and at most one $x_0^{(2)}(y)$. It is clear that $x_0^{(1)}(y)$ exists for $H(F) < y < H(D)$, and $x_0^{(2)}(y)$ exists for $H(C) < y < H(A)$.

Define $k(x)$, $x \in R^2$, as follows.

$$k(x) = \begin{cases} i, & \text{if } x_0^{(1)}(H(x)) \text{ exists and } x \text{ belongs to the domain of} \\ & \text{attraction of } x_0^{(i)}(H(x)), i = 1, 2; \\ 1, & \text{if } H(x) \geq H(D), \text{ or if } x \text{ belongs to the arc } CMD; \\ 2, & \text{if } H(x) \leq H(C), \text{ or if } x \text{ belongs to the arc } ANF. \end{cases}$$

It is easy to see that $k(x)$ is a first integral for the system shown in Fig. 26(a).

Let now $\tilde{X}_t^\varepsilon$ be the perturbed process defined by (7.2) for the dynamical system shown in Fig. 26(a). Let X_t^ε be the corresponding rescaled process: $X_t^\varepsilon = \tilde{X}_{t/\varepsilon^2}^\varepsilon$.

Consider the graph Γ in Fig. 26(b). The pairs (i, H), $H \geq H(0) = 0$, $i \in \{1, 2\}$, can be used as coordinates on Γ. The points $(1, H(F))$ and $(2, H(B))$ as well as $(1, H(E))$ and $(2, H(A))$ are identified.

Define the mapping $Y(x) : R^2 \to \Gamma$ by $Y(x) = (k(x), H(x))$, and let $Y_t^\varepsilon = Y(X_t^\varepsilon)$.

It is easy to see that, when $H(X_t^\varepsilon) \notin [H(C), H(D)]$, the fast component of the process X_t^ε has distribution on $C(H(X_t^\varepsilon))$ close to the unique invariant measure of the nonperturbed system on the corresponding level set of $H(x)$. When $H(x_t^\varepsilon) \in (H(C),\ H(B)) \cup (H(A), H(D))$, the distribution of the fast component is close to the measure concentrated at the stable rest point of the nonperturbed system on the corresponding level set. But when $H(X_t^\varepsilon) \in (H(B), H(A))$, the fast coordinate of the X_t^ε is concentrated near one of the points $x_0^{(1)}(H(X_t^\varepsilon))$ or $x_0^{(2)}(H(X_t^\varepsilon))$, depending upon the side from which $H(X_t^\varepsilon)$ last entered the segment $[H(B), H(A)]$. If $\varepsilon \ll 1$, the process X_t^ε "jumps" from F to B along a deterministic trajectory. Similarly, X_t^ε "jumps" from A to E.

Using these arguments, one can check that the processes Y_t^ε converge weakly as $\varepsilon \to 0$ in the space of continuous functions on $[0, T]$, $T < \infty$, with the values in Γ to a diffusion process Y_t on Γ. The process Y_t is defined as follows. Let $T(y) = \oint_{C(y)} [|\beta(x)| \nabla H(x)|||]^{-1}\, d\ell$ and

$$A_1(y) = \begin{cases} [T(y)]^{-1} \displaystyle\oint_{C(y)} |\nabla H(x)| \frac{d\ell}{|\beta(x)|}, & y > H(D), \\ |\nabla H(x_0^{(1)}(y))|^2, & H(F) \leq y \leq H(D); \end{cases}$$

$$B_1(y) = \begin{cases} (2T(y))^{-1} \displaystyle\oint_{C(y)} \frac{\Delta H(x)\, d\ell}{|\beta(x) \nabla H(x)|}, & y > H(D), \\ \frac{1}{2} \Delta H(x_0^{(1)}(y)), & H(F) \leq y \leq H(D); \end{cases}$$

$$A_2(y) = \begin{cases} [T(y)]^{-1} \displaystyle\oint_{C(y)} |\nabla H(x)| \frac{d\ell}{|\beta(x)|}, & y < H(C), \\ |\nabla H(x_0^{(2)}(y))|^2, & H(C) \leq y \leq H(A); \end{cases}$$

$$B_2(y) = \begin{cases} (2T(y))^{-1} \displaystyle\oint_{C(y)} \frac{\Delta H(x)\, d\ell}{|\beta(x) \nabla H(x)|}, & y < H(C), \\ \frac{1}{2} \Delta H(x_0^{(2)}(y)), & H(C) \leq y \leq H(A). \end{cases}$$

Define the operators L_i, $i = 1, 2$,

$$L_i = \frac{1}{2} A_i(y) \frac{d^2}{dy^2} + B_i(y) \frac{d}{dy}.$$

The process Y_t on Γ is governed by the operator L_1 on the upper part of the graph $\Gamma (i=1)$, and by L_2 on the lower part $(i=2)$. The operator L_2 degenerates at the point $(0,2) \in \Gamma$, so that this point is inaccessible for the process Y_t. To define the process Y_t for all $t \geq 0$ in a unique way, one should add the gluing conditions at the points $(1, H(F))$ identified with $(2, H(B))$ and at $(1, H(E))$ identified with $(2, H(A))$. These gluing conditions define the behavior of the process at the vertices. As we have already seen, the gluing conditions describe the domain of definition of the generator for the process Y_t. Let Y_t be the process on Γ such that its generator is defined for functions $f(i,y)$, $(i,y) \in \Gamma$, which are continuous on Γ, smooth on $\{(i,y) \in \Gamma :\ y \geq H(F), i=1\}$ and on $\{(i,y) \in \Gamma : y \leq H(A), i=2\}$, and satisfy $L_1 f(1, H(F)) = L_2 f(2, H(B))$ and $L_1 f(1, H(E)) = L_2 f(2, H(A))$. Obviously, these gluing conditions have the form described earlier in this chapter with one of $\beta_{ki} = 0$ for $k = 1,2$. The generator coincides with L_1 at the points of the set $\{(i,y) \in \Gamma : y > H(F), i=1\}$ and with L_2 at the points of $\{(i,y) \in \Gamma : y < H(A), i=2\}$. The limiting process Y_t on Γ is determined by these conditions in a unique way.

2. We have seen that the graphs associated with the Hamiltonian systems in R^2 always have a structure of a tree. If we consider dynamical systems in the phase space of a more complicated topological structure, the corresponding graph, on which the slow motion of the perturbed system should be considered, can have loops. Consider, for example, the dynamical system on the two-dimensional torus T^2 with Hamiltonian $H(x)$ defined as follows. Let the torus be imbedded in R^3 and the symmetry axis of the torus be parallel to a plane Π (see Fig. 27(a)), and $H(x)$ be equal to the distance of a point $x \in T^2$ from Π. The trajectories of the system then lie in the planes parallel to Π as shown in Fig. 27(a). The set of connected components of the level sets of the Hamiltonian H is homeomorphic to the graph Γ shown in Fig. 27(b). This graph has a loop. One can describe the slow motion for the diffusion process resulting from small white noise perturbations of the dynamical system in the same way as in the case of the systems in R^2.

The dynamical system or this example is not a generic Hamiltonian system on T^2. Generic Hamiltonian systems on two-dimensional tori were studied in Arnold [2] and Sinai and Khanin [1]. In particular, it was shown there that a generic dynamical system on T^2 preserving the area has the following structure. There are finitely many loops in T^2 such that inside each loop the system behaves as a Hamiltonian system in a finite part of the plane. Each trajectory lying in the exterior of all loops is dense in the exterior. These trajectories form one ergodic class. There are two such loops in the example shown in Fig. 28.

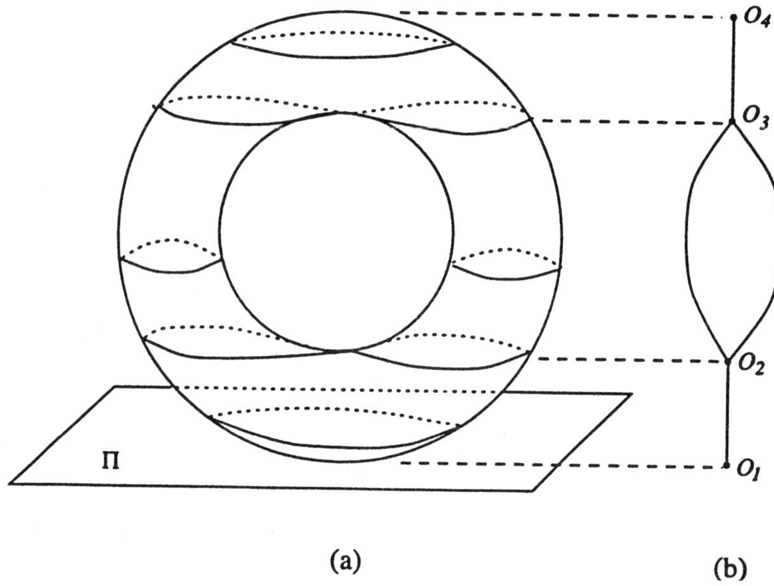

Figure 27.

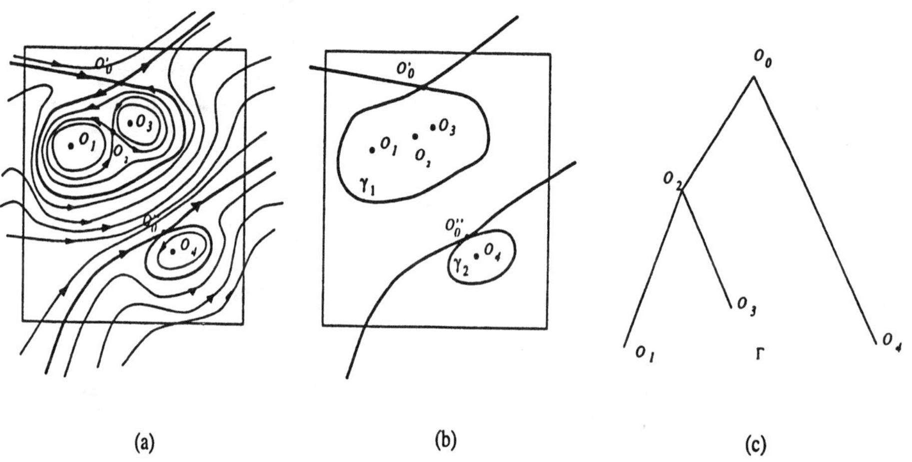

Figure 28.

The Hamiltonian H of such a system on T^2 is multivalued, but its gradient $\nabla H(x)$, $x \in T^2$, is a smooth vector field on the torus. Consider small white noise perturbation of such a system:

$$\dot{\tilde{X}}_t^\varepsilon = \bar{\nabla} H(\tilde{x}_t^\varepsilon) + \varepsilon \dot{\tilde{W}}_t, \quad X_t^\varepsilon = \tilde{x}_{t/\varepsilon^2}^\varepsilon.$$

To study the slow component of the perturbed process, we should identify some points of the phase space. First, as we did before, we should identify all the points of each periodic trajectory. But here we have an additional identification: since the trajectories of the dynamical system outside the loops are dense in the exterior of the loops, all the points of the exterior should be glued together. The set of all the connected components of the level sets of the Hamiltonian, provided with the natural topology, is homeomorphic in this case to a graph Γ having the following structure: the exterior of the loops corresponds to one vertex O_0. The number of branches connected with O_0 is equal to the number of loops. Each branch describes the set of connected components of the Hamiltonian level sets inside the corresponding loop. For example, the left branch of the graph Γ shown in Fig. 28(c), consisting of the edges O_0O_2, O_2O_1, O_2O_3, describes the set of connected components inside the loop γ_1.

Let $H_1, H_2, \ldots, H_N$ be the values of the Hamiltonian function on the loops $\gamma_1, \ldots, \gamma_N$. Denote by $\overline{H}(x)$, $x \in T^2$, the function equal to zero outside the loops and equal to $H(x) - H_k$ inside the kth loop, $k = 1, \ldots, N$. Let all the edges of Γ be indexed by the numbers $1, \ldots, n$. The pairs (k, y), where k is the number of the edge containing a point $z \in \Gamma$ and y is the value of $\overline{H}(x)$ on the level set component corresponding to z, form a coordinate system on Γ. Consider the mapping $Y : T^2 \mapsto \Gamma$, $Y(x) = (k(x), \overline{H}(x))$.

Then one can expect that the processes $Y_t^\varepsilon = Y(X_t^\varepsilon)$, $0 \le t \le T$, $T < \infty$, converge weakly in the space of continuous functions on $[0, T]$ with the values in Γ to a diffusion process Y_t on Γ. One can prove in the same way as in the case of Hamiltonian systems in R^2, that the limiting process is governed inside the edges by the operators obtained by averaging with respect to the invariant density of the fast motion. The gluing conditions at all the vertices besides O_0 have the same form as before. But the gluing condition at O_0 is special. The uniform distribution on T^2 is invariant for any process X_t^ε, $\varepsilon > 0$. Thus, the time each of them spends in the exterior of the loops is proportional to the area of the exterior. Therefore, the limiting process Y_t spends a positive amount of time at the vertex O_0, that corresponds to the exterior of the loops. The process Y_t spends zero time at all other vertices.

To calculate the gluing conditions at O_0, assume that the Hamiltonian $H(x)$ inside each loop has just one extremum (like inside the loop γ_2 in Fig. 28). Since the gluing conditions at O_0 are determined by the behavior of the process X_t^ε in a neighborhood of the loops, we can assume this without loss of generality. Let the area of the torus T^2 be equal to one, and the area of the exterior of the loops be equal to α. Denote by μ the measure on Γ such that $\mu\{O_0\} = \alpha$, $(d\mu/dH)(k, H) = T_k(H)$ inside the kth edge, where

$$T_k(H) = \oint_{C_k(H)} |\nabla H(x)|^{-1}\, d\ell,$$

$C_k(H)$ is the component of the level set $C(H) = \{x \in T^2 : \overline{H}(x) = H\}$ inside the kth loop. It is easy to check that $T_k(\overline{H}) = |dS_k(\overline{H})/d\overline{H}|$, where $S_k(\overline{H})$ is the area of the domain bounded by $C_k(\overline{H})$. Taking this into account, one can check that μ is the normalized invariant measure for the processes $Y_t^\varepsilon = Y(X_t^\varepsilon)$ on Γ for any $\varepsilon > 0$, and it is invariant for the limiting process Y_t as well.

On the other hand, a diffusion process on Γ with the generator A has an invariant measure μ if and only if

$$\int_\Gamma Au(k, y)\, d\mu = 0 \tag{7.3}$$

for any continuous function $u(z)$, $z \in \Gamma$, that is smooth at the interior points of the edges and such that $Au(z)$ is also continuous. Taking into account that the operator A for the process Y_t coincides with the differential operator

$$L_k = \frac{1}{2T_k(\overline{H})}\frac{d}{dH}\left(a_k(\overline{H})\frac{d}{dH}\right), \qquad a_k(\overline{H}) = \oint_{C_k(\overline{H})} |\nabla H(x)|\, d\ell,$$

inside any edge I_k of the graph Γ, we conclude from (7.3) that the measure μ is invariant for the process Y_t only if

$$\alpha Au(O_0) = \frac{1}{2}\sum_{k=1}^{N}(\pm 1)a_k(0)\frac{du}{dH}(k, O_0). \tag{7.4}$$

Here $(du/dH)(k, O_0)$ means the derivative of the function $u(z)$, $z \in \Gamma$ along the kth edge $I_k \sim O_0$ at the point O_0. Sign "+" should be taken in the kth term if $\overline{H}(x) < 0$ on I_k, and sign "−" if $\overline{H}(x) > 0$ on I_k. Equality (7.4) gives us the gluing condition at the vertex O_0.

3. Consider an oscillator with one degree of freedom,

$$\ddot{p}_t + f(p_t) = 0. \tag{7.5}$$

Let the force $f(p)$ be a smooth generic function, $F(p) = \int_0^p f(z)\, dz$. Assume that $\lim_{|p|\to\infty} F(p) = \infty$. One can introduce the Hamiltonian $H(p, q) = \frac{1}{2}q^2 + F(p)$ of system (7.5) and rewrite (7.5) in the Hamiltonian form,

$$\begin{aligned}\dot{p}_t &= \frac{\partial H}{\partial_q}(p_t, q_t) \equiv q_t\\ \dot{q}_t &= -\frac{\partial H}{\partial_q}(p_t, q_t) \equiv -f(p_t).\end{aligned} \tag{7.6}$$

If we denote by x the point (p, q) of the phase space, equations (7.6) have the form $\dot{x}_t = \overline{\nabla} H(x_t)$.

Consider now perturbations of (7.5) by white noise,

$$\ddot{\tilde{p}}_t^{\varepsilon} + f(\tilde{p}_t^{\varepsilon}) = \varepsilon \dot{\tilde{W}}_t. \tag{7.7}$$

Here $\tilde{W}_t$ is the Wiener process in R^1, $0 < \varepsilon \ll 1$. One can rewrite (7.7) as a system:

$$\begin{aligned} \dot{\tilde{p}}_t^{\varepsilon} &= \tilde{q}_t^{\varepsilon}, \\ \dot{\tilde{q}}_t^{\varepsilon} &= -f(\tilde{p}_t^{\varepsilon}) + \varepsilon \dot{\tilde{W}}_t. \end{aligned} \tag{7.8}$$

Now we have the degenerate process $(0, \varepsilon \tilde{W}_t)$ in R^2 as the perturbation. Although the general construction and the scheme of reasoning in this case are more or less the same as in the nondegenerate case, some additional estimates are needed, especially when we prove the Markov property and calculate the gluing conditions for the limiting process. On the other hand, we show that the characteristics of the limiting process, in the case under consideration, have simple geometric sense. Here we follow Freidlin and Weber [1].

As in the case of nondegenerate perturbations, the process $(\tilde{p}_t^{\varepsilon}, \tilde{q}_t^{\varepsilon})$ in R^2 defined by (7.8) has, roughly speaking, fast and slow components. The fast component corresponds to the motion along the nonperturbed periodic trajectories, and the slow component corresponds to the motion transversal to the deterministic trajectories. The fast component can be characterized by the invariant density of the nonperturbed system on the corresponding periodic trajectory.

To describe the slow component, it is convenient to rescale the time: put $p_t^{\varepsilon} = \tilde{p}_{t/\varepsilon^2}^{\varepsilon}$, $q_t^{\varepsilon} = \tilde{q}_{t/\varepsilon^2}^{\varepsilon}$. Then the function, p_t^{ε}, q_t^{ε} satisfies the equations

$$\begin{aligned} \dot{p}_t^{\varepsilon} &= \frac{1}{\varepsilon^2} q_t^{\varepsilon}, \\ \dot{q}_t^{\varepsilon} &= -\frac{1}{\varepsilon^2} f(p_t^{\varepsilon}) + \dot{W}_t, \end{aligned} \tag{7.9}$$

where W_t is a Wiener process in R^1.

First, let us consider the case of the force $f(p)$ having just one zero, say at $p = 0$. Then the slow component of the process defined by (7.9) can be characterized by how $H(p_t^{\varepsilon}, q_t^{\varepsilon})$ changes. Using the Itô formula, we have

$$H(p_t^{\varepsilon}, q_t^{\varepsilon}) - H(p_0, q_0) = \int_0^t H_q'(p_s^{\varepsilon}, q_s^{\varepsilon})\, dW_s + \frac{1}{2} \int_0^t H_{qq}''(p_s^{\varepsilon}, p_s^{\varepsilon})\, ds.$$

We made use of the orthogonality of $\nabla H(p,q)$ and $\bar{\nabla} H(p,q)$. Applying the standard averaging procedure with respect to the invariant density of the fast motion with the frozen slow component, one can easily prove that for any $T > 0$ the processes $Y_t^\varepsilon = H(p_t^\varepsilon, q_t^\varepsilon)$ converge weakly in the space of continuous functions on $[0, T]$ to the diffusion process Y_t on $[0, \infty)$, corresponding to the operator

$$L = \frac{1}{2} A(y) \frac{d^2}{dy^2} + B(y) \frac{d}{dy},$$
$$A(y) = \frac{1}{T(y)} \oint_{C(y)} \frac{H_q^2(p,q)\, d\ell}{|\nabla H(p,q)|}, \tag{7.10}$$
$$B(y) = \frac{1}{2T(y)} \oint_{C(y)} \frac{H_{qq}\, d\ell}{|\nabla H(p,q)|},$$

where $T(y) = \oint_{C(y)} (d\ell / |\nabla H(p,q)|)$ is the period of the nonperturbed oscillations with the energy $H(p,q) = y$. Here, as before, $C(y) = \{(p,q) \in R^2 : H(p,q) = y\}$; $d\ell$ is the length element on $C(y)$. The point 0 is inaccessible for the process Y_t on $[0, \infty)$. The proof of this convergence can be carried out in the same way as in the nondegenerate case.

Using the Gauss formula, one can check that

$$\frac{d}{dy}[A(y)T(y)] = B(y)T(y).$$

Thus,

$$L = \frac{1}{2T(y)} \frac{d}{dy} \left(a(y) \frac{d}{dy} \right),$$

where $a(y) = A(y)T(y)$.

Now, since $H(p,q) = \frac{1}{2}q^2 + F(p)$, the contour $C(y)$ can be described as follows.

$$C(y) = \{(p,q) \in R^2 : \alpha(y) \leq p \leq \beta(y), q = \pm \sqrt{2(y - F(p))}\}, \tag{7.11}$$

where $\alpha(y)$, $\beta(y)$ are the roots of theequation $F(p) = y$. Note, that since $F'(p) = f(p)$ has just one zero at $p = 0$, and $\lim_{|p| \to \infty} F(p) = \infty$, the equation $F(p) = y$ has exactly two roots for any $y > 0$. Using (7.11), we calculate:

$$d\ell = dp \sqrt{1 + \frac{f^2(p)}{2(y - F(p))}}; \qquad H_q^2(p,q) = q^2 = 2(y - F(p))$$

and $|\nabla H(p,q)| = f^2(p) + 2(y - F(p))$ for $(p,q) \in C(y)$. Thus,

$$a(y) = \oint_{C(y)} \frac{H_q^2(p,q)\,d\ell}{|\nabla H(p,q)|} = 2\int_{\alpha(y)}^{\beta(y)} dp\sqrt{2(y - F(p))} = S(y),$$

where $S(y)$ is the area of the domain bounded by $C(y)$. Taking into account that the period $\lambda(y) = dS(y)/dy$, one can rewrite the operator defined by (7.10) in the form

$$L = \frac{1}{2S'(y)}\frac{1}{dy}\left(S(y)\frac{d}{dy}\right). \tag{7.12}$$

Now let $f(p)$ have more than one zero. Then the level sets of the Hamiltonian $H(p,q) = \frac{1}{2}q^2 + F(p)$ may have more than one component. Let $\Gamma = \{O_1, \ldots, O_m; I_1, \ldots, I_n\}$ be the graph homeomorphic to the set of connected components of the level sets of $H(x,y)$, provided with the natural topology. Let $Y : R^2 \to \Gamma$ be the mapping introduced earlier in this chapter, $Y(x) = (k(x), H(x))$, where $k(x) \in \{1, \ldots, n\}$ is the number of the edge containing the point of Γ corresponding to the nonperturbed trajectory starting at $x \in R^2$. The pair (k, H) forms the coordinates on Γ.

If $f(p)$ has just one zero, the graph Γ is reduced to one edge $[0,\infty)$. But if the number of zeros is greater than one, the function $k(x)$ is not a constant, and $k(x)$ is an additional, independent of $H(x)$, first integral of the nonperturbed system. This implies that the process $H(p_t^\varepsilon, q_t^\varepsilon)$ in the case of many zeros no longer converges to a Markov process (if $f(p)$ has no special symmetries; but remember that we assume that $f(p)$ is generic). To have in the limit a Markov process, we should, as in the case of the nondegenerate perturbations, extend the phase space by inclusion of the additional first integral. In other words, we should consider the processes $Y_t^\varepsilon = Y(p_t^\varepsilon, q_t^\varepsilon)$ on the graph Γ related to the Hamiltonian $H(p,q)$.

To describe the limiting process on Γ, define a function $S(z)$, $z = (k,y) \in \Gamma$: put $S(k,y)$ equal to the area bounded by the component $C_k(y)$ of the level set $C(y)$, corresponding to the point $z = (k,y) \in \Gamma \backslash \{O_1, \ldots, O_m\}$. If $z = (k,y)$ is a vertex, corresponding to the extremum of $H(p,q)$ (exterior vertex), then $S(z) = 0$. If $z = O_\ell$ is a vertex corresponding to a saddle point of $H(x)$ (interior vertex), then $S(z)$ has a discontinuity at O_ℓ. Each interior vertex is an end for three edges, say, I_{i_1}, I_{i_2}, and I_{i_3}. It is easy to see that $\lim_{z\to O_\ell,\, z\in I_{i_r}} S(z) = S_{i_r}(O_\ell)$, $r = 1,2,3$ exist. These limits $S_{i_r}(O_\ell)$, in general, are different for $r = 1,2,3$, and one of them is equal to the sum of the other two.

Theorem 7.1 *Let $f(p) \in C^\infty$, $F(p) = \int_0^p f(s)\,ds$, $\lim_{|p|\to\infty} F(p) = \infty$. Assume that $f(p)$ has a finite number of simple zeros and the values of $F(p)$ at different* *critical points are different. Then for any $T > 0$, the process $Z_t^\varepsilon = Y(p_t^\varepsilon, q_t^\varepsilon)$ on Γ converges weakly in the space of continuous functions on $[0, T]$ with values in Γ to a Markov diffusion process Z_t as $\varepsilon \downarrow 0$. The limiting process Z_t is governed by the operator*

$$L_k = \frac{1}{2S_y'(k,y)} \frac{d}{dy}\left(S(k,y)\frac{d}{dy}\right)$$

inside the edge $I_k \subset \Gamma$, $k = 1, \ldots, n$, and by the gluing conditions at the vertices: a bounded continuous on Γ and smooth inside the edges function $g(z)$, $z \in \Gamma$, belongs to the domain of definition of the generator A of the limiting process Z_t if and only if Ag is continuous on Γ, and at any interior (corresponding to a saddle point of $H(p,q)$) vertex O_ℓ of Γ, the following equality holds.

$$S_{i_1}(O_\ell)\frac{d_{i_1}g}{dy}(O_\ell) + S_{i_2}(O_\ell)\frac{d_{i_2}g}{dy}(O_\ell) = S_{i_3}(O_\ell)\frac{d_{i_3}g}{dy}(O_\ell).$$

Here $S_{i_k}(O_\ell) = \lim_{\substack{z \to O_\ell \\ z \in I_{i_k}}} S(z)$, $I_{i_k} \sim O_\ell$, $k = 1, 2, 3$; we assume that $H(Y^{-1}(z)) < H(Y^{-1}(O_\ell))$ or $H(Y^{-1}(z)) > H(Y^{-1}(O_\ell))$ simultaneously for $z \in I_{i_1} \cup I_{i_2}$.

The operators L_k, $k = 1, \ldots, n$ and the gluing conditions define the limiting process Z_t in a unique way.

The proof of this theorem is carried out, in general, using the same scheme as in the case of nondegenerate perturbations. The most essential difference arises in the proof of the Markov property (step 4) in the plan described in §8.1). We use here a priori bounds of Hörmander type instead of Krylov and Safonov [1]. The detailed proof of Theorem 7.1 can be found in Freidlin and Weber [1]. The degenerate perturbations of a general Hamiltonian system with one degree of freedom are considered in that paper as well.

4. Consider the diffusion process (X_t, P_x) in R^r corresponding to the operator

$$L = \frac{1}{2}\sum_{i,j=1}^{r} A^{ij}(x)\frac{\partial^2}{\partial x^i \partial x^j} + \sum_{i=1}^{r} B^i(x)\frac{\partial}{\partial x^i}.$$

We say that a function $H(x)$ is a first integral for the process (X_t, P_x), if

$$P_x\{H(X_t) = H(x)\} = 1, \quad x \in R^r.$$

If the function $H(x)$, $x \in R^r$, is smooth, then $H(x)$ is a first integral for the process X_t if and only if for any $x \in R^r$,

$$\sum_{i,j=1}^{r} A^{ij}(x) \frac{\partial H}{\partial x^i}(x) \frac{\partial H}{\partial x^j}(x) = 0, \quad LH(x) = 0. \tag{7.13}$$

This follows immediately from the Itô formula. Of course, just degenerate diffusion processes can have a nontrivial first integral. If all diffusion coefficients are equal to zero, the process turns into a dynamical system defined by the vector field $B(x) = (B^1(x), \dots, B^r(x))$, and conditions (7.13) into equality $B(x) \cdot \nabla H(x) = 0$, $x \in R^r$.

Trajectories of the process governed by the operator L can be described by the equation

$$dX_t = \sigma_0(X_t)\,dW_t + B(X_t)\,dt, \quad X_0 = x \in R^r,$$

where $\sigma_0(x)\sigma_0^*(x) = (A^{ij}(x))$ and W_t is the Wiener process. Consider now random perturbations of the process X_t:

$$d\tilde{X}_t^\varepsilon = \sigma_0(\tilde{X}_t^\varepsilon)\,dW_t + B(X_t)\,dt + \varepsilon\sigma_1(\tilde{X}_t^\varepsilon)\,d\tilde{W}_t + \varepsilon^2 b(\tilde{X}_t^\varepsilon)\,dt;$$

here $\sigma_1(x)\sigma_1^*(x) = (a^{ij}(x))$, $b(x) = (b^1(x), \dots, b^r(x))$, $\tilde{W}_t$ is a Wiener process independent of W_t. Let us rescale the time: $X_t^\varepsilon = \tilde{X}_{t/\varepsilon^2}^\varepsilon$. The generator of the new process X_t^ε is

$$L^\varepsilon = \frac{1}{\varepsilon^2} L + L_1, \quad L_1 = \frac{1}{2} \sum_{i,j=1}^{r} a^{ij}(x) \frac{\partial^2}{\partial x^i \partial x^j} + \sum_{i=1}^{r} b^i(x) \frac{\partial}{\partial x^i}.$$

Assume that the operator L_1 is uniformly elliptic with smooth bounded coefficients. Suppose the nonperturbed process X_t has a smooth first integral $H(x)$. Assume that $H(x)$ is a generic function. It has a finite number of nondegenerate critical points, and each level set contains at most one critical point. Moreover, let $\lim_{|x|\to\infty} H(x) = \infty$, $\min_{x \in R^2} H(x) = 0$. Let $C(y) = \{x \in R^2 : H(x) = y\}$, $y \geq 0$. The set $C(y)$ is compact and consists of a finite number $n(y)$ of connected components $C_k(y)$, $C(y) = \bigcup_{k=1}^{n(y)} C_k(y)$.

The process X_t is degenerate since it has a nontrivial first integral. But we assume that $\sum_{i,j=1}^{r} A^{ij}(x)e_i e_j \geq a_k(y)|e|^2$ for any $e = (e_1, \dots, e_r)$, $e \cdot \nabla H(x) = 0$, $x \in C_k(y)$, with some $a_k(y) > 0$, if $C_k(y)$ has no critical points. The last assumption means that the process X_t is not degenerate if considered on a nonsingular manifold $C_k(y)$. The manifold $C_k(y)$ is compact. Thus, the process X_t has on any nonsingular $C_k(y)$ a unique invariant density $M_y^k(x)$. If $C_k(y)$ contains a critical point O, we assume that the process X_t considered on $C_k(y)$ has just one invariant probability measure concentrated at the point O.

One can introduce fast and slow components of the perturbed process (X_t^ε, P_x): the fast motion along the level sets of the first integral and the slow motion in the transversal direction. The slow motion can be described, at least in a neighborhood of a nonsignular level set $C_k(y)$, by the evolution of $H(X_t^\varepsilon)$. The fast component near the manifold $C_k(y)$ has a distribution close to the distribution with the density $M_y^k(y)$ if $0 < \varepsilon \ll 1$.

To describe the evolution of the slow component, let us apply the Itô formula to $H(X_t^\varepsilon)$:

$$H(X_t^\varepsilon) - H(x) = \frac{1}{\varepsilon}\int_0^t \nabla H(X_s^\varepsilon)\cdot\sigma_0(X_s^\varepsilon)\,dW_s + \frac{1}{\varepsilon^2}\int_0^t LH(X_s^\varepsilon)\,ds + \int_0^t \nabla H(X_s^\varepsilon)\cdot\sigma_1(X_s^\varepsilon)\,d\tilde{W}_s + \int_0^t L_1H(X_s^\varepsilon)\,ds. \tag{7.14}$$

Since $H(x)$ is a smooth first integral, condition (7.13) is fulfilled. Therefore, the first two integrals on the right-hand side of (7.14) having large factors vanish:

$$H(X_t^\varepsilon) - H(x) = \int_0^t \nabla H(X_s^\varepsilon)\cdot\sigma_1(X_s^\varepsilon)\,d\tilde{W}_s + \int_0^t L_1H(X_s^\varepsilon)\,ds. \tag{7.15}$$

Consider now the graph Γ homeomorphic to the set of connected components of the level sets of the function $H(x)$. Let Γ consist of the edges $I_1, \dots, I_n$ and vertices $O_1, \dots, O_m$. Let $Y : R^r \to \Gamma$ be the mapping defined earlier in this chapter in the case $r = 2$: $Y(x)$ is the point of Γ corresponding to the connected component of $C(H(x))$ containing the point $x \in R^r$; $Y(x) = (k(x), H(x))$, where (k, H) are the coordinates on Γ. The function $k(x)$, as well as $H(x)$, is a first integral for the process X_t.

Consider the random processes $Y_t^\varepsilon = Y(X_t^\varepsilon)$ on Γ, $\varepsilon > 0$. These processes converge weakly to a diffusion process Y_t on Γ as $\varepsilon \downarrow 0$. To calculate the characteristics of the limiting process, consider, first, an interior point (y, k) of the edge $I_k \subset \Gamma$. Let $C_k(y)$ be the corresponding level set component. Then, using (7.15) and the fact that the process X_t on $C_k(y)$ is ergodic and $M_y^k(x)$ is its limiting density, one can prove that the process Y_t inside I_k is governed by the operator

$$\bar{L}_k = \frac{1}{2}\bar{a}_k(y)\frac{d^2}{dy^2} + \bar{b}_k(y)\frac{d}{dy},$$

$$\bar{a}_k(y) = \oint_{C_k(y)} \sum_{i,j=1}^r a^{ij}(x)\frac{\partial H(x)}{\partial x^i}\frac{\partial H(x)}{\partial x^j} M_y^k(x)\,dx,$$

$$\bar{b}_k(y) = \oint_{C_k(y)} L_1H(x)M_y^k(x)\,dx.$$

To determine the limiting process Y_t for all $t \geq 0$, the behavior of the process after touching the vertices should be described. One can prove that the exterior vertices are inaccessible. To calculate the gluing conditions at the interior vertices, one can, sometimes, use the same approach as we used before. Assume that the processes in R^r corresponding to the operators L and L_1 have the same invariant density $M(x)$. Then the density $M(x)$ is invariant for the process (X_t^ε, P_x) for any $\varepsilon > 0$. Define the measure $\bar{\mu}$ on Γ as the projection on Γ of the measure μ, $\mu(D) = \int_D M(x)\,dx$, $D \subset R^r$:

$$\bar{\mu}(\gamma) = \mu(Y^{-1}(\gamma));$$

γ is a Borel set in Γ. Then the measure $\bar{\mu}$ will be invariant for the processes Y_t^ε, $\varepsilon > 0$, and for the limiting process Y_t. Now, if we know that Y_t is a continuous Markov process on Γ, the gluing conditions at a vertex O_k are described by a set of nonnegative constants $\alpha_k, \beta_{k1}, \dots, \beta_{k\ell}$, where ℓ is the number of edges connected with O_k. These constants can be chosen in a unique way so that the process Y_t has the prescribed invariant measure.

For example, one can use such an approach if

$$L = \hat{L} + B(x) \cdot \nabla, \quad L_1 = \hat{L}_1 + b(x) \cdot \nabla,$$

where $\hat{L}$ and $\hat{L}_1$ are self-adjoint operators and the vector fields $B(x)$ and $b(x)$ are divergence free. The Lebesgue measure is invariant for L and L_1 in this case.

Note that if the process (X_t, P_x) degenerates in a dynamical system, to repeat our construction and arguments, we should assume that the dynamical system has a unique invariant measure on each connected component $C_k(y)$. This assumption, in the case when the dimension of the manifolds $C_k(y)$ is greater than 1, is rather restrictive. One can expect that the convergence to a diffusion process on the graph still holds under a less restrictive assumption, namely, that the dynamical system on $C_k(y)$ has a unique invariant measure stable with respect to small random perturbations.

5. The nonperturbed process (X_t, P_x) (or the dynamical system) in R^r governed by the operator L may have several smooth first integrals $H_1(x), \dots, H_\ell(x)$; $\ell < r$. Let

$$C(y) = \{x \in R^r : H_1(x) = y_1, \dots, H_\ell(x) = y_\ell\}, \quad y = (y_1, \dots, y_\ell) \in R^\ell.$$

The set $C(y)$ may be empty for some $y \in R^\ell$. Let $C(y)$ consist of $n(y)$ connected components: $C(y) = \bigcup_{k=1}^{n(y)} C_k(y)$. Assume that at least one of the first integrals, say $H_1(x)$, tends to $+\infty$ as $|x| \to \infty$. Then each $C_k(y)$ is compact.

A point $x \in R^r$ is called nonsingular if the matrix $(\partial H_i(x)/\partial x^j)$, $1 \leq i \leq \ell$, $1 \leq j \leq r$, has the maximal rank ℓ. Assume that the nonperturbed process at a nonsingular point $x \in R^r$ is nondegenerate, if considered on the manifold $C(H_1(x), \ldots, H_\ell(x))$. This means that

$$\sum_{i,j} A^{ij}(x) e_i e_j \geq a(x)|e|^2, \qquad a(x) > 0,$$

for any $e = (e_1, \ldots, e_r)$ such that $e \cdot \nabla H_k(x) = 0$, $k = 1, \ldots, \ell$. Then, the diffusion process X_t on each $C_k(y)$, consisting of nonsingular points, has a unique invariant measure. Let $M_y^k(x)$, $y \in R^\ell$, $y \in C_k(y)$, be the density of that measure.

The collection of all connected components of the level sets $C(y)$, $y \in R^\ell$, provided with the natural topology is homeomorphic to a set Γ consisting of glued ℓ-dimensional pieces. The interior points of these pieces correspond to the nonsingular components $C_k(y)$.

Define the mapping $Y : R^r \mapsto \Gamma : Y(x)$, $x \in R^r$, is the point of Γ corresponding to the connected component of $C(H_1(x), \ldots, H_\ell(x))$ containing x. One can expect that the stochastic processes $Y_t^\varepsilon = Y(X_t^\varepsilon)$, where X_t^ε corresponds to $L^\varepsilon = (1/\varepsilon^2)L + L_1$, converge weakly (in the space of continuous functions on $[0, T]$ with values in Γ, provided with the uniform topology) to a diffusion process Y_t on Γ as $\varepsilon \to 0$.

Inside an ℓ-dimensional piece $\gamma_k \subset \Gamma$, the matrix $(\partial H_i(x)/\partial x^j)$ has rank ℓ, and the values of the first integrals $H_1(x), \ldots, H_\ell(x)$ can be used as coordinates. The process Y_t in these coordinates is governed by an operator

$$\bar{L}_k = \frac{1}{2} \sum_{i,j=1}^{\ell} \bar{a}_k^{ij}(y) \frac{\partial^2}{\partial y^i \partial y^j} + \sum_{i=1}^{\ell} \bar{b}_k^i(y) \frac{\partial}{\partial y^i}.$$

The coefficients $\bar{a}_k^{ij}(y)$ and $\bar{b}_k^i(y)$ can be calculated by the averaging procedure with respect to the density $M_y^k(x)$:

$$\bar{a}_k^{ij}(y) = \oint_{C_k(y)} \sum_{m,n=1}^{r} A^{mn}(x) \frac{\partial H_i(x)}{\partial x^m} \frac{\partial H_j(x)}{\partial x^n} M_y^k(x)\, dx,$$

$$\bar{b}_k^i(y) = \oint_{C_k(y)} L H_i(x) M_y^k(x)\, dx.$$

To determine the limiting process, Y_t for all $t \geq 0$, one should supplement the operators $\bar{L}_k$ governing the process inside the ℓ-dimensional pieces, by the gluing conditions in the places where several ℓ-dimensional pieces are glued together. These gluing conditions should be, in a sense, similar to the boundary conditions for the multidimensional process (see Wentzell [8]).

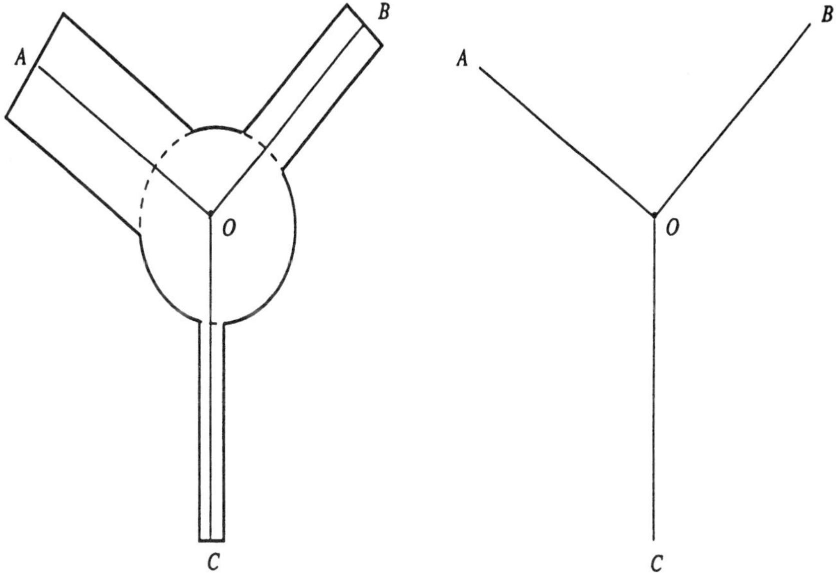

Figure 29.

6. In the conclusion of this section, we would like to mention one more asymptotic problem leading to a process on a graph.

Consider the Wiener process X_t^ε in the domain $G^\varepsilon \subset R^r$, shown in Fig. 29(a), with the normal reflection on the boundary. The domain G^ε consists of a ball of the radius ρ^ε and of three cylinders of radii εr_1, εr_2, εr_3, respectively. Let the axes of these cylinders intersect at center O of the ball. It is natural to expect that the projection of the process X_t^ε on the "skeleton" Γ of the domain G^ε, shown in Fig. 29(b), converges to a stochastic process on Γ, as $\varepsilon \downarrow 0$. One can check that, inside each edge of the graph, the limiting process will be the one-dimensional Wiener process. Assuming that near the points A, B, C, the boundary of G^ε consists of a piece of the plane orthogonal to the axis of the corresponding cylinder, it is easy to see that the limiting process has instantaneous reflection at the points $A, B, C \in \Gamma$. To determine the limiting process for all $t \geq 0$, we should describe its behavior at the vertex O.

The gluing condition at O depends on the relation between ρ^ε and ε. If $\varepsilon \max_k r_k \leq \rho^\varepsilon \ll \varepsilon^{(r-1)/r}$, the gluing condition has the form

$$\sum_{i=1}^{3} r_i^{r-1} \frac{du}{dy^i}(0) = 0, \tag{7.16}$$

where d/dy^i means the differentiation along the ith edge and y_i is the distance from O. Condition (7.16) means that the limiting process spends time zero at the vertex. If $\rho^\varepsilon \gg \varepsilon^{(r-1)/r}$, then the point O is a trap for the process: after touching O, the trajectory stays there forever. The gluing condition in this case is

$$u''(0) = 0,$$

independently of the edge along which this derivative is calculated.

If $\lim_{\varepsilon \downarrow 0} \rho^\varepsilon \varepsilon^{(r-1)/r} = \kappa$, then the domain of the generator of the limiting process consists of continuous functions $u(x)$, $x \in \Gamma$, for which $u''(x)$ is also continuous and

$$\sum_{i=1}^{3} r_i^{r-1} \frac{du}{dy_i}(0) = \frac{\kappa^r \Gamma(\frac{1}{2}) \Gamma\left(\frac{r+1}{2}\right)}{\Gamma\left(\frac{r+2}{2}\right)} u''(0).$$

The limiting process in this case spends a positive time at O.

This problem and some other problems leading to the processes on graphs were considered in Freidlin and Wentzell [4] in Freidlin [21].

Chapter 9

Stability Under Random Perturbations

§1. Formulation of the Problem

In the theory of ordinary differential equations much work is devoted to the study of stability of solutions with respect to small perturbations of the initial conditions or of the right side of an equation. In this chapter we consider some problems concerning stability under random perturbations. First we recall the basic notions of classical stability theory. Let the dynamical system

$$\dot{x}_t = b(x_t) \tag{1.1}$$

in R^r have an equilibrium position at the point $O: b(O) = 0$.

The equilibrium position O is said to be stable (Lyapunov stable) if for every neighborhood U_1 of O there exists a neighborhood U_2 of O such that the solutions of equation (1.1) with initial condition $x_0 = x \in U_2$ do not leave U_1 for positive t. If, in addition, $\lim_{t\to\infty} x_t = O$ for trajectories issued from points $x_0 = x$ sufficiently close to O, then the equilibrium position O is said to be asymptotically stable.

With stability with respect to perturbations of the initial conditions there is closely connected the problem of stability under continuously acting perturbations. To clarify the meaning of this problem, along with equation (1.1), we consider the equation

$$\dot{\tilde{x}}_t = b(\tilde{x}_t) + \zeta_t, \tag{1.2}$$

where ζ_t is a bounded continuous function on the half-line $[0, \infty)$ with values in R^r. The problem of stability under continuously acting perturbations can be formulated in the following way: under what conditions on the field $b(x)$ does the solution of problem (1.2) with initial condition $\tilde{x}_0 = x$ converge uniformly on $[0, \infty)$ to the constant solution $\tilde{x}_t \equiv O$, as $|x - O| + \sup_{0 \le t < \infty} |\zeta_t| \to 0$. It can be proved that if the equilibrium position is stable, in a sufficiently strong sense, with respect to small perturbations of the initial conditions, then it is also stable under continuously acting perturbations.

For example, if the equilibrium position is asymptotically stable, i.e.,

$$\lim_{t \to \infty} x_t = O$$

uniformly in $x_0 = x$ belonging to some neighborhood of O, then O is also stable with respect to small continuously acting perturbations (cf. Malkin [1]).

The situation changes essentially if we omit the condition of boundedness of ζ_t on $[0, \infty)$. In this case the solutions of equation (1.2) may, in general, leave any neighborhood of the equilibrium position even if the equilibrium position is stable with respect to perturbations of the initial conditions in the strongest sense. Actually, the very notion of smallness of continuously acting perturbations needs to be adjusted in this case.

Now let ζ_t in (1.2) be a random process: $\zeta_t = \zeta_t(\omega)$. If the perturbations $\zeta_t(\omega)$ are uniformly small in probability, i.e., $\sup_{0 \le t < \infty} |\zeta_t(\omega)| \to 0$ in probability, then the situation is not different from the deterministic case: if the point O is stable for system (1.1) in a sufficiently strong sense, then as

$$\sup_{0 \le t < \infty} \zeta_t(\omega)$$

converges to zero in probability and the initial condition $x_0 = x$ converges to O, we have: $\mathsf{P}\{\sup_{0 \le t < \infty} |x_t - O| \ge \delta\} \to 0$ for any $\delta > 0$. Nevertheless, the assumption that ζ_t converges to zero uniformly on the whole half-line $[0, \infty)$ is too stringent in a majority of problems. An assumption of the following kind is more natural: $\sup_t \mathsf{M}|\zeta_t|^2$ or some other characteristic of the process ζ_t converges to zero, which makes large values of $|\zeta_t|$ unlikely at every fixed time t but allows the functions $|\zeta_t(\omega)|$ to assume large values at some moments of time depending on ω. Under assumptions of this kind, the trajectories of the process $\tilde{x}_t$ may, in general, leave any neighborhood of the equilibrium position sooner or later. For example, if ζ_t is a stationary Gaussian process, then the trajectories of $\tilde{x}_t$ have arbitrarily large deviations from the equilibrium position with probability one regardless of how small $\mathsf{M}|\zeta_t|^2 = \alpha \neq 0$ is.

To clarify the character of problems, involving stability under random perturbations, to be discussed in this chapter, we consider an object whose state can be described by a point $x \in R^r$. We assume that in the absence of random perturbations, the evolution of this object can be described by equation (1.1), and random perturbations $\zeta_t(\omega)$ lead to an evolution which can be described by the equation

$$\dot{X}_t = b(X_t, \zeta_t). \tag{1.3}$$

As ζ_t we also allow some generalized random processes for which equation (1.3) is meaningful (there exists a unique solution of equation (1.3) for any initial point $x_0 = x \in R^r$). For example, we shall consider the case where $b(x, y) = b(x) + \sigma(x)y$ and ζ_t is a white noise process or the derivative of a Poisson process.

We assume that there exists a domain $D \subset R^r$ such that as long as the phase point characterizing the state of our object belongs to D, the object does not undergo essential changes and when the phase point leaves D, the object gets destroyed. Such a domain D will be called a *critical domain.*

Let X_t^x be the solution of equation (1.3) with initial condition $X_0^x = x$. We introduce the random variable $\tau^x = \min\{t: X_t^x \notin D\}$—the time elapsed until the destruction of the object. As the measure of stability of the system with respect to random perturbations ζ_t relative to the domain D, it is natural to choose various theoretical probability characteristics of τ^x).[1] Hence if we are interested in our object only over a time interval $[0, T]$, then as the measure of instability of the system, we may choose $\mathsf{P}\{\tau^x \leq T\}$. If the time interval is not given beforehand, then stability can be characterized by $\mathsf{M}\tau^x$. In cases where the sojourn of the phase point outside D does not lead to destruction of the object but is only undesirable, as the measure of stability, we may choose the value of the invariant measure of the process for the complement of D (if there exists an invariant measure). This measure will characterize the ratio of the time spent by the trajectory X_t^x outside D.

However, a precise calculation of these theoretical probability characteristics is possible only rarely, mainly when X_t^x turns out to be a Markov process or a component of a Markov process in a higher dimensional space. As is known, in the Markov case the probabilities and mean values under consideration are solutions of some boundary value problems for appropriate equations. Even in those cases where equations and boundary conditions can be written down, it is not at all simple to obtain useful information from the equations. So it is of interest to determine various kinds of asymptotics of the mentioned theoretical probability characteristics as one parameter or another, occurring in the equation, converges to zero.

In a wide class of problems we may assume that the intensity of the noise is small in some sense compared with the deterministic factors determining the evolution of the system. Consequently, a small parameter appears in the problem. If O is an asymptotically stable equilibrium position of the unperturbed system, then exit from a domain D takes place due to small random perturbations, in spite of the deterministic constituents. As we know, in such situations estimates of $\mathsf{P}\{\tau^x \leq T\}$ and $\mathsf{M}\tau^x$ are given by means of an action functional. Although in this way we can only calculate the rough, logarithmic, asymptotics, this is usually sufficient in many problems. For example, the logarithmic asymptotics enables us to compare various critical

[1] We note that τ^x may turn out to be equal to ∞ with positive probability. If the random perturbations vanish at the equilibrium position O itself, then $\mathsf{P}\{\tau^x < \infty\}$ may converge to 0 as $x \to O$. For differential equations with perturbations vanishing as the equilibrium position is approached, a stability theory close to the classical theory can be created (cf. Khas'minskii [1]).

domains, various vector fields $b(x, y)$ and also enables us to solve some problems of optimal stabilization.

This approach is developed in the article by Wentzell and Freidlin [5].

We introduce several different formulations of stability problems and outline methods of their solution.

Let X_t^h, $t \geq 0$, be the family of random processes in R^r, obtained as a result of small random perturbations of system (1.1); the probabilities and mathematical expectations corresponding to a given value of the parameter and a given initial point will be denoted by P_x^h, M_x^h, respectively. Let $\lambda(h)S_{0T}(\varphi)$ be the action functional for the family of processes X_t^h with respect to the metric $\rho_{0T}(\varphi, \psi) = \sup_{0 \leq t \leq T} |\varphi_t - \psi_t|$ as $h \downarrow 0$. It is clear that S_{0T} vanishes for trajectories of system (1.1) and only for them.

Let O be a stable equilibrium position of system (1.1), let D be a domain containing O, and let $\tau_D = \inf\{t: X_t^h \notin D\}$ be the time of first exit of the process from D. We are interested in the stability of the system on a finite time interval $[0, T]$. We shall characterize stability by the asymptotics of $\mathsf{P}_x^h\{\tau_D \leq T\}$ as $h \downarrow 0$.

It is appropriate to introduce the following *measure of stability* of our system with respect to the given random perturbations and domain D:

$$V_D^{T,x} = \inf\left\{S_{0T}(\varphi): \varphi \in \bigcup_{0 \leq t \leq T} \bigcup_{y \notin D} H_{xy}(t)\right\},$$

where the set $H_{xy}(t)$ consists of all functions φ_s defined for $s \in [0, T]$, such that $\varphi_0 = x$ and $\varphi_t = y$. The sense of this measure of stability is the following: if the infima of S_{0T} over the closure and interior of $\bigcup_{0 \leq t \leq T} \bigcup_{y \notin D} H_{xy}(t)$ coincide, then $\mathsf{P}_x^h\{\tau_D \leq T\}$ is logarithmically equivalent to

$$\exp\{-\lambda(h) V_D^{t,x}\}.$$

We note that for the coincidence of the infima over the closure and the interior it is sufficient (in the case of a functional S_{0T} of the form considered in Chs. 4–5) that D coincide with the interior of its closure.

The measure of stability $V_D^{T,x}$ is given by $\inf_{0 \leq t \leq T,\, y \notin D} u(t, x, y)$, where $u(t, x, y) = \inf\{S_{0T}(\varphi): \varphi \in H_{x,y}(t)\}$ can be determined from the Jacobi equation.

The problem in which there is no fixed time interval $[0, T]$ of observation is characterized by another measure of stability. We define μ_D as the infimum of the values of the functional S_{0T} of the functions φ_t defined on intervals $[0, T]$ of any length, such that $\varphi_0 = O$, $\varphi_T \notin D$. We may calculate μ_D as the infimum, over $y \notin D$, of the quasipotential

$$V(O, y) = \inf\{S_{0T}(\varphi): \varphi_0 = O, \varphi_T = y; 0 < T < \infty\},$$

which can be calculated (Theorem 4.3 of Ch. 5) as the solution of an appropriate problem for a partial differential equation of the first order. According to results of Chs. 4–7, under suitable assumptions on the processes X_t^h and the domain D, the mean exit time $\mathsf{M}_x^h \tau_D$ of D is logarithmically equivalent to $\exp\{\lambda(h)\mu_D\}$ for all points x belonging to D for which the trajectory of system (1.1) issued from x converges to the equilibrium position O, without leaving D (Theorems 4.1 of Ch. 4, 5.3 of Ch. 6 and 6.1 of Ch. 7). In this case the mathematical expectation represents a typical value of the exit time to within logarithmic equivalence. Namely, for any $\gamma > 0$ we have

$$\lim_{h \downarrow 0} \mathsf{P}_x^h\{\exp\{\lambda(h)[\mu_D - \gamma]\} < \tau_D < \exp\{\lambda(h)[\mu_D + \gamma]\}\} = 1$$

(Theorem 4.2 of Ch. 4). Further, the value of the normalized invariant measure of X_t^h for the set $R^r \setminus D$ is logarithmically equivalent to

$$\exp\{-\lambda(h)\mu_D\}$$

(Theorem 4.3 of Ch. 4 and 4.2 of Ch. 5). Consequently, if the time interval is not fixed beforehand, then the constant μ_D is, in some sense, a universal measure of stability for perturbations and critical domain of the kinds being considered. If the critical domain is not given, then such a universal characteristic of stability of the equilibrium position is the quasipotential $V(O, y)$ of random perturbations.

The "most dangerous point" on the boundary of the critical domain can be expressed in terms of $V(O, y)$: under certain assumptions, the "destruction" of the object takes place, with overwhelming probability for small h, near the points $y \in \partial D$ where $V(O, y)$ attains its infimum over ∂D.

Now we consider the problem of selecting an optimal critical domain. We assume that for domains D containing the equilibrium position O of the unperturbed system, a monotone functional $H(D)$ is defined: for the sake of definiteness, we assume that this functional has the form $\int_D h(x)\,dx$, where $h(x)$ is a positive function. From the domains D with a given value H_0 of $H(D)$ we try to select one with the smallest probability of exit from the domain over a given time T or with the largest mathematical expectation of the exit time. The optimal critical domain depends on h in general. We shall seek an *asymptotic* solution of the problem, i.e., we shall try to construct a domain which is better than any other domain (independent of h) for sufficiently small h.

It is clear that the problem is reduced to maximization of the corresponding measure of stability, $V_D^{T,x}$ or μ_D. It is easy to see that it has to be solved in the following way: we choose domains D_c of the form

$$\left\{y: \inf_{0 \le t \le T} u(t, x, y) < c\right\}$$

(or $\{y: V(O, y) < c\}$, respectively); from the increasing family of these domains we choose that one for which $H(D_c) = H_0$. If the function

$$\inf_{0 \le t \le T} u(t, x, y)$$

(or $V(O, y)$) is smooth in y, then the good properties of D_c are guaranteed. Any domain whose boundary is not a level surface of the function

$$\inf_{0 \le t \le T} u(t, x, y)$$

($V(O, y)$, respectively) can be made smaller with preservation of the value of $\inf_{y \notin D} \inf_{0 \le t \le T} u(t, x, y)$ (or $\inf_{y \notin D} V(O, y)$), and then be replaced by a larger domain D_c with the former value of H.

For example for the system $\dot{x}_t = Ax_t$ with a normal matrix A and perturbations of the type of a "white noise" (i.e., $\dot{X}_t^\varepsilon = AX_t^\varepsilon + \varepsilon w_t$), the optimal critical domain for the problem without a fixed time interval is an ellipsoid (cf. Example 3.2, Ch. 4).

We pass to problems of optimal stabilization.

We assume that the perturbed equation (1.3) contains a parameter (or several parameters) which can be controlled. A choice of the way of control of the process consists of a choice of the form of dependence of the controlling parameter on values of the controlled random process. We introduce the following restriction on the character of this dependence. To every form a of dependence of the controlling parameter on values of the process let there correspond a family of random processes $X_t^{a,h}$. For all of them we assume that there exists an action functional $\lambda(h)S^a(\varphi)$ in which the normalizing coefficient $\lambda(h)$ does not depend on the choice of control.

The solution of certain problems of optimal control of a process for small h is connected with the functional $S^a(\varphi) = S_{0T}^a(\varphi)$ and the quasipotential $V^a(x, y) = \inf\{S_{0T}^a(\varphi): \varphi_0 = x, \varphi_T = y; 0 \le T < \infty\}$. In particular, it is plausible that the problem of the choice of control maximizing the mean exit time of a domain D (asymptotically as $h \downarrow 0$) is connected with the function

$$\overline{V}(x, y) = \sup V^a(x, y), \tag{1.4}$$

where the supremum is taken over all admissible ways of control.

We restrict ourselves to controlled Markov processes, homogeneous in time. As admissible controls we shall consider those in which the value of the controlling parameter at a given time t is a definite function $a(X_t)$ of the value of the process at the same moment of time. Such controls lead to Markov processes $X_t^{a,h}$, homogeneous in time (cf. Krylov [2]); at every point x their local characteristics depend on x and on the value of the controlling parameter $a(x)$ at x.

Hence let the class of admissible controls consist of functions $a(x)$ whose values at every point x belong to the set $\Pi(x)$ of admissible controls at x (and which are subject to some regularity conditions). To every admissible function there corresponds a Markov process $(X_t^{a,h}, \mathsf{P}_x^{a,h})$ (the corresponding mathematical expectation, normalized action functional, and quasipotential will be denoted by $\mathsf{M}_x^{a,h}$, $S_{0T}^a(\varphi)$, and $V^a(x, y)$). We shall consider the following problem: we seek a function a which maximizes

$$\lim_{h \downarrow 0} \lambda(h)^{-1} \ln \mathsf{M}_x^{h,a} \tau_D.$$

As we have already mentioned, the solution of this problem is most likely connected with the function $\overline{V}(x, y)$ defined by formula (1.4). A series of questions arises here: How can the maximal quasipotential $\overline{V}(x, y)$ be determined? How can the optimal control problem be solved by means of it? In the following sections we consider these questions (for a certain class of controlled Markov processes) with examples.

§2. The Problem of Optimal Stabilization

The class of families of random processes for which we shall consider the problem is the family of locally infinitely divisible processes considered in Ch. 5. We assume that at every $x \in R^r$ we are given a set $\Pi(x) \subseteq R^l$. To every pair $x \in R^r$, $a \in \Pi(x)$ let there correspond: a vector $b(x, a) = (b^1(x, a), \ldots, b^r(x, a))$, a symmetric nonnegative definite matrix $(a^{ij}(x, a))$ of order r, and a measure $\mu(x, a, \cdot)$ on $R^r \setminus \{0\}$, such that $\int |\beta|^2 \mu(x, a, d\beta) < \infty$.

We define the process $(X_t^{h,a}, \mathsf{P}_x^{h,a})$ corresponding to the value h of the parameter (from $(0, \infty)$) and the choice $a(x)$ of the control function ($a(x)$ belongs to $\Pi(x)$ for every x) to be the Markov process with infinitesimal generator defined for twice continuously differentiable functions with compact support by the formula

$$\begin{aligned} A^{h,a} f(x) &= \sum b^i(x, a(x)) \frac{\partial f(x)}{\partial x^i} + \frac{h}{2} \sum a^{ij}(x, a(x)) \frac{\partial^2 f(x)}{\partial x^i \partial x^j} \\ &\quad + h^{-1} \int_{R^r \setminus \{0\}} \left[f(x + h\beta) - f(x) - h \sum \frac{\partial f(x)}{\partial x^i} \beta^i \right] \mu(x, a(x), d\beta). \end{aligned} \tag{2.1}$$

Of course, it may turn out that no Markov process corresponds to a given function $a(x)$ for some $h > 0$; in this case we shall consider the control function inadmissible. We denote by $\mathfrak{A}$ the class of admissible controls.

A large number of publications is devoted to the problem of finding conditions guaranteeing that to a given set of local characteristics there corresponds a locally infinitely divisible process; this problem has been studied especially extensively in the case of diffusion processes, i.e., where

$\mu \equiv 0$. For example, it is sufficient that the diffusion matrix be uniformly nondegenerate and continuous and the drift vector be bounded and measurable (cf. Krylov [1], Stroock and Varadhan [1]). The case of a μ, different from zero, has been considered in Komatsu [1], Stroock [1], Lepeltier and Marchal [1], etc.

For a varying choice of $a(x)$, the probabilities of unlikely events for $(X_t^{h,a}, \mathsf{P}_x^{h,a})$ are connected with the function of three variables

$$H(x, a, \alpha) = \sum b^i(x, a)\alpha_i + \frac{1}{2}\sum a^{ij}(x, a)\alpha_i\alpha_j + \int_{R^r \setminus \{0\}} [\exp\{\sum \beta^i \alpha_i\} - 1 - \sum \beta^i \alpha_i]\mu(x, a, d\beta), \quad (2.2)$$

which, together with its derivatives with respect to α_i, we shall assume to be finite and continuous in all arguments. Under the hypotheses of Theorem 2.1 of Ch. 5, the action functional for the family of processes $(X_t^{h,a}, \mathsf{P}_x^{h,a})$ as $h \downarrow 0$ is given by the formula

$$h^{-1}S_{0T}^{\alpha}(\varphi) = h^{-1}\int_0^T L(\varphi_t, a(\varphi_t), \dot{\varphi}_t)\, dt, \quad (2.3)$$

where $L(x, a, \beta)$ is the Legendre transform of $H(x, a, \alpha)$ in the third argument.

In Ch. 5 we required that the functions $H(x, \alpha)$, $L(x, \beta)$ be continuous in x. Therefore, it would be desirable to consider only continuous controlling functions $a(x)$. However, it is known that in problems of optimal control we can only rarely get along with continuous controls. It turns out that the necessary results can be carried over to the case where the controls are discontinuous at one point (or a finite number of points) but the results cannot be preserved in general if discontinuity occurs on a curve (or on a surface). Therefore, we introduce the following class of functions. Let $a(x)$ be defined on R^r with values in R^l. We shall write $a \in \mathbf{\Pi}$ if a is continuous for all $x \in R^r$ except, maybe, one point and for every x the value of a belongs to the set $\Pi(x)$ of admissible controls at x.

Theorems 2.1–2.3 for diffusion processes are contained in Wentzell and Freidlin [5].

Theorem 2.1. *Let τ_D be the time of exit of the process $X_t^{h,a}$ from a bounded domain D whose boundary coincides with the boundary of its closure. For any control function $a \in \mathbf{\Pi} \cap \mathfrak{A}$ we have*

$$\varlimsup_{h \downarrow 0} h \ln \mathsf{M}_x^{h,a}\tau_D \le V_0 = \max_{x_0 \in D} \min_{y \in \partial D} \sup_{a \in \mathbf{\Pi}} V^a(x_0, y) \quad (2.4)$$

uniformly in $x \in D$.

We note that V^a is in turn the infimum of the values of $S^a(\varphi)$ for functions leading from x_0 to y, so that the right side of (2.4) contains a quite complicated combination of maxima and minima: max min sup inf.

The proof of the theorem can be carried out in the following way. For any $\gamma > 0$ we choose $T > 0$ such that for any control function $a \in \mathbf{\Pi}$ and any $x_0 \in D$ there exists a function φ_t, $0 \leq t \leq T$, such that $\varphi_0 = x_0$, φ_t leaves D for some $t \in [0, T]$ and $S^a_{0T}(\varphi) \leq V_0 + \gamma$. Then we prove the lemma below.

Lemma 2.1. *For any $a \in \mathbf{\Pi} \cap \mathfrak{A}$ we have*

$$\mathsf{P}_x^{h,a}\{\tau_D \leq T + 1\} \geq \exp\{-h^{-1}(V_0 + 2\gamma)\} \tag{2.5}$$

for sufficiently small h and all $x \in D$.

The proof can be carried out as that of part (a) of Theorem 4.1 of Ch. 4: the function φ_t is extended beyond D; we use the lower estimate given by Theorem 2.1 of Ch. 5, of the probability of passing through a tube. The only difference is that the hypotheses of Theorem 2.1 of Ch. 5 are not satisfied in the neighborhood of the point x_* where $a(x)$ is discontinuous. Nevertheless, the corresponding segment of the function φ_t can be replaced by a straight line segment and the lower estimate gives the following lemma.

Lemma 2.2. *Let $(X_t^{h,a}, \mathsf{P}_x^{h,a})$ be a family of locally infinitely divisible processes with infinitesimal generator of the form (2.1) and let the corresponding function $H(x, a(x), \alpha)$, along with its first and second derivatives with respect to α, be bounded in any finite domain of variation of α (but it may be arbitrarily discontinuous as a function x). Then for any $\gamma > 0$ and any $\delta > 0$ there exists $\rho_0 > 0$ such that*

$$\mathsf{P}_x^{h,a}\{\rho_{0t_0}(X^{h,a}, \varphi) < \delta\} \geq \exp\{-\gamma h^{-1}\} \tag{2.6}$$

for sufficiently small $h > 0$, for all x, y such that $|y - x| \leq \rho_0$, where $\varphi_t = x + t[(y - x)/(|y - x|)]$, $0 \leq t \leq t_0 = |y - x|$.

The proof copies the corresponding part of the proof of Theorem 2.1 of Ch. 5 but without using the continuity of H or L.

After the proof of (2.5), the proof of Theorem 2.1 can be completed by applying the Markov property:

$$\mathsf{P}_x^{h,a}\{\tau_D > n(T + 1)\} \leq [1 - \exp\{-h^{-1}(V_0 + 2\gamma)\}]^n;$$

$$\begin{aligned}\mathsf{M}_x^{h,a}\tau_D &\leq (T + 1) \sum_{n=0}^{\infty} \mathsf{P}_x^{h,a}\{\tau_D > n(T + 1)\} \\ &\leq (T + 1)\exp\{h^{-1}(V_0 + 2\gamma)\}.\end{aligned}$$

Now we put

$$H(x, \alpha) = \inf_{a \in \Pi(x)} H(x, a, \alpha).$$

Theorem 2.2. *Let $\overline{V}(x)$ be the solution of problem $\mathbf{R}_{x_0}$ for the equation*

$$H(x, \nabla \overline{V}(x)) = 0 \tag{2.7}$$

in a domain D. For any function a belonging to $\mathbf{\Pi}$ (i.e., a is continuous everywhere except one point and $a(x) \in \mathbf{\Pi}(x)$ for every x) *for the quasipotential we have*

$$V^a(x_0, x) \le \overline{V}(x)$$

for all x in the set

$$B = \left\{x \in D \cup \partial D : \overline{V}(x) \le \inf_{y \in \partial D} \overline{V}(y)\right\}.$$

Moreover, suppose that there exists a function $a(x, \alpha)$, continuous on the set $\{(x, \alpha): x \ne x_0, \alpha \ne 0, H(x, \alpha) = 0\}$, such that $a(x, \alpha) \in \Pi(x)$ and

$$H(x, a(x, \alpha), \alpha) = H(x, \alpha) = 0.$$

Then

$$\sup_{a \in \mathbf{\Pi}} V^a(x_0, x) = \overline{V}(x)$$

for all $x \in B$ and the supremum is attained for the function $a(x) = a(x, \nabla \overline{V}(x))$.

Proof. Suppose that for some $a \in \mathbf{\Pi}$, $x \in B$ and $\varepsilon > 0$ we have

$$V^\alpha(x_0, x) > \varepsilon + (1 + \varepsilon)\overline{V}(x).$$

We may assume that $\varepsilon \ne \varepsilon_* = [V^a(x_0, x_*) - \overline{V}(x_*)]/[1 + \overline{V}(x_*)]$, where x_* is the point where $a(x)$ is discontinuous.

We consider the set $A = B \cap \{x: V^a(x_0, x) > \varepsilon + (1 + \varepsilon)\overline{V}(x)\}$; this set is open in B. We put $V_0 = \inf\{\overline{V}(x): x \in A\}$. The infimum V_0 is not attained; let x_∞ be a limit point of A on the level surface $\{x : \overline{V}(x) = V_0\}$. Let $x_1, \ldots,$ $x_n, \ldots$ be a sequence of points of A converging to c_∞ (Fig. 30). It is clear that the points of the surface $\{x: \overline{V}(x) = V_0\}$ do not belong to A and $\overline{V}(x_n) > V_0$. We note that by virtue of the choice of $\varepsilon \ne \varepsilon_*$, the point x_∞ does not coincide with the point x_* of discontinuity of a.

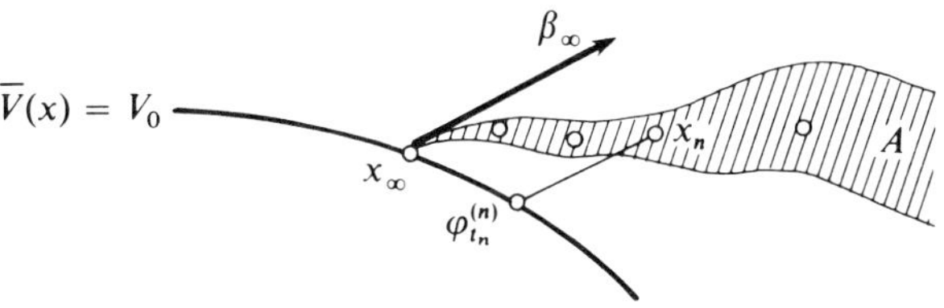

Figure 30.

We consider the vectors $\nabla \bar{V}(x_\infty)$ and $\beta_\infty = \nabla_\alpha H(x_\infty, a(x_\infty), \nabla \bar{V}(x_\infty))$. By virtue of the properties of the Legendre transformation,

$$(\nabla \bar{V}(x_\infty), \beta_\infty) = L(x_\infty, a(x_\infty), \beta_\infty) + H(x_\infty, a(x_\infty), \nabla \bar{V}(x_\infty)). \tag{2.8}$$

The function $L(x_\infty, a(x_\infty), \beta)$ is nonnegative everywhere and vanishes only at $\beta = \nabla_\alpha H(x_\infty, a(x_\infty), 0)$. On the other hand, $\nabla \bar{V}(x_\infty) \neq 0$ (because $\bar{V}$ is the solution of problem $\mathbf{R}_{x_0}$). Therefore, the first term in (2.8) is positive. The second term is not less than $H(x_\infty, \nabla \bar{V}(x_\infty)) = 0$. Therefore, the scalar product $(\nabla \bar{V}(x_\infty), \beta_\infty)$ is positive, i.e., the vector β_∞ is directed outside the surface $\{x: \bar{V}(x) = V_0\}$ at x_∞. By virtue of the continuity of $\nabla \bar{V}(x)$, the situation is the same at points close to x_∞.

For every point x_n we define the function

$$\varphi_t^{(n)} = x_n + t\beta_\infty, \qquad t \leq 0.$$

For x_n sufficiently close to x_∞ (i.e., for sufficiently large n), the straight line $\varphi_t^{(n)}$ intersects the surface $\{x: \bar{V}(x) = V_0\}$ for a small negative value t_n, where

$$|t_n| \sim \frac{\bar{V}(x_n) - V_0}{(\nabla \bar{V}(x_\infty), \beta_\infty)}$$

as $n \to \infty$. The denominator here is not less than $L(x_\infty, a(x_\infty), \beta_\infty)$, so that

$$|t_n| \leq \frac{\bar{V}(x_n) - V_0}{L(x_\infty, a(x_\infty), \beta_\infty)}(1 + o(1))$$

as $n \to \infty$.

We estimate the value of the functional S^a of the function $\varphi_t^{(n)}$ for $t_n \leq t \leq 0$:

$$\begin{aligned} S_{t_n 0}^a(\varphi^{(n)}) &= \int_{t_n}^0 L(\varphi_t^{(n)}, a(\varphi_t^{(n)}), \beta_\infty)\, dt \\ &\sim |t_n| L(x_\infty, a(x_\infty), \beta_\infty) \leq (\bar{V}(x_n) - V_0)(1 + o(1)) \end{aligned} \tag{2.9}$$

as $n \to \infty$. By virtue of the definition of the quasipotential V^a we have

$$V^a(x_0, x_n) \le V^a(x_0, \varphi_{t_n}^{(n)}) + S_{t_n 0}^a(\varphi^{(n)}).$$

The first term here does not exceed $\varepsilon + (1 + \varepsilon)\bar{V}(\varphi_{t_n}^{(n)}) = \varepsilon + (1 + \varepsilon)V_0$; the second term can be estimated by formula (2.9). Consequently, the inequality

$$V^a(x_0, x_n) < \varepsilon + (1 + \varepsilon)V_0 + (1 + \varepsilon)(\bar{V}(x_n) - V_0) = \varepsilon + (1 + \varepsilon)\bar{V}(x_n)$$

is satisfied for sufficiently large n. On the other hand, this contradicts the fact that $x_n \in A$, i.e., that $V^a(x_0, x_n) > \varepsilon + (1 + \varepsilon)\bar{V}(x_n)$. It follows that A is empty. The first part of the theorem is proved.

For the proof of the second part it is sufficient to apply Theorem 4.3 of Ch. 5 to the functions $H(x, a(x), \alpha) \leftrightarrow L(x, a(x), \beta)$, continuous for $x = x_0$, where $a(x) = a(x, \nabla\bar{V}(x))$. □

Now we assume that there exists a function $a(x, \alpha)$ mentioned in the hypothesis of Theorem 2.2, for every $x_0 \in D$ there exists a function $\bar{V}(x) = \bar{V}_{x_0}(x)$ satisfying the hypotheses of Theorem 2.2 and for every x_0, the function $a_{x_0}(x) = a(x, \nabla_x \bar{V}_{x_0}(x))$ belongs to the class $\mathfrak{A}$ of admissible controls. These conditions imply, in particular, that for any two points $x_0, x_1 \in D$, sufficiently close to each other, there exists a control function $a(x)$ such that there is a "most probable" trajectory from x_1 to x_0—a solution of the equation $\dot{x}_t = b(x_t, a(x_t))$. We introduce the additional requirement that 0 is an isolated point of the set $\{\alpha: H(x, \alpha) = 0\}$ for every x. Then x_0 can be reached from x_1 over a finite time and it is easy to prove that the same remains true for arbitrary $x_0, x_1 \in D$, not only for points close to each other.

Theorem 2.3. *Let the conditions just formulated be satisfied. We choose a point x_0 for which $\min_{y \in \partial D} \bar{V}_{x_0}(y)$ attains the maximum (equal to V_0). We choose the control function $\bar{a}(x)$ in the following way: in the set*

$$B_{x_0} = \left\{x: \bar{V}_{x_0}(x) \le \min_{y \in \partial D} \bar{V}_{x_0}(y)\right\}$$

we put $\bar{a}(x) = a(x, \nabla_x \bar{V}_{x_0}(x))$; for the remaining x we define $\bar{a}(x)$ in an arbitrary way, only ensuring that $\bar{a}(x)$ is continuous and that from any point of D the solution of the system $\dot{x}_t = b(x_t, a(x_t))$ reaches the set B_{x_0} for positive t, remaining in D. Then

$$\lim_{h \downarrow 0} h \ln \mathsf{M}_x^{h, \bar{a}} \tau_D = V_0. \tag{2.10}$$

for any $x \in D$.

Together with Theorem 2.1, this theorem means that the function $\bar{a}$ is a solution of our optimal stabilization problem.

The proof can be carried out in the following way: for the points of B_{x_0}, the function $\bar{V}_{x_0}(x)$ is a Lyapunov function for the system $\dot{x}_t = b(x_t, \bar{a}(x_t))$. This, together with the structure of $\bar{a}(x)$ for the remaining x, shows that x_0 is the unique stable equilibrium position, which attracts the trajectories issued from points of the domain. Now (2.10) follows from Theorem 4.1 of Ch. 4, generalized as indicated in §4, Ch. 5.

§3. Examples

At the end of Chapter 5 we considered the example of calculating the quasi-potential and the asymptotics of the mean exit time of a neighborhood of a stable equilibrium position and of the invariant measure for the family of one-dimensional processes jumping distance h to the right and to the left with probabilities $h^{-1}r(x)\,dt$ and $h^{-1}l(x)\,dt$ over time dt. By the same token, we determined the characteristics of the stability of an equilibrium position. This example admits various interpretations; in particular, the process of division and death of a large number of cells (cf. §2, Ch. 5). The problem of stability of an equilibrium position of such a system, i.e., the problem of determining the time over which the number of cells remains below a given level may represent a great interest, especially if we consider, for example, the number of cells of a certain kind in the blood system of an organism rather than the number of bacteria in a culture.

Another concrete interpretation of the same scheme is a system consisting of a large number of elements N, which go out of work independently of each other after an exponential time of service. These elements start to be repaired; the maintenance time is exponential with coefficient μ depending on the ratio of the elements having gone out of work. The change of this ratio x with time is a process of the indicated form with $h = N^{-1}$, $r(x) = (1 - x) \cdot \lambda$ (λ is the coefficient of the distribution of the time of service) and $l(x) = x\mu(x)$.

A nonexponential maintenance time or a nonexponential time between subsequent cell divisions lead to other schemes, cf. Freidlin [8] and Levina, Leontovich, and Pyatetskii-Shapiro [1].

We consider examples of optimal stabilization.

Example 3.1. Let the family of controlled processes have the same structure at all points; in other words, let the function H and the set of admissible controls at a given point be independent of x:

$$H(x, a, \alpha) \equiv H(a, \alpha): \qquad \Pi(x) \equiv \Pi.$$

In this case the function H is also independent of x:

$$H(x, \alpha) \equiv H(\alpha) = \inf_{a \in \Pi} H(a, \alpha)$$

and equation (2.7) turns into

$$H(\nabla \bar{V}(x)) = 0. \tag{3.1}$$

The function $H(\alpha)$ vanishes only for one vector in every direction (except $\alpha = 0$). The locus of the terminal points of these vectors will be denoted by A. Equation (3.1) may be rewritten in the form $\nabla \bar{V}(x) \in A$.

Equation (3.1) has an infinite set of solutions—the r-dimensional planes of the form $\bar{V}_{x_0\alpha_0}(x) = (\alpha_0, x - x_0)$, $\alpha_0 \in A$—but none of them is a solution of problem $\mathbf{R}_{x_0}$ for the equation. To find this solution, we note that the solution planes depend on the $(r - 1)$-dimensional parameter $\alpha_0 \in A$ and a one-dimensional parameter independent of α_0 on the scalar product (α_0, x_0). This family is a complete integral of equation (3.1). As is known (cf. Courant [1], p. 111), the envelope of any $(r - 1)$-parameter family of these solutions is also a solution. If for a fixed x_0, the family of planes $\bar{V}_{x_0\alpha_0}(x)$ has a smooth envelope, then this envelope is the desired solution $\bar{V}_{x_0}(x)$ of problem $\mathbf{R}_{x_0}$.

This solution is given by a conic surface at any rate (i.e., $\bar{V}_{x_0}(x)$ is a positively homogeneous function of degree one in $x - x_0$); it is convenient to define it by means of its $(r - 1)$-dimensional level surface

$$U_1 = \{x - x_0 : \bar{V}_{x_0}(x) = 1\}$$

(it is independent of x_0). The surface U_1 has one point β_0 on every ray emanating from the origin. We shall see how to find this point.

Let the generator of the cone $\bar{V}_{x_0}(x)$, corresponding to a point β_0, be a line of tangency of the cone with the plane $\bar{V}_{x_0\alpha_0}(x)$ (α_0 is determined uniquely, because the direction of this vector is given: it is the direction of the exterior normal to U_1 at β_0). The intersection with the horizontal plane at height 1 is the $(r - 1)$-dimensional plane $\{\beta : (\alpha_0, \beta) = 1\}$, which is tangent to U_1 at β_0 (Fig. 31). The point of this plane which is the closest to the origin is situated at distance $|\alpha_0|^{-1}$ in the same direction from the origin as α_0, i.e.,

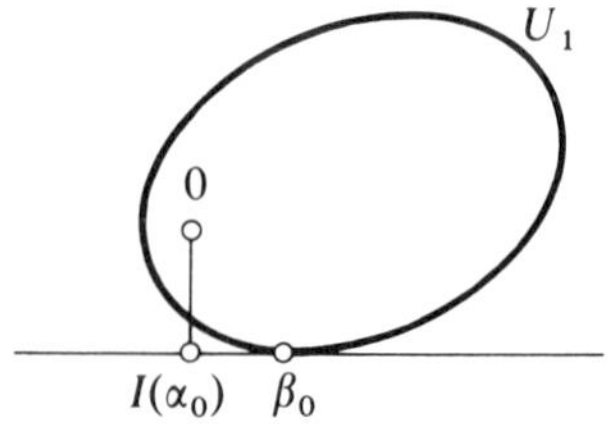

Figure 31.

it is the point $I(\alpha_0)$ obtained from α_0 by inversion with respect to the unit sphere with center at 0.

Hence to find the level surface U_1, we have to invert the surface A; through every point of the surface $I(A)$ thus obtained we consider the plane orthogonal to the corresponding radius and take the envelope of these planes. This geometric transformation does not always lead to a smooth convex surface: "corners" or cuspidal edges may appear. Criteria may be given for the smoothness of U_1 in terms of the centers of curvature of the original surface A.

If the surface U_1 turns out to be smooth (continuously differentiable), then the solution of the optimal control problem may be obtained in the following way. In D we inscribe the largest figure homothetic to U_1 with positive coefficient of homothety c, i.e., the figure $x_0 + cU_1$, where x_0 is a point of D (not defined in a unique way in general). Inside this figure, the optimal control field $\bar{a}(x)$ is defined in the following way: we choose a point $\beta_0 \in U_1$ situated on the ray in the direction of $x - x_0$; we determine the corresponding point α_0 of A; as $\bar{a}(x)$ we choose that value a_0 of the controlling parameter in Π for which $\min_{a \in \Pi} H(a, \alpha_0)$ is attained (this minimum is equal to $H(\alpha_0) = 0$; we assume that the minimum is attained and that a_0 depends continuously on α_0 belonging to A).

Consequently, the value of the control parameter $\bar{a}(x)$ is constant on every radius issued from x_0. At points x lying outside $x_0 + cU_1$, the field $\bar{a}(x)$ may be defined almost arbitrarily, it only has to drive all points inside the indicated surface.

The mean exit time is logarithmically equivalent to $\exp\{ch^{-1}\}$ as $h \downarrow 0$.

In Wentzell and Freidlin [5], a special case of this example was considered, where the subject was the control of a diffusion process with small diffusion by means of the choice of the drift.

Example 3.2. We consider a dynamical system perturbed by a small white noise; it can be subjected to control effects whose magnitude is under our control but which themselves may contain a noise. The mathematical model of the situation is as follows:

$$\dot{X}_t^{\varepsilon, a} = b(X_t^{\varepsilon, a}) + \varepsilon\sigma(X_t^{\varepsilon, a})\dot{w}_t + a(X_t^{\varepsilon, a})[\tilde{b}(X_t^{\varepsilon, a}) + \varepsilon\tilde{\sigma}(X_t^{\varepsilon, a})\dot{\tilde{w}}_t],$$

where w_t, $\tilde{w}_t$ are independent Wiener processes and $a(x)$ is a control function, which is chosen within the limits of the set $\Pi(x)$.

We consider the simplest case: the process is one-dimensional, $\tilde{\sigma}$, $\tilde{b}$, and σ are positive constants, $b(x)$ is a continuous function, D is the interval (x_1, x_2) and the controlling parameter $a(x)$ varies within the limits

$$\pm\left[2\tilde{b}^{-1} \max_{[x_1, x_2]} |b(x)| + \tilde{\sigma}^{-1}\sigma\right]$$

at every point (or within the limits of any larger segment $\Pi(x)$).

We have

$$H(x, a, \alpha) = (b(x) + a\tilde{b})\alpha + \tfrac{1}{2}(\sigma^2 + a^2\tilde{\sigma}^2)\alpha^2;$$

$$H(x, \alpha) = \min_a H(x, a, \alpha) = b(x)\alpha - \frac{\tilde{b}^2}{2\tilde{\sigma}^2} + \tfrac{1}{2}\sigma^2\alpha^2$$

for $|\alpha| \geq [2\tilde{\sigma}^2\tilde{b}^{-2}\max_{[x_1, x_2]}|b(x)| + \sigma\tilde{\sigma}\tilde{b}^{-1}]^{-1}$. The minimum here is attained for $a = -\tilde{b}\tilde{\sigma}^{-2}\alpha^{-1}$. The function $H(x, \alpha)$ vanishes for

$$\alpha = \alpha_1(x) = -\frac{b(x)}{\sigma^2} - \sqrt{\frac{b(x)^2}{\sigma^4} + \frac{\tilde{b}^2}{\sigma^2\tilde{\sigma}^2}} < 0$$

and for

$$\alpha = \alpha_2(x) = -\frac{b(x)}{\sigma^2} + \sqrt{\frac{b(x)^2}{\sigma^4} + \frac{\tilde{b}^2}{\sigma^2\tilde{\sigma}^2}} > 0,$$

and also for $\alpha = 0$, which is, by the way, immaterial for the determination of the optimal quasipotential (it can be proved easily that $|\alpha_1(x)|$ and $|\alpha_2(x)|$ surpass the indicated boundary for $|\alpha|$). Problem $\mathbf{R}_{x_0}$ for the equation $H(x, \bar{V}'_{x_0}(x)) = 0$ reduces to the equation

$$\bar{V}'_{x_0}(x) = \begin{cases} \alpha_1(x) & \text{for } x_1 \leq x < x_0, \\ \alpha_2(x) & \text{for } x_0 < x \leq x_2 \end{cases}$$

with the additional condition $\lim_{x \to x_0} \bar{V}_{x_0}(x) = 0$. It is easy to see that $\min(\bar{V}_{x_0}(x_1), \bar{V}_{x_0}(x_2))$ attains its largest value for an x_0 for which $\bar{V}_{x_0}(x_1)$ and $\bar{V}_{x_0}(x_2)$ are equal to each other. This leads to the following equation for x_0:

$$\int_{x_1}^{x_0}\left[\sqrt{\frac{b(x)^2}{\sigma^4} + \frac{\tilde{b}^2}{\sigma^2\tilde{\sigma}^2}} + \frac{b(x)}{\sigma^2}\right]dx = \int_{x_0}^{x_2}\left[\sqrt{\frac{b(x)^2}{\sigma^4} + \frac{\tilde{b}^2}{\sigma^2\tilde{\sigma}^2}} - \frac{b(x)}{\sigma^2}\right]dx,$$

which has a unique solution. After finding the optimal equilibrium position x_0, we can determine the optimal control according to the formula

$$\bar{a}(x) = \begin{cases} -\tilde{b}\tilde{\sigma}^{-2}\alpha_1(x)^{-1} & \text{to the left of } x_0, \\ -\tilde{b}\tilde{\sigma}^{-2}\alpha_2(x)^{-1} & \text{to the right of } x_0. \end{cases}$$

Chapter 10

Sharpenings and Generalizations

§1. Local Theorems and Sharp Asymptotics

In Chapters 3, 4, 5 and 7 we established limit theorems on large deviations, involving the rough asymptotics of probabilities of the type $\mathsf{P}\{X^h \in A\}$. There arises the following question: Is it possible to obtain subtler results for families of random processes (similar to those obtained for sums of independent random variables)—local limit theorems on large deviations and theorems on sharp asymptotics? There is some work in this direction; we give a survey of the results in this section.

If X_t^h is a family of random processes, a local theorem on large deviations may involve the asymptotics, as $h \downarrow 0$, of the density $p_t^h(y)$ of the distribution of the value of the process being considered at time t, where the point y is different from the "most probable" value of $x(t)$ for small h (we take into account that the density is not defined uniquely in general; we have to indicate that we speak of, for example, the continuous version of density). We may prove local theorems involving the joint density $p_{t_1,\ldots,t_n}^h(y_1, \ldots, y_n)$ of random variables $X_{t_1}^h, \ldots, X_{t_n}^h$. We may also consider the asymptotics of the density of the distribution of a functional $F(X^h)$ at points different from the "most probable" value of $F(x(\cdot))$.

However, obtaining these kinds of results involving a large class of functionals F and families of random processes X^h is unrealistic for the time being, at least because we need to obtain results on the existence of a continuous density of the distribution of $F(X^h)$ beforehand. For the same reason, positive results in the area of local theorems on large deviations involving densities of values of a process at separate moments of time up to now are restricted to families of diffusion processes, for which the problem of the existence of a density is well studied and solved in the affirmative (sense) under insignificant restrictions, in the case of a nonsingular diffusion matrix. The transition probability density $p^h(t, x, y)$ for the value of a diffusion process X_t^h under the assumption that $X_0^h = x$ has the meaning of the fundamental solution of the corresponding parabolic differential equation (or that of the Green's function for the corresponding problem in the case of a diffusion process in a domain with attainable boundary). This provides a base for both an additional method of study of the density p^h and the area of possible applications.

We begin with Friedman's work [1] because its results are connected more directly with what has been discussed (although it appeared a little later than Kifer's work [1] containing stronger results). Let $p^h(t, x, y)$ be the fundamental solution of the equation $\partial u/\partial t = L^h u$ in the r-dimensional space, where

$$L^h u = \frac{h}{2} \sum a^{ij}(x) \frac{\partial^2 u}{\partial x^i \, \partial x^j} + \sum b^i(x) \frac{\partial u}{\partial x^i}; \tag{1.1}$$

in other words, $p^h(t, x, y)$ is the continuous version of the transition probability density of the corresponding diffusion process (X_t^h, P_x^h). (Here it is more convenient to denote the small parameter in the diffusion matrix by h rather than ε^2, as had been done beginning with Ch. 4.) We put

$$V(t, x, y) = \min\{S_{0t}(\varphi): \varphi_0 = x, \varphi_t = y\}, \tag{1.2}$$

where $h^{-1} S_{0t}$ is the action functional for the family of processes (X_t^h, P_x^h). The functional S_{0t} is given, as we know, by the formula

$$S_{0t}(\varphi) = \frac{1}{2} \int_0^t \sum a_{ij}(\varphi_s)(\dot{\varphi}_s^i - b^i(\varphi_s))(\dot{\varphi}_s^j - b^j(\varphi_s)) \, ds, \tag{1.3}$$

where $(a_{ij}(x)) = (a^{ij}(x))^{-1}$. It can be proved that

$$\lim_{h \downarrow 0} h \ln p^h(t, x, y) = -V(t, x, y). \tag{1.4}$$

This rough local theorem on large deviations is obtained by applying rough (integral) theorems of Wentzell and Freidlin [4] and estimates from Aronson [1]:

$$p^h(t, x, y) \le \frac{A_0}{(ht)^{r/2}} \exp\left\{-\frac{c_0 |y - x(t, x)|^2}{ht}\right\},$$

$$p^h(t, x, y) \ge \frac{A_1}{(ht)^{r/2}} \exp\left\{-\frac{c_1 |y - x(t, x)|^2}{ht}\right\} - \frac{A_2}{(ht)^{r/2 - \alpha}} \exp\left\{-\frac{c_2 |(y - x(t, x))|^2}{ht}\right\}$$

for sufficiently small t, where the A_i, c_i and α are positive constants and $x(t, x)$ is the solution of the system $\dot{x} = b(x)$ with initial condition $x(0, x) = x$.

Further, if (X_t^h, P_x^h) is the diffusion process corresponding to L^h in a domain D with smooth boundary ∂D, vanishing upon reaching the boundary, then its transition probability density $q^h(t, x, y)$ is the Green's function for the equation $\partial u/\partial t = L^h u$ with boundary condition $u = 0$ on ∂D. It is proved that

$$\lim_{h \downarrow 0} h \ln q^h(t, x, y) = -V_D(t, x, y), \tag{1.5}$$

where $V_D(t, x, y)$ is defined as the infimum of $S_{0T}(\varphi)$ for curves φ connecting x and y over time t, without leaving D. This implies in particular that

$$\lim_{h \downarrow 0} \frac{q^h(t, x, y)}{p^h(t, x, y)} = 0 \tag{1.6}$$

if all extremals for which the minimum (1.2) is attained leave $D \cup \partial D$. The following opposite result can also be proved: if all extremals pass inside D, then

$$\lim_{h \downarrow 0} \frac{q^h(t, x, y)}{p^h(t, x, y)} = 1. \tag{1.7}$$

(According to M. Kac's terminology, the process X_t^h "does not feel" the boundary for small h.)

In the scheme being considered we may also include, in a natural way (and almost precisely), questions involving the asymptotics, for small values of time, of the transition density of a diffusion process (of the fundamental solution of a parabolic equation) not depending on a parameter. Indeed, let $p(t, x, y)$ be the transition probability density of the diffusion process (X_t, P_x) corresponding to the elliptic operator

$$Lu = \frac{1}{2} \sum a^{ij}(x) \frac{\partial^2 u}{\partial x^j \partial x^j} + \sum b^i(x) \frac{\partial u}{\partial x^i}.$$

We consider the family of diffusion processes $X_t^h = X_{ht}$, $h > 0$. The transition density $p(h, x, y)$ of X_t over time h is equal to the transition density $p^h(1, x, y)$ of X_t^h over time 1. The process (X_t^h, P_x^h) is governed by the differential operator

$$L^h u = hLu = \frac{h}{2} \sum a^{ij}(x) \frac{\partial^2 u}{\partial x^i \partial x^j} + h \sum b^i(x) \frac{\partial u}{\partial x^i}. \tag{1.8}$$

The drift coefficients $h \cdot b^i(x)$ converge to zero as $h \downarrow 0$, so that the family of operators (1.8) is almost the same as the family (1.1) with the $b^i(x)$ replaced by zero. In any event, the action functional may be written out without difficulty: it has the form $h^{-1} S_{0t}(\varphi)$, where

$$S_{0t}(\varphi) = \frac{1}{2} \int_0^t \sum a_{ij}(\varphi_s) \dot{\varphi}_s^i \dot{\varphi}_s^j \, ds. \tag{1.9}$$

The problem of finding the minimum of this functional can be solved in terms of the following Riemannian metric $\rho(x, y)$ connected with the matrix (a_{ij}):

$$\rho(x, y) = \min_{\varphi_0 = x,\, \varphi_t = y} \int_0^t [\sum a_{ij}(\varphi_s)\dot{\varphi}_s^i \dot{\varphi}_s^j]^{1/2}\, ds.$$

It is easy to prove that the minimum of the functional (1.9) for all parametrizations φ_s, $0 \le s \le t$, of a given curve is equal to the square of the Riemannian length of the curve multiplied by $(2t)^{-1}$. Consequently, the minimum (1.2) is equal to $(2t)^{-1}\rho(x, y)^2$.

Application of formula (1.4) yields

$$\lim_{h \downarrow 0} h \ln p(h, x, y) = -\tfrac{1}{2}\rho(x, y)^2. \tag{1.10}$$

Accordingly, for the transition density of the diffusion process which vanishes on the boundary of D we obtain

$$\lim_{h \downarrow 0} h \ln q(h, x, y) = -\tfrac{1}{2}\rho_D(x, y)^2, \tag{1.11}$$

where $\rho_D(x, y)$ is the infimum of the Riemannian lengths of the curves connecting x and y, without leaving D. There are also results, corresponding to (1.6), (1.7), involving the ratio of the densities q and p for small values of the time argument. In particular, the "principle of not feeling of the boundary" assumes the form

$$\lim_{t \downarrow 0} \frac{q(t, x, y)}{p(t, x, y)} = 1 \tag{1.12}$$

if all shortest geodesics connecting x and y lie entirely in D.

The results (1.10)–(1.12) were obtained in Varadhan [2], [3] (the result (1.12) in the special case of a Wiener process was obtained in Ciesielski [1]).

In Kifer [1], [3] and Molchanov [1] sharp versions of these results were obtained. In obtaining them, an essential role is played by the corresponding rough results (it is insignificant whether in local or integral form): they provide an opportunity to exclude from the consideration all but a neighborhood of an extremal (extremals) of the action functional; after this, the problem becomes local.

The sharp asymptotics of the transition probability density from x to y turns out to depend on whether x and y are conjugate or not on an extremal connecting them (cf. Gel'fand and Fomin [1]). In the case where x and y are nonconjugate and the coefficients of the operator smooth, not only can the asymptotics of the density up to equivalence be obtained but an asymptotic expansion in powers of the small parameter can also be obtained for the family of processes corresponding to the operators (1.1) we have

$$p^h(t, x, y) = (2\pi ht)^{-r/2} \exp\{-h^{-1}V(t, x, y)\}[K_0(t, x, y) + hK_1(t, x, y) + \cdots + h^m K_m(t, x, y) + o(h^m)] \tag{1.13}$$

as $h \downarrow 0$ (Kifer [3]); for the process with generator (1.8) we have

$$p(t, x, y) = (2\pi t)^{-r/2} \exp\{-\rho(x, y)^2/2t\}[K_0(x, y) + tK_1(x, y) + \cdots + t^m K_m(x, y) + o(t^m)] \tag{1.14}$$

as $t \downarrow 0$ (Molchanov [1]). Methods in probability theory are combined with analytic methods in these publications. We outline the proof of expansion (1.13) in the simplest case, moreover for $m = 0$, i.e., the proof of the existence of the finite limit

$$\lim_{h \downarrow 0} (2\pi ht)^{r/2} p^h(t, x, y) \exp\{h^{-1}V(t, x, y)\}. \tag{1.15}$$

Let the operator L^h have the form

$$L^h u = \frac{h}{2}\Delta u + \sum b^i(x)\frac{\partial u}{\partial x^i}. \tag{1.16}$$

The process corresponding to this operator may be given by means of the stochastic equation

$$\dot{X}_s^h = b(X_s^h) + h^{1/2}\dot{w}_s, \qquad X_0^h = x, \tag{1.17}$$

where w_s is an r-dimensional Wiener process. Along with X_s^h, we consider the diffusion process Y_s^h, nonhomogeneous in time, given by the stochastic equation

$$\dot{Y}_s^h = \dot{\varphi}_s + h^{1/2}\dot{w}_s, \qquad Y_0^h = x, \tag{1.18}$$

where φ_s, $0 \le s \le t$, is an extremal of the action functional from x to y over time t (we assume that it is unique). The density of the distribution of Y_t^h can be written out easily: it is equal to the density of the normal distribution with mean φ_t and covariance matrix htE; at y it is equal to $(2\pi ht)^{-r/2}$.

The ratio of the probability densities of X_t^h and Y_t^h is equal to the limit of the ratio

$$\mathsf{P}\{X_t^h \in y + D\}/\mathsf{P}\{Y_t^h \in y + D\} \tag{1.19}$$

as the diameter of the neighborhood D of the origin of coordinates converges to zero. We use the fact that the measures in $\mathbf{C}_{0t}(R^r)$, corresponding to the random processes X_s^h and Y_s^h, are absolutely continuous with respect to each other; the density has the form

$$\frac{d\mu_{X^h}}{d\mu_{Y^h}}(Y^h) = \exp\left\{h^{-1/2}\int_0^t (b(Y_s^h) - \dot{\varphi}_s, dw_s) - (2h)^{-1}\int_0^t |b(Y_s^h) - \dot{\varphi}_s|^2\, ds\right\}. \tag{1.20}$$

The absolute continuity enables us to express the probability of any event connected with X^h in the form of an integral of a functional of Y^h; in particular,

$$\mathsf{P}\{X_t^h \in y + D\} = \mathsf{M}\left\{Y_t^h \in y + D; \frac{d\mu_{X^h}}{d\mu_{Y^h}}(Y^h)\right\}.$$

Expression (1.19) assumes the form

$$\begin{aligned}
&\mathsf{M}\left\{\exp\left\{h^{-1/2}\int_0^t (b(Y_s^h) - \dot{\varphi}_s, dw_s)\right.\right.\\
&\qquad\left.\left.- (2h)^{-1}\int_0^t |b(Y_s^h) - \dot{\varphi}_s|^2\, ds\right\} \middle|\, Y_t^h \in y + D\right\}\\
&= \mathsf{M}\left\{\exp\left\{h^{-1/2}\int_0^t (b(\varphi_s + h^{1/2}w_s) - \dot{\varphi}_s, dw_s)\right.\right.\\
&\qquad\left.\left.- (2h)^{-1}\int_0^t |b(\varphi_s + h^{1/2}w_s) - \dot{\varphi}_s|^2\, ds\right\} \middle|\, w_t \in h^{-1/2}D\right\}
\end{aligned}$$

(we use the fact that $w_0 = 0$ and $Y_s^h = \varphi_s + h^{1/2}w_s$ for $0 \le s \le t$).

If D (and together with it, $h^{-1/2}D$) shrinks to zero, we obtain the conditional mathematical expectation under the condition $w_t = 0$. Hence

$$\begin{aligned}
\frac{p^h(t, x, y)}{(2\pi h t)^{-r/2}} &= \mathsf{M}\left\{\exp\left\{h^{-1/2}\int_0^t (b(\varphi_s + h^{1/2}w_s) - \dot{\varphi}_s, dw_s)\right.\right.\\
&\qquad\left.\left.- (2h)^{-1}\int_0^t |b(\varphi_s + h^{1/2}w_s) - \dot{\varphi}_s|^2\, ds\right\} \middle|\, w_t = 0\right\}.
\end{aligned} \tag{1.21}$$

(Formula (1.21) has already been obtained by Hunt [1]).

To establish the existence of the limit (1.15) as $h \downarrow 0$, first of all we truncate the mathematical expectation (1.21) by multiplying the exponential expression by the indicator of the event $\{\max_{0 \le s \le t} |h^{1/2}w_s| < \delta\}$. The circumstance that the omitted part may be neglected as $h \downarrow 0$ may be established by means of rough estimates connected with the action functional. Then we transform the exponent by taking account of the smoothness of b. In the first integral we take the expansion of $b(\varphi_s + h^{1/2}w_s)$ in powers of $h^{1/2}w_s$ up to terms of the first order and in the second integral up to terms of the second order. The principal term arising in the second integral is equal to

$$-(2h)^{-1}\int_0^t |b(\varphi_s) - \dot{\varphi}_s|^2\,ds,$$

and upon substitution into (1.15) it cancels with $h^{-1}V(t, x, y)$. If we verify that the terms of order $h^{-1/2}$ also cancel each other, it only remains to be proved that the terms of order 1 and the infinitely small terms as $h \downarrow 0$ do not hinder convergence (this can be done in the case where x and y are not conjugate).

We discuss the terms of order $h^{-1/2}$, arising from the first integral in (1.21). We integrate by parts:

$$h^{-1/2}\int_0^t (b(\varphi_s) - \dot{\varphi}_s, dw_s) = h^{-1/2}(b(\varphi_t) - \dot{\varphi}_t, w_t)$$
$$- h^{-1/2}\int_0^t \left(w_s, \frac{d}{ds}(b(\varphi_s) - \dot{\varphi}_s)\right) ds.$$

The integrated term vanishes by virtue of the condition $w_t = 0$. The expression $(d/ds)(b(\varphi_s) - \dot{\varphi}_s)$ can be transformed by taking account of Euler's equation for an extremal. It is easy to see that the integral cancels with the terms of order $h^{-1/2}$, arising from the second integral in (1.21).

It can be seen from the proof outlined here that the coefficient $K_0(t, x, y)$ in the expansion (1.13) has the meaning of the conditional mathematical expectation of the exponential function of a quadratic functional of a Wiener process under the condition $w_t = 0$.

The situation is not more complicated in the case where x and y are connected with a finite number of extremals and they are not conjugate to each other on any of these extremals. In the case of conjugate points the transition density over time t can be expressed by means of the Chapman–Kolmogorov equation in terms of a transition density over a shorter time. In the integral thus obtained, the main role is played by densities at non-conjugate points and the asymptotics is obtained by applying Laplace's method (the finite-dimensional one). For the density $p(t, x, y)$ as $t \downarrow 0$, this is done in Molchanov [1]; in particular, for various structures of the set of minimal geodesics connecting x and y, there arise asymptotic expressions of the form $p(t, x, y) \sim Ct^{-\alpha}e^{-\rho(x,y)^2/2t}$ with varying α.

Concerning sharp asymptotics in problems involving large deviations not reducing to one-dimensional or finite-dimensional distributions, little has been done yet. The results obtained in this area do not relate to probabilities $\mathsf{P}^h\{X^h \in A\}$ or densities but rather to mathematical expectations

$$\mathsf{M}^h \exp\{h^{-1}F(X^h)\}, \tag{1.22}$$

where F is a smooth functional (the normalizing coefficient is assumed to be equal to h^{-1}). The consideration of problems of this kind is natural as a first step, since even in the case of large deviations for sums of independent two-dimensional random vectors, the sharp asymptotics of integrals analogous to (1.22) can be found much easier than the sharp asymptotics of the probability of hitting a domain.

The expression (1.22) is logarithmically equivalent to

$$\exp\{h^{-1} \max[F - S]\},$$

where S is the normalized action functional. If the extremal φ providing this maximum is unique, then the mathematical expectation (1.22) differs from

$$\mathsf{M}^h\{\rho(X^h, \varphi) < \delta; \exp\{h^{-1}F(X^h)\}\} \tag{1.23}$$

by a number which is exponentially small compared with (1.22) or (1.23). This enables us to localize the problem.

The plan of further study is analogous, to a great degree, to what has been done in Cramér [1]: the generalized Cramér transformation is performed, which transforms the measure P^h into a new probability measure $\hat{\mathsf{P}}^h$ such that the "most probable" trajectory of X_t^h with respect to $\hat{\mathsf{P}}^h$ turns out to be the extremal φ_t for h small. With respect to the new probability measure, the random process $h^{-1/2}[X_t^h - \varphi_t]$ turns out to be asymptotically Gaussian with characteristics which are easy to determine. If the functionals F and S are twice differentiable at φ (the requirement of smoothness of S can be reduced to smoothness requirements of local characteristics of the family of processes X_t^h) and the quadratic functional corresponding to the second derivative $(F'' - S'')(\varphi)$ is strictly negative definite, then for the means (1.23) and (1.22) we obtain the following sharp asymptotics:

$$\mathsf{M}^h \exp\{h^{-1}F(X^h)\} \sim K_0 \exp\{h^{-1}[F(\varphi) - S(\varphi)]\} \tag{1.24}$$

as $h \downarrow 0$. The constant K_0 can be expressed as the mathematical expectation of a certain functional of a Gaussian random process.

If F and S are $\nu + 2$ times differentiable, then for (1.22), (1.23) we obtain an asymptotic expansion of the form

$$\exp\{h^{-1}[F(\varphi) - S(\varphi)]\}(K_0 + K_1 h + \cdots + K_{[\nu/2]}h^{[\nu/2]} + o(h^{\nu/2})). \tag{1.25}$$

Analogous results may be obtained for the mathematical expectations of functionals of the form $G(X^h)\exp\{h^{-1}F(X^h)\}$.

This program has been realized in Schilder [1] for the family of random processes $X_t^h = h^{-1/2}w_t$, where w_t is a Wiener process and in Dubrovskii [1], [2], [3] for the families of locally infinitely divisible Markov processes, considered by us in Ch. 5. We note that in the case considered in Schilder [1], the part connected with the asymptotic Gaussianness of $h^{-1/2}[X_t^h - \varphi_t]$ with respect to $\hat{\mathsf{P}}^h$ falls out of our scheme outlined here (because of the triviality of that part). In exactly the same way, in this simple situation we do not need to be aware of the connection of the employed method with H. Cramér's method.

§2. Large Deviations for Random Measures

We consider a Wiener process ξ_t on the interval $[0, 1]$ with reflection at the endpoints. For every Borel set $\Gamma \subseteq [0, 1]$ we may consider a random variable $\pi_T(\Gamma)$, the proportion of time spent by ξ_t in Γ for $t \in [0, T]$:

$$\pi_T(\Gamma) = \frac{1}{T}\int_0^T \chi_\Gamma(\xi_s)\,ds. \tag{2.1}$$

The random variable $\pi_T(\Gamma) = \pi_T(\Gamma, \omega)$ is a probability measure in Γ. In the space of measures on $[0, 1]$, let a metric ρ be given, for example, the metric defined by the equality

$$\rho(\mu, \nu) = \sup_{0<x<1} |\mu[0, x] - \nu[0, x]|. \tag{2.2}$$

It is known that for any initial point $\xi_0 = x \in [0, 1]$, the measure π_T converges to Lebesgue measure l on $[0, 1]$ with probability 1 as $T \to \infty$. If A is a set in the space of measures, at a positive distance from Lebesgue measure l, then $\mathsf{P}_x\{\pi_T \in A\} \to 0$ as $T \to \infty$, i.e., the event $\{\omega : \pi_T \in A\}$ is related to large deviations.

Of course, formula (2.1) defines a random measure $\pi_T(\Gamma)$ for every measurable random process $\xi_t(\omega)$, $t \geq 0$. If this process is, for example, stationary and ergodic, then by Birkhoff's theorem, $\pi_T(\Gamma)$ converges to a nonrandom measure $m(\Gamma)$ (to the one-dimensional distribution of the process) as $T \to \infty$ and the deviations of π_T from m which do not converge to zero belong to the area of large deviations.

Families of random measures, converging in probability to a nonrandom measure, also arise in other situations, in particular, in problems concerning the intersection of a level. For example, let ξ_t be a stationary process with sufficiently regular trajectories and let $\eta_t^T = \xi_{Tt}(T > 0)$. Along with ξ_t we may consider the family of random measures π^T on $[0, 1]$, where $\pi^T(\Gamma)$ is the number, normalized by dividing by T, of intersections, taking place for $t \in \Gamma \subseteq [0, 1]$, of a given level by η_t^T. Under certain regularity conditions on ξ_t, the measure π^T converges, as $T \to \infty$, to Lebesgue measure multiplied by the average number of intersections over the time unit.

We may consider limit theorems of various kinds for random measures. Here we discuss some results concerning the behavior of probabilities of large deviations. The rough asymptotics of probabilities of large deviations for measures (2.1), connected with Markov processes, was studied in a series of publications by Donsker and Varadhan [1], [2]. Independently of them, general results were obtained by Gärtner [2], [3]. Gärtner also applied these results to Markov and diffusion processes and considered a series of examples. Our exposition is close to Gärtner's work.

Let $(E, \mathscr{B})$ be a measurable space. We denote by **B** the space of bounded measurable functions on E and by **V** the space of finite countably additive set functions (charges) on the σ-algebra $\mathscr{B}$. We introduce the notation $\langle \mu, f \rangle = \int_E f(x)\mu(dx)$.

In **V** we may consider the **B***-weak topology (cf. Dunford and Schwartz [1]) given by neighborhoods of the form

$$\{\mu\colon |\langle \mu - \mu_0, f_i \rangle| < \delta, i = 1, \ldots, n\},$$

$f_i \in \mathbf{B}$. Along with it, we shall consider a metric ρ and the corresponding topology in **V**.

We shall only consider metrics given in the following way. A system $\mathfrak{M}$ of functions $f \in \mathbf{B}$ bounded in absolute value by one is fixed and we put

$$\rho(\mu, \nu) = \sup_{f \in \mathfrak{M}} |\langle \mu, f \rangle - \langle \nu, f \rangle|; \qquad \mu, \nu \in V. \tag{2.3}$$

Of course, in order that equality (2.3) define a metric, we need to take a sufficiently rich supply of functions f: every charge $\mu \in \mathbf{V}$ need to be defined uniquely by the integrals $\langle \mu, f \rangle$ for $f \in \mathfrak{M}$.

Let a finite measure m be fixed on $(E, \mathscr{B})$. We shall say that a metric ρ, defined by (2.3) in **V**, satisfies condition (1) if for every $\delta > 0$ there exist finite systems $\mathfrak{A} = \mathfrak{A}_\delta$, $\mathfrak{B} = \mathfrak{B}_\delta$ of measurable functions on E, bounded in absolute value by one, such that $\langle m, w \rangle \leq \delta$ for $w \in \mathfrak{B}$ and for every $f \in \mathfrak{M}$ there exist $v \in \mathfrak{A}$ and $w \in \mathfrak{B}$ for which $|f - v| \leq w$.

Condition (1) enables us to reduce the study of large deviations for measures to the study of large deviations for random vectors in a finite-dimensional space and to use the results of §1, Ch. 5.

The metric (2.2) in $\mathbf{V} = \mathbf{V}([0, 1])$ is a special case of the metric (2.3) with the system $\mathfrak{M}$ consisting of the indicators of all intervals $[0, x]$; it satisfies condition (1) if as m we choose Lebesgue measure. As $\mathfrak{A}_\delta$ we may choose the indicators of the intervals of the form $[0, k\delta]$ and as $\mathfrak{B}_\delta$, the indicators of the intervals $[k\delta, (k + 1)\delta]$. Another example of a metric satisfying condition

(1): E is a compactum, $\mathscr{B}$ is the σ-algebra of its Borel subsets, ρ is the metric corresponding to the family $\mathfrak{M}$ of functions bounded in absolute value by one and satisfying a Lipschitz condition with constant 1 (this metric corresponds to $\mathbf{C}^*$-weak convergence in the space $\mathbf{V}$, considered usually in probability theory). If m is any measure with $m(E) = 1$, then as $\mathfrak{A}_\delta$ we may choose a finite δ-net in $\mathfrak{M}$ and as $\mathfrak{B}_\delta$ we may choose the singleton consisting of the constant δ function.

We fix a complete probability space $\{\Omega, \mathscr{F}, \mathsf{P}\}$. A mapping $\pi: \Omega \times B \to R^1$ is called a random measure if $\pi(\cdot, A)$ is a random variable for every $A \subset \mathfrak{B}$ and $\pi(\omega, \cdot)$ is a measure for almost all ω.

In what follows we shall often consider sets of the form

$$\{\omega: \rho(\pi(\omega, \cdot), \mu) < \delta\}.$$

In order that these sets be measurable for any $\mu \in \mathbf{V}$ and $\delta > 0$, it is sufficient to assume that in $\mathfrak{M}$ there exists a countable subset $\mathfrak{M}_0$ such that for every function $f \in \mathfrak{M}$ and every measure μ there exists a bounded sequence of elements f_n of $\mathfrak{M}_0$, converging to f almost everywhere with respect to μ. We shall always assume that this condition is satisfied.

Now let us be given a family of random measures π^h, depending on a parameter h. For the sake of simplicity we assume that h is a positive numerical parameter.

Our fundamental assumption consists in the following. There exists a function $\lambda(h)$ converging to $+\infty$ as $h \downarrow 0$ and such that the finite limit

$$H(f) = \lim_{h \downarrow 0} \lambda(h)^{-1} \ln \mathsf{M} \exp\{\lambda(h)\langle \pi^h, f\rangle\} \tag{2.4}$$

exists for every $f \in \mathbf{B}$. As in §1, Ch. 5, it can be proved that $H(f)$ is a convex functional. We make the following assumptions concerning this functional:

A.1. The functional $H(f)$ is Gâteau differentiable, i.e., the function $h(\gamma) = H(f + \gamma g)$, $\gamma \in R^1$, is differentiable for every $f, g \in \mathbf{B}$.

A.2. If a bounded sequence $\{f_n\}$ of measurable functions on E converges to an $f \in \mathbf{B}$ in measure with respect to the measure m (m is the measure singled out on $(E, \mathscr{B})$), then $H(f_n) \to H(f)$.

We denote by $S(\mu)$ the Legendre transform of the functional $H(f)$:

$$S(\mu) = \sup_{f \in \underline{B}} [\langle \mu, f\rangle - H(f)], \qquad \mu \in V. \tag{2.5}$$

It is easy to see that $S(\mu)$ is a convex functional on $\mathbf{V}$, assuming nonnegative values, possibly including $+\infty$.

We list the basic properties of S. In the meanwhile we introduce some notation to be used in what follows.

B.1. In order that $S(\mu) < \infty$, it is necessary that the charge μ be a measure. If $\pi^h(E) = 1$ with probability 1, then $S(\mu) < \infty$ only for measures μ with $\mu(E) = 1$. The functional S is lower semicontinuous with respect to $\mathbf{B}^*$-weak convergence.

B.2. For any $s \geq 0$, the measures μ such that $S(\mu) \leq s$ are uniformly bounded and uniformly absolutely continuous with respect to m.

B.3. If condition (1) is satisfied, then S is lower semicontinuous in the topology induced by the metric ρ and the set

$$\Phi(s) = \{\mu \in \mathbf{V}: S(\mu) \leq s\},$$

$s < \infty$, is compact in this topology.

B.4. Let $\mathfrak{A}$ be a finite system of functions belonging to $\mathbf{B}$. We write

$$\rho_{\mathfrak{A}}(\mu, \nu) = \sup_{v \in \mathfrak{A}} |\langle \mu, v \rangle - \langle \nu, v \rangle|,$$

$$S_{\mathfrak{A}}(\mu) = \sup_{f \in \mathscr{L}(\mathfrak{A})} [\langle \mu, f \rangle - H(f)],$$

where $\mathscr{L}(\mathfrak{A})$ is the linear hull of $\mathfrak{A}$. Moreover, we write $\Phi_{\mathfrak{A}}(s) = \{\mu \in \mathbf{V}: S_{\mathfrak{A}}(\mu) \leq s\}$, $s \in [0, \infty)$. (We note that $\rho_{\mathfrak{A}}(\mu, \nu) = 0$ does not necessarily imply that $\mu = \nu$.) For any $\delta > 0$ and $s \geq 0$ we have the inequality

$$\inf\{S(\nu): \rho_{\mathfrak{A}}(\nu, \mu) < \delta\} \leq S_{\mathfrak{A}}(\mu) \tag{2.6}$$

and the inclusion

$$\Phi_{\mathfrak{A}}(s) \subseteq \{\Phi(s)\}^{\rho_{\mathfrak{A}}}_{+\delta} = \{\mu \in \mathbf{V}: \rho_{\mathfrak{A}}(\mu, \Phi(s)) < \delta\}. \tag{2.7}$$

We outline the proofs of these properties (complete proofs may be found in Gärtner [3]).

B.1 may be proved very simply. We do not even have to use assumptions A.1, A.2. For example: if $\mu(A) < 0$ for some $A \in \mathscr{B}$, then

$$S(\mu) \geq \sup_{\gamma \leq 0} [\langle \mu, \gamma\chi_A \rangle - H(\gamma\chi_A)] \geq \sup_{\gamma \leq 0} \gamma\mu(A) = \infty.$$

Here we have used the fact that $H(f) \leq 0$ for $f \leq 0$.

B.2. If $S(\mu) \leq s$, then we use the fact that the supremum (2.5) is not smaller than the value of the expression in square brackets for $f = \gamma\chi_A$, $\gamma > 0$. We obtain

$$\mu(A) \leq \frac{H(\gamma\chi_A) + s}{\gamma}.$$

The boundedness follows from this immediately. To verify the uniform absolute continuity, it is sufficient to prove that for any $\varepsilon > 0$ there exists $\delta > 0$ such that $m(A) < \delta$ implies $\mu(A) < \varepsilon$ for all measures $\mu \in \Phi(s)$. We choose γ such that s/γ is smaller than $\varepsilon/2$; condition A.2 implies that there exists $\delta > 0$ such that $H(\gamma\chi_A) < \gamma\varepsilon/2$ if $m(A) < \delta$.

B.3. The compactness of $\Phi(s)$ in the $\mathbf{B}^*$-weak topology follows from B.1 and B.2. It only remains to be shown that μ_n, $\mu \in \Phi(s)$, $\mu_n \to \mu$ in the $\mathbf{B}^*$-weak topology imply that $\rho(\mu_n, \mu) \to 0$. This can be deduced from (1) by using the uniform absolute continuity of the μ_n and μ.

B.4. According to A.1 the derivative $H'(f)(h) = \lim_{\gamma \to 0} \gamma^{-1}(Hf + \gamma h) - H(f))$ exists, which is a linear functional of $h \in \mathbf{B}$. The equality $\mu^f(A) = H'(f)(\chi_A)$ defines a finite positive countably additive (by virtue of assumption A.2) measure on $(E, \mathscr{B})$. It follows from the properties of convex functions (cf. Rockafellar [1]; we have already used these properties in the proof of Lemma 5.2 of Ch. 7) that it is sufficient to prove inequality (2.6) for measures μ for which the supremum in the definition of $S_A(\mu)$ is attained for some function $f_0 \in \mathscr{L}(\mathfrak{A})$. Then for all $h \in \mathscr{L}(\mathfrak{A})$ we have: $\langle \mu, h \rangle = H'(f_0)(h) = \langle \mu^{f_0}, h \rangle$. It is easy to deduce from the definition of μ^{f_0} that the upward convex functional $a(f) = \langle \mu^{f_0}, f \rangle - H(f)$, $f \in \mathbf{B}$, attains its maximum for $f = f_0$. From this we obtain $S(\mu^{f_0}) = \langle \mu^{f_0}, f_0 \rangle - H(f_0) = \langle \mu, f_0 \rangle - H(f_0) = S_{\mathfrak{A}}(\mu)$, where $\rho_{\mathfrak{A}}(\mu^{f_0}, \mu) = 0$ ($\rho_{\mathfrak{A}}$ is a semimetric!). This proves inequality (2.6) and along with it, inclusion (2.7).

Now we formulate the fundamental result of this section.

Theorem 2.1. *Let the metric ρ satisfy condition* (1), *let the limit* (2.4) *exist, and let the functional $H(f)$ satisfy conditions* A.1 *and* A.2. *Then $\lambda(h)S(\mu)$ is the action functional for the family of random measures π^h in the metric space (V, ρ) as $h \downarrow 0$; i.e., for any γ, δ, $s > 0$ and $\mu \in V$ we have*

$$\mathsf{P}\{\rho(\pi^h, \mu) < \delta\} \geq \exp\{-\lambda(h)[S(\mu) + \gamma]\}, \tag{2.8}$$

$$\mathsf{P}\{\rho(\pi^h, \Phi(s)) \geq \delta\} \leq \exp\{-\lambda(h)[s - \gamma]\} \tag{2.9}$$

for sufficiently small h, where $\Phi(s) = \{\mu \in \mathbf{V} : S(\mu) \leq s\}$.

Proof. First we obtain estimate (2.8). If $S(\mu) = +\infty$, there is nothing to be proved. Therefore, we assume that $S(\mu) < \infty$. We use condition (1) for $\delta_1 > 0$. We obtain the estimate

$$\rho(\pi^h, \mu) \leq \rho_{\mathfrak{A}}(\pi^h, \mu) + \max_{w \in \mathfrak{B}} \langle \pi^h, w \rangle + \max_{w \in \mathfrak{B}} \langle \mu, w \rangle. \tag{2.10}$$

Since by B.2, the measure μ is absolutely continuous with respect to m, the last term on the right side of (2.10) is smaller than $\delta/4$ for sufficiently small δ_1. Consequently,

$$\mathsf{P}\{\rho(\pi^h, \mu) < \delta\} \geq \mathsf{P}\{\rho_{\mathfrak{A}}(\pi^h, \mu) < \delta/2\} - \mathsf{P}\left\{\max_{w \in \mathfrak{B}} \langle \pi^h, w \rangle \geq \delta/4\right\}. \tag{2.11}$$

Applying Theorem 1.2 of Ch. 5 to the family of finite-dimensional vectors $\eta^h = \{\pi^h(v)\}_{v \in \mathfrak{A}}$, $h \downarrow 0$, and taking into account that in a finite-dimensional space all norms are equivalent, we obtain the estimate

$$\mathsf{P}\{\rho_{\mathfrak{A}}(\pi^h, \mu) < \delta/2\} \geq \exp\{-\lambda(h)[S_{\mathfrak{A}}(\mu) + \gamma/2]\} \geq \exp\{-\lambda(h)[S(\mu) + \gamma/2]\}. \tag{2.12}$$

Now we estimate the subtrahend in (2.11). The exponential Chebyshev inequality yields

$$\begin{aligned}\mathsf{P}\{\langle \pi^h, w\rangle > \delta/4\} &\leq \exp\left\{-\varkappa\lambda(h)\frac{\delta}{4}\right\}\mathsf{M}\exp\{\varkappa\lambda(h)\langle\pi^h, w\rangle\} \\ &= \exp\left\{-\lambda(h)\left[\varkappa\frac{\delta}{4} - \lambda(h)^{-1}\ln \mathsf{M}\exp\{\lambda(h)\langle\pi^h, \varkappa w\rangle\}\right]\right\}\end{aligned}$$

for any $\varkappa > 0$. The expression in square brackets converges to $\varkappa(\delta/4) - H(\varkappa w)$ as $h \downarrow 0$. If we choose $\varkappa$ sufficiently large and then decrease δ_1 so that $H(\varkappa w)$ be sufficiently small for all $w \in \mathfrak{B}$ (this can be done by virtue of condition A.2), we obtain $\varkappa(\delta/4) - H(\varkappa w) > S(\mu) + \gamma$ for all $w \in \mathfrak{B}$. This implies that

$$\mathsf{P}\{\langle\pi^h, w\rangle \geq \delta/4\} \leq \exp\{-\lambda(h)[S(\mu) + \gamma]\} \tag{2.13}$$

for sufficiently small $h > 0$. Estimate (2.8) follows from (2.11)–(2.13).

Now we deduce (2.9). We use condition (1) again. By virtue of B.2 we have $\max_{w \in \mathfrak{B}} \langle \mu, w\rangle < \delta/4$ for sufficiently small positive δ_1 and for all $\mu \in \Phi(s)$. From this and (2.10) we conclude that

$$\mathsf{P}\{\rho(\pi^h, \Phi(s)) \geq \delta\} \leq \mathsf{P}\{\rho_{\mathfrak{A}}(\pi^h, \Phi(s)) \geq \delta/2\} + \mathsf{P}\left\{\max_{w \in \mathfrak{B}} \langle\pi^h, w\rangle \geq \delta/4\right\}. \tag{2.14}$$

Let the functions $v_1, \ldots, v_n \in \mathfrak{A}$ form a basis of the linear space $\mathscr{L}(\mathfrak{A})$. To estimate the first term on the right side of (2.14), we use the inclusion $\Phi_{\mathfrak{A}}(s) \subseteq \{\Phi(s)\}^{\rho_{\mathfrak{A}}}_{+\delta/4}$ (cf. formula (2.7)) and apply Theorem 1.1 of Ch. 5 to the family of finite-dimensional vectors $\eta^h = (\pi^h(v_1), \ldots, \pi^h(v_n))$:

$$\mathsf{P}\{\rho_{\mathfrak{A}}(\pi^h, \Phi(s)) \geq \delta/2\} \leq \mathsf{P}\{\rho_{\mathfrak{A}}(\pi_h, \Phi_{\mathfrak{A}}(s)) \geq \delta/4\} \leq \exp\{-\lambda(h)(s - \gamma/2)\}.$$

The second term in (2.14) can be estimated by means of (2.13) and we obtain estimate (2.9) for small h. □

We consider an example. Let (ξ_t, P_x) be a diffusion process on a compact manifold E of class $\mathbf{C}^{(\infty)}$, controlled by an elliptic differential operator L

with infinitely differentiable coefficients. Let us consider the family of random measures π_T defined by formula (2.1). We verify that this family satisfies the condition that the limit (2.4) exists, where instead of $h \downarrow 0$, the parameter T goes to ∞ and as the function λ we take T: for any function $f \in \mathbf{B}$, the finite limit

$$H(f) = \lim_{T \to \infty} T^{-1} \ln \mathsf{M}_x \exp\left\{\int_0^T f(\xi_s)\, ds\right\} \tag{2.15}$$

exists. As in the proof of Theorem 4.2 of Ch. 7, for this we note that the family of operators

$$T_t^f g(x) = \mathsf{M}_x \exp\left\{\int_0^T f(\xi_s)\, ds\right\} g(\xi_t)$$

forms a semigroup acting in $\mathbf{B}$. If $g(x)$ is a nonnegative function belonging to $\mathbf{B}$, assuming positive values on some open set, then the function $T_t^f g(x)$ is strictly positive for any $t > 0$. As in the proof of Theorem 4.2 of Ch. 7, we may deduce from this that the limit (2.15) exists and is equal to $\lambda(f)$, the logarithm of the spectral radius of T_1^f (cf., for example, Kato [1]). The fulfilment of conditions A.1 and A.2 follows from results of perturbation theory. Moreover, it is easy to prove that the supremum in the definition of $S(\mu)$ may be taken for not all functions belonging to $\mathbf{B}$ but only continuous ones:

$$S(\mu) = \sup_{f \in \mathbf{C}} [\langle \mu, f \rangle - \lambda(f)]. \tag{2.16}$$

For continuous functions f the logarithm of the spectral radius of T_1^f coincides with that eigenvalue of the infinitesimal generator A^f of the semigroup T_1^f which has the largest real part: this eigenvalue is real and simple. For smooth functions, A^f coincides with the elliptic differential operator $L + f$. This implies that by $\lambda(f)$ in formula (2.16) we may understand the maximal eigenvalue of $L + f$.

The random measure π_T is a probability measure for every $T > 0$. According to B.1, the functional $S(\mu)$ is finite only for probability measures μ. Since $\lambda(f + c) = \lambda(f) + c$ for any constant c, the supremum in (2.16) may be taken only for those $f \in \mathbf{C}$ for which $\lambda(f) = 0$. On the other hand, the functions f for which $\lambda(f) = 0$ are exactly those functions which can be represented in the form $-Lu/u$, $u > 0$. Consequently, for probability measures μ, the definition of $S(\mu)$ may be rewritten in the form

$$S(\mu) = -\inf_{u > 0} \left\langle \mu, \frac{Lu}{u} \right\rangle. \tag{2.17}$$

For measures μ admitting a smooth positive density with respect to the Riemannian volume m induced by the Riemannian metric connected with the principal terms of L, we may write out Euler's equation for an extremal of problem (2.17) and obtain an expression for $S(\mu)$, not containing the infimum. In particular, if L is self-adjoint with respect to m, then the equation for an extremal can be solved explicitly. The infimum in (2.17) is attained for $u = \sqrt{d\mu/dm}$ and $S(\mu)$ can be written in the form

$$S(\mu) = \frac{1}{8}\int_E \frac{\left|\nabla \dfrac{d\mu}{dm}\right|^2}{\dfrac{d\mu}{dm}}\, dm.$$

The functional $S(\mu)$ may be calculated analogously for measures π_T in the case where (ξ_t, P_x) is a process in a bounded domain with reflection at the boundary (Gärtner [2]).

§3. Processes with Small Diffusion with Reflection at the Boundary

In Chapters 2–6 we considered the application of methods in probability theory to the first boundary value problem for differential equations with a small parameter at the derivatives of highest order. There naturally arises the question of what can be done for the second boundary value problem or for a mixed boundary value problem, where Dirichlet conditions are given on some parts of the boundary and Neumann conditions on other parts.

Let the smooth boundary of a bounded domain D consist of two components, $\partial_1 D$ and $\partial_2 D$. We shall consider the boundary value problem

$$\begin{gathered} L^\varepsilon u^\varepsilon(x) \equiv \frac{\varepsilon^2}{2}\sum a^{ij}(x)\frac{\partial^2 u^\varepsilon}{\partial x^i\,\partial x^j} + \sum b^i(x)\frac{\partial u^\varepsilon}{\partial x^i} = 0, \qquad x \in D, \\ \left.\frac{\partial u^\varepsilon(x)}{\partial l}\right|_{\partial_1 D} = 0, \qquad \left.u^\varepsilon(x)\right|_{\partial_2 D} = f(x), \end{gathered} \tag{3.1}$$

where $\partial/\partial l$ is the derivative in some nontangential direction; the coefficients of the equation, the direction l and the function f are assumed to be sufficiently smooth functions of x. This problem has a unique solution, which can be written in the form $u^\varepsilon(x) = \mathsf{M}_x^\varepsilon f(X_{\tau^\varepsilon}^\varepsilon)$, where $(X_t^\varepsilon, \mathsf{P}_x^\varepsilon)$ is the diffusion process in $D \cup \partial_1 D$, governed by L^ε in D, undergoing reflection in the direction l on the part $\partial_1 D$ of the boundary; $\tau^\varepsilon = \min\{t : X_t^\varepsilon \in \partial_2 D\}$ (Fig. 32). The asymptotics of the solution u^ε may be deduced from results involving

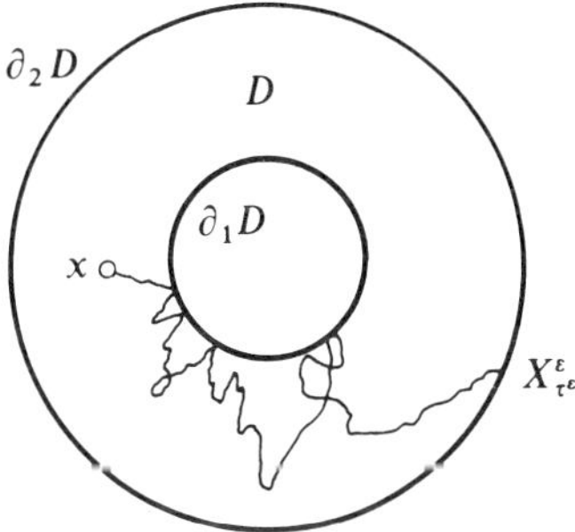

Figure 32.

the limit behavior of X_t^ε for small ε. We begin with results of the type of laws of large numbers.

Along with X_t^ε, we consider the dynamical system $\dot{x}_t = b(x_t)$ in D, obtained from X_t^ε for $\varepsilon = 0$. It follows from results of Ch. 2 that with probability close to one for small ε, the trajectory of X_t^ε, beginning at a point $x \in D$, is close to the trajectory $x_t(x)$ of the dynamical system until the time of exit of $x_t(x)$ to the boundary (if this time is finite). From this we obtain the following: if $x_t(x)$ goes to $\partial_2 D$ sooner than to $\partial_1 D$ and at the place $y(x)$ of exit, the field b is directed strictly outside the domain, then the value $u^\varepsilon(x)$ of the solution of problem (3.1) at the given point $x \in D$ converges to $f(y(x))$ as $\varepsilon \to 0$.

If X_t^ε begins at a point $x \in \partial_1 D$ (or, moving near the trajectory of the dynamical system, reaches $\partial_1 D$ sooner than $\partial_2 D$), then its most probable behavior depends on whether $b(x)$ is directed strictly inside or outside the domain. In the first case, X_t^ε will be close to the trajectory, issued from $x \in \partial_1 D$, of the same dynamical system $\dot{x}_t = b(x_t)$. From this we obtain the following result.

Theorem 3.1. *Let the field b be directed strictly inside D on $\partial_1 D$ and strictly outside D on $\partial_2 D$. Let all trajectories of the dynamical system $\dot{x}_t = b(x_t)$, beginning at points $x \in D \cup \partial_1 D$, exit from D (naturally, through $\partial_2 D$). Then the solution $u^\varepsilon(x)$ of problem* (3.1) *converges to the solution $u^0(x)$ of the degenerate problem*

$$\sum b^i(x)\frac{\partial u^0}{\partial x^i} = 0, \qquad x \in D,$$
$$u^0(x)|_{\partial_2 D} = f(x), \tag{3.2}$$

uniformly for $x \in D \cup \partial D$ as $\varepsilon \to 0$.

If b is directed strictly outside D on $\partial_1 D$, then the trajectory of the dynamical system goes out of $D \cup \partial_1 D$ through $\partial_1 D$ and X_t^ε is deprived of the opportunity of following it. It turns out that in this case X_t^ε begins to move along a trajectory of the system $\dot{x}_t = \hat{b}(x_t)$ on $\partial_1 D$, where $\hat{b}$ is obtained by projecting b parallel to the direction onto the tangential direction. This result also admits a formulation in the language of partial differential equations but in a situation different from the case of a boundary consisting of two components, considered here (cf. Freidlin [2]).

In the case where the trajectories of the dynamical system do not leave D through $\partial_2 D$ but instead enter D through $\partial_2 D$, results of the type of laws of large numbers are not sufficient for the determination of $\lim_{\varepsilon \to 0} u^\varepsilon(x)$. A study of large deviations for diffusion processes with reflection and of the asymptotics of solutions of the corresponding boundary value problems was carried out in Zhivoglyadova and Freidlin [1] and Anderson and Orey [1]. We present the results of these publications briefly.

We discuss Anderson's and Orey's [1] construction enabling us to construct a diffusion process with reflection by means of stochastic equations, beginning with a Wiener process. We restrict ourselves to the case of a process in the half-plane $R^2_+ = \{(x^1, x^2): x^1 \geq 0\}$ with reflection along the normal direction of the boundary, i.e., along the x^1-direction. We introduce a mapping Γ of $\mathbf{C}_{0T}(R^2)$ into $\mathbf{C}_{0T}(R^2_+)$: for $\xi \in \mathbf{C}_{0T}(R^2)$ we define the value $\eta_t = \Gamma_t(\xi)$ of the function $\eta = \Gamma(\xi)$ by the equalities

$$\eta_t = (\eta_t^1, \eta_t^2), \qquad \eta_t^2 = \zeta_t^2,$$
$$\eta_t^1 = \zeta_t^1 - \min\left(0, \min_{0 \leq s \leq t} \zeta_s^1\right). \tag{3.3}$$

It is clear that $\Gamma(\xi) \in \mathbf{C}_{0T}(R^2_+)$, the mapping Γ is continuous and

$$|\eta_t - \eta_s| \leq |\zeta_t - \zeta_s|.$$

We would like to construct a diffusion process with reflection at the boundary of R^2_+, with diffusion matrix $(a^{ij}(x))$ and drift $b(x) = (b^1(x), b^2(x))$. We represent the diffusion matrix in the form $(a^{ij}(x)) = \sigma(x)\sigma^*(x)$, where $\sigma(x)$ is a bounded matrix-valued function with entries satisfying a Lipschitz condition (such a matrix exists if the $a^{ij}(x)$, together with their derivatives, are bounded and continuous). We define the functionals $\sigma_t(\zeta)$ and $b_t(\zeta)$ on $\mathbf{C}_{0T}(R^2)$ by putting

$$\sigma_t(\zeta) = \sigma(\Gamma_t(\zeta)), \qquad b_t(\zeta) = b(\Gamma_t(\zeta)).$$

We consider the stochastic equation

$$\tilde{X}_t = x + \int_0^t \sigma_s(\tilde{X})\, dw_s + \int_0^t b_s(\tilde{X})\, ds, \qquad x \in R^2_+.$$

The existence and uniqueness theorem for such equations can be proved in the same way as for standard stochastic differential equations. It turns out that the random process $X_t = \Gamma_t(\tilde{X})$ is exactly the diffusion process with reflection in R^2_+, having the given diffusion and drift coefficients.

Now let us assume that there is a small parameter in the equation. For the sake of simplicity, we restrict ourselves to the case where σ is the identity matrix multiplied by ε:

$$\tilde{X}_t^\varepsilon = x + \varepsilon w_t + \int_0^t b_s(\tilde{X}^\varepsilon)\, ds; \tag{3.4}$$

the corresponding process with reflection will also be equipped with the index ε: $X_t^\varepsilon = \Gamma_t(\tilde{X}^\varepsilon)$. It can be proved that the solution of equation (3.4) can be obtained by applying a continuous mapping B_x to the function $\varepsilon w \in \mathbf{C}_{0T}(R^2)$ (the proof is the same as that of Lemma 1.1 of Ch. 4). Hence the process X^ε with reflection can be obtained from εw by means of the composition of two continuous mappings: $X^\varepsilon = \Gamma(B_x(\varepsilon w))$. Using general properties of the action functional (§3, Ch. 3), we obtain that the action functional for the family of processes X_t^ε in $\mathbf{C}_{0T}$ as $\varepsilon \to 0$ has the form $\varepsilon^{-2} S^+_{0T}(\varphi)$, where

$$S^+_{0T}(\varphi) = \min_{\chi:\, \Gamma(B_x(\chi)) = \varphi} \frac{1}{2} \int_0^T |\dot{\chi}_s|^2\, ds. \tag{3.5}$$

On the other hand, B_x is invertible:

$$(B_x^{-1}\psi)_t = \psi_t - x - \int_0^t b_s(\psi)\, ds;$$

taking account of this, expression (3.5) can be rewritten as

$$S^+_{0T}(\varphi) = \min_{\psi:\, \Gamma(\psi) = \varphi} \frac{1}{2} \int_0^T |\dot{\psi}_s - b(\varphi_s)|^2\, ds \tag{3.6}$$

(we use the fact that $b_s(\psi) = b(\Gamma_s(\psi)) = b(\varphi_s)$).

It is easy to verify that the minimum (3.6) is equal to

$$S^+_{0T}(\varphi) = \frac{1}{2} \int_0^T |\dot{\varphi}_s - \hat{b}(\varphi_s)|^2\, ds, \tag{3.7}$$

where $\hat{b}(x)$ is the field coinciding with $b(x)$ everywhere except at those points of $\partial_1 D$ at which $b(x)$ is directed outside D; at these points $\hat{b}(x)$ is defined as the projection of $b(x)$ onto the direction of the boundary. The minimum is attained for the function ψ defined by the equalities

$$\psi_t = (\psi_t^1, \psi_t^2), \qquad \psi_t^2 = \varphi_t^2,$$

$$\psi_t^1 = \varphi_t^1 + \int_0^t \chi_{\{0\}}(\varphi_s^1) \min(0, b^1(0, \varphi_s^2))\, ds. \tag{3.8}$$

A formula analogous to (3.7) also holds for a varying matrix $(a^{ij}(x))$ and for processes in an arbitrary domain D with reflection along any smooth field l, nontangent to the boundary, as well.

Results involving the action functional imply, in particular, that for $\varepsilon \to 0$ the trajectory of X_t^ε converges in probability to a function at which S^+ vanishes, i.e., to a solution of the system $\dot{x}_t = \hat{b}(x_t)$.

Now we present those results involving the asymptotics of solutions of problem (3.1) which follow from the above result, (cf. Zhivoglyadova and Freidlin [1]). Let D be the ring in the plane, having the form $\{(r, \theta): 1 < r < 2\}$ in the polar coordinates (r, θ). We consider the problem

$$L^\varepsilon u^\varepsilon \equiv \frac{\varepsilon^2}{2}\left(\frac{\partial^2 u^\varepsilon}{\partial r^2} + \frac{\partial^2 u^\varepsilon}{\partial \theta^2}\right) + b_r(r, 0)\frac{\partial u^\varepsilon}{\partial r} + b_\theta(r, \theta)\frac{\partial u^\varepsilon}{\partial \theta} = 0, \qquad x \in D,$$

$$\frac{\partial u^\varepsilon}{\partial r}(1, \theta) = 0, \qquad u^\varepsilon(2, \theta) = f(\theta). \tag{3.9}$$

We shall say that condition 1 is satisfied if the trajectories, beginning in D, of the dynamical system $\dot{x}_t = b(x_t)$ go to $\partial_1 D = \{r = 1\}$ sooner than to $\partial_2 D = \{r = 2)$ and $b_r(1, \theta) < 0$ for all θ.

We assume that on the interval $[0, 2\pi]$ the function $b_\theta(1, \theta)$ has a finite number of zeros. Let $K_1, K_2, \ldots, K_l$ be those of them at which $b_\theta(1, \theta)$ changes its sign from plus to minus as θ increases.

We consider

$$V(K_i, y) = \inf\{S_{0T}^+(\varphi): \varphi_0 = K_i, \varphi_T = y; T > 0\}$$

and assume that for any i, the minimum of $V(K_i, y)$ for $y \in \partial_2 D$ is attained at a unique point $y_i = (2, \theta_i)$. We put $V(K_i, \partial_2 D) = V(K_i, y_i)$.

Theorem 3.2. *Let the above conditions be satisfied. Let g^* be the unique $\{\partial_2 D\}$-graph over the set of symbols $\{K_1, \ldots, K_l, \partial_2 D\}$ for which the minimum of $\sum_{(\alpha \to \beta) \in g} V(\alpha, \beta)$ is attained over all $\{\partial_2 D\}$-graphs g. Let the trajectory, beginning at a point $x \in D$, of the system $\dot{x}_t = b(x_t)$ go out to the circle $\partial_1 D$ at the point $(1, \theta_x)$; for $x \in \partial_1 D$, as θ_x we choose the angular coordinate of x. Let K_i be the point to which the solution of the equation $\dot{\theta}_t = b_\theta(1, \theta_t)$ with initial condition θ_x is attracted as $t \to \infty$. Then $\lim_{\varepsilon \to 0} u^\varepsilon(x) = f(\theta_k)$, where $K_k \to \partial_2 D$ is the last arrow in the path leading from K_i to $\partial_2 D$ in g^*.*

The proof of this theorem can be carried out in the same way as that of Theorem 5.2 of Ch. 6.

§4. Wave Fronts in Semilinear PDEs and Large Deviations

Consider the Cauchy problem

$$\begin{aligned} \frac{\partial u(t,x)}{\partial t} &= \frac{D}{2}\frac{\partial^2 u}{\partial x^2} + cu, \qquad t > 0, x \in R^1, \\ u(0,x) &= \chi^-(x) = \begin{cases} 1, & x \le 0, \\ 0, & x > 0. \end{cases} \end{aligned} \tag{4.1}$$

Here c and D are positive constants. Let X_t be the process in R^1 corresponding to $(D/2)(d^2/dx^2) : \dot{X}^t = \sqrt{D}\dot{W}_t$. Then we have:

$$\begin{aligned} u(t,x) &= \mathsf{M}_x \chi^-(X_t) e^{ct} \\ &= e^{ct} P\{x + \sqrt{D} W_t \le 0\} \\ &= e^{ct} \frac{1}{\sqrt{2\pi t}} \int_{-\infty}^{-x/\sqrt{D}} e^{-y^2/2t}\, dy. \end{aligned}$$

One can easily derive from this equality that for any $\alpha > 0$

$$u(t,\alpha t) \approx \exp\left\{ t\left(c - \frac{\alpha^2}{2D}\right)\right\}, \qquad t \to \infty.$$

If $\alpha^* = \sqrt{2cD}$, then

$$\begin{aligned} \lim_{t\to\infty} \sup_{x \ge (\alpha^* + h)t} u(t,x) &= 0 \\ \lim_{t\to\infty} \inf_{x \le (\alpha^* - h)t} u(t,x) &= \infty \end{aligned} \tag{4.2}$$

for any $h > 0$. This means that the region of large values of $u(t,x)$ propagates, roughly speaking, with the speed $\alpha^* = \sqrt{2cD}$.

Now let the constant c in (4.1) be replaced by a smooth function $c(u)$ such that

$$\begin{aligned} c(u) &> 0, \quad \text{for } u < 1, \\ c(u) &< 0, \quad \text{for } u > 1, \\ c = c(0) &= \max_{u \ge 0} c(u); \end{aligned} \tag{4.3}$$

$$\frac{\partial u(t,x)}{\partial t} = \frac{D}{2}\frac{\partial^2 u}{\partial x^2} + f(u), \qquad t > 0, x \in R^1,$$
$$u(0,x) = \chi^-(x), \tag{4.4}$$

where $f(u) = c(u)u$.

Applying the Feynman–Kac formula, one can obtain the following equation for $u(t,x)$

$$u(t,x) = \mathsf{M}_x \chi^-(X_t) \exp\left\{\int_0^t c(u(t-s, X_s))\, dx\right\}. \tag{4.5}$$

Since, according to (4.3), $c(u) \leq c(0) = c$, we conclude from (4.5),

$$0 \leq u(t,x) \leq e^{ct} P_x\{X_t \leq 0\},$$

and thus the first of the equalities (4.2) holds for the solution of the nonlinear problem. The second equality of (4.2), of course, does not hold for the solution of problem (4.4). Since $c(u) < 0$ for $u > 1$, the solution never exceeds 1. But as we show later, one can prove using the large deviation estimates that $\lim_{t\to\infty} \inf_{x \leq (\alpha^* - h)t} u(t,x) = 1$ for any $h > 0$, where $u(t,x)$ is the solution of (4.4).

Equation (4.4) appeared, first, in Fisher [1] and in Kolmogorov, Petrovskii, and Piskunov [1]. The rigorous results for nonlinearities $f(u) = c(u)u$ satisfying (4.3) were obtained in the second of these papers. Therefore, this nonlinearity is called KPP-type nonlinearity. It was shown in Kolmogorov, Petrovskii, and Piskunov [1] that the solution of problem (4.4) with a KPP-type nonlinear term behaves for large t, roughly speaking, as a running wave:

$$u(t,x) \approx v(x - \alpha^* t), \qquad t \to \infty,$$

where $\alpha^* = \sqrt{2Dc}$, $c = f'(0)$, and $v(z)$ is the unique solution of the problem

$$\frac{D}{2} v''(z) + \alpha^* v'(z) + f(v(z)) = 0, \qquad -\infty < z < \infty,$$
$$v(-\infty) = 1, \qquad v(+\infty) = 1, \qquad v(0) = \tfrac{1}{2}.$$

Thus, the asymptotic behavior of $u(t,x)$ for large t is characterized by the asymptotic speed α^* and by the shape of the wave $v(z)$.

Instead of considering large t and x, one can introduce a small parameter $\varepsilon > 0$: put $u^\varepsilon(t,x) = u(t/\varepsilon, x/\varepsilon)$. Then the function $u^\varepsilon(t,x)$ is the solution of the problem

$$\frac{\partial u^\varepsilon(t,x)}{\partial t} = \frac{\varepsilon D}{2}\frac{\partial^2 u^\varepsilon}{\partial x^2} + \frac{1}{\varepsilon} f(u^\varepsilon),$$
$$u^\varepsilon(0,x) = \chi^-(x). \tag{4.6}$$

One can expect that $u^\varepsilon(t,x) = u(t/\varepsilon, x/\varepsilon) \approx v((x-\alpha^* t)/\varepsilon) \approx \chi^-(x-\alpha^* t)$ as $\varepsilon \downarrow 0$. This means that the first approximation has to do only with the asymptotic speed α^*, and does not involve the asymptotic shape $v(z)$.

Consider now a generalization of problem (4.6) (see Freidlin [12], [14], [17], [18]):

$$\frac{\partial u^\varepsilon(t,x)}{\partial t} = \frac{\varepsilon}{2}\sum_{i,j=1}^r a^{ij}(x)\frac{\partial^2 u^\varepsilon}{\partial x^i \partial x^j} + \frac{1}{\varepsilon} f(x,u^\varepsilon), \qquad t>0, \qquad x \in R^r,$$
$$u^\varepsilon(0,x) = g(x) \geq 0. \tag{4.7}$$

A number of interesting problems, like the small diffusion asymptotics for reaction-diffusion equations (RDEs) with slowly changing coefficients, can be reduced to (4.7) after a proper time and space rescaling. We assume that the coefficients $a^{ij}(x)$ and $f(x,u)$ are smooth enough, $\underline{a}\sum_{i=1}^r \lambda_i^2 \leq \sum a^{ij}(x)\lambda_i\lambda_j \leq \bar{a}\sum_{i=1}^r \lambda_i^2$ for some $0<\underline{a}<\bar{a}$; the function $f(x,u) = c(x,u)u$, where $c(x,u)$ satisfies conditions (4.3) for any $x \in R^r$ (in other words, $f(x,u)$ is of KPP-type for any $x \in R^r$). Denote by G_0 the support of the initial function $g(x)$. We assume that $0 \leq g(x) \leq \bar{g} < \infty$ and that the closure $[G_0]$ of G_0 coincides with the closure of the interior (G_0) of $G : [(G_0)] = [G_0]$. Moreover, assume that $g(x)$ is continuous in (G_0).

Denote by (X_t^ε, P_x) the diffusion process in R^r governed by the operator

$$L^\varepsilon = \frac{\varepsilon}{2}\sum_{i,j=1}^r a^{ij}(x)\frac{\partial^2}{\partial x^i \partial x^j}.$$

The Feynman–Kac formula yields an equation for $u^\varepsilon(t,x)$,

$$u^\varepsilon(t,x) - \mathsf{M}_x g(X_t^\varepsilon)\exp\left\{\frac{1}{\varepsilon}\int_0^t c(X_s^\varepsilon, u^\varepsilon(t-s, X_s^\varepsilon))\,ds\right\}. \tag{4.8}$$

First, assume that $c(x,0) = c =$ constant. Then, taking into account that $c(x,u) \leq c(x,0) = c$, we derive from (4.8),

$$0 \leq u^\varepsilon(t,x) \leq M_x g(X_t^\varepsilon)e^{ct/\varepsilon} \leq \bar{g}e^{ct/\varepsilon}P_x\{X_t^\varepsilon \in G_0\}. \tag{4.9}$$

The action functional for the family of processes X_s^ε, $0 \le s \le t$, in the space of continuous functions is equal to

$$\frac{1}{\varepsilon} S_{0t}(\phi) = \frac{1}{2\varepsilon} \int_0^t \sum_{i,j=1}^r a_{ij}(\phi_s) \dot{\phi}_s^i \dot{\phi}_s^j \, ds,$$

where ϕ is absolutely continuous, $(a_{ij}(x)) = (a^{ij}(x))^{-1}$, and

$$\lim_{\varepsilon \downarrow 0} \varepsilon \ln P_x\{X_t^\varepsilon \in G_0\} = -\inf\{S_{0t}(\phi) : \phi \in C_{0t}, \phi_0 = x, \phi_t \in G_0\}. \tag{4.10}$$

Denote by $\rho(x, y)$ the Riemannian distance in R^r corresponding to the form $ds^2 = \sum_{i,j=1}^r a_{ij}(x)\, dx^i\, dx^j$. One can check that the infimum in (4.10) is actually equal to $(1/2t)\rho^2(x, G_0)$. We derive from (4.9) and (4.10),

$$0 \le u^\varepsilon(t, x) \le \bar{g} e^{ct} P_x\{X_t^\varepsilon \in G_0\} \asymp \exp\left\{\frac{1}{\varepsilon}\left(ct - \frac{\rho^2(x, G_0)}{2t}\right)\right\}.$$

It follows from the last bound that

$$\lim_{\varepsilon \downarrow 0} u^\varepsilon(t, x) = 0, \qquad \text{if } \rho(x, G_0) > t\sqrt{2c}, \qquad t > 0. \tag{4.11}$$

We now outline the proof of the equality

$$\lim_{\varepsilon \downarrow 0} u^\varepsilon(t, x) = 1, \quad \text{if } \rho(x, G_0) < t\sqrt{2c}, \quad t > 0. \tag{4.12}$$

Let us first prove that

$$\lim_{\varepsilon \downarrow 0} \varepsilon \ln u^\varepsilon(t, x) = 0, \tag{4.13}$$

if $\rho(x, G_0) = t\sqrt{2c}$, $t > 0$. If K is a compact set in $\{t > 0, x \in R^r\}$, then the convergence in (4.13) is uniform in $(t, x) \in K \cap \{\rho(x, G_0) = t\sqrt{2c}\}$. One can derive from the maximum principle that

$$\overline{\lim_{\varepsilon \downarrow 0}}\, u^\varepsilon(t, x) \le 1, \qquad t > 0, x \in R^r. \tag{4.14}$$

Therefore, to prove (4.13) it is sufficient to show that

$$\underline{\lim_{\varepsilon \downarrow 0}}\, \varepsilon \ln u^\varepsilon(t, x) \ge 0, \tag{4.15}$$

if $\rho(x, G_0) = t\sqrt{2c}$, $t > 0$.

Let ϕ_s^*, $0 \le s \le t$, be the shortest geodesic connecting x and $[G_0]$ with the parameter s proportional to the Riemannian distance from x. For any small $\delta > 0$, choose $z \in (G_0)$, $|z - \phi_t^*| < \delta$, and define

$$\phi_s^\delta = \begin{cases} x, & 0 \le s \le \delta, \\ \phi^*_{t(s-\delta)/(t-\delta)}, & \delta \le s \le t-\delta, \\ h_s, & t-\delta \le s \le t, \end{cases}$$

where h_s is the linear function such that $h_{t-\delta} = \phi^*_{t(t-2\delta)/(t-\delta)}$ and $h_t = z$. Denote by $\mathscr{E}_\mu$, $\mu > 0$, the μ-neighborhood in C_{0t} of the function ϕ_s^δ, and by χ_μ the indicator of this set in C_{0t}. Let $\mu > 0$ be chosen so small that the ball in R^r of radius 2μ centered at z belongs to (G_0) and for any $\psi \in \mathscr{E}_\mu$

$$\min_{\delta \le s \le t-\delta} d(\psi_s, G_s) = d_0 > 0,$$

where $d(\cdot,\cdot)$ is the Euclidean distance in R^r and $G_s = \{x \in R^r,\ \rho(x, G_0) \le s\sqrt{2c}\}$. Such a choice of μ is possible since $d(\phi_s^\delta, G_s) \ge d_1 > 0$ for $\delta \le s \le t-\delta$. Then (4.11) implies that $u^\varepsilon(t-s, \phi_s) \to 0$ as $\varepsilon \downarrow 0$ for $\delta \le s \le t-\delta$ and $\psi \in E_\mu$, and thus $c(\psi_s, u^\varepsilon(t-s, \psi_s)) \to c(\psi_s, 0) = c$ as $\varepsilon \downarrow 0$ for such ψ and s.

Since $S_{0t}(\phi^*) = \rho^2(x, G_0)/2t$, one can derive from the definition of ϕ_s^δ that for any $h > 0$ there exists $\delta > 0$ so small that

$$S_{0t}(\phi^\delta) - \frac{\rho^2(x, G_0)}{2t} < h.$$

Let $g_0 = \min\{g(x) : |x - z| \le \mu\}$. It follows from (4.8) that

$$\begin{aligned} u^\varepsilon(t,x) &\ge \mathsf{M}_x g(X_t^\varepsilon)\chi_\mu(X^\varepsilon)\exp\left\{\frac{1}{\varepsilon}\int_0^t c(X_s^\varepsilon, u^\varepsilon(t-s, X_s^\varepsilon))\,ds\right\} \\ &\ge g_0 e^{(c(t-3\delta))/\varepsilon}\mathsf{P}_x\{\mathscr{E}_\mu\} \\ &\ge g_0 e^{c(t-3\delta)/\varepsilon}\exp\left\{-\frac{1}{\varepsilon}\left(\frac{\rho^2(x, G_0)}{2t} + 2h\right)\right\} \end{aligned} \tag{4.16}$$

for $\varepsilon > 0$ small enough. In (4.16) we used the lower bound from the definition of the action functional. Taking into account that $\rho(x, G_0) = t\sqrt{2c}$, we derive (4.15) from (4.16).

We now prove (4.12). Suppose it is not true. Then, taking into account (4.14), $x_0 \in R^r$ and $t_0 > 0$ exist such that

$$\rho(x_0, G_0) < t_0\sqrt{2c}, \qquad \varliminf_{\varepsilon \downarrow 0} u^\varepsilon(t_0, x_0) < 1 - 2h \tag{4.17}$$

for some $h > 0$. Let $U = U^\varepsilon$ be the connected component of the set

$$\{(t, x) : t > 0, x \in R^r, \rho(x_0, G_0) < t\sqrt{2c}, u^\varepsilon(t, x) < 1 - h\},$$

containing the point (t_0, x_0). The boundary ∂U of the set $U \subset R^{r+1}$ consists of sets ∂U_1 and ∂U_2:

$$\partial U_1 = \partial U \cap \{(t, x) : t > 0, x \in R^r, u^\varepsilon(t, x) = 1 - h\},$$

$$\partial U_2 = \partial U \cap (\{(t, x) : t > 0, \rho(x, G_0) = t\sqrt{2c}\} \cup \{(t, x) : t = 0, x \in G_0\}).$$

Define $\tau^\varepsilon = \min\{s : (t - s, X_s^\varepsilon) \in U\}$ and let χ_i^ε, $i \in \{1, 2\}$, be the indicator function of the set of trajectories X^ε for which $(t - \tau^\varepsilon, X_{\tau^\varepsilon}^\varepsilon) \in \partial U_i$. Let λ_1 be the distance in R^{r+1} between (t_0, x_0) and the surface $\{(t, x) : t > 0, \rho(x, G_0) = t\sqrt{2c}\}$, $\lambda = \lambda_1 \wedge t_0$. Let χ_0^ε denote the indicator function of the set of trajectories X^ε, such that $\max_{0 \le s \le t} |X_s^\varepsilon - x| \le \lambda$.

Using the strong Markov property, one can derive from (4.8),

$$\begin{aligned} u^\varepsilon(t_0, x_0) &= \mathsf{M}_{x_0} u^\varepsilon(t_0 - \tau^\varepsilon, X_{\tau^\varepsilon}^\varepsilon) \exp\left\{\frac{1}{\varepsilon}\int_0^{\tau^\varepsilon} c(X_s^\varepsilon, u^\varepsilon(t_0 - s, X_s^\varepsilon))\, ds\right\} \\ &\le \sum_{i=1}^{2} \mathsf{M}_{x_0} \chi_i^\varepsilon \chi_0^\varepsilon u^\varepsilon(t_0 - \tau^\varepsilon, X_{\tau^\varepsilon}^\varepsilon) \exp\left\{\frac{1}{\varepsilon}\int_0^{\tau^\varepsilon} c(X^\varepsilon, u^\varepsilon, (t_0 - s, X_s^\varepsilon))\, ds\right\}. \end{aligned} \tag{4.18}$$

The first term of the sum in the right-hand side of (4.18) can be bounded from below by $(1 - h)\mathsf{M}_{x_0}\chi_1^\varepsilon\chi_0^\varepsilon$, since $u^\varepsilon(t, x) = 1 - h$ on ∂U_1 and $c(x, u) \ge 0$. To estimate the second term, note that $\tau^\varepsilon > \lambda$ and

$$\min_{0 \le s \le \tau^\varepsilon} c(X_s^\varepsilon, u^\varepsilon(t_0 - s, X_s^\varepsilon)) \ge \min_{\substack{0 \le u \le 1-h \\ |x - x_0| \le \lambda}} c(x, u) = c_0 > 0,$$

if $(t_0 - \tau^\varepsilon, X_{\tau^\varepsilon}^\varepsilon) \in \partial U_2$. Thus, the second term in the right-hand side of (4.18) is greater than

$$\mathsf{M}_{x_0}\chi_0^\varepsilon\chi_2^\varepsilon e^{(c_0\lambda)/\varepsilon} u^\varepsilon(t_0 - \tau^\varepsilon, X_{\tau^\varepsilon}^\varepsilon).$$

Using bound (4.13), we conclude that the last expression is bounded from below by $\mathsf{M}_{x_0}\chi_0^\varepsilon\chi_2^\varepsilon$ if $\varepsilon > 0$ is small enough.

Combining these bounds together, one can derive from (4.18):

$$\begin{aligned} u^\varepsilon(t_0, x_0) &\ge (1 - h)\mathsf{M}_{x_0}\chi_1^\varepsilon\chi_0^\varepsilon + \mathsf{M}_{x_0}\chi_2^\varepsilon\chi_0^\varepsilon \\ &\ge 1 - h - \mathsf{M}_{x_0}(1 - \chi_0^\varepsilon) \end{aligned}$$

for ε small enough. Since $M_{x_0}\chi_0^\varepsilon \to 1$ as $\varepsilon \downarrow 0$, the last inequality implies that $\varepsilon_0 > 0$ exists such that

$$u^\varepsilon(t_0, x_0) \geq 1 - \frac{3h}{2} \quad \text{for } 0 < \varepsilon < \varepsilon_0.$$

This bound contradicts (4.17) and thus (4.12) holds.

Equalities (4.11) and (4.12) mean that, in the case $c(x, 0) = c = $ constant, the motion of the interface separating areas where $u^\varepsilon(t, x)$ is close to 0 and to 1 for $0 < \varepsilon \ll 1$, can be described by the Huygens principle with the velocity field $v(x, e), x, e \in R^r, |e| = 1$, which is homogeneous and isotropic if calculated with respect to the Riemannian metric $\rho(\cdot, \cdot)$ corresponding to the form $ds^2 = \sum_{i,j=1}^2 a_{ij}(x)\, dx^i\, dx^j$, $(a_{ij}(x)) = (a^{ij}(x))^{-1}$, and $|v(x, e)| = \sqrt{2c}$. If $G_t = \{x \in R^r : \lim_{\varepsilon \downarrow 0} u^\varepsilon(t, x) = 1\}$, $t > 0$, then

$$G_{t+h} = \{x \in R^r : \rho(x, G_t) < h\sqrt{2c}\}, \qquad h > 0.$$

In particular, if $G_0 = G_0^\varepsilon$ is a ball of radius ε with the center at a point $z \in R^r$, and $g(x) = g^\varepsilon(x)$ is the indicator function of $G_0^\varepsilon \subset R^r$, then G_t is the Riemannian ball of radius $t\sqrt{2c}$ with the center at z.

Now let $c(x, 0) = \max_{0 \leq u} c(x, u) = c(x)$ not be a constant. Put

$$V(t, x) = \inf\left\{\int_0^t \left[c(\phi_s) - \frac{1}{2}\sum_{i,j=1}^r a_{ij}(\phi_s)\dot{\phi}_s^i\dot{\phi}_s^j\right] ds : \phi \in C_{0t}, \phi_0 = x, \phi_t \in G_0\right\}.$$

One can check that $V(t, x)$ is a Lipschitz continuous function increasing in t. It follows from (4.8) and (4.3) that

$$u^\varepsilon(t, x) \leq \mathsf{M}_x g(X_t^\varepsilon) \exp\left\{\frac{1}{\varepsilon}\int_0^t C(X_s^\varepsilon)\, ds\right\}.$$

The expectation in the right-hand side of the last inequality, as follows from §3.3, is logarithmically equivalent to $\exp\{(1/\varepsilon)V(t, x)\}$, $\varepsilon \downarrow 0$. Thus,

$$\lim_{\varepsilon \downarrow 0} u^\varepsilon(t, x) = 0, \quad \text{if } V(t, x) < 0. \tag{4.19}$$

If we could prove that

$$\lim_{\varepsilon \downarrow 0} u^\varepsilon(t, x) = 1, \quad \text{if } V(t, x) > 0, \tag{4.20}$$

then the equation $V(t, x) = 0$ would give us the position of the interface (wavefront) separating areas where $u^\varepsilon(t,x)$ is close to 0 and to 1 at time t. But simple examples show that, in general, this is not true; the area where $\lim_{\varepsilon\downarrow 0} u^\varepsilon(t,x) = 0$ can be larger than prescribed by (4.19). This, actually, depends on the behavior of the extremals in the variational problem defining $V(t,x)$.

We say that condition (N) is fulfilled if for any (t,x) such that $V(t,x) = 0$,

$$V(t,x) = \sup\left\{ \int_0^t \left[c(\phi_s) - \frac{1}{2}\sum_{i,j=1}^r a_{ij}(\phi_s)\dot\phi_s^i\dot\phi_s^j \right] ds : \phi \in C_{0t}, \right.$$

$$\left. \phi_0 = x, \phi_t \in [G_t], V(t-s,\phi_s) < 0 \text{ for } 0 < s < t \right\}.$$

Theorem 4.1. *Let $u^\varepsilon(t,x)$ be the solution of problem* (4.7) *with a KPP-type nonlinear term $f(x,y) = c(x,u)u$. Then*

$$\lim_{\varepsilon\downarrow 0} u^\varepsilon(t,x) = 0$$

uniformly in (t,x) changing in any compact subset of $\{(t,x) : t > 0, x \in R^r, V(t,x) < 0\}$.

If the condition (N) holds,

$$\lim_{\varepsilon\downarrow 0} u^\varepsilon(t,x) = 1$$

uniformly in any compact subset of $\{(t,x) : t > 0, x \in R^r, V(t,x) > 0\}$, and the equation $V(t,x) = 0$ describes the position of the wavefront at time t.

The first statement of this theorem is a small refinement of (4.19). The proof of the second statement is similar to the proof of (4.12) in the case $c(x) = c =$ constant; the condition (N) allows us to check that $\lim_{\varepsilon\downarrow 0} \varepsilon \ln u^\varepsilon(t,x) = 0$, if $t > 0$ and $V(t,x) = 0$. The rest of the proof is based on (4.8), strong Markov property, and condition (4.3), as in the case $c(x) =$ constant. For the detailed proof, see Freidlin [15].

The motion of the front in the case of variable $c(x) = c(x,0)$ has a number of special features, which cannot be observed if $c(x) = c =$constant.

Consider, for example, the one-dimensional case

$$\frac{\partial u^\varepsilon(t,x)}{\partial t} = \frac{\varepsilon}{2}\frac{\partial^2 u^\varepsilon}{\partial x^2} + \frac{1}{\varepsilon} c(x)u^\varepsilon(1-u^\varepsilon), \qquad t > 0, x \in R^1, \tag{4.21}$$

$$u^\varepsilon(0,x) = \chi^-(x-\beta).$$

Here $c(x)$ is a smooth increasing positive function; $\beta > 0$ is a constant. One can check that the condition (N) is fulfilled in this case, and the equation $V(t,x) = V_\beta(t,x) = 0$ describes the position of the front at time t. If $x > \beta$, then

$$V_\beta(t,x) = \sup\left\{\int_0^t \left[c(\phi_s) - \frac{1}{2}\dot{\phi}_s^2\right] ds, \phi_0 = x, \phi_t = \beta\right\}.$$

Let, first, $c(x) = 1$ for $x < 0$ and $c(x) = 1 + x$ for $x > 0$. Then the function $V_\beta(t,x)$ can be calculated explicitly:

$$V_\beta(t,x) = \frac{t^3}{24} + t\left(1 + \frac{\beta + x}{2}\right) - \frac{(\beta - x)^2}{2t}, \qquad x > \beta,$$

$$V_\beta(t,x) > 0, \qquad x < \beta.$$

The interface between areas where $u^\varepsilon(t,x) \to 1$ and $u^\varepsilon(t,x) \to 0$ as $\varepsilon \downarrow 0$ moves in the positive direction, and solving the equation $V_\beta(t,x) = 0$, we define its position $x_\beta(t)$ at time $t > 0$,

$$x_\beta(t) = \frac{t^2}{2} + \beta + \sqrt{\frac{t^4}{3} + 2t^2(1+\beta)}.$$

The interface motion in this case has some particularities which are worth mentioning. Evaluate $\tilde{\beta} = x_0(1)$, $x_{\tilde{\beta}}(1)$, $x_0(2)$:

$$x_0(1) = \tilde{\beta} = \frac{1}{2} + \sqrt{\frac{7}{3}} \approx 2.0,$$

$$x_0(2) = 2 + \sqrt{\frac{40}{3}} \approx 5.6,$$

$$x_{\tilde{\beta}}(0) = 1 + \sqrt{\frac{7}{3}} + \sqrt{\frac{10}{3} + \sqrt{\frac{28}{3}}} \approx 5.1.$$

We see that $x_0(2) > x_{\tilde{\beta}}(1)$. This means that the motion of the front does not satisfy the Markov property (i.e., the semigroup property): given the position of the front at time 1, the behavior of the front before time 1 can influence the behavior of the front after time 1. The evolution of the function $u^\varepsilon(t,x)$ satisfies, of course, the semigroup property: if $u^\varepsilon(s,x)$ is known for $x \in R^1$, then $u^\varepsilon(t,x)$, $t > s$, can be calculated in a unique way. However, it turns out that for $c(x) \neq$ constant, the evolution of the main term of $u^\varepsilon(t,x)$ as $\varepsilon \downarrow 0$, which is a step function with values 0 and 1, already loses the semigroup property. To preserve this property, one must extend the phase space. The phase space should include not just the point where $V_\beta(t,x)$ is equal to zero, but the whole function $v_\beta(t,x) = V_\beta(t,x) \wedge 0$.

The loss of the semigroup property also shows that the motion of the front cannot be described by a Huygens principle.

One more property of the front motion in this example: Let $u_0^\varepsilon(t,x)$ and $u_1^\varepsilon(t,x)$ be the solutions of equation (4.21) with different initial functions $u_0^\varepsilon(0,x) = \chi^-(x)$, $u_1^\varepsilon(0,x) = 1 - \chi^-(x-1)$. In the first case, the front moves from $x = 0$ in the positive direction and its position at time t is given by $x_0(t)$. In the second case, the front moves from $x = 1$ in the negative direction of the x-axis. Condition (N) is not satisfied in the last case, but one can prove that the position $\tilde{X}(t)$ of the front at time t is the solution of the equation (Freidlin [15]),

$$\dot{\tilde{X}}(t) = -\sqrt{c(\tilde{X}_t)}, \qquad \tilde{X}_0 = 1. \tag{4.22}$$

In our example, $c(x) = 1 + x$ for $x > 0$. Solving (4.22), one can find that the front comes to $x = 0$ at time $t_{1\to 0} \approx 1.22$ in the second case. On the other hand, solving the equation $x_\beta(t) = 1$ for $\beta = 0$, we see that in the first case, the front needs time $t_{0\to 1} \approx 0.57$ to come from 0 to 1. Thus, the motion in the direction in which $c(x)$ increases is faster than in the opposite direction. This property is also incompatible with the Huygens principle.

We have seen in this example that if $c(x)$ is linearly increasing, the front moves at an increasing, but finite, speed. It turns out that if $c(x)$ increases fast enough on a short interval, remaining smooth, the front may have a jump. For example, consider problem (4.21) with $\beta = 0$ and take as $c(x)$ a smooth increasing function which coincides with the step function $\bar{c}(x) = 1 + 3(1 - \chi^-(x-h))$, $h > 0$, everywhere except a δ-neighborhood of the point $x = h$. One can check that the condition (N) is satisfied in this case. But if δ is small enough, the solution $x_0 = x_0(t)$ of the equation $V(t, x_0) = 0$ has the form shown in Fig. 33.

This means that for $0 < t < T_0$, the front moves continuously. But at time $t = T_0$ (see Fig. 33) a new source arises at a point h' close to h if δ is small enough. The region where $u^\varepsilon(t,x)$ is close to 1 for $0 < \varepsilon \ll 1$ propagates from this source in both directions, so that $u^\varepsilon(t,x)$ is close to 1 at time $t_1 \in (T_0, T_1)$ on $[0, x_0(t_1))$ and on $(y_1(t_1), y_2(t_1))$. Outside the closure of these intervals $u^\varepsilon(t_1, x) \to 0$ as $\varepsilon \downarrow 0$. At time T_1 both components meet. These and some other examples are considered in more detail in Freidlin [14].

The condition (N) is not always satisfied. For example, if in problem (4.21) $c(x)$ is a decreasing function, then (N) is not fulfilled. To describe the behavior of the wavefront in the general case, introduce the function

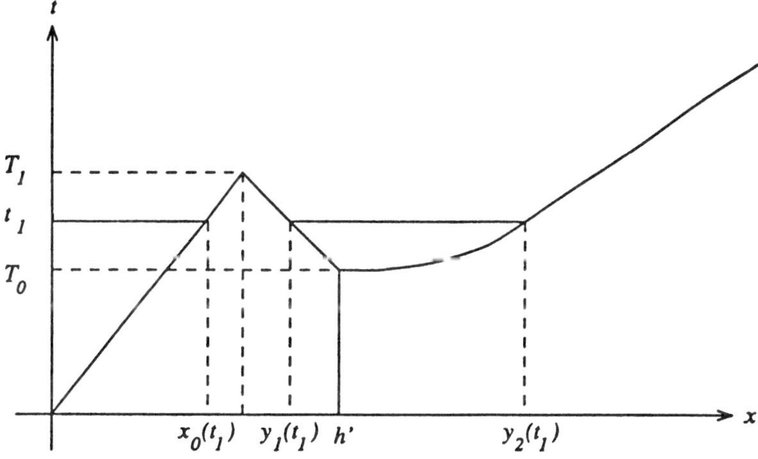

Figure 33.

$$W(t,x) = \sup\left\{ \min_{0 \le a \le t} \int_0^a \left[c(\phi_s) - \frac{1}{2}\sum_{i,j=1}^{r} a_{ij}(\phi_s)\dot{\phi}_s^i\dot{\phi}_s^j \right] ds : \right.$$
$$\left. \phi \in C_{0t}, \phi_0 = x, \phi_t \in G_0 \right\}.$$

One can prove that the function $W(t,x)$ is Lipschitz, continuous, and nonpositive. If the condition (N) is satisfied, then $W(t,x) = V(t,x) \wedge 0$.

Theorem 4.2. *Let $u^\varepsilon(t,x)$ be the solution of problem* (4.7) *with a KPP-type nonlinear term $f(x,u) = c(x,u)u$, Then*

(i) $\lim_{\varepsilon\downarrow 0} u^\varepsilon(t,x) = 0$, *uniformly in $(t,x) \in F_1$, where F_1 is a compact subset of the set $\{(t,x) : t > 0, x \in R^r, W(t,x) < 0\}$;*

(ii) $\lim_{\varepsilon\downarrow 0} u^\varepsilon(t,x) = 1$, *uniformly in $(t,x) \in F_2$, where F_2 is a compact subset of the interior of the set $\{(t,x) : t > 0, x \in R^r, W(t,x) = 0\}$.*

The proof of this theorem and various examples can be found in Freidlin [17] and Freidlin and Lee [1].

Various generalizations of problem (4.7) for KPP-type nonlinear terms can be found in Freidlin [18], [20], [22], and Freidlin and Lee [1]. In particular, one can consider RDE-systems of the KPP-type, equations with nonlinear boundary conditions, and degenerate reaction-diffusion equations. An analytic approach to some of these problems is available at present as well; corresponding references can be found in Freidlin and Lee [1]. Wavefront propagation in periodic and in random media is studied in Gärtner and Freidlin [1], Freidlin [15], and Gärtner [5].

Consider one-dimensional dynamical system $\dot{u} = f(u)$. If $f(u)$ is of the KPP-1 type, the dynamical system has two equilibrium points: one unstable point at $u = 0$ and one stable point at $u = 1$. One can consider problem (4.7) with such a nonlinear term that the dynamical system has two stable rest points separated by the unstable third one. Such nonlinear terms are called bistable. Problem (4.7) with a nonlinear term that is bistable for each $x \in R^r$ is of great interest. One can prove that $u^\varepsilon(t, x)$ also converges in this case to a step function as $\varepsilon \downarrow 0$, and the evolution of this step function can be described by a Huygens principle. These results can be found in Gärtner [4].

§5. Random Perturbations of Infinite-Dimensional Systems

We considered, in the previous chapters of this book, random perturbations of finite-dimensional dynamical systems. Similar questions for infinite-dimensional systems are of interest as well. From a general point of view, we face in the infinite-dimensional case, again, problems of the law-of-large-numbers type, of the central-limit-theorem type, and of the large-deviation type. But many new interesting and rather delicate effects appear in this case.

A rich class of infinite-dimensional dynamical systems and semiflows is described by evolutionary differential equations. For example, one can consider the dynamical system associated with a linear hyperbolic equation or a semiflow associated with the reaction-diffusion equation

$$\begin{aligned} \frac{\partial u(t,x)}{\partial t} &= D\Delta u + f(x,u), \qquad t > 0, x \in G \subset R^r, \\ u(0,x) &= g(x), \\ \left.\frac{\partial u(t,x)}{\partial n}\right|_{\partial G} &= 0. \end{aligned} \tag{5.1}$$

Here D is a positive constant, $f(x, u)$ is a Lipschitz continuous function, and $n = n(x)$ is the interior normal to the boundary ∂G of the domain G. The domain G is assumed to be regular enough. Problem (5.1) is solvable for any continuous bounded $g(x)$. The corresponding solution $u(t, x)$ will be differentiable in $x \in G$ for any $t > 0$, even if $g(x)$ is just continuous. This shows that problem (5.1) cannot be solved, in general, for $t < 0$. Therefore, equation (5.1) defines a semiflow $T_t g(x) = u(t, x)$, $t \geq 0$, but not a flow (dynamical system), in the space of continuous functions.

Of course, there are many ways to introduce perturbations of the semiflow $T_t g$. One can consider various kinds of perturbations of the equation, perturbations of the domain, and perturbations of the boundary or initial conditions.

Let us start with additive perturbations of the equation. We restrict ourselves to the case of one spatial variable and consider the periodic problem, so that we do not have to care about the boundary conditions:

$$\frac{\partial u^\varepsilon(t,x)}{\partial t} = D\frac{\partial^2 u^\varepsilon}{\partial x^2} + f(x,u^\varepsilon) + \varepsilon\zeta(t,x), \qquad t>0, x\in S^1,$$
$$u^\varepsilon(0,x) = g(x). \tag{5.2}$$

We assume that the functions $f(x,u)$, $g(x)$, $\zeta(t,x)$ are 2π-periodic in x, so that they can be considered for x belonging to the circle S^1 of radius 1, and we consider 2π-periodic in x solutions of problem (5.2). When $\varepsilon = 0$, problem (5.2) defines a semiflow in the space C_{S^1} of continuous functions on S^1 provided with the uniform topology.

What kind of noise $\zeta(t,x)$ will we consider? Of course, the smoother $\zeta(t,x)$, the smoother the solution of (5.2). But on the other hand, the smoothness makes statistical properties of the noise and of the solution more complicated. The simplest (from the statistical point of view) noise $\zeta(t,x)$ is the space–time white noise

$$\zeta(t,x) = \frac{\partial^2 W(t,x)}{\partial t\partial x}, \qquad t\geq 0, x\in[0,2\pi).$$

Here $W(t,x)$, $t\geq 0$, $x\in R^1$, is the Brownian sheet, that is, the continuous, mean zero Gaussian random field with the correlation function $\mathsf{M}W(s,x)W(t,y) = (s\wedge t)(x\wedge y)$. One can prove that such a random field exists, but $\partial^2 W(t,x)/\partial t\partial x$ does not exist as a usual function. Therefore, we have to explain the solution of problem (5.2) with $\zeta(t,x) = \partial^2 W(t,x)/\partial t\partial x$.

A measurable function $u^\varepsilon(t,x)$, $t\geq 0$, $x\in S^1$, is called the generalized solution of (5.2) with $\zeta(t,x) = \partial^2 W(t,x)/\partial t\partial x$, if

$$\begin{aligned}\int_{S_1} u^\varepsilon(t,x)\phi(x)\,dx - \int_{S_1} g(x)\phi(x)\,dx\\ = \int_0^t\int_{S^1}[u^\varepsilon(s,x)D\phi''(x) - f(x,u^\varepsilon(s,x))\phi(x)]\,dx\,ds\\ + \varepsilon\int_{S^1}\phi'(x)W(t,x)\,dx\end{aligned} \tag{5.3}$$

for any $\phi\in C^\infty_{S^1}$ with probability 1. It is proved in Walsh [1] that such a solution exists and is unique, if $f(x,u)$ is Lipschitz continuous and $g\in C_{S^1}$. Some properties of the solution were established in that paper as well. In particular, it was shown that the solution is Hölder continuous and defines a Markov process $u^\varepsilon_t = u^\varepsilon(t,0)$ in C_{S^1}. The asymptotic behavior of u^ε_t as $\varepsilon\downarrow 0$ was studied in Faris and Jona-Lasinio [1] and in Freidlin [16].

Consider, first, the linear case,

$$\frac{\partial v^\varepsilon(t,x)}{\partial t} = D\frac{\partial^2 v^\varepsilon}{\partial x^2} - \alpha v^\varepsilon + \varepsilon\frac{\partial^2 W(t,x)}{\partial t\partial x}, \qquad t>0, x\in S^1$$
$$v^\varepsilon(0,x) = g(x). \tag{5.4}$$

Here D, $\alpha > 0$, $W(t,x)$ is the Brownian sheet. The solution of (5.4) with $g(x)\equiv 0$ is denoted by $v_0^\varepsilon(t,x)$, and let $v_g^0(t,x)$ be the solution of (5.4) with $\varepsilon = 0$; $v_g^0(t,x)$ is not random.

It is clear that $v^\varepsilon(t,x) = v_g^0(t,x) + v_0^\varepsilon(t,x)$.

Due to the linearity of the problem, $v_0^\varepsilon(t,x)$ is a Gaussian mean zero random field. One can solve problem (5.4) using the Fourier method. The eigenvalues of the operator $Lh(x) = Dh''(x) - \alpha h(x)$, $x\in S^1$, are $\lambda_k = Dk^2+\alpha$, $k = 0,1,2,\dots$. The normalized eigenfunctions corresponding to λ_k are $\pi^{-1/2}\cos kx$ and $\pi^{-1/2}\sin kx$. Then

$$v^\varepsilon(t,x) = v_g^0(t,x) + \frac{\varepsilon}{\sqrt{2\pi}}A_0(t) + \frac{\varepsilon}{\sqrt{\pi}}\sum_{k=1}^\infty (A_k(t)\sin kx + B_k(t)\cos kx), \tag{5.5}$$

where $A_k(t)$ and $B_k(t)$ are independent Ornstein–Uhlenbleck processes, satisfying the equations

$$dA_k(t) = dW_k(t) - \lambda_k A_k(t)\,dt, \qquad A_k(0) = 0,$$
$$dB_k(t) = d\tilde{W}_k(t) - \lambda_k B_k(t)\,dt, \qquad B_k(0) = 0.$$

Here $W_k(t)$ and $\tilde{W}_k(t)$ are independent Wiener processes. The correlation function of the field $v^\varepsilon(t,x)$ has the form

$$\begin{aligned}\mathsf{M}&(v^\varepsilon(s,x) - Mv^\varepsilon(s,x))(v^\varepsilon(t,y) - Mv^\varepsilon(t,y))\\ &= \varepsilon^2 B(s,x,t,y)\\ &= \frac{\varepsilon^2}{2\pi}\sum_{k=0}^\infty \frac{\cos(x-y)}{\lambda_k}(e^{-\lambda_k(t-s)} - e^{-\lambda_k(t+s)}); \qquad x,y\in S^1, 0\le s\le t.\end{aligned} \tag{5.6}$$

One can view $v_t^\varepsilon = v^\varepsilon(t,\cdot)$, $t\ge 0$, as a Markov process in C_{S^1} (generalized Ornstein–Uhlenbeck process). Equalities (5.5) and (5.6) imply that the process v_t^ε has a limiting distribution μ^ε as $t\to\infty$. The distribution μ^ε in C_{S^1} is mean zero Gaussian with the correlation function

$$\varepsilon^2 B(x,y) = \frac{\varepsilon^2}{2\pi}\sum_{k=0}^\infty \frac{\cos k(x-y)}{\lambda_k}; \tag{5.7}$$

μ^ε is the unique invariant measure of the process v_t^ε in C_{S^1}.

Consider now the random field $v^\varepsilon(t,x)$, $0 \le t \le T$, $x \in S^1$, for some $T \in (0,\infty)$. It is easy to see that $v^\varepsilon(t,x)$ converges to $v_g^0(t,x)$ as $\varepsilon \to 0$, but with a small probability $v^\varepsilon(t,x)$ will be in a neighborhood of any function $\phi(t,x) \in C_{[0,T]\times S^1}$ such that $\phi(0,x) = g(x)$, if $\varepsilon > 0$. The probabilities of such deviations are characterized by an action functional. One can derive from Theorem 3.4.1 and (5.6) that the action functional for the family $v^\varepsilon(t,x)$ in $L^2_{[0,T]\times S^1}$ as $\varepsilon \to 0$ is $\varepsilon^{-2}S^v(\phi)$, where

$$S^v(\phi) = \begin{cases} \frac{1}{2}\int_{S^1}\int_0^T |\phi_t'(t,x) - D\phi_{xx}''(t,x) + \alpha\phi(t,x)|^2\,dt\,dx, & \phi \in W_2^{1,2} \\ +\infty, \quad \text{for the rest of } L^2_{[0,T]\times S^1}. \end{cases}$$

Here $W_2^{1,2}$ is the Sobolev space of functions on $[0,T] \times S^1$ possessing generalized square integrable derivatives of the first order in t and of the second order in x.

Note that there is a continuous imbedding of $W^{1,2}$ in $C_{[0,T]\times S^1}$. We preserve the same notation for the restriction of $S^v(\phi)$ on $C_{[0,T]\times S^1}$. Using Theorem 3.4.1 and Fernique's bound for the probability of exceeding a high level by a continuous Gaussian field (Fernique [1]), one can prove that $\varepsilon^{-2}S^v(\phi)$ is the action functional for $v^\varepsilon(t,x)$ in $C_{[0,T]\times S^1}$ as $\varepsilon \to 0$.

Consider now equation (5.2) with $\zeta(t,x) = \partial^2 W(t,x)/\partial t\partial x$. Let $F(x,u) = \int_0^u f(x,z)\,dz$, and

$$U(\phi) = \int_0^{2\pi}\left[\frac{D}{2}\left(\frac{d\phi}{dx}\right)^2 + F(x,\phi(x))\right]dx, \qquad \phi \in W_2^1,$$

where W_2^1 is the Sobolev space of functions on S^1 having square integrable first derivatives. It is readily checked that the variational derivative $\delta U(\phi)/\delta\phi$ has the form

$$\frac{\delta U(\phi)}{\delta \phi} = -D\frac{d^2\phi}{dx^2} - f(x,\phi).$$

So the functional $U(\phi)$ is the potential for the semiflow in C_{S^1} defined by (5.1):

$$\frac{\partial u(t,x)}{\partial t} = -\frac{\delta U(u(t,x))}{\delta u}.$$

We have seen in Chapter 4 that for finite-dimensional dynamical systems, potentiality leads to certain simplifications. In particular, the density $m^\varepsilon(x)$

of the invariant measure of the process

$$\dot{x}_t^\varepsilon = -\nabla U(x_t^\varepsilon) + \varepsilon \dot{W}_t$$

in R^r can be written down explicitly:

$$m^\varepsilon(x) = C_\varepsilon e^{-2\varepsilon^{-2}U(x)}, \qquad x \in R^r,$$

$$C_\varepsilon^{-1} = \int_{R^r} e^{-2\varepsilon^{-2}U(x)}\,dx.$$

Convergence of the last integral is the necessary and sufficient condition for existence of a finite invariant measure. One could expect that a similar formula could be written down for the invariant density $m^\varepsilon(\phi)$ of the process u_t^ε:

$$m^\varepsilon(\phi) = \text{constant} \times \exp\left\{-\frac{2}{\varepsilon^2}\,U(\phi)\right\}. \tag{5.8}$$

The difficulty involved here, first of all, is due to the fact that there is no counterpart of Lebesgue measure in the corresponding space of functions. One can try to avoid this difficulty in various ways.

First, for (5.8) to make sense, one can do the following. Denote by $\mathscr{E}_\delta(\phi)$ the δ-neighborhood of the function ϕ in C_{S^1}-norm. If ν is the normalized invariant measure for the process u_t^ε, then one can expect that

$$P_g\left\{\lim_{T\to\infty}\frac{1}{T}\int_0^T \chi_{\mathscr{E}_\delta(\phi)}(u_t^\varepsilon)\,dt = \nu(\mathscr{E}_\delta(\phi))\right\} = 1,$$

where $\chi_{\mathscr{E}_\delta(\phi)}$ is the indicator of the set $\mathscr{E}_\delta(\phi)$. Then it is natural to call $\exp\{-(2/\varepsilon^2)U(\phi)\}$ the nonnormalized density function of the stationary distribution of the process u_t^ε (with respect to nonexisting uniform distribution) provided

$$\lim_{\delta\downarrow 0}\lim_{T\to\infty}\frac{\int_0^T \chi_{\mathscr{E}_\delta(\phi_1)}(u_t^\varepsilon)\,dt}{\int_0^T \chi_{\mathscr{E}_\delta(\phi_2)}(u_t^\varepsilon)\,dt} = \exp\left\{-\frac{2}{\varepsilon^2}(U(\phi_1) - U(\phi_2))\right\}. \tag{5.9}$$

In (5.9), ε is fixed. It is intuitively clear from (5.9) that the invariant measure concentrates as $\varepsilon \to 0$ near the points where $U(\phi)$ has the absolute minimum. One can try to give exact meaning not to (5.8) but to its intuitive implications, which characterize the behavior of u_t^ε for $\varepsilon \ll 1$.

It is also possible to try to write down a formula for the density of the invariant measure of the process u_t^ε with respect to an appropriate Gaussian measure correctly defined in C_{S^1}. Of course, the form of the density with respect to this reference measure will differ from (5.8). We commence with this last approach.

As the reference measure in C_{S^1}, we choose the Gaussian measure μ_α^ε which is the invariant measure for the process v_t^ε in C_{S^1} defined by (5.4). The measure μ^ε has zero mean and the correlation function (5.7); α is a positive constant. Denote by $\mathsf{M}^{\alpha,\varepsilon}$ the expectation with respect to the measure μ_α^ε.

Theorem 5.1. *Assume that for some* $\alpha > 0$

$$A_\varepsilon = \mathsf{M}^{\alpha,\varepsilon} \exp\left\{ -\frac{2}{\varepsilon^2} \int_0^{2\pi} [F(x, \phi(x)) - \tfrac{1}{2}\alpha\phi^2(x)]\, dx \right\} < \infty.$$

Denote by ν^ε *the measure in* C_{S^1} *such that*

$$\frac{d\nu^\varepsilon}{d\mu_\alpha^\varepsilon}(\phi) = A_\varepsilon^{-1} \exp\left\{ -\frac{2}{\varepsilon^2} \int_0^{2\pi} [F(x, \phi(x)) - \tfrac{1}{2}\alpha\phi^2(x)]\, dx \right\}.$$

Then ν^ε *is the unique normalized invariant measure of the process* u_t^ε *in* C_{S^1} *defined by* (5.3). *For any Borel set* $\Gamma \subset C_{S^1}$ *and* $g \in C_{S^1}$,

$$\mathsf{P}_g\left\{ \lim_{T\to\infty} \frac{1}{T} \int_0^T \chi_\Gamma(u_t^\varepsilon)\, dt = \nu^\varepsilon(\Gamma) \right\} = 1.$$

The proof of this theorem can be found in Freidlin [16].

Equality (5.9) can be deduced from this theorem.

One can use Theorem 5.1 for studying the behavior of the invariant measure ν^ε as $\varepsilon \to 0$. In particular, if the potential $U(\phi)$ has a unique point $\hat{\phi} \in C_{S^1}$ of the absolute minimum, then the measure ν^ε is concentrated in a neighborhood of $\hat{\phi}$ if $0 < \varepsilon \ll 1$: for any $\delta > 0$,

$$\lim_{\varepsilon\to 0} \nu^\varepsilon\left\{ \phi \in C_{S^1} : \max_{x\in S^1} |\phi(x) - \hat{\phi}(x)| > \delta \right\} = 0.$$

If the absolute minimum of $U(\phi)$ is attained at finitely many m points of C_{S^1}, then the limiting invariant measure is distributed between these m points, and under certain assumptions, the weight of each of these points can be calculated.

One can check that the field $u^\varepsilon(t,x)$ can be obtained from $v^\varepsilon(t,x)$ with the help of a continuous invertible transformation B of the space $C_{[0,T]\times S^1} : u(t,x) = B[v](t,x)$. Therefore, by Theorem 3.3.1, the action functional for the family $u^\varepsilon(t,x)$ as $\varepsilon \to 0$ in $C_{[0,T]\times S^1}$ has the form

$$\varepsilon^{-2}S^u(\psi) = \varepsilon^{-2}S^v(B^{-1}(\psi)), \qquad \psi \in C_{[0,T]\times S^1},$$

where $\varepsilon^{-2}S^v$ is the action functional for $v^\varepsilon(t,x)$ calculated above. This gives us the following expression for $S^u(\psi)$.

$$S^u(\psi) = \begin{cases} \frac{1}{2}\int_0^T\int_0^{2\pi}\left|\frac{\partial\psi(t,x)}{\partial t} - D\,\frac{\partial^2\psi(t,x)}{\partial x^2} - f(x,\psi(t,x))\right|^2 dx\,dt, & \psi \in W_2^{1,2}, \\ +\infty, \quad if\, \phi \in C_{[0,T]\times S^1} \setminus W_2^{1,2}. \end{cases} \tag{5.10}$$

One can use this result, for example, for calculation of the asymptotics in the exit problems for the process u_t^ε.

Consider now a system of reaction-diffusion equations perturbed by the space–time white noise,

$$\begin{aligned} \frac{\partial u_k^\varepsilon(t,x)}{\partial t} &= D_k\,\frac{\partial^2 u_k^\varepsilon}{\partial x^2} + f_k(x;u_1^\varepsilon,\ldots,u_n^\varepsilon) + \varepsilon\,\frac{\partial^2 W_k(t,x)}{\partial t\partial x} \\ u_k^\varepsilon(0,x) &= g_k(x), \qquad t>0, x \in S^1, k \in \{1,2,\ldots,n\}. \end{aligned} \tag{5.11}$$

Here $W_k(t,x)$ are independent Brownian sheets. If the vector field $(f_1(x,u),\ldots,f_n(x,u)) = f(x,u)$, $u \in R^n$, for any $x \in S^1$ is potential, that is, $f_k(x,u) = -\partial F(x,u)/\partial u_k$ for some potential $F(x,u)$, $k \in \{1,\ldots,n\}$, $x \in S^1$, then the semiflow $u_t^0 = (u_1^0(t,\cdot),\ldots,u_n^0(t,\cdot))$ in $[C_{S^1}]^n$ is potential:

$$\frac{\partial u_k^0}{\partial t} = -\frac{\delta U(u^0)}{\delta u_k^0},$$

$$U(\phi) = \int_0^{2\pi}\left[\frac{1}{2}\sum_{k=1}^{h} D_k\left(\frac{d\phi_k(x)}{dx}\right)^2 + F(x,\phi(x))\right]dx,$$

$$\phi = (\phi^1(x),\ldots,\phi^n(x)).$$

One can give a representation for the invariant measure in this case similar to the case of a single RDE. If the field $f(x,u)$ is not potential, the semiflow (5.11) is not potential as well. But even in the nonpotential case, one can calculate the action functional for the family $u^\varepsilon(t,x)$ defined by (5.11). It has a form similar to (5.10). For example, this action functional allows one to calculate the limiting behavior of the invariant measures as $\varepsilon \to 0$ in the nonpotential case.

If the spatial variable in (5.1) has dimension greater than 1, one cannot consider perturbations of (5.1) by a space–time white noise; the corresponding equation has no solution in the case of such irregular perturbations. One can add to (5.1) the noise $\varepsilon\zeta(t,x)$, where $\zeta(t,x)$ is the mean zero Gaussian field with a correlation function $R(s,x,t,y) = \delta(t-s)r(x,y)$, where $r(x,y)$ is smooth. Then the perturbed process exists, but its statistical properties are more complicated than in the white noise case. It is interesting to consider here the double asymptotic problem, when $r(x,y) = (1/\lambda^d)\tilde{r}((x-y)/\lambda)$, where d is the dimension of the space, and both ε and λ tend to zero; $r(u)$ is assumed to be a smooth function such that $\lim_{|u|\to\infty} \tilde{r}(u) = 0$.

The additive perturbations of the equation do not, of course, exhaust all interesting ways to perturb a differential equation. Consider, for example, the following linear Cauchy problem.

$$\frac{\partial u^\varepsilon(t,x)}{\partial t} = a^\varepsilon(x)\frac{\partial^2 u^\varepsilon}{\partial x^2} + b^\varepsilon(x)\frac{\partial u^\varepsilon}{\partial x} = L^\varepsilon u^\varepsilon, \qquad t>0, x \in R^1,$$
$$u^\varepsilon(0,x) = g(x). \tag{5.12}$$

Here $a^\varepsilon(x)$ and $b^\varepsilon(x)$ are random fields depending on a parameter $\varepsilon > 0$ and $g(x)$ is a continuous bounded function. Let, for example, $a^\varepsilon(x) = a(x/\varepsilon)$, $b^\varepsilon(x) = b(x/\varepsilon)$, where $(a(x), b(x))$ is an ergodic stationary process. Then one can expect that the semiflow u_t^ε defined by problem (5.12) converges as $\varepsilon \downarrow 0$ to the semiflow defined by the equation

$$\frac{\partial u^0(t,x)}{\partial t} = \hat{a}\,\frac{\partial^2 u^0}{\partial x^2} + \hat{b}\,\frac{\partial u^0}{\partial x}, \qquad u^0(0,x) = g(x),$$

where $\hat{a}$ and $\hat{b}$ are appropriate constants. This is an example of the homogenization problem. One can ask a more general question: what are the conditions on the coefficients $(a^\varepsilon(x), b^\varepsilon(x))$ in (5.12) that ensure convergence of corresponding semiflows.

It is useful to consider a generalization of problem (5.12). Let $u(x)$, $v(x)$ be monotone increasing functions on R^1, and let $u(x)$ be continuous. One can consider a generalized differential operator $D_v D_u$ (see Feller [2]), where D_u is defined as

$$D_u h(x) = \lim_{\Delta\to 0} \frac{h(x+\Delta) - h(x)}{u(x+\Delta) - u(x)},$$

and D_v is defined in a similar way. The operator L^ε in (5.12) also can be written in the form $D_{v^\varepsilon} D_{u^\varepsilon}$ with

$$u^\varepsilon(x) = \int_0^x dy \cdot \exp\left\{-\int_0^y \frac{b^\varepsilon(z)\,dz}{a^\varepsilon(z)}\right\},$$
$$v^\varepsilon(x) = \int_0^x \frac{dy}{a^\varepsilon(y)} \exp\left\{\int_0^y \frac{b^\varepsilon(z)\,dz}{a^\varepsilon(z)}\right\}.$$

Note that the pairs (u, v), $(u + c_1, v + c_2)$ and $(au, a^{-1}v)$, where c_1, c_2, a are constants, $a > 0$, correspond to the same operator $D_v D_u$.

Consider the semiflow W_t^ε, $\varepsilon \geq 0$, defined by the equation

$$\frac{\partial W^\varepsilon(t,x)}{\partial t} = D_{v^\varepsilon} D_{u^\varepsilon} W^\varepsilon(t,x),$$

$$u^\varepsilon(0,x) = g(x).$$

Then $\lim_{\varepsilon\downarrow 0}(W_t^\varepsilon g - W_t^0 g) = 0$ for any continuous bounded $g(x)$ if and only if $u^\varepsilon(x) \to u^0(x)$, $v^\varepsilon(x) \to v^0(x)$ as $\varepsilon \downarrow 0$ at any continuity point of the limiting function (after a proper choice of the functions u^0, v^0 having to do with the above-mentioned nonuniqueness) (Freidlin and Wentzell [4]). In particular, all the homogenization results for equation (5.12) can be deduced from that statement. Such a result, which is complete, in a sense, is known only in the case of one spatial variable. Results concerning homogenization in the multidimensional case were obtained by Kozlov [1] and Papanicolaou and Varadhan [1].

Some large deviation results for equation (5.12) with rapidly oscillating coefficients, which are useful, in particular, when wavefronts in random media are studied, were obtained in Gärtner and Freidlin [1] (see also Freidlin [15]). The corresponding multidimensional problem is still open.

Perturbations of semiflows defined by PDEs caused by random perturbations of the boundary conditions are considered in Freidlin and Wentzell [1]. Initial-boundary problems for domains with many small holes are considered in Papanicolaou and Varadhan [2].

References

Anderson, R. F. and Orey, S.

[1] Small random perturbations of dynamical systems with reflecting boundary. *Nagoya Math. J.*, **60** (1976), 189–216.

Anosov, D. V.

[1] Osrednenie v sistemakh obyknovennykh differentsial'nykh uravenenii s bystro kolebIyushchimisya resheniyami. *Izv. Akad. Nauk SSSR Ser. Mat.*, **24**, No. 5 (1960), 721–742.

Arnold, V. I.

[1] *Matematicheskie Metody Klassicheskoi Mekhaniki.* Nauka: Moscow, 1975.
English translation: *Mathematical Methods of Classical Mechanics*, Graduate Texts in Mathematics, 60. Springer-Verlag: New York, Heidelberg, Berlin, 1978.

[2] Topological and ergodic properties of closed 1-forms with incommensurable period. *Funct. Analysis and Appl.*, **23**, No. 2 (1991), 1–12.

Aronson, D. G.

[1] The fundamental solution of a linear parabolic equation containing a small parameter. *Ill. J. Math.*, **3** (1959), 580–619.

Baier, D. and Freidlin, M. I.

[1] Teoremy o bol'shikh ukloneniyakh i ustoichivost' pri sluchainykh vozmushcheniyakh. *Dokl. Akad. Nauk SSSR*, **235**, No. 2 (1977), 253–256.
English translation: Baier, D. and Freidlin, M. I., Theorems on large deviations and stability under random perturbations. *Soviet Math. Dokl.*, **18**, No. 4 (1977), 905–909.

Baxter, J. R. and Chacon, R. V.

[1] The equivalence of diffusions on networks to Brownian motion. *Cont. Math.*, **26** (1984), 33–49.

Bernstein, S.

[1] Sur l'équation différentiel de Fokker–Planck. *C. R. Acad. Sci. Paris*, **196** (1933), 1062–1064.

[2] Sur l'extension du théorème limite du calcul des probabilités aux sommes de quantités dépendantes. *Math. Ann.*, **97** (1926), 1–59.

Blagoveshchenskii, Yu. N.

[1] Diffuzionnye protsessy, zavisyashchie ot malogo parametra. *Teor. Veroyatnost. i Primenen.*, **7**, No. 2 (1962), 135–152.
English translation: Diffusion processes depending on a small parameter. *Theory Probab. Appl.*, **7**, No. 2 (1962), 130–146.

Blagoveshchenskii, Yu. N. and Freidlin, M. I.

[1] Nekotorye svoistva diffuzionnykh protsessov, zavisyashchikh ot parametra. *Dokl. Akad. Nauk SSSR*, **138**, No. 3 (1961), 508–511.
English translation: Blagoveščenskii, Yu. N. and Freidlin, M. I., Certain properties of diffusion processes depending on a parameter. **2**, No. 3 (1961), 633–636.

Bogolyubov, N. M. and Mitropol'skii, Y. A.

[1] *Asymptotic Methods in the Theory of Nonlinear Oscillations*, 2nd ed. Gordon & Breach: New York, Delhi, 1961.

Bogolyubov, N. N. and Zubarev, D. N.
[1] Metod asimptoticheskogo priblizheniya dlya sistem s vrashchayuschcheisya fazoi i ego primenenie k dvizheniyu zaryazhennykh chastits v magnitnom pole. *Ukrain. Mat. Zh.*, **7**, No. 7 (1955).

Borodin, A. N.
[1] Predel'naya teorema dlya reshenii differentsial'nykh uravnenii so sluchainoi pravoi chast'yu. *Teor. Veroyatnost. i Primenen.*, **22**, No. 3 (1977), 498–512.
English translation: A limit theorem for solutions of differential equation with random right-hand side. *Theory Probab. Appl.*, **22**, No. 3 (1977), 482–497.

Borodin, A. N. and Freidlin, M. I.
[1] Fast oscillating random perturbations of dynamical systems with conservation laws. *Ann. Inst. H. Poincaré*, **31**, No. 3 (1995), 485–525.

Borovkov, A. A.
[1] Granichnye zadachi dlya sluchainykh bluzhdanii i bol'shie ukloneniya v funktsional'nykh prostranstvakh. *Teor. Veroyatnost. i Primenen.*, **12**, No. 4 (1967), 635–654.
English translation: Boundary-value problems for random walks and large deviations in function spaces. *Theory Probab. Appl.*, **12**, No. 4 (1967), 575–595.

Burdzeiko, V. P., Ignatov, F. F., Khas'minskii, R. Z., and Shakhgil'dyan, V. Z.
[1] Statisticheskii analiz odnoi sistemy fazovoi sinkhronizacii. *Proceedings of the Conference on Information Theory, Tbilisi, 1979*, 64–67.

Ciesielski, Z.
[1] Heat conduction and the principle of not feeling the boundary. *Bull. Acad. Polon. Sci. Ser. Math.*, **14**, No. 8 (1966), 435–440.

Coddington, E. A. and Levinson, N.
[1] *Theory of Ordinary Differential Equations*. McGraw-Hill: New York, 1955.

Courant, R. (Lax, P.)
[1] *Partial Differential Equations*. New York University. Institute of Mathematical Sciences: New York, 1952.

Cramér, H.
[1] Sur un nouveau théorème limite de la théorie des probabilités, *Acta Sci. et Ind.*, (1938), 736.

Da Prato, G. and Zabczyk, J.
[1] *Stochastic Equations in Infinite Dimensions*. Cambridge University Press: New York, 1992.

Day, M. V.
[1] Recent progress on the small parameter exit problem. *Stoch. Stoch. Rep.*, **20** (1987), 121–150.
[2] Large deviation results for exit problem with characteristic boundary. *J. Math. Anal. Appl.*, **147**, No. 1 (1990), 134–153.

Dembo, A. and Zeitouni, O.
[1] *Large Deviations Techniques and Applications*. Jones and Bartlett; Boston, 1992.

Deuschel, J.-D. and Stroock, D. W.
[1] *Large Deviations*. Academic Press, 1989.

Devinatz, A., Ellis, R., and Friedman, A.
[1] The asymptotic behavior of the first real eigenvalue of the second-order elliptic operator with a small parameter in the higher derivatives, II. *Indiana Univ. Math. J.*, (1973/74), 991–1011.

Donsker, M. D. and Varadhan, S. R. S.
[1] Asymptotic evaluation of certain Markov process expectations for large time, I: *Comm. Pure Appl. Math.* **28**, No. 1 (1975), 1–47; II: *ibid.*, **28**, No. 2 (1975), 279–301; III: *ibid.*, **29**, No. 4 (1976), 389–461.
[2] Asymptotics for the Wiener sausage. *Comm. Pure Appl. Math.*, **28**, No. 4 (1975), 525–565.

Doob, J. L.
[1] *Stochastic Processes*. Wiley: New York, 1953.

Dubrovksii, V. N.

[1] Asimptoticheskaya formula laplasovskogo tipa dlya razryvnykh markovskikh protsessov. *Teor. Veroyatnost. i Primenen.*, **21**, No. 1 (1976), 219–222.
English translation: The Laplace asymptotic formula for discontinuous Markov processes. *Theory Probab. Appl.*, **21**, No. 1 (1976), 213–216.

[2] Tochnye asimptoticheskie formuly laplasovskogo tipa dlya markovskikh protsessov. *Dokl. Akad. Nauk SSSR*, **226**, No. 5 (1976), 1001–1004.
English translation: Dubrovskiĭ, V. N., Exact asymptotic formulas of Laplace type for Markov processes. *Soviet Math. Dokl.*, **17**, No. 1 (1976), 223–227.

[3] Tochnye asimptoticheskie formuly laplasovskogo tipa dlya markovskikh protsessov. Kandidatskaya dissertatsiya, Moskva, 1976.

Dunford, N. and Schwartz, J. T.

[1] *Linear Operators*, Vols. I–III. Wiley: New York, London, Sydney, Toronto, 1958–1971.

Dynkin, E. B.

[1] *Osnovaniya Teorii Markovskikh Protsessov*. Fizmatgiz: Moskva, 1959.
German translation: *Die Grundlagen der Theorie der Markoffschen Prozesse*. Die Grundlehren der mathematischen Wissenschaften in Einzeldarstellungen, Band 108. Springer-Verlag: Berlin, Heidelberg, New York, 1961.

[2] *Markovskie Protsessy*. Fizmatgiz: Moskva, 1963.
English translation: *Markov Processes*. Die Grundlehren der mathematischen Wissenschaften in Einzeldarstellungen, Band 121–122. Springer-Verlag: Berlin, Heidelberg, New York, 1965.

Evgrafov, M. A.

[1] *Asimptoticheskie Otsenki i Tselye Funktsii*. Gostekhizdat: Moskva, 1957.
English translation: *Asymptotic Estimates and Entire Functions*. Gordon & Breach: New-York, 1961.

Faris, W. and Jona-Lasinio, G.

[1] Large deviations for nonlinear heat equation with noise. *J. Phys.* A, **15** (1982), 3025–3055.

Feller, W.

[1] The parabolic differential equations and the associated semi-groups of transformations. *Ann. Math.*, **55** (1952), 468–519.

[2] Generalized second-order differential operators and their lateral conditions. *Ill. J. Math.*, **1** (1957), 459–504.

Fernique, X., Conze, J., and Gani, J.

[1] *École d'été de Saint-Flour IV—1974*. P. L. Hennequin, Ed., *Lecture Notes in Math.*, 400, Springer, 1975.

Feynman, R. P. and Hibbs, A. R.

[1] *Quantum Mechanics and Path Integrals*. McGraw-Hill: New York, 1965.

Fisher, R. A.

[1] The wave of advance of advantageous genes. *Ann. Eugen.*, **7** (1937), 355–359.

Freidlin, M. I.

[1] O stokhasticheskikh uravneniyakh Ito i vyrozhdayushchikhsya ellipticheskikh uravneniyakh. *Izv. Akad. Nauk SSSR Ser. Mat.*, **26**, No. 5 (1962), 653–676.

[2] Smeshannaya kraevaya zadacha dlya ellipticheskikh uravnenii vtorogo poryadka s malym parametrom. *Dokl. Akad. Nauk SSSR*, **143**, No. 6 (1962), 1300–1303.
English translation: A mixed boundary value problem for elliptic differential equations of second order with a small parameter. *Soviet Math. Dokl.*, **3**, No. 2 (1962), 616–620.

[3] Diffuzionnye protsessy s otrazheniem i zadacha s kosoi proizvodnoi na mnogoobrazii s kraem. *Teor. Veroyatnost. i Primenen.*, **8**, No. 1 (1963), 80–88.
English translation: Diffusion processes with reflection and problems with a directional derivative on a manifold with a boundary. *Theory Probab. Appl.*, **8**, No. 1 (1963), 75–83.

[4] Ob apriornykh otsenkakh reshenii vyrozhdayushchikhsya ellipticheskikh uravnenii. *Dokl. Akad. Nauk SSSR*, **158**, No. 2 (1964), 281–283.
English translation: Freidlin, M. I., A priori estimates of solutions of degenerating elliptic equations. *Soviet Math. Dokl.*, **5**, No. 5 (1964), 1231–1234.

[5] O faktorizatsii neotritsatel'no opredelennykh matrits. *Teor. Veroyatnost. i Primenen.*, **13**, No. 2 (1968), 375–378.
English translation: On the factorization of non-negative definite matrices, *Theory Probab. Appl.*, **13**, No. 2 (1968), 354–356.

[6] O gladkosti reshenii vyrozhdayushchikhsya ellipticheskikh uravnenii. *Izv. Akad. Nauk SSSR Ser. Mat.*, **32**, No. 6 (1968), 1391–1413.
English translation: Freidlin, M. I., On the smoothness of solutions of degenerate elliptic equations. *Math. USSR-Izv.*, **32**, No. 6 (1968), 1337–1359.

[7] Funktsional deistviya dlya odnogo klassa sluchainykh protsessov. *Teor. Veroyatnost. i Primenen.*, **17**, No. 3 (1972), 536–541.
English translation: The action functional for a class of stochastic processes. *Theory Probab. Appl.*, **17**, No. 3 (1972), 511–515.

[8] O stabil'nosti vysokonadezhnykh sistem. *Teor. Veroyatnost. i Primenen.*, **20**, No. 3 (1975), 584–595.
English translation: On the stability of highly reliable systems. *Theory Probab. Appl.*, **20**, No. 3 (1975), 572–583.

[9] Fluktuatsii v dinamicheskikh sistemakh s usredneniem. *Dokl. Akad. Nauk SSSR*, **226**, No. 2 (1976), 273–276.
English translation: Freidlin, M. I., Fluctuations in dynamical systems with averaging. *Soviet Math. Dokl.*, **17**, No. 1 (1976), 104–108.

[10] Subpredel'nye raspredeleniya i stabilizatsiya reshenii parabolicheskikh uravnenii s malym parametrom. *Dokl. Akad. Nauk SSSR*, **235**, No. 5 (1977), 1042–1045.
English translation: Freidlin, M. I., Sublimiting distributions and stabilization of solutions of parabolic equations with a small parameter. *Soviet Math. Dokl.*, **18**, No. 4 (1977), 1114–1118.

[11] Printsip usredneniya i teoremy o bol'shikh ukloneniyakh. *Uspekhi Mat. Nauk*, **33**, No. 5 (1978), 107–160.
English translation: The averaging principle and theorems on large deviations. *Russian Math. Surveys*, **33**, No. 5 (1978), 117–176.

[12] Propagation of concentration waves due to a random motion connected with growth. *Soviet Math. Dokl.*, **246** (1979), 544–548.

[13] On wave front propagation in periodic media. *Stochastic Analysis and Applications*, M. A. Pinsky, Ed., Marcel Decker: New York: 1984, 147–166.

[14] Limit theorems for large deviations and reaction-diffusion equations. *Ann. Probab.*, **13**, No. 3 (1985), 639–676.

[15] *Functional Integration and Partial Differential Equations*. Princeton Univ. Press: Princeton, NJ, 1985.

[16] Random perturbations of RDE's. *TAMS*, **305**, No. 2 (1988), 665–697.

[17] Coupled reaction-diffusion equations. *Ann. Probab.*, **19**, No. 1 (1991), 29–57.

[18] Semi-linear PDE's and limit theorems for large deviations. *École d'été de Probab. de Saint-Flour XX—1990. Lecture Notes in Math.* 1527, Springer, 1991.

[19] Random perturbations of dynamical systems: Large deviations and averaging. *Math. J. Univ. of São-Paulo*, **1**, No. 2/3 (1994), 183–216.

[20] Wave front propagation for KPP-type equations. In *Surveys in Applied Mathematics*, 2, J. B. Keller, D. M. McLaughlin, and G. Papanicolaou, Ed., Plenum: New York, 1995, 1–62.

[21] *Markov Processes and Differential Equations: Asymptotic Problems*. Birkhäuser, 1996.

Freidlin, M. I. and Lee, T.-Y.

[1] Wave front propagation and large deviations for diffusion-transmutation processes. *Probab. Theory Related Fields*, **106**, No. 1 (1996), 39–70.

Freidlin, M. I. and Weber, M.

[1] Random perturbations of nonlinear oscillator, Technical Report—1997, Dresden Technical University, 1997.

Freidlin, M. I. and Wentzell, A. D.

[1] Reaction-diffusion equations with randomly perturbed boundary conditions. *Ann. Probab.*, **20**, No. 2 (1992), 963–986.

[2] Diffusion processes on graphs and the averaging principle, *Ann. Probab.*, **21**, No. 4 (1993), 2215–2245.

[3] Random perturbations of Hamiltonian systems. In *Memoirs of American Mathematical Society*, No. 523, May 1994.

[4] Necessary and sufficient conditions for weak convergence of one-dimensional Markov processes. In *The Dynkin Festschrift. Markov Processes and their Applications*, M. I. Freidlin, Ed., Birkhäuser, 1994. 95–110.

Friedman, A.
[1] Small random perturbations of dynamical systems and applications to parabolic equations. *Indiana Univ. Math. J.*, **24**, No. 6 (1974), 533–553; Erratum to this article, *ibid.*, **24**, No. 9 (1975).

Gantmakher, F. R.
[1] *Teoriya Matrits*. Nauka: Moskva, 1967.
English translation: Gantmakher, F. R., *Theory of Matrices*, Chelsea: New York, 1977.

Gärtner, J.
[1] Teoremy o bol'shikh ukloneniyakh dlya nekotorogo klassa sluchainykh protsessov. *Teor. Veroyatnost. i Primenen.*, **21**, No. 1 (1976), 95–106.
English translation: Theorems on large deviations for a certain class of random processes. *Theory Probab. Appl.*, **21**, No. 1 (1976), 96–307.
[2] O logarifmicheskoi asimptotike veroyatnostei bol'shikh uklonenii. Kandidatskaya dissertatsiya, Moskva, 1976.
[3] O bol'shikh ukloneniyakh ot invariantnoi mery. *Teor. Veroyatnost. i Primenen.*, **22**, No. 1 (1977), 27–42.
English translation: On large deviations from the invariant measure. *Theory Probab. Appl.*, **22**, No. 1 (1977), 24–39.
[4] Nonlinear diffusion equations and excitable media. *Soviet Math. Dokl.*, **254** (1980), 1310–1314.
[5] On wave front propagation. *Math. Nachr.*, **100** (1981), 271–296.

Gärtner, J. and Freidlin, M. I.
[1] On the propagation of concentration waves in periodic and random media. *Soviet Math. Dokl.*, **249** (1979), 521–525.

Gel'fand, I. M. and Fomin, S. V.
[1] *Variatsionnoe Ischislenie*. Fizmatgiz: Moskva, 1961.
English translation: Gel'fand, I. M. and Fomin, S. V., *Calculus of Variations*, revised English edition, Prentice-Hall: Englewood Cliffs, NJ, 1963.

Gikhman, I. I.
[1] Po povodu odnoi teoremy N. I. Bogolyubova. *Ukrain. Mat. Zh.*, **4**, No. 2 (1952), 215–218.

Gikhman, I. I. and Skorokhod, A. V.
[1] *Vvedenie v Teoriyu Sluchainykh Protsessov*. Nauka: Moskva, 1965.
English translation: *Introduction to the Theory of Random Processes*, W. B. Saunders: Philadelphia, 1969.
[2] *Sluchainye Protsessy*. Nauka: Moskva, 1971.
English translation: Gikhman, I. I. and Skorokhod, A. V., *The Theory of Stochastic Processes*. Die Grundlehren der mathematischen Wissenschaften in Einzeldarstellungen, Band 210. Springer-Verlag: Berlin, Heidelberg, New York, 1974–1979.

Girsanov, I. V.
[1] O preobrazovanii odnogo klassa sluchainykh protsessov s pomoshch'yu absolyutno nepreryvnoi zameny mery. *Teor. Veroyatnost. i Primenen.*, **5**, No. 3 (1960), 314–330.
English translation: On transforming a certain class of stochastic processes by absolutely continuous substitution of measures. *Theory Probab. Appl.*, **5**, No. 3 (1960), 285–301.

Grigelionis, B.
[1] O strukture plotnostei mer, sootvetstvuyushchikh sluchainym protsessam. *Litovsk. Mat. Sb.*, **13**, No. 1, (1973), 71–78.

Grin', A. G.
[1] O vozmushcheniyakh dinamicheskikh sistem regulyarnymi gaussovskimi protsessami. *Teor. Veroyatnost. i Primenen.*, **20**, No. 2 (1975), 456–457.
English translation: On perturbations of dynamical systems by regular Gaussian processes. *Theory Probabl. Appl.*, **20**, No. 2 (1975), 442–443.

[2] Nekotorye zadachi, kasayushchiesya ustoichivosti pri malykh sluchainykh vozmushcheniyakh. Kandidatskaya dissertatsiya, Moskva, 1976.
[3] O malykh sluchainykh impul'snykh vozmuschcheniyakh dinamicheskikh sistem. *Teor. Veroyatnost. i Primenen.*, **20**, No. 1 (1975), 150–158.
English translation: On small random impulsive perturbations of dynamical systems, **20**, No. 1 (1975), 150–158.

Gulinsky, O. V. and Veretennikov, A. Yu.
[1] *Large Deviations for Discrete-Time Processes with Averaging*. VSP: Utrecht, 1993.

Holland, C. J.
[1] Singular perturbations in elliptic boundary value problems. *J. Differential Equations*, **20**, No. 1 (1976), 248–265.

Hunt, G. A.
[1] Some theorems concerning Brownian motion. *Trans. Amer. Math. Soc.*, 81 (1956), 294–319.

Ibragimov, I. A. and Linnik, Yu. V.
[1] *Nezavisimye i Statsionarno Svyazannye Velichiny*. Nauka: Moskva, 1965.
English translation: *Independent and Stationary Sequences of Random Variables*. Wolters-Nordhoff: Groningen, 1971.

Ibragimov, I. A. and Rozanov, Yu. A.
[1] *Gaussian Random Processes*. Springer, 1978.

Ikeda, N.
[1] On the construction of two-dimensional diffusion processes satisfying Wentzell's boundary conditions and its application to boundary value problems. *Mem. Coll. Sci. Kyoto. Ser. A*, **33**, No. 3 (1961), 367–427.

Ioffe, A. D. and Tikhomirov, V. M.
[1] *Teoriya Ekstremalnykh Zadach*. Nauka: Moskva, 1974.
English translation: Ioffe, A. D. and Tikhomirov, V. M., *Theory of Extremal Problems*. Studies in Mathematics and its Applications, Vol. 6. North-Holland: Amsterdam, New York, 1979.

Kato, T.
[1] *Perturbation Theory for Linear Operators*, 2nd ed. Springer-Verlag: Berlin, Heidelberg, New York, 1976.

Khas'minskii, R. Z.
[1] *Ustoichivost' Sistem Differentsial'nykh Uravnenii Pri Sluchainykh Vozmushcheniyakh Ikh Parametrov*. Nauka: Moskva, 1969.
English translation: Khas'minskii, R. Z., *Stochastic Stability of Differential Equations*, 2nd ed. Sijthoff & Noordhoff: Alphen aan den Rijn, Netherlands and Rockville, MD, 1980.
[2] O polozhitelnykh resheniyakh uravneniya $\mathfrak{A}u + V \cdot u = 0$. *Teor. Veroyatnost. i Primenen.*, **4**, No. 3 (1959), 332–341.
English translation: On positive solutions of the equation $\mathfrak{A}u + V \cdot u = 0$. *Theory Probab. Appl.*, **4**, No. 3 (1959), 309–318.
[3] O printsipe usredneniya dlya parabolicheskikh i ellipticheskikh differentsial'nykh uravnenii i markovskikh protsessov s maloi diffuziei. *Teor. Veroyatnost. i Primenen.*, **8**, No. 1 (1963), 3–25.
English translation: Principle of averaging for parabolic and elliptic differential equations and for Markov processes with small diffusion. *Theory Probab. Appl.*, **8**, No. 1 (1963), 1–21.
[4] O sluchainykh protsessakh, opredelyaemykh differentsial'-nymi uravneniyami s malym parametrom. *Teor. Veroyatnost. i Primenen.*, **11**, No. 2 (1966), 240–259.
English translation: On stochastic processes defined by differential equations with a small parameter. *Theory Probab. Appl.*, **11**, No. 2 (1966), 211–228.
[5] Predel'naya teorema dlya reshenii differentsial'nykh uravnenii so sluchainoi pravoi chast'yu. *Teor. Veroyatnost. i Primenen.*, **11**, No. 3 (1966), 444–462.
English translation: A limit theorem for the solutions of differential equations with random right-hand sides. *Theory Probab. Appl.*, **11**, No. 3 (1966), 390–406.

[6] O printsipe usredneniya dlya stokhasticheskikh differentsial'nykh uravnenii Ito. *Kybernetika*, **4**, No. 3 (1968), 260–279.
[7] Ergodic properties of recurrent diffusion processes and stabilization of the solutions of the Cauchy problem for parabolic equations, *Teor. veroyatnost. i primenen.*, **5** (1960), 179–196.

Kifer, Yu. I.
[1] Nekotorye rezul'taty, kasayushchiesya malykh sluchaynykh vozmushchenii dinamicheskikh sistem. *Teor. Veroyatnost. i Primenen.*, **19**, No. 2 (1974), 514–532.
English translation: Certain results concerning small random perturbations of dynamical systems. *Theory Probab. Appl.*, **19**, No. 2 (1974), 487–505.
[2] O malykh sluchaynikh vozmushcheniyakh nekotorykh gladkikh dinamicheskikh sistem. *Izv. Akad. Nauk SSSR Ser. Mat.*, **38**, No. 5 (1974), 1091–1115.
English translation: Kifer, Ju. I., On small random perturbations of some smooth dynamical systems. *Math. USSR-Izv.*, **8**, No. 5 (1974), 1083–1107.
[3] Ob asimptotike perekhodnykh plotnostei protsessov s maloi diffuziei. *Teor. Veroyatnost. i Primenen.*, **21**, No. 3 (1976), 527–536.
English translation: On the asymptotics of the transition density of processes with small diffusion. *Theory Probab. Appl.*, **21**, No. 3 (1976), 513–522.
[4] Large deviations in dynamical systems and stochastic processes. *Invent. Math.*, **110** (1992), 337–370.
[5] Limit theorems on averaging for dynamical systems, *Ergodic Theory Dynamical Syst.*, **15** (1995), 1143–1172.

Kolmogorov, A., Petrovskii, I., and Piskunov, N.
[1] Étude de l'équation de la diffusion avec croissnence de la matière et son application à un problèmne biologique. *Moscow Univ. Buul. Math.*, **1** (1937), 1–25.

Kolmogorov, A. N.
[1] Zur Umkehrbarkeit der statistischen Naturgesetze. *Math. Ann.*, **113** (1937), 766–772.

Kolmogorov, A. N. and Fomin, S. V.
[1] *Elementy Teorii Funktsii i Funktsional'nogo Analiza*. Nauka: Moskva, 1968.
English translation: *Elements of the Theory of Functions and Functional Analysis*. Graylock Press: Rochester, NY, 1957.

Komatsu, T.
[1] Markov processes associated with certain integro-differential equations, *Osaka J. Math.*, **10** (1973), 271–303.

Kozlov, S. M.
[1] The averaging of random operators. *Math. USSR Sbornik*, **37**, No. 2 (1979), 167–180.

Krasnosel'skii, M. A. and Krein, S. G.
[1] O printsipe usredneniya v nelineinoi mekhanike. *Uspekhi Mat. Nauk*, **10**, No. 3 (1955), 147–152.

Krylov, N. V.
[1] O kvazidiffuzionnykh protsessakh. *Teor. Veroyatnost. i Primenen.*, **11**, No. 3 (1966), 424–443.
English translation: On quasi-diffusional processes. *Theory Probab. Appl.*, **11**, No. 3 (1966), 373–389.
[2] *Upravlyaemye Diffuzionnye Protsessy*. Nauka: Moskva, 1977.
English translation: *Controlled Diffusion Processes*. Springer-Verlag: New York, Heidelberg, Berlin, 1980.

Krylov, N. V. and Safonov, M. V.
[1] On a problem suggested by A. D. Wentzell. In *The Dynkin Festschrift. Markov Processes and their Applications*, M. I. Freidlin, Ed., Birkhäuser, 1994, 209–220.

Kunita, H. and Watanabe, S.
[1] On square integrable martingales. *Nagoya Math. J.*, **30** (1967), 209–245.

Labkovskii, V. A.

[1] Novye predel'nye teoremy o vremeni pervogo dostizheniya granitsy tsep'yu Markova. *Soobshch. Akad. Nauk Gruzin. SSR*, **67**, No. 1 (1972), 41–44.

Ladyzhenskaya, O. A. and Ural'tseva, N. N.

[1] *Lineinye i Kvazilineinye Uravneniya Ellipticheskogo Tipa*. Nauka: Moskva, 1964. English translation: Ladyzhenskaya, O. A. and Ural'tseva, N. N., *Linear and Quasilinear Elliptic Equations*. Mathematics in Science and Engineering, vol. 46. Academic Press: New York, 1968.

Lepeltier, J. P. and Marchal, B.

[1] Problème des martingales et équations différentielles stochastiques associés à un opérateur intégro-différentiel. *Ann. Inst. H. Poincaré*, **B12**, No. 1 (1976), 43–103.

Levina, L. V., Leontovich, A. M., and Pyatetskii-Shapiro, I. I.

[1] Ob odnom reguliruemom vetvyashchemsya protsesse. *Problemy Peredachi Informatsii*, **4**, No. 2 (1968), 72–83.
English translation: A controllable branching process. *Problems Inform. Transmission*, **4**, No. 2 (1968), 55–64.

Levinson, N.

[1] The first boundary value problem for equation $\varepsilon\Delta u + Au_x + Bu_k + Cu = D$ for small ε. *Ann. Math.*, **51**, No. 2 (1950), 428–445.

Liptser, R. S.

[1] Large deviations for occupation measures of Markov processes. *Theory Probab. Appl.*, **1** (1996), 65–88.

[2] Large deviations for two-scaled diffusion. *Probab. Theory Related Fields*, **1** (1996), 71–104.

Malkin, I. G.

[1] *Theory of Stability of Motion*. United States Atomic Energy Commission: Washington, DC, 1952.

Mandl, P.

[1] *Analytic Treatment of One-Dimensional Markov Processes*. Springer: Prague, Academia, 1968.

McKean, H. P., Jr.

[1] *Stochastic integrals. Probability and Mathematical Statistics*, vol. 5. Academic Press: New York, 1969.

Mogul'skii, A. A.

[1] Bol'shie ukloneniya dlya traektorii mnogomernykh sluchainykh bluzhdanii. *Teor. Veroyatnost. i Primenen.*, **21**, No. 2 (1976), 309–323.
English translation: Large deviations for trajectories of multi-dimensional random walks. *Theory Probab. Appl.*, **21**, No. 2 (1976), 300–315.

Molchanov, S. A.

[1] Diffuzionnye protsessy i rimanova geometriya. *Uspekhi Mat. Nauk*, **30**, No. 1 (1975), 3–59.
English translation: Diffusion processes and Riemannian geometry. *Russian Math. Surveys*, **30**, No. 1 (1975), 1–63.

Neishtadt, A. I.

[1] Ob osrednenii v mnogochastotnykh sistemakh, I: *Dokl. Akad. Nauk SSSR*, **223**, No. 2 (1975), 314–317; II: *ibid.*, **226**, No. 6 (1976), 1296–1298.
English translation: Averaging in multifrequency systems, I: *Soviet Physics. Doklady*, **20**, No. 7 (1975), 492–494; II: *ibid.*, 21, No. 2 (1976), 80–82.

Nevel'son, M. B.

[1] O povedenii invariantnoi mery diffuzionnogo protsessa s maloi diffuziei na okruzhnosti. *Teor. Veroyatnost. i Primenen.*, **9**, No. 1 (1964), 139–146.
English translation: On the behavior of the invariant measure of a diffusion process with small diffusion on a circle. *Theory Probab. Appl.*, **9**, No. 1 (1964), 125–131.

Nguen Viet Fu

[1] K odnoi zadache ob ustoichivosti pri malykh sluchainykh vozmushcheniyakh. *Vestnik Moskov. Univ. Ser. Mat. Mekh.*, **1974**, No. 5, 8–13.

[2] Malye gaussovskie vozmushcheniya i uravneniya Eilera. *Vestnik Moskov. Univ. Ser. Mat. Mekh.*, **1974**, No. 6, 12–18.

Papanicolaou, G. and Varadhan, S. R. S.
[1] Boundary-value problems with rapidly oscillating random coefficients. In *Proceedings of the Conference on Random Fields, Esztergon, Hungary, Colloquia Math. Soc. Janos Bolyai*, **27** (1981), 835–873.
[2] Diffusion in regions with many small holes. In *Proceedings of the Conference Stochastic Differential Systems and Control, Vilnius, 1978*, 1979.

Petrov, V. V.
[1] *Summy Nezavisimykh Sluchainykh Velichin.* Nauka: Moskva, 1972.
English translation: *Sums of Independent Random Variables.* Ergebnisse der Mathematik und ihrer Grenzgebiete, 82. Springer-Verlag: Berlin, Heidelberg, New York, 1975.

Pontryagin, L. S., Andronov, A. A., and Vitt, A. A.
[1] O statisticheskom rassmotrenii dinamicheskikh sistem. *Zh. Eksper. Teoret. Fiz.*, **3** (1933), No. 3, 165–180.

Prokhorov, Yu. V.
[1] Skhodimost' sluchainykh protsessov i predelnye teoremy teorii veroyatnostei. *Teor. Veroyatnost. i Primenen.*, **1** (1956), No. 2, 177–238.
English translation: Convergence of random processes and limit theorems in probability theory. *Theory Probab. Appl.*, **1**, No. 2 (1956), 157–214.

Riesz, F. and Szökefalvi-Nagy, B.
[1] *Functional Analysis.* Translation of second French edition. Ungar: New York, 1955.

Rockafellar, R. T.
[1] *Convex Analysis.* Princeton Mathematical Series, 28. Princeton University Press: Princeton, NJ, 1970.

Rozanov, Yu. A.
[1] *Statsionarnye Sluchainye Protsessy.* Fizmatgiz: Moskva, 1963.
English translation: *Stationary Random Processes.* Holden-Day: San Francisco, 1967.

Sarafyan, V. V., Safaryan, R. G., and Freidlin, M. I.
[1] Vyrozhdennye diffuzionnye protsessy i diffuzionnye uravneniya s malym parametrom. *Uspekhi Mat. Nauk*, **33**, No. 6 (1978), 233–234.
English translation: Degenerate diffusion processes and differential equations with a small parameter. *Russian Math. Surveys*, **33**, No. 6 (1978), 257–260.

Schilder, M.
[1] Some asymptotic formulas for Wiener integrals. *Trans. Amer. Math. Soc.*, **125**, No. 1 (1966), 63–85.

Sinai, Ya. G.
[1] Gibbovskie mery v ergodicheskoi teorii. *Uspekhi Mat. Nauk*, **27**, No. 4 (1972), 21–64.
English translation: Gibbs measures in ergodic theory. *Russian. Math. Surveys*, **27**, No. 4 (1972), 21–69.

Sinai, Ya. G. and Khanin, K. M.
[1] Mixing for some classes of special flows over a circle rotation. *Funct. Anal. Appl.*, **26**, No. 3 (1992), 1–21.

Skorokhod, A. V.
[1] *Sluchainye Protsessy s Nezavisimymi Prirashchemiyami.* Nauka: Moskva, 1964.

Sobolev, S. L.
[1] Nekotorye primeneniya funktsional'nogo analiza v matematicheskoi fizike. Izd. Len. gos. universiteta, Leningrad, 1950.
English translation: *Applications of Functional Analysis in Mathematical Physics.* Translations of Mathematical Monographs, 7. American Mathematical Society: Providence, RI, 1963.

Stratonovich, R. L.

[1] *Uslovnye Markovskie Protsessy i Ikh Primenenie k Teorii Optimal'nogo Upravleniya*. Izd. MGU: Moskva, 1966.
English translation: *Conditional Markov Processes and Their Application to the Theory of Optimal Control*. Modern Analytic and Computational Methods in Science and Mathematics, 7. American Elsevier: New York, 1968.

Stroock, D. W.

[1] Diffusion processes with Lévy generators. *Z. Wahrsch. Verw. Gegiete*, **32** (1975), 209–244.

Stroock, D. W. and Varadhan, S. R. S.

[1] Diffusion processes with continuous coefficients, I: *Comm. Pure Appl. Math.*, **22**, No. 3 (1969), 345–400; II: *ibid.*, **22**, No. 4 (1969), 479–530.

[2] *Multidimensional Diffusion Processes*. Springer, 1979.

Sytaya, G. N.

[1] O nekotorykh lokal'nykh predstavleniyakh dlya gaussovoi mery v gil'bertovom prostranstve. Tezisy dokladov Mezhdunarodnoi konferentsii po teorii veroyatnostei i matematicheskoi statistike. Vilnius, 1973, Vol. II, 267–268.

Varadhan, S. R. S.

[1] Asymptotic probabilities and differential equations. *Comm. Pure Appl. Math.*, **19**, No. 3 (1966), 261–286.

[2] On the behavior of the fundamental solution of the heat equation with variable coefficients. *Comm. Pure Appl. Math.*, **20**, No. 2 (1967), 431–455.

[3] Diffusion processes in a small time interval. *Comm. Pure Appl. Math.*, **20**, No. 4 (1967), 659–685.

[4] *Large Deviations and Application. In SIAM, CBMS-NSF, Regional Conferences series in Applied Mathematics*, **46** (1984).

Volosov, V. M.

[1] Usrednenie v sistemakh obyknovennykh differentsial'nykh uravnenii. *Uspekhi Mat. Nauk*, **17**, No. 6 (1962), 3–126.
English translation: Averaging in systems of ordinary differential equations. *Russian Math. Surveys*, **17**, No. 6 (1962), 1–126.

Walsh, J.

[1] An introduction to stochastic partial differential equations. In *École d'été de Probabilité de Saint-Flour XIV—1984*, P. L. Hennequin, Ed., *Lecture Notes in Mathematics*, 1180, Springer, 1984, 265–439.

Wentzell, A. D.

[1] *Kurs Teorii Sluchainykh Protsessov*. Nauka: Moskva, 1975.
English translation: Wentzell, A. D., *A Course in the Theory of Stochastic Processes*. McGraw-Hill: New York, London, 1980.

[2] Ob asimptotike naibol'shego sobstvennogo znacheniya ellipticheskogo differentsial'nogo operatora s malym parametrom pri starshikh proizvodnykh. *Dokl. Akad. Nauk SSSR*, **202**, No. 1 (1972), 19–21.
English translation: Wentzell. A. D., On the asymptotic behavior of the greatest eigenvalue of a second-order elliptic differential operator with a small parameter in the higher derivatives. *Soviet Math. Dokl.*, **13**, No. 1 (1972), 13–17.

[3] Ob asimptotike sobstvennykh znachenii matrits s elementami proyadka $\exp\{-V_{ij}/2\varepsilon^2\}$. *Dokl. Akad. Nauk SSSR*, **202**, No. 2 (1972), 263–266.
English translation: Wentzell, A. D., On the asymptotics of eigenvalues of matrices with elements of order $\exp\{-V_{ij}/(2\varepsilon^2)\}$. *Soviet Math. Dokl.*, **13**, No. 1 (1972), 65–68.

[4] Teoremy, kasayushchiesya funktsionala deistviya dlya gaussovskikh sluchainykh funktsii. *Teor. Veroyatnost. i Primenen.*, **17**, No. 3 (1972), 541–544.
English translation: Wentzell, A. D., Theorems on the action functional for Gaussian random functions. *Theory Probab. Appl.*, **17**, No. 3 (1972), 515–517.

[5] Formuly dlya sobstvennykh funktsii i mer, svyazannykh s markovskim protsessom. *Teor. Veroyatnost. i Primenen.*, **18**, No. 1 (1973), 3–29.
English translation: Wentzell, A. D., Formulae for eigenfunctions and eigenmeasures associated with a Markov process. *Theory Probab. Appl.*, **18**, No. 1 (1973), 1–26.

[6] Ob asimptotike pervogo sobstvennogo znacheniya differentsial'nogo operatora vtorogo poryadka s malym parametrom pri starshikh proizvodnykh. *Teor. Veroyatnost. i Primenen.*, **20**, No. 3 (1975), 610–613.
English translation: Wentzell, A. D., On the asymptotic behavior of the first eigenvalue of a second-order differential operator with small parameter in higher derivatives. *Theory Probab. Appl.*, **20**, No. 3 (1975), 599–602.

[7] Grubye predel'nye teoremy o bol'shikh ukloneniyakh dlya markovskikh sluchainykh protsessov, I: *Teor. Veroyatnost. i Primenen.*, **21**, No. 2 (1976), 231–251; II: *ibid.*, **21**, No. 3 (1976), 512–526.
English translation: Wentzell, A. D., Rough limit theorems on large deviations for Markov stochastic processes, I: *Theory Probab. Appl.*, **21**, No. 2 (1976), 227–242, II: *ibid.*, **21**, No. 3 (1976), 499–512.

[8] On lateral conditions for multidimensional diffusion processes. *Teor. Veroyatn.i Primen.*, **4**, No. 2 (1959), 172–185.

[9] *Limit Theorems on Large Deviations for Markov Stochastic Processes.* Kluwer: Dordrecht, 1991.

Wentzell, A. D. and Freidlin, M. I.

[1] Malye sluchainye vozmushcheniya dinamicheskoi sistemy s ustoichivym polozheniem ravnovesiya. *Dokl. Akad. Nauk SSSR*, **187**, No. 3 (1969), 506–509.
English translation: Wentzell, A. D. and Freidlin, M. I., Small random perturbations of a dynamical system with a stable equilibrium position. *Soviet Math. Dokl.*, **10**, No. 4 (1969), 886–890.

[2] O predel'nom povedenii invariantnoi mery pri malykh sluchainykh vozmusheniyakh dinamicheskikh sistem. *Dokl. Akad. Nauk SSSR*, **188**, No. 1 (1969), 13–16.
English translation: Wentzell, A. D. and Freidlin, M. I., On the limiting behavior of an invariant measure under small perturbations of a dynamical system. *Soviet Math. Dokl.*, **10**, No. 5 (1969); 1047–1051.

[3] O dvizhenii diffundiruyushchei chastitsy protiv techeniya. *Uspekhi Mat. Nauk*, **24**, No. 5 (1969), 229–230.

[4] O malykh sluchainykh vozmushcheniyakh dinamicheskikh sistem. *Uspekhi Mat. Nauk*, **25**, No. 1 (1970), 3–55.
English translation: Wentzell, A. D. and Freidlin, M. I., On small random perturbations of dynamical systems. *Russian Math. Surveys*, **25**, No. 1 (1970), 1–55.

[5] Nekotorye zadachi, kasayushchiesya ustoichivosti pri malykh sluchainykh vozmushcheniyakh. *Teor. Veroyatnost. i Primenen.*, **17**, No. 2 (1972), 281–295.
English translation: Wentzell, A. D. and Freidlin, M. I., Some problems concerning stability under small random perturbations. *Theory Probab. Appl.*, **17**, No. 2 (1972), 269–283.

Wolansky, G.

[1] Stochastic perturbations to conservative dynamical systems on the plane. *Trans. Amer. Math. Soc.*, **309** (1988), 621–639.

[2] Limit theorems for a dynamical system in the presence of resonances and homoclinic orbits. *J. Differential Equations*, **83** (1990), 300–335.

Zhivoglyadova, L. V. and Freidlin, M. I.

[1] Kraevye zadachi s malym parametrom dlya diffuzionnogo protsessa s otrazheniem. *Uspekhi Mat. Nauk*, **31**, No. 5 (1976), 241–242.

Index

Action function 80
for families of distributions in a finite-dimensional space 138, 140
for the invariant measure 190–191
normalized 80
Action functional 73, 81, 413–414
connection with Feynman's approach to quantum mechanics 73
for diffusion processes with small diffusion 104, 154, 155
for diffusion processes with small diffusion and reflection 395
for a dynamical system with Gaussian perturbations 133
for families of locally infinitely divisible processes 146
for families of random measures 387, 391
for Gaussian random processes and fields 92
normalized 74
for stochastic equations with averaging 277
for systems with averaging 235–236, 254
for the Wiener process with a small parameter 74
Asymptotic expansion
for the solution of the perturbed system 2, 52
for the solution of a stochastic equation with a small parameter 56
for the time and place of the first exit from a domain 58
Averaging principle 213, 217, 269, 286

Brownian sheet 409

Cramér transformation 90, 148, 384

Diffusion processes on graphs 289

Eigenvalue problems 204
Equivalence relation connected with the action functional 169
Exterior vertex 295

First integral
for a dynamical system 283
for a Markov process 354

Gluing conditions 289, 295
Graph corresponding to a function 287

Hamiltonian system 283
Hierarchy of cycles 198
Homogenization 415
Huygens principle 403

Interior vertex 295

KPP-equation 398–399

Laplace method 70, 121, 383
Legendre transformation of convex functions 136
Limit behavior
of the solution of the Dirichlet problem 64, 67, 69, 115, 158, 176, 198, 273–274, 281–282, 297–298
of the solution of the mixed problem 392, 396
of the solution of parabolic equations 60, 105–108, 280–281

Limit distribution. *See* Problem of the invariant measure
Logarithmic equivalence 11

Measure of stability 364–365

Normal deviations. *See* Results of the type of the central limit theorem
Normalizing coefficient 80

Perturbations of dynamical systems
  by a white noise 8, 45
  nonrandom 1
  random 3
Principle of not feeling the boundary 379
Problem of the asymptotics of the mean exit time 124, 127, 134, 160, 249
Problem of exit from a domain 108, 158, 160, 176, 192, 300
Problem of the invariant measure 128, 158, 160, 185, 412–413
Problem of optimal stabilization 366–367
Problem of selecting an optimal critical domain 365

Quasipotential 108, 118, 159

Reaction-diffusion equation (RDE) 408, 414
Regularity of a set with respect to a function (a functional) 85
Results of the type of laws of large numbers 5
  for perturbations of the white noise type 45
  for stochastic equations with averaging 269
  for systems with averaging 216
Results of the type of the central limit theorem 5
  for perturbations of the white noise type 57
  for systems with averaging 221
Results on large deviations 5, 377. *See also* Action functional

Sets stable with respect to the action functional 188
Sublimit distributions 198

Van der Pol equation 214, 261

W-graphs 177
Wave front 404

Grundlehren der mathematischen Wissenschaften

Continued from page ii

258. Smoller: Shock Waves and Reaction–Diffusion Equations, 2nd Edition
259. Duren: Univalent Functions
260. Freidlin/Wentzell: Random Perturbations of Dynamical Systems, Second Edition
261. Bosch/Güntzer/Remmert: Non Archimedian Analysis—A System Approach to Rigid Analytic Geometry
262. Doob: Classical Potential Theory and Its Probabilistic Counterpart
263. Krasnosel'skiĭ/Zabreĭko: Geometrical Methods of Nonlinear Analysis
264. Aubin/Cellina: Differential Inclusions
265. Grauert/Remmert: Coherent Analytic Sheaves
266. de Rham: Differentiable Manifolds
267. Arbarello/Cornalba/Griffiths/Harris: Geometry of Algebraic Curves, Vol. I
268. Arbarello/Cornalba/Griffiths/Harris: Geometry of Algebraic Curves, Vol. II
269. Schapira: Microdifferential Systems in the Complex Domain
270. Scharlau: Quadratic and Hermitian Forms
271. Ellis: Entropy, Large Deviations, and Statistical Mechanics
272. Elliott: Arithmetic Functions and Integer Products
273. Nikol'skiĭ: Treatise on the Shift Operator
274. Hörmander: The Analysis of Linear Partial Differential Operators III
275. Hörmander: The Analysis of Linear Partial Differential Operators IV
276. Liggett: Interacting Particle Systems
277. Fulton/Lang: Riemann-Roch Algebra
278. Barr/Wells: Toposes, Triples and Theories
279. Bishop/Bridges: Constructive Analysis
280. Neukirch: Class Field Theory
281. Chandrasekharan: Elliptic Functions
282. Lelong/Gruman: Entire Functions of Several Complex Variables
283. Kodaira: Complex Manifolds and Deformation of Complex Structures
284. Finn: Equilibrium Capillary Surfaces
285. Burago/Zalgaller: Geometric Inequalities
286. Andrianov: Quadratic Forms and Hecke Operators
287. Maskit: Kleinian Groups
288. Jacod/Shiryaev: Limit Theorems for Stochastic Processes
289. Manin: Gauge Field Theory and Complex Geometry
290. Conway/Sloane: Sphere Packings, Lattices and Groups
291. Hahn/O'Meara: The Classical Groups and K-Theory
292. Kashiwara/Schapira: Sheaves on Manifolds
293. Revuz/Yor: Continuous Martingales and Brownian Motion
294. Knus: Quadratic and Hermitian Forms over Rings
295. Dierkes/Hildebrandt/Küster/Wohlrab: Minimal Surfaces I
296. Dierkes/Hildebrandt/Küster/Wohlrab: Minimal Surfaces II
297. Pastur/Figotin: Spectra of Random and Almost-Periodic Operators
298. Berline/Getzler/Vergne: Heat Kernels and Dirac Operators
299. Pommerenke: Boundary Behaviour of Conformal Maps
300. Orlik/Terao: Arrangements of Hyperplanes
301. Loday: Cyclic Homology
302. Lange/Birkenhake: Complex Abelian Varieties
303. DeVore/Lorentz: Constructive Approximation
304. Lorentz/v. Golitschek/Makovoz: Constructive Approximation. Advanced Problems
305. Hiriart-Urruty/Lemaréchal: Convex Analysis and Minimization Algorithms I. Fundamentals
306. Hiriart-Urruty/Lemaréchal: Convex Analysis and Minimization Algorithms II. Advanced Theory and Bundle Methods
307. Schwarz: Quantum Field Theory and Topology
308. Schwarz: Topology for Physicists
309. Adem/Milgram: Cohomology of Finite Groups
310. Giaquinta/Hildebrandt: Calculus of Variations I: The Lagrangian Formalism
311. Giaquinta/Hildebrandt: Calculus of Variations II: The Hamiltonian Formalism

312. Chung/Zhao: From Brownian Motion to Schrödinger's Equation
313. Malliavin: Stochastic Analysis
314. Adams/Hedberg: Function Spaces and Potential Theory
315. Bürgisser/Clausen/Shokrollahi: Algebraic Complexity Theory
316. Saff/Totik: Logarithmic Potentials with External Fields
317. Rockafellar/Wets: Variational Analysis
318. Kobayashi: Hyperbolic Complex Spaces